Molecular Pathways of Estrogen Receptor Action

Molecular Pathways of Estrogen Receptor Action

Special Issue Editor

Farzad Pakdel

MDPI • Basel • Beijing • Wuhan • Barcelona • Belgrade

MDPI

Special Issue Editor
Farzad Pakdel
University of Rennes
France

Editorial Office
MDPI
St. Alban-Anlage 66
Basel, Switzerland

This is a reprint of articles from the Special Issue published online in the open access journal *International Journal of Molecular Sciences* (ISSN 1422-0067) from 2017 to 2018 (available at: https://www.mdpi.com/journal/ijms/special_issues/estrogen)

For citation purposes, cite each article independently as indicated on the article page online and as indicated below:

LastName, A.A.; LastName, B.B.; LastName, C.C. Article Title. *Journal Name* **Year**, *Article Number*, Page Range.

ISBN 978-3-03897-296-9 (Pbk)
ISBN 978-3-03897-297-6 (PDF)

Contents

About the Special Issue Editor

Farzad Pakdel obtained his Ph.D. in 1989 in molecular biology from the University of Rennes in France. He carried out his postdoctoral work in the United States at the University of Illinois (1989–1992). He is currently director of research at the CNRS at IRSET, research institute on health, environment, and occupational environment, UMR 1085 Inserm, at the University of Rennes in France. His work focuses on the molecular pathways of estrogen receptor action, in gene transcription, in breast cancer, and during development. Dr. Pakdel is also interested in understanding the transcriptional and epigenetic effects of endocrine disruptors and phytochemicals interacting with estrogen receptors. In addition, his lab has an interest in understanding signaling crosstalk and the molecular mechanisms of breast cancer progression. It was among the first labs that developed in vivo and in vitro bioassays to screen the hormonal activity and mechanisms of actions of environmental chemicals.

International Journal of
Molecular Sciences

MDPI

Editorial

Molecular Pathways of Estrogen Receptor Action

Farzad Pakdel

Research Institute in Health, Environment and Occupation (Irset), Inserm U1085, Transcription, Environment and Cancer Group, University of Rennes 1, 35000 Rennes, France; farzad.pakdel@univ-rennes1.fr

Received: 17 July 2018; Accepted: 30 August 2018; Published: 31 August 2018

The estrogen receptors (ERs) are typical members of the superfamily of nuclear receptors that includes the receptors that mediate the effects of steroid hormones, thyroid hormones, retinoid and vitamin D, as well as numerous orphan receptors. ERs, as other steroid receptors, mainly function as ligand-inducible transcription factors which bind chromatin, as homodimers, at specific response elements. It should also be noted that a tight reciprocal coupling between rapid 'non-genomic' and 'genomic' biological responses to estrogen occurs in many physiological processes. ERs have long been evaluated for their roles in controlling the expression of genes involved in vital cellular processes such as proliferation, apoptosis and differentiation. Given the various and pleiotropic functions of ERs, the dysregulation of their pathways contributes to several diseases such as, the hormone-dependent breast, endometrial and ovarian cancers as well as neurodegenerative diseases, cardiovascular diseases and osteoporosis. Several classes of ER ligands with agonist or antagonist activities in different E2-target tissues have been characterized. Moreover, ER ligands that efficiently block tumor growth and kill cancer cells have been developed.

In this special issue, "Molecular Pathways of Estrogen Receptor Action", promising results in understanding the mechanisms underlying ER-mediated effects in various pathophysiological processes are represented, covering different roles of ER pathways in the tumorigenesis, the resistance to endocrine therapy, the dynamics of 3D genome organization and the cross-talk with other signaling pathways.

A key step in the physiological processes is the regulation of the transcriptional dynamics of gene networks. The article by Le Dily and Beato [1] summarizes the restructuration and chromatin folding during steroid hormone exposure, as well as the influence of three-dimensional genome organization in the response to steroid hormones. Deciphering these events may particularly be important to understand cell transformation and its progression in cancers where the genome is often rearranged during tumorigenesis. In addition, Yang et al. [2] update the effect of hypoxia on ER function in breast cancer. They focus on the link between ERs, the hypoxia inducible factor 1 and the histone lysine demethylase KDM4B, an important epigenetic modifier in cancer. Additionally, Saito and Cui [3] describe a possible cross-talk via transcriptional regulation between ERs and the estrogen-related receptors (ERRs) that partially share common target genes. Moreover, ERs can directly regulate the expression of genes encoding ERRs through the estrogen-response element within the promoter region. As ERRα is at the center of the coordination of transcriptional networks for neuronal and adaptive responses, this can potentially explain estrogenic actions in social behavior. Further, Hsu et al. [4] provide an overview of the possible role of ERs in lung cancer. Different aspects of the disease development, clinical studies, effects of tobacco smoking and environmental estrogens as well as ER activation and interactions with EGFR (epidermal growth factor receptor) are discussed. A critical review on the natural human anti-ERα antibodies capable of inducing estrogenic responses in breast cancer cells is given by Guy Leclercq [5]. These observations, not much mentioned previously, were recently confirmed and have been extended to autoimmune diseases. These data will open new paths to develop new strategies and to combine immunological and endocrine approaches for the management of breast cancer. The mechanism of action of these antibodies is also addressed.

In addition to cancerous cells, the non-cancer cells including tumor microenvironment (TME) are critical mediators of tumor progression. Besides the intracellular signaling, the interactions between cancer cells, stromal cells, immune cells, and extracellular molecules within the TME greatly impact antitumor immunity and the immunotherapeutic response. The potential role of estrogen signaling pathway, as a regulator of tumor immune responses, in the tumor microenvironment is discussed and reviewed by Rothenberger et al. [6]. Radiation therapy is widely used as one of the most common and effective therapeutic strategies. Nevertheless, the effect of ionizing radiation on the expression of ERs and ER signaling pathways in cancerous tissues, as well as on the endocrine therapy is not well-known. This topic is reviewed and discussed by Rong et al. [7]. They also summarize basic, pre-clinical and clinical studies that assess the consequences of anti-estrogen treatments in combination with radiotherapy in cancer.

There is an important link between estrogen signaling pathways and the regulation of the cardiovascular and immune systems. Trenti et al. [8] review the current understanding of the protective effects of estrogen on the cardiovascular system, including promoting endothelial healing and angiogenesis. They also describe the actions of estrogens in the immune function of the monocyte-macrophage system, through different pathways and in particular with regard to the production of cytokines. Recent studies have also suggested that estrogens exert their vascular protective effects, at least in part, through microRNA activity. Pérez-Cremades et al. [9] focus on the recent progress in determining the roles of estrogen-regulated microRNAs and their contribution in vascular biology. They summarize the microRNAs involved in estrogen action and the major role played by miR-23a and miR-22. However, further works focused on characterizing the role of estradiol-mediated miRNAs involved in vascular function are needed. Wnuk and Kajta [10] highlight the role of steroid and xenobiotic receptor signaling in apoptosis and autophagy of the central nervous system, and their potential implications in brain diseases. Finally, Lecomte et al. [11] discuss and summarize the in vitro and in vivo effects of phytochemicals interacting with ERs and their potential role in human health. The diversity of the mechanisms of action and the subtle balance between beneficial and harmful biological outcomes are also given.

In addition to the reviews mentioned above, eight research articles are included in this special issue. A clinical study reported by Matta et al. [12] describes a substantial variability in DNA repair capacity among breast cancer subtypes and suggests lowest repairs in triple negative breast cancer. Cardoso et al. [13] report estrogen metabolism-associated CYP2D6 and IL6-174G/C polymorphisms in Schistosoma haematobium Infection. From a primary culture approach, Kranc et al. [14] analyze the expression profile of genes regulating steroid biosynthesis and metabolism in human ovarian granulosa cells. An in vivo study conducted by d'Adesky et al. [15] indicates that nicotine modifies ER-β-regulated inflammasome activity and aggravates ischemic brain damage in female rats. The study conducted by Casanova-Nakayama et al. [16] examines the immune-specific expression and estrogenic regulation of the four ER isoforms in female rainbow trout. Alexandre-Pires et al. [17] evaluate functional aspects of sheep inguinal sinus gland and the mRNA and protein expressions of several hormone receptors including ERs. An in vivo and in vitro study conducted by Hinfray et al. [18] provides evidence regarding antagonistic effects of estradiol and genistein in combination using mixture concentration-response modeling in zebrafish. Serra et al. [19] report that triclosan lacks (anti-)estrogenic effects in zebrafish cells but alters estrogen response in zebrafish embryos.

While much remains to be learned, this special issue provides a background of the molecular mechanisms of ERs that is needed in clinical studies against estrogen-related diseases. Lastly, I would like to thank all the authors and referees for their efforts in supporting this special issue.

Conflicts of Interest: The author declares no conflict of interest.

References

1. Le Dily, F.; Beato, M. Signaling by Steroid Hormones in the 3D Nuclear Space. *Int. J. Mol. Sci.* **2018**, *19*, 306. [CrossRef] [PubMed]
2. Yang, J.; Harris, A.L.; Davidoff, A.M. Hypoxia and Hormone-Mediated Pathways Converge at the Histone Demethylase KDM4B in Cancer. *Int. J. Mol. Sci.* **2018**, *19*, 240. [CrossRef] [PubMed]
3. Saito, K.; Cui, H. Emerging Roles of Estrogen-Related Receptors in the Brain: Potential Interactions with Estrogen Signaling. *Int. J. Mol. Sci.* **2018**, *19*, 1091. [CrossRef] [PubMed]
4. Hsu, L.; Chu, N.; Kao, S. Estrogen, Estrogen Receptor and Lung Cancer. *Int. J. Mol. Sci.* **2017**, *18*, 1713. [CrossRef] [PubMed]
5. Leclercq, G. Natural Anti-Estrogen Receptor Alpha Antibodies Able to Induce Estrogenic Responses in Breast Cancer Cells: Hypotheses Concerning Their Mechanisms of Action and Emergence. *Int. J. Mol. Sci.* **2018**, *19*, 411. [CrossRef] [PubMed]
6. Rothenberger, N.J.; Somasundaram, A.; Stabile, L.P. The Role of the Estrogen Pathway in the Tumor Microenvironment. *Int. J. Mol. Sci.* **2018**, *19*, 611. [CrossRef] [PubMed]
7. Rong, C.; Meinert, E.F.R.C.; Hess, J. Estrogen Receptor Signaling in Radiotherapy: From Molecular Mechanisms to Clinical Studies. *Int. J. Mol. Sci.* **2018**, *19*, 713. [CrossRef] [PubMed]
8. Trenti, A.; Tedesco, S.; Boscaro, C.; Trevisi, L.; Bolego, C.; Cignarella, A. Estrogen, Angiogenesis, Immunity and Cell Metabolism: Solving the Puzzle. *Int. J. Mol. Sci.* **2018**, *19*, 859. [CrossRef] [PubMed]
9. Pérez-Cremades, D.; Mompeón, A.; Vidal-Gómez, X.; Hermenegildo, C.; Novella, S. miRNA as a New Regulatory Mechanism of Estrogen Vascular Action. *Int. J. Mol. Sci.* **2018**, *19*, 473. [CrossRef] [PubMed]
10. Wnuk, A.; Kajta, M. Steroid and Xenobiotic Receptor Signalling in Apoptosis and Autophagy of the Nervous System. *Int. J. Mol. Sci.* **2017**, *18*, 2394. [CrossRef] [PubMed]
11. Lecomte, S.; Demay, F.; Ferrière, F.; Pakdel, F. Phytochemicals Targeting Estrogen Receptors: Beneficial Rather Than Adverse Effects? *Int. J. Mol. Sci.* **2017**, *18*, 1381. [CrossRef] [PubMed]
12. Matta, J.; Ortiz, C.; Encarnación, J.; Dutil, J.; Suárez, E. Variability in DNA Repair Capacity Levels among Molecular Breast Cancer Subtypes: Triple Negative Breast Cancer Shows Lowest Repair. *Int. J. Mol. Sci.* **2017**, *18*, 1505. [CrossRef] [PubMed]
13. Cardoso, R.; Lacerda, P.C.; Costa, P.P.; Machado, A.; Carvalho, A.; Bordalo, A.; Fernandes, R.; Soares, R.; Richter, J.; Alves, H.; et al. Estrogen Metabolism-Associated CYP2D6 and IL6-174G/C Polymorphisms in Schistosoma haematobium Infection. *Int. J. Mol. Sci.* **2017**, *18*, 2560. [CrossRef] [PubMed]
14. Kranc, W.; Brązert, M.; Ożegowska, K.; Nawrocki, M.J.; Budna, J.; Celichowski, P.; Dyszkiewicz-Konwińska, M.; Jankowski, M.; Jeseta, M.; Pawelczyk, L.; et al. Expression Profile of Genes Regulating Steroid Biosynthesis and Metabolism in Human Ovarian Granulosa Cells—A Primary Culture Approach. *Int. J. Mol. Sci.* **2017**, *18*, 2673. [CrossRef] [PubMed]
15. D'adesky, N.D.; de Rivero Vaccari, J.P.; Bhattacharya, P.; Schatz, M.; Perez-Pinzon, M.A.; Bramlett, H.M.; Raval, A.P. Alters Estrogen Receptor-Beta-Regulated Inflammasome Activity and Exacerbates Ischemic Brain Damage in Female Rats. *Int. J. Mol. Sci.* **2018**, *19*, 1330. [CrossRef] [PubMed]
16. Casanova-Nakayama, A.; Wernicke von Siebenthal, E.; Kropf, C.; Oldenberg, E.; Segner, H. Immune-Specific Expression and Estrogenic Regulation of the Four Estrogen Receptor Isoforms in Female Rainbow Trout (Oncorhynchus mykiss). *Int. J. Mol. Sci.* **2018**, *19*, 932. [CrossRef] [PubMed]
17. Alexandre-Pires, G.; Martins, C.; Galvão, A.M.; Miranda, M.; Silva, O.; Ligeiro, D.; Nunes, T.; Ferreira-Dias, G. Understanding the Inguinal Sinus in Sheep (Ovis aries)—Morphology, Secretion, and Expression of Progesterone, Estrogens, and Prolactin Receptors. *Int. J. Mol. Sci.* **2017**, *18*, 1516. [CrossRef] [PubMed]
18. Hinfray, N.; Tebby, C.; Piccini, B.; Bourgine, G.; Aït-Aïssa, S.; Porcher, J.M.; Pakdel, F.; Brion, F. Mixture Concentration-Response Modeling Reveals Antagonistic Effects of Estradiol and Genistein in Combination on Brain Aromatase Gene (cyp19a1b) in Zebrafish. *Int. J. Mol. Sci.* **2018**, *19*, 1047. [CrossRef] [PubMed]
19. Serra, H.; Brion, F.; Porcher, J.M.; Budzinski, H.; Aït-Aïssa, S. Triclosan Lacks (Anti-)Estrogenic Effects in Zebrafish Cells but Modulates Estrogen Response in Zebrafish Embryos. *Int. J. Mol. Sci.* **2018**, *19*, 1175. [CrossRef] [PubMed]

International Journal of
Molecular Sciences

MDPI

Review

Signaling by Steroid Hormones in the 3D Nuclear Space

François Le Dily [1,2] and Miguel Beato [1,2,*]

[1] Gene Regulation, Stem Cells and Cancer Program, Centre for Genomic Regulation (CRG),
    The Barcelona Institute of Science and Technology (BIST), Doctor Aiguader 88,
    08003 Barcelona, Spain; francois.ledily@crg.es
[2] Universitat Pompeu Fabra (UPF), 08003 Barcelona, Spain
*   Correspondence: miguel.beato@crg.es; Tel.: +34-93-316-0119; Fax: +34-93-316-0099

Received: 29 December 2017; Accepted: 19 January 2018; Published: 23 January 2018

Abstract: Initial studies showed that ligand-activated hormone receptors act by binding to the proximal promoters of individual target genes. Genome-wide studies have now revealed that regulation of transcription by steroid hormones mainly depends on binding of the receptors to distal regulatory elements. Those distal elements, either enhancers or silencers, act on the regulation of target genes by chromatin looping to the gene promoters. In the nucleus, this level of chromatin folding is integrated within dynamic higher orders of genome structures, which are organized in a non-random fashion. Terminally differentiated cells exhibit a tissue-specific three-dimensional (3D) organization of the genome that favors or restrains the activity of transcription factors and modulates the function of steroid hormone receptors, which are transiently activated upon hormone exposure. Conversely, integration of the hormones signal may require modifications of the 3D organization to allow appropriate transcriptional outcomes. In this review, we summarize the main levels of organization of the genome, review how they can modulate the response to steroids in a cell specific manner and discuss the role of receptors in shaping and rewiring the structure in response to hormone. Taking into account the dynamics of 3D genome organization will contribute to a better understanding of the pleiotropic effects of steroid hormones in normal and cancer cells.

Keywords: chromatin conformation; estrogen receptor; steroid receptors; topological domains; transcription regulation

1. Introduction

Similarly to other steroids, Estrogens (e.g., Estradiol, E2) exert their action by binding to their cognate nuclear receptors, the Estrogen Receptor (ER), which mainly functions as ligand-activated transcription factor [1,2]. Upon activation, ER translocates to the nucleus and converges to chromatin together with effectors of signaling pathways activated at the plasma membrane through non-genomic pathways [3]. ER binds directly to DNA through Estrogen Responsive Elements (ERE), which correspond to palindromic repeats [4], as well as indirectly through protein-protein interactions with other transcription factors [5,6]. It has been initially proposed that the effects E2 exerts on transcription depend on the binding of the ER to response elements located within the proximal promoters of the target genes. There, activated receptors orchestrate the recruitment of co-regulators, chromatin remodeling complexes and general transcription factors [1,2,7]. Although such mechanisms have been described in details for model estrogen responsive genes [8,9], the emergence of high-throughput technologies challenged this view: genome-wide analysis of transcripts levels by micro-arrays or RNA-Seq showed that the hormone modulate the expression of several hundreds of targets genes, many without direct binding of the ER at the proximal promoter [10–12]. Indeed, ChIP-Seq experiments targeting the ER in model estrogen-responsive cells showed that the receptors bind to DNA in an unexpected genome-wide fashion. The majority of the binding sites are located in intergenic regions, frequently far away from genes, and rather correspond to enhancer regions [13–15]. Similar observations have been made for other steroid receptors, such as the Glucocorticoid (GR) and Progesterone (PR) receptors, suggesting a shared mode of action [16,17].

Enhancers classically regulate transcription through chromatin looping to bring the regulatory machinery in close proximity to the promoters they target [18]. This level of chromatin folding and the formation of regulatory loops is embedded in more complex levels of organization of the genome. Indeed, it is becoming evident that the genome is organized in a highly compartmentalized and non-random fashion in the nucleus in interphase [19–21]. Such three-dimensional (3D) structures, in part cell specific, constrain the treatment of the genetic information in processes such as replication or transcription [22–25].

In terminally differentiated cells, 3D genome organization may constitute an epigenetic level controlling signal-induced modifications of transcription, as in the case of the rapid response induced by steroid hormones [26–29]. In the context of this review, we give an overview of the recent advances on understanding genome folding in eukaryotic cells and describe how it can interfere with the activity of transcription factors and the response to external cues. We further discuss observations of steroid dependent reorganization of the 3D genome architecture at local or more global scales.

2. Genome 3D Organization

Increasing experimental evidences support of a highly compartmentalized organization of the genome within the nucleus in interphase. Both cytological approaches such as Fluorescent In Situ Hybridization (FISH), or biochemical methods deriving from the chromosome conformation capture (3C) technique [30], have demonstrated that chromosomes do not decondense in a random way but rather organize following hierarchical order of structures [19–21]. The emergence of high-throughput 3C-derivatives, in particular Hi-C [31], allowed analysis of genome organization at various scales: individual chromosomes are organized in chromosome territories and are segmented in domains of preferential local contacts known as Topologically Associating Domains (TADs), which belong to functionally and epigenetically distinct chromatin compartments [24,25,31–33].

2.1. Chromosome Territories

The use of fluorescent whole chromosome paint probes permitted to confirm the hypothesis that chromosomes do not decompact in an unorganized way after mitosis but rather occupy a discrete area within the nuclear space [34,35]. These structures, known as chromosome territories, show limited intermingling between them and appear to distribute at preferential positions within the nucleus. In human cells, for example, long gene poor chromosomes are preferentially observed at the periphery, close to the nuclear lamina, while small gene rich chromosomes are frequently located within the central part of the nucleus. This suggests a functional radial positioning of chromosomes in relation to their transcriptional activity [35,36]. If the chromosome territories were originally observed by cytological approaches, this level of structure has also been confirmed by Hi-C experiments, which showed that most of the contacts detected were occurring in *cis* (i.e., intra-chromosomal contacts—Figure 1A) and that the *trans*, inter-chromosomal, interactions were reflecting the relative positioning of chromosomes [31].

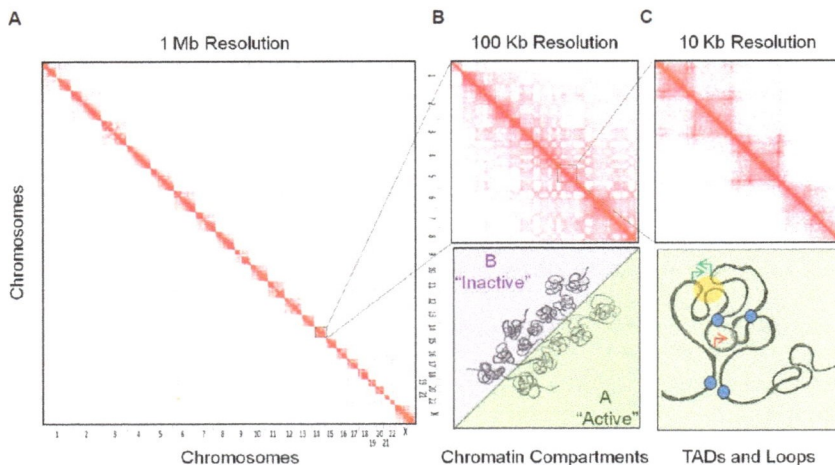

Figure 1. Hierarchical organization of the genome. Hi-C permits genome-wide detection of pair-wise contacts between genomic loci. They could be summarized as contact matrices where the color scale highlights the frequency of ligations events observed between any pairs of loci in the genome (from white to red, low to high frequencies, respectively). At different scales of resolution (e.g., 1 Mb, 100 or 10 Kb), higher orders of structure emerge: (**A**) chromosome territories, (**B**) chromatin compartments, (**C**) Topologically Associating Domains (TADs) and loops. (**B,C**) Bottom panels correspond to possible interpretations of the contact matrices: (**B**) active and inactive chromatin segregate spatially in two distinct chromatin compartments (A and B, respectively). (**C**) Architectural proteins (blue circles), such as CTCF (CCCTC-binding Factor), participate in the partitioning of the genome in TADs and generate sub-megabase structures, which can bring together specific loci or exclude genes from the activity of distal regulatory regions (orange circle: active enhancer; green arrows: expressed genes; red arrows: silenced genes).

2.2. Chromatin Compartments

In addition to support the existence of territories, the contact matrices obtained by Hi-C show a striking "plaid" or "chess" pattern (Figure 1B), which corresponds to the engagement of preferential long-range associations by non-contiguous genomic domains [31]. Such arrangement reflects the segregation of two types of genomic domains, which tend to not intermingle between them. Correlation of this pattern with epigenetic marks and transcription data demonstrated that the two types of regions corresponded mainly to the active and inactive parts of the genome (Figure 1B), also referred to as A and B chromatin compartments, respectively [31]. The existence of these two spatially segregated chromatin compartments has been confirmed by high-resolution microscopy using specific oligo-paint FISH probes [37]. This approach allowed the distinguishing of the two compartments spatially polarized in single chromosomes and further suggested that compartmentalization differs with the transcriptional activity [37]. This bimodal chromatin compartmentalization was initially observed based on Hi-C contact maps at resolutions between 0.1 to 1Mb. Recent high coverage Hi-C studies further demonstrated that the segregation of chromatin domains could be defined in a finer way, with the A and B compartments being subdivided in sub-compartments in correlation with their activity [33,38]. Importantly, this spatial segregation of chromatin compartments appears largely cell specific. Through the process of differentiation, chromosomal domains can dynamically switch from one to the other compartment, in correlation with changes of expression of tissue specific genes [23,39]. Conversely, in the process of dedifferentiation or cell reprogramming, chromosomal domains can change dynamically their association to one or the other compartment, in some cases prior to corresponding transcriptional modifications, supporting an instructive role of the 3D structure on transcription [40].

2.3. Topologically Associating Domains

At a resolution of 100 Kb or below, Hi-C chromosomal contact maps show that chromosomes are segmented in domains of high local interactions separated from each other by sharp boundaries (Figure 1C). These megabase-sized domains were referred to as Topological Domains, Topologically Associating Domains or TADs [24,32]. The boundaries between TADs are characterized by the presence of highly expressed housekeeping genes as well as by enrichments in epigenetic marks (e.g., H3K4me3, H3K36me3) linked to gene activation and in binding sites for architectural proteins such as CCCTC-binding Factor (CTCF) and cohesins [24,25,41,42]. In contrast to what is observed at the level of chromatin compartments, boundaries between TADs are largely conserved between cell types and through evolution. This suggests that TADs are important structural levels of organization [23,24]. However, a recent study based on a multi-scale analysis of insulation between genomic domains suggested that, rather to be a structurally favored level of organization, TADs represent an optimal functional level of folding for the establishment of specific interactions [43]. Such organization will notably facilitate the coordinated regulation of genes by facilitating the organization of specific wiring between genes promoters and regulatory elements [43]. In this view, TADs can be considered as epigenetic domains characterized by relatively homogeneous epigenetic features, suggesting that the border between them could limit the spreading of epigenetic marks [24,25,32]. In addition, TADs can behave as transcription units where genes are co-regulated under the control of specific regulatory elements during differentiation [32,43,44] or in response to steroid hormones [28].

2.4. Sub-Domains and Chromatin Loops

If the boundaries between TADs are conserved between cell types, their internal organization appears more dynamic and cell specific. At higher resolutions, the contact maps show cell specific internal sub-TADs or sub-domains (Figure 1C), which correspond to structures generated by the interactions between specific elements, either structural and/or related to gene activity [33,38,45,46]. In particular, during differentiation, differential binding of CTCF and cohesins, together with subunits of the mediator complex lead to the formation of such cell specific sub-domains [45]. This sub-megabase level of organization leads some authors to propose that chromosome-neighborhoods, which correspond to the establishment of specific CTCF loops that embed genes together with or without regulatory elements, might be the functional minimal unit of organization of the genome [33,38,47]. Additionally, other zinc finger proteins able to form homodimers, such as YY1, can mediate enhancer-promoter loops and are essential for specific gene regulation [48].

In summary, the genome is organized in a hierarchy of structures that have been correlated with the processing of the genetic information. Although these preferential structures can be observed in cell populations, it is important to keep in mind that, in single cells, the underlying spatial interactions remain highly dynamic and rather stochastic, as exemplified by results obtained in single cell Hi-C [49] and by the frequent discrepancies observed between 3C derived population results and direct visualization in single cells by FISH [50]. Globally however, genome-wide contact datasets suggest a cell type specific organization, which could participate in the integration of the different signals received by the cell. Degron-mediated knock-down of the levels of CTCF or subunits of the cohesin complex lead to a loss of the organization in loops and TADs without affecting the segregation of chromatin compartments [51–53]. These studies confirmed that architectural proteins are essential for the maintenance of cell specific organization of TADs and suggest that the different levels of structure are partially uncoupled. However, these proteins probably act in combinations with other regulatory factors but the precise role of tissue-specific transcription factors in organizing different levels of organization remains largely unexplored.

3. 3D Genome Folding Modulates the Response to Steroids

Comparative ChiP-Seq studies demonstrated that the landscape of binding of steroid receptors varies quantitatively and qualitatively from cell type to cell type, even between cells lines of similar origins [17,54]. This cell specificity probably participates in the regulation of distinct subsets of responsive genes, explaining the differences observed in different cell lines in response to the same stimulus [55,56]. The different levels of structures described above are part of the cell identity and one can reasonably hypothesize that they act as an epigenetic level to condition the activity of transcription factors. In particular, in the case of steroid receptors, which activity is regulated by external signals in terminally differentiated cells, this 3D organization can participate in demarcating the sets of regulatory elements potentially bound by the receptors as well as in restricting the genes that will be targeted.

3.1. Steroid Receptors Cistrome

Although variables depending on cell types, time of treatment and detection approaches, results from transcriptomic studies showed that steroid hormones elicit genome-wide changes in gene expression, with between hundreds to thousands of genes being either up- or down-regulated upon hours of treatment [10–12,16]. Some of these changes rely on indirect secondary regulations; nevertheless, the use of Global-Run-On method (GRO-Seq) confirmed these broad effects of E2 on transcription [12]. In addition, GRO-Seq permitted to highlight rapid changes in transcription not only of protein coding genes but also of many non-annotated, non-coding transcripts [12]. Rather than giving a direct explanation to these broad transcriptional changes, analysis of ER binding by ChIP-on-chip or later on by ChIP-Seq experiments largely modified the classical view of the mechanisms involved in the cellular response to steroids [13,15,57]. For instance, ChiP-Seq experiments performed in MCF-7 cells after treatment with E2 demonstrated an unexpected genome-wide binding of the ER, with more than 14,000 binding sites detected; much more than the number of genes actually regulated by the hormone in these cells [15]. The location of these binding sites throughout the genome was also unexpected: only a small proportion was located within the proximal regulatory regions of targets genes; the majority of sites were rather broadly distributed, with particular enrichment in distal inter-genic regions. This suggests that, in addition to act at the levels of promoters, ER exert their actions from distal regulatory regions. Similar behaviors were observed in other cell types and for other nuclear receptors, such as the GR and the PR [16,17]. For instance, upon 30 min exposure to progestins, PR bind to more than 25,000 sites characterized by enhancer marks and located at more than 30 Kb away from promoters [16]. This suggests a shared mode of actions of steroid receptors in acting as regulators of the activity of distal, enhancer or silencer, regions.

3.2. Differential Accessibility of Hormone Response Elements

Despite the large number of steroid receptors binding sites detected by ChIP-Seq, they represent only a small fraction of the potential Hormone Response Elements (HRE) identified by searching for consensus sequences in the genome [58], suggesting that a large fraction of HRE is not accessible for binding. This differential accessibility represents the first level of epigenetic regulation of steroid receptors function as transcription factors and can consequently dictates the cell specific target genes. Several mechanisms can explain why the receptors can only bind to part of their consensus element on DNA. ER, for example, binds preferentially to DNA elements that are not protected by nucleosomes, and requires the activity of pioneer factors. Pioneer factors are able to bind to nucleosomal DNA where they orchestrate local arrangements of the chromatin and therefore act cooperatively to facilitate the binding of activated receptors [59]. For example, Forkhead Box Protein A1 (FOXA1) has been shown to mark the sites that will be bound by ER after exposure to the hormone [60]. The PR is able to bind to sites occupied by nucleosomes [16] and other mechanisms, such as methylation of DNA, potentially prevent its binding to some of its responsive elements [61].

More generally, the precise mechanisms by which transcription factors find their DNA targets within the nuclear environment remains mainly unknown [62]. In addition to epigenetic modifications of DNA and nucleosomes, the way the chromatin is spatially organized in the nucleus can interfere with, either promote or limit, the accessibility of chromatin to nucleoplasmique factors. Several models of transcription factor search strategies have been proposed, which notably include facilitated diffusion through chromatin. In this model, transcription factors binding to specific chromatin sites occurs through a combination of diffusion, linear tracking and jumps, or hopping, along the chromatin fiber [63]. The fact that the genome folds readily in a non-random manner may favor such hopping from regions far away on the linear genome and may direct transcription factors to specific sites (Figure 2A). In addition, pre-existing chromatin loops can favor the residence time of transcription factors and lead to the nucleation of transcription factors enriched environments [64]. Finally, active and inactive (A/B) chromatin compartments are characterized by epigenetic signatures associated with open or compact chromatin, respectively. Transcription factors may therefore preferentially access (and bind to) the accessible active chromatin compartment (Figure 2A). The fact that domains belonging to the same compartment are found preferentially together within the nuclear space can delimit the space that transcription factors have to visit prior to reach functional sites. In this context, the cell specific compartmentalization of chromatin can contribute to the establishment of a cell specific landscape for binding of transcription factors. In line with this hypothesis, differential binding of the ER in the breast cancer cells MCF-7 and T47D is observed in genomic domains that belong to distinct chromatin compartments (Figure 2B).

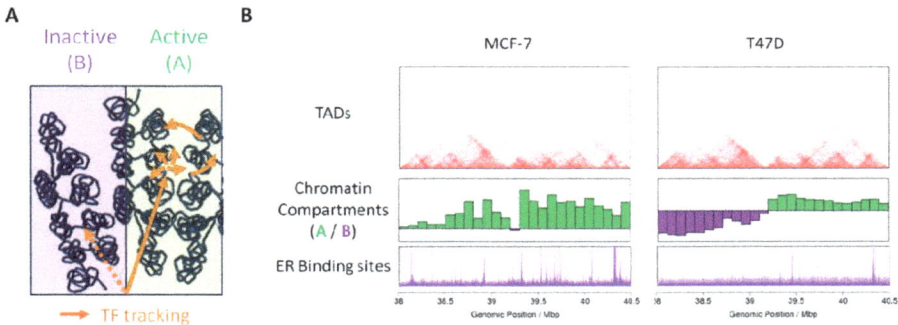

Figure 2. 3D genome folding modulates the binding of steroid receptors. (**A**) Active and inactive chromatin compartments can favor or limit, respectively, the diffusion of transcription factors within the nuclear space. Long-range folding of the chromatin fiber can facilitate the tracking and permit local enrichment of factors in given nuclear environments. (**B**) In contrast to the conservation of borders between TADs (top panel—frequencies of contacts in red), chromosomal domains belong to the A (green) or B (purple) compartment in a cell specific manner (middle panel). These differences correlate with the extent of binding of the ER as shown by ChIP-Seq experiments in MCF-7 and T47D (bottom panel).

3.3. Topological Restraint of Promoter-Enhancer Looping

An important, still opened, question is to determine whether all the sites detected by ChIP-Seq are functional and participate in transcriptional regulation or whether they are reflecting non-productive binding or chromatin interactions involved in other processes. The use of ChIA-PET, a method allowing the analysis of the spatial contacts between loci bound by a specific factor [65], showed that many of the ER bound loci were interacting together to form complex loops anchoring distal and proximal ER sites [27]. These observations support the concept that the inter-genic ER binding sites correspond to enhancer or silencer regions, which could act on distal target genes by chromatin looping. In this context, higher levels of organization may constrain the activity of the receptors by delimiting the targets they can reach. Most of the interactions between enhancers and promoters occur within TADs [47,66,67]. By limiting the space to be explored, this level of organization favors the activity of regulatory elements on the genes lying within the same domain, independently of the genomic distance that separate them (Figure 3). Conversely, the boundaries that separate contiguous domains can act as barriers and impede ectopic action of regulatory elements on genes located outside of the domain (Figure 3). The importance of these boundaries in demarcating the targets of a given regulatory region is supported by experiments where the borders between TADs were specifically deleted [68]. The use of CRISR-Cas9 approach to engineer borders between contiguous TADs induces structural modifications, with establishment of novel interactions between enhancers and promoters. This rewiring is accompanied with a misexpression of the genes located outside of the original domain, supporting a role for TADs in demarcating the range of action of regulatory elements [68]. Within a given TAD, additional cell specific loops can generate sub-domains, or chromosome neighborhoods [33,38,45,47]. These local structures can serve to isolate given genes from the activity of regulatory elements in order to maintain them in a silenced state (Figure 3) or conversely to favor the regulation of a given set of genes [47].

The organization of the genome in chromosomal domains is therefore both limiting the activity of given regulatory elements to a specific set of target genes and demarcating the 3D space that has to be explored by a regulatory element to engage specific looping with its targets. This can explain the frequent genomic clustering of steroid responsive genes as well as their co-regulation within TADs in response to hormones [28,69–71].

Figure 3. Structural segmentation of the genome restrains the range of action of regulatory elements. The enrichment in proteins with insulator function at the borders between TADs can prevent contact with regulatory elements activated by steroid receptors with promoters located outside of the domains. The natural tendency of contacts within a TAD limits the space to be explored by an activated enhancer and favor the stability of promoter-enhancer contacts within the domain independently of the genomic distance. Intra-domain loops established in a cell specific manner can serve to isolate genes from the activity of distal regulatory elements or conversely, favor contacts between enhancers and specific sets of genes.

3.4. Influence of Architectural Proteins in Steroid Response

There are evidences that architectural proteins, notably CTCF and cohesins, are directly involved in the response to E2 [72,73]. Depletion of CTCF affects the transcriptional response to E2 of model responsive genes [74]. ER binding sites frequently co-localize with CTCF sites in a cell specific manner, suggesting that CTCF can direct the ER to specific regions, potentially acting upstream of FOXA1 or other pioneer factors [73]. ER is also found at sites occupied by cohesins in a CTCF independent manner and those sites are associated with responsive genes and more prone to establish chromatin loops [75]. In this line, depletion of CTCF or RAD21 Cohesin Complex Component by small interfering

RNA (siRNA) led to destabilization of the loops engaged by ER binding sites [74]. In line with the models described above, those architectural proteins can participate in licensing ER binding in a cell specific manner and/or can facilitate the establishment of specific regulatory chromatin loops. In link with its function as insulator, CTCF also demarcates specific chromosome neighborhoods [47] or regulatory units, which could restrain the range of interactions of a given ER bound site [76].

4. Steroid Receptors Mediated Genome (re)-Organization

As described above, the conformation of the genome can restrain the response to steroid hormones by modulating the binding of the receptors to chromatin as well as by demarcating the target genes that could be regulated in a given cell type. Once bound to their DNA elements, the receptors recruit a plethora of co-factors, which act as nucleosomes remodeling machinery or histones modifying enzymes [7,77–79]. Recruitment of these co-regulators leads to either local or long-range modifications of the chromatin fiber, which can consequently reshape the 3D organization and may be essential to set the stage for subsequently acting transcription regulators.

4.1. Steroid Receptors Dependent Promoter-Enhancer Loops

The existence of functional chromatin loops between distal steroid receptor binding sites and promoters during the activation of model genes, such as *TFF1* and *CTSD*, have been demonstrated by 3C [27,74,80]. In addition to loops between enhancer and promoters, there are also evidences for steroid induced looping involved in gene repression [81]. These experiments, performed in absence or presence of ligand, showed increased frequency of contacts of the distal regions with proximal promoters upon binding of the receptors at the regulatory sites, suggesting that the binding of the receptors has an instructive role on the looping. However, how receptors actually reorganize the folding of these loci remains mainly hypothetical. Remodeling of nucleosomes as well as modifications of histones tails, at or around receptors binding sites, lead to an opening of chromatin. This could provide more flexibility to the chromatin fiber, increasing the probability for the looping to occur. In addition, ER and other steroid receptors are known to interact with components of the mediator complex, such as the Mediator Complex Subunit 1 (MED1), which could help in the stabilization of the contacts between distal transcription machineries [82–84]. It has also been observed that enhancers bound by ER are sites of transcription of short RNA products known as enhancer RNA (eRNA)—[12,85,86]. These RNAs may participate in the stabilization of the contacts between distal elements and/or in the recruitment of additional co-factors [87]. However, it is still not clear whether their production is necessary and sufficient to induce the looping between enhancers and target promoters. Inhibition of the production of the eRNA did not prevent the formation of loops between enhancer and promoter when MCF-7 cells were co-treated with E2 and flavopiridol [86]. In other models, the expression of non-coding RNA was necessary to the establishment of productive looping [87].

4.2. Dynamic and/or Pre-Settled Organization of Steroid Responsive Hubs

Genome-wide studies of chromosome conformation highlighted that contacts between enhancers and promoters occur in a complex fashion: enhancers frequently contact multiple targets and a given gene can be submitted to the activity of several enhancers [67]. The existence of such chromatin hubs is supported by ChIA-PET data where ER binding sites were observed to be engaged in multiple interactions, generating a complex network of loops and anchors after exposure to hormone [27]. Establishment of loops between multiple enhancers and promoters was observed for other nuclear receptors [26,28,88]. It remains unclear however whether those hubs are relatively stable structures or whether they are established de novo upon binding of the activated receptors (Figure 4). ChIA-PET per se does not permit to determine whether the contacts observed depend or not on the binding of the receptors. However, in the same study, various alternative approaches (3C, 4C and FISH) were used to confirm that the looping network was dependent on hormone exposure [27]. In response to progestins, changes in transcription are associated with concomitant changes in chromatin structure and conformation of TADs [28], arguing for hormone induced rewiring of local spatial contacts (Figure 4A). On another hand, some contacts between poised enhancers and promoters have also been observed prior to exposure to the signal (Figure 4B). In this case, activated receptors bind within a pre-established structure and the binding does not dramatically modify the existing contacts [46]. Steroid receptors themselves could participate in maintaining those pre-existing loops in the absence of hormones. Indeed, large regulatory regions of clustering of ER and PR after exposure to the hormones are frequently already occupied by the unliganded receptors in basal conditions. The structure of TADs exhibiting binding of unliganded ER or PR within these regions largely differ between cells expressing or not the receptors, suggesting a direct role for the unliganded receptors in maintaining a structure that could facilitate further binding of the receptors after activation [88]. Active looping induced by the hormones or binding within pre-existing structures are probably not mutually exclusive models of action (Figure 4C). For instance, in the case of the response to glucocorticoids, the use of ChIA-PET targeting p300 showed that a large fraction of enhancers bound by the GR upon exposure to the hormone were already bound by p300 and engaged in interactions with gene promoters prior to exposure to the hormone. In parallel, GR also brought de novo the histone acetyl-transferase to a significant fraction of sites, which interactions with promoters and pre-existing sites were dynamically modified by the hormone [26].

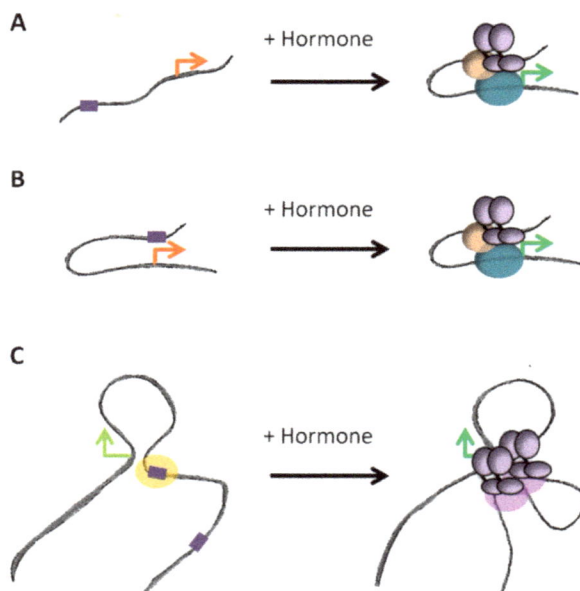

Figure 4. Different dynamics of promoter-enhancer looping. (**A**) Binding of the receptors upon exposure to the hormone induces the formation of loop between promoter and enhancer in an active process. (**B**) In other cases, the loop is established prior binding of the receptors, which activate the enhancer region in an already favorable conformation. (**C**) A promoter could be regulated by several enhancers, which require or not de novo chromatin looping. Binding of steroid receptors in the absence of hormones (unliganded receptors) might serve in maintaining such structures prior exposure to the hormone. Orange and green arrows correspond to paused or activated promoters, respectively.

4.3. *Steroid Induced Changes at Higher Levels of Organization*

As mentioned above the loops between regulatory elements and promoters appear to be limited to loci laying in the same TAD and exposure to steroid hormones can lead to local restructuration of TADs, reflecting a rewiring of the contacts [26–28]. Some observations also suggest that steroid responsive loci located further away are found together within the nuclear space. For instance, clusters of binding sites of ER and PR located in different TADs can establish long-range interactions between them [88] and genes responsive to glucocorticoids have been observed to cluster within nuclear hubs [89]. These observations could suggest the existence of specific hubs or transcription factories specialized in the response to steroids. In addition, steroids have been shown to induce large-scale remodeling of chromatin [90,91]. This can occur through spreading of chromatin modifications over large chromatin domains or even more globally within the nucleus [28,81,92]. Whether these large-scale modifications are accompanied with more global changes of the 3D genome structure remains unclear. Despite the important changes of chromatin induced by progestins, dynamic changes of chromosomal organization were mainly observed at the level of TADs [28]. Similarly, the long-range clustering of glucocorticoid responsive genes in nuclear hubs was not modified upon exposure to the hormones [89]. In contrast, exposure to E2 have been proposed to induce large-scale reorganization of the structure of chromosomes [93] and some authors proposed that E2-responsive genes could be actively brought together within the nuclear space upon exposure to the hormone, facilitating their transcription in specific hubs [94]. However, such active clustering of E2-responsive genes was not observed by others in similar models [95].

Although such long-range organizations may be favored, they probably remain highly stochastic and it will be important to determine to what extent they facilitate the transcriptional response. Additional studies of their dynamic will be also necessary to determine to what extent they are reflected by functional modifications of higher levels of structure of the genome.

5. Future Directions

In summary, the way the genome organizes can modulate the activity of steroid receptors and other transcription factors at several levels: accessibility to their binding sites, topological restraint of their potential targets as well as facilitation of effective enhancer/promoter looping. Conversely, the modifications that the receptors exert on chromatin upon exposure to their cognate signal lead to specific restructuration of the chromatin folding, which are probably important for fine-tuning the transcriptional response. Together these observations support a role for steroid receptors as genome organizers, not only at local but also at global scale.

In addition to their direct effects on chromatin, it is also important to consider that steroids exert important so-called non-genomic actions, which can actively participate in the final transcriptional output [96,97]. For instance, estrogens and progestins can rapidly activate protein kinases [96–99] and chromatin effectors such as Poly(ADP-Ribose) Polymerase 1 (PARP1) [100] that are both important for their transcriptional effects. Activation of PARP1 upon exposure to hormones leads to the formation of Poly-ADP-ribose (PAR), which directly acts on chromatin structure through histone H1 displacement [100] as well as on the synthesis of nuclear ATP required for the response to estrogen and progestins [101]. Interestingly, PAR has been associated with phase transition mechanisms [102] and ATP acts as an hydrotrope for liquid droplet formation [103]. Such processes have been recently highlighted for their potential role in the distribution of proteins and chromatin within the nuclear space. More profound analysis of these non-genomic processes and their consequences on genome structure will be important to better understand the pleiotropic effects that steroid hormones have on gene transcription and cell fate.

Finally, if the 3D structure of the genome plays a role in transcription regulation in normal cells, one could easily hypothesize that its modifications may favor inappropriate responses. Indeed, chromosomal rearrangements could have dramatic influence on gene expression by modifying the normal landscape of action of enhancers [68,104]. Juxtaposition of specific chromosomes or genomic domains in the nucleus in normal cells reflects the preferential breakpoints that lead to oncogenic fusion proteins in some cancers [105]. Hormone induced double strand breaks upon binding of the Androgen Receptor (AR) has been involved in such processes in prostate cancers [106]. A better understanding of how these events occur is of particular importance in diseases associated with response to steroid hormones, such as breast or ovarian cancers, where the genome is frequently rearranged during the process of transformation.

Acknowledgments: We thank all members of the Chromatin and Gene Expression group (CRG, Barcelona) and the members of the 4DGenome project (CRG and CNAG-CRG, Barcelona) for helpful discussions. Research in the Beato's laboratory receives funding from the European Research Council under the European Union's Seventh Framework Programme (FP7/2007-2013)/ERC Synergy grant agreement 609989 (4DGenome). The content of this manuscript reflects only the author's views and the Union is not liable for any use that may be made of the information contained therein. We also acknowledge support of the Spanish Ministry of Economy and Competitiveness, "Centro de Excelencia Severo Ochoa 2013-2017" and Plan Nacional (SAF2016-75006-P), as well as support of the CERCA Programme/Generalitat de Catalunya.

Conflicts of Interest: The authors declare no conflicts of interest.

References

1. Nilsson, S.; Makela, S.; Treuter, E.; Tujague, M.; Thomsen, J.; Andersson, G.; Enmark, E.; Pettersson, K.; Warner, M.; Gustafsson, J.A. Mechanisms of estrogen action. *Physiol. Rev.* **2001**, *81*, 1535–1565. [CrossRef] [PubMed]

2. Mangelsdorf, D.J.; Thummel, C.; Beato, M.; Herrlich, P.; Schutz, G.; Umesono, K.; Blumberg, B.; Kastner, P.; Mark, M.; Chambon, P.; et al. The nuclear receptor superfamily: The second decade. *Cell* **1995**, *83*, 835–839. [CrossRef]

3. Levin, E.R. Plasma membrane estrogen receptors. *Trends Endocrinol. Metab. TEM* **2009**, *20*, 477–482. [CrossRef] [PubMed]

4. Kumar, V.; Chambon, P. The estrogen receptor binds tightly to its responsive element as a ligand-induced homodimer. *Cell* **1988**, *55*, 145–156. [CrossRef]

5. Safe, S. Transcriptional activation of genes by 17β-estradiol through estrogen receptor-sp1 interactions. *Vitam. Horm.* **2001**, *62*, 231–252. [PubMed]

6. Webb, P.; Nguyen, P.; Valentine, C.; Lopez, G.N.; Kwok, G.R.; McInerney, E.; Katzenellenbogen, B.S.; Enmark, E.; Gustafsson, J.A.; Nilsson, S.; et al. The estrogen receptor enhances ap-1 activity by two distinct mechanisms with different requirements for receptor transactivation functions. *Mol. Endocrinol.* **1999**, *13*, 1672–1685. [CrossRef] [PubMed]

7. Perissi, V.; Rosenfeld, M.G. Controlling nuclear receptors: The circular logic of cofactor cycles. *Nat. Rev. Mol. Cell Biol.* **2005**, *6*, 542–554. [CrossRef] [PubMed]

8. Metivier, R.; Penot, G.; Hubner, M.R.; Reid, G.; Brand, H.; Kos, M.; Gannon, F. Estrogen receptor-α directs ordered, cyclical, and combinatorial recruitment of cofactors on a natural target promoter. *Cell* **2003**, *115*, 751–763. [CrossRef]

9. Shang, Y.; Hu, X.; DiRenzo, J.; Lazar, M.A.; Brown, M. Cofactor dynamics and sufficiency in estrogen receptor-regulated transcription. *Cell* **2000**, *103*, 843–852. [CrossRef]

10. Cicatiello, L.; Scafoglio, C.; Altucci, L.; Cancemi, M.; Natoli, G.; Facchiano, A.; Iazzetti, G.; Calogero, R.; Biglia, N.; De Bortoli, M.; et al. A genomic view of estrogen actions in human breast cancer cells by expression profiling of the hormone-responsive transcriptome. *J. Mol. Endocrinol.* **2004**, *32*, 719–775. [CrossRef] [PubMed]

11. Kininis, M.; Isaacs, G.D.; Core, L.J.; Hah, N.; Kraus, W.L. Postrecruitment regulation of RNA polymerase ii directs rapid signaling responses at the promoters of estrogen target genes. *Mol. Cell. Biol.* **2009**, *29*, 1123–1133. [CrossRef] [PubMed]

12. Hah, N.; Danko, C.G.; Core, L.; Waterfall, J.J.; Siepel, A.; Lis, J.T.; Kraus, W.L. A rapid, extensive, and transient transcriptional response to estrogen signaling in breast cancer cells. *Cell* **2011**, *145*, 622–634. [CrossRef] [PubMed]

13. Carroll, J.S.; Liu, X.S.; Brodsky, A.S.; Li, W.; Meyer, C.A.; Szary, A.J.; Eeckhoute, J.; Shao, W.; Hestermann, E.V.; Geistlinger, T.R.; et al. Chromosome-wide mapping of estrogen receptor binding reveals long-range regulation requiring the forkhead protein foxa1. *Cell* **2005**, *122*, 33–43. [CrossRef] [PubMed]

14. Welboren, W.J.; van Driel, M.A.; Janssen-Megens, E.M.; van Heeringen, S.J.; Sweep, F.C.; Span, P.N.; Stunnenberg, H.G. CHIP-Seq of ERα and RNA polymerase ii defines genes differentially responding to ligands. *EMBO J.* **2009**, *28*, 1418–1428. [CrossRef] [PubMed]

15. Ross-Innes, C.S.; Stark, R.; Holmes, K.A.; Schmidt, D.; Spyrou, C.; Russell, R.; Massie, C.E.; Vowler, S.L.; Eldridge, M.; Carroll, J.S. Cooperative interaction between retinoic acid receptor-α and estrogen receptor in breast cancer. *Genes Dev.* **2010**, *24*, 171–182. [CrossRef] [PubMed]

16. Ballare, C.; Castellano, G.; Gaveglia, L.; Althammer, S.; Gonzalez-Vallinas, J.; Eyras, E.; Le Dily, F.; Zaurin, R.; Soronellas, D.; Vicent, G.P.; et al. Nucleosome-driven transcription factor binding and gene regulation. *Mol. Cell* **2013**, *49*, 67–79. [CrossRef] [PubMed]

17. John, S.; Sabo, P.J.; Thurman, R.E.; Sung, M.H.; Biddie, S.C.; Johnson, T.A.; Hager, G.L.; Stamatoyannopoulos, J.A. Chromatin accessibility pre-determines glucocorticoid receptor binding patterns. *Nat. Genet.* **2011**, *43*, 264–268. [CrossRef] [PubMed]

18. Harmston, N.; Lenhard, B. Chromatin and epigenetic features of long-range gene regulation. *Nucleic Acids Res.* **2013**, *41*, 7185–7199. [CrossRef] [PubMed]

19. Cavalli, G.; Misteli, T. Functional implications of genome topology. *Nat. Struct. Mol. Biol.* **2013**, *20*, 290–299. [CrossRef] [PubMed]

20. Dekker, J.; Mirny, L. The 3d genome as moderator of chromosomal communication. *Cell* **2016**, *164*, 1110–1121. [CrossRef] [PubMed]

21. Gibcus, J.H.; Dekker, J. The hierarchy of the 3d genome. *Mol. Cell* **2013**, *49*, 773–782. [CrossRef] [PubMed]

22. Pope, B.D.; Ryba, T.; Dileep, V.; Yue, F.; Wu, W.; Denas, O.; Vera, D.L.; Wang, Y.; Hansen, R.S.; Canfield, T.K.; et al. Topologically associating domains are stable units of replication-timing regulation. *Nature* **2014**, *515*, 402–405. [CrossRef] [PubMed]

23. Dixon, J.R.; Jung, I.; Selvaraj, S.; Shen, Y.; Antosiewicz-Bourget, J.E.; Lee, A.Y.; Ye, Z.; Kim, A.; Rajagopal, N.; Xie, W.; et al. Chromatin architecture reorganization during stem cell differentiation. *Nature* **2015**, *518*, 331–336. [CrossRef] [PubMed]

24. Dixon, J.R.; Selvaraj, S.; Yue, F.; Kim, A.; Li, Y.; Shen, Y.; Hu, M.; Liu, J.S.; Ren, B. Topological domains in mammalian genomes identified by analysis of chromatin interactions. *Nature* **2012**, *485*, 376–380. [CrossRef] [PubMed]

25. Sexton, T.; Yaffe, E.; Kenigsberg, E.; Bantignies, F.; Leblanc, B.; Hoichman, M.; Parrinello, H.; Tanay, A.; Cavalli, G. Three-dimensional folding and functional organization principles of the drosophila genome. *Cell* **2012**, *148*, 458–472. [CrossRef] [PubMed]

26. Kuznetsova, T.; Wang, S.Y.; Rao, N.A.; Mandoli, A.; Martens, J.H.; Rother, N.; Aartse, A.; Groh, L.; Janssen-Megens, E.M.; Li, G.; et al. Glucocorticoid receptor and nuclear factor κ-b affect three-dimensional chromatin organization. *Genome Biol.* **2015**, *16*, 264. [CrossRef] [PubMed]

27. Fullwood, M.J.; Liu, M.H.; Pan, Y.F.; Liu, J.; Xu, H.; Mohamed, Y.B.; Orlov, Y.L.; Velkov, S.; Ho, A.; Mei, P.H.; et al. An oestrogen-receptor-α-bound human chromatin interactome. *Nature* **2009**, *462*, 58–64. [CrossRef] [PubMed]

28. Le Dily, F.; Bau, D.; Pohl, A.; Vicent, G.P.; Serra, F.; Soronellas, D.; Castellano, G.; Wright, R.H.; Ballare, C.; Filion, G.; et al. Distinct structural transitions of chromatin topological domains correlate with coordinated hormone-induced gene regulation. *Genes Dev.* **2014**, *28*, 2151–2162. [CrossRef] [PubMed]

29. Le Dily, F.; Beato, M. Tads as modular and dynamic units for gene regulation by hormones. *FEBS Lett.* **2015**, *589*, 2885–2892. [CrossRef] [PubMed]

30. Dekker, J.; Rippe, K.; Dekker, M.; Kleckner, N. Capturing chromosome conformation. *Science* **2002**, *295*, 1306–1311. [CrossRef] [PubMed]

31. Lieberman-Aiden, E.; van Berkum, N.L.; Williams, L.; Imakaev, M.; Ragoczy, T.; Telling, A.; Amit, I.; Lajoie, B.R.; Sabo, P.J.; Dorschner, M.O.; et al. Comprehensive mapping of long-range interactions reveals folding principles of the human genome. *Science* **2009**, *326*, 289–293. [CrossRef] [PubMed]

32. Nora, E.P.; Lajoie, B.R.; Schulz, E.G.; Giorgetti, L.; Okamoto, I.; Servant, N.; Piolot, T.; van Berkum, N.L.; Meisig, J.; Sedat, J.; et al. Spatial partitioning of the regulatory landscape of the x-inactivation centre. *Nature* **2012**, *485*, 381–385. [CrossRef] [PubMed]

33. Rao, S.S.; Huntley, M.H.; Durand, N.C.; Stamenova, E.K.; Bochkov, I.D.; Robinson, J.T.; Sanborn, A.L.; Machol, I.; Omer, A.D.; Lander, E.S.; et al. A 3D map of the human genome at kilobase resolution reveals principles of chromatin looping. *Cell* **2014**, *159*, 1665–1680. [CrossRef] [PubMed]

34. Cremer, T.; Cremer, C.; Schneider, T.; Baumann, H.; Hens, L.; Kirsch-Volders, M. Analysis of chromosome positions in the interphase nucleus of chinese hamster cells by laser-UV-microirradiation experiments. *Hum. Genet.* **1982**, *62*, 201–209. [CrossRef] [PubMed]

35. Postberg, J.; Lipps, H.J.; Cremer, T. Evolutionary origin of the cell nucleus and its functional architecture. *Essays Biochem.* **2010**, *48*, 1–24. [CrossRef] [PubMed]

36. Heride, C.; Ricoul, M.; Kieu, K.; von Hase, J.; Guillemot, V.; Cremer, C.; Dubrana, K.; Sabatier, L. Distance between homologous chromosomes results from chromosome positioning constraints. *J. Cell Sci.* **2010**, *123*, 4063–4075. [CrossRef] [PubMed]

37. Wang, S.; Su, J.H.; Beliveau, B.J.; Bintu, B.; Moffitt, J.R.; Wu, C.T.; Zhuang, X. Spatial organization of chromatin domains and compartments in single chromosomes. *Science* **2016**, *353*, 598–602. [CrossRef] [PubMed]

38. Rowley, M.J.; Nichols, M.H.; Lyu, X.; Ando-Kuri, M.; Rivera, I.S.M.; Hermetz, K.; Wang, P.; Ruan, Y.; Corces, V.G. Evolutionarily conserved principles predict 3d chromatin organization. *Mol. Cell* **2017**, *67*, 837–852. [CrossRef] [PubMed]

39. Bonev, B.; Mendelson Cohen, N.; Szabo, Q.; Fritsch, L.; Papadopoulos, G.L.; Lubling, Y.; Xu, X.; Lv, X.; Hugnot, J.P.; Tanay, A.; et al. Multiscale 3d genome rewiring during mouse neural development. *Cell* **2017**, *171*, 557–572. [CrossRef] [PubMed]

40. Stadhouders, R.; Vidal, E.; Serra, F.; Di Stefano, B.; Le Dily, F.; Quilez, J.; Gomez, A.; Collombet, S.; Berenguer, C.; Cuartero, Y.; et al. Transcription factors orchestrate dynamic interplay between genome topology and gene regulation during cell reprogramming. *Nat. Genet.* **2018**. [CrossRef] [PubMed]

41. Hou, C.; Li, L.; Qin, Z.S.; Corces, V.G. Gene density, transcription, and insulators contribute to the partition of the drosophila genome into physical domains. *Mol. Cell* **2012**, *48*, 471–484. [CrossRef] [PubMed]

42. Van Bortle, K.; Corces, V.G. The role of chromatin insulators in nuclear architecture and genome function. *Curr. Opin. Genet. Dev.* **2013**, *23*, 212–218. [CrossRef] [PubMed]

43. Zhan, Y.; Mariani, L.; Barozzi, I.; Schulz, E.G.; Bluthgen, N.; Stadler, M.; Tiana, G.; Giorgetti, L. Reciprocal insulation analysis of Hi-C data shows that tads represent a functionally but not structurally privileged scale in the hierarchical folding of chromosomes. *Genome Res.* **2017**, *27*, 479–490. [CrossRef] [PubMed]

44. Neems, D.S.; Garza-Gongora, A.G.; Smith, E.D.; Kosak, S.T. Topologically associated domains enriched for lineage-specific genes reveal expression-dependent nuclear topologies during myogenesis. *Proc. Natl. Acad. Sci. USA* **2016**, *113*, E1691–E1700. [CrossRef] [PubMed]

45. Phillips-Cremins, J.E.; Sauria, M.E.; Sanyal, A.; Gerasimova, T.I.; Lajoie, B.R.; Bell, J.S.; Ong, C.T.; Hookway, T.A.; Guo, C.; Sun, Y.; et al. Architectural protein subclasses shape 3d organization of genomes during lineage commitment. *Cell* **2013**, *153*, 1281–1295. [CrossRef] [PubMed]

46. Jin, F.; Li, Y.; Dixon, J.R.; Selvaraj, S.; Ye, Z.; Lee, A.Y.; Yen, C.A.; Schmitt, A.D.; Espinoza, C.A.; Ren, B. A high-resolution map of the three-dimensional chromatin interactome in human cells. *Nature* **2013**, *503*, 290–294. [CrossRef] [PubMed]

47. Dowen, J.M.; Fan, Z.P.; Hnisz, D.; Ren, G.; Abraham, B.J.; Zhang, L.N.; Weintraub, A.S.; Schujiers, J.; Lee, T.I.; Zhao, K.; et al. Control of cell identity genes occurs in insulated neighborhoods in mammalian chromosomes. *Cell* **2014**, *159*, 374–387. [CrossRef] [PubMed]

48. Weintraub, A.S.; Li, C.H.; Zamudio, A.V.; Sigova, A.A.; Hannett, N.M.; Day, D.S.; Abraham, B.J.; Cohen, M.A.; Nabet, B.; Buckley, D.L.; et al. Yy1 is a structural regulator of enhancer-promoter loops. *Cell* **2017**, *171*, 1573–1588. [CrossRef] [PubMed]

49. Nagano, T.; Lubling, Y.; Yaffe, E.; Wingett, S.W.; Dean, W.; Tanay, A.; Fraser, P. Single-cell Hi-C for genome-wide detection of chromatin interactions that occur simultaneously in a single cell. *Nat. Protoc.* **2015**, *10*, 1986–2003. [CrossRef] [PubMed]

50. Williamson, I.; Berlivet, S.; Eskeland, R.; Boyle, S.; Illingworth, R.S.; Paquette, D.; Dostie, J.; Bickmore, W.A. Spatial genome organization: Contrasting views from chromosome conformation capture and fluorescence in situ hybridization. *Genes Dev.* **2014**, *28*, 2778–2791. [CrossRef] [PubMed]

51. Schwarzer, W.; Abdennur, N.; Goloborodko, A.; Pekowska, A.; Fudenberg, G.; Loe-Mie, Y.; Fonseca, N.A.; Huber, W.; Haering, C.H.; Mirny, L.; et al. Two independent modes of chromatin organization revealed by cohesin removal. *Nature* **2017**, *551*, 51–56. [CrossRef] [PubMed]

52. Nora, E.P.; Goloborodko, A.; Valton, A.L.; Gibcus, J.H.; Uebersohn, A.; Abdennur, N.; Dekker, J.; Mirny, L.A.; Bruneau, B.G. Targeted degradation of ctcf decouples local insulation of chromosome domains from genomic compartmentalization. *Cell* **2017**, *169*, 930–944. [CrossRef] [PubMed]

53. Rao, S.S.P.; Huang, S.C.; Glenn St Hilaire, B.; Engreitz, J.M.; Perez, E.M.; Kieffer-Kwon, K.R.; Sanborn, A.L.; Johnstone, S.E.; Bascom, G.D.; Bochkov, I.D.; et al. Cohesin loss eliminates all loop domains. *Cell* **2017**, *171*, 305–320. [CrossRef] [PubMed]

54. Ross-Innes, C.S.; Stark, R.; Teschendorff, A.E.; Holmes, K.A.; Ali, H.R.; Dunning, M.J.; Brown, G.D.; Gojis, O.; Ellis, I.O.; Green, A.R.; et al. Differential oestrogen receptor binding is associated with clinical outcome in breast cancer. *Nature* **2012**, *481*, 389–393. [CrossRef] [PubMed]

55. Creighton, C.J.; Cordero, K.E.; Larios, J.M.; Miller, R.S.; Johnson, M.D.; Chinnaiyan, A.M.; Lippman, M.E.; Rae, J.M. Genes regulated by estrogen in breast tumor cells in vitro are similarly regulated in vivo in tumor xenografts and human breast tumors. *Genome Biol.* **2006**, *7*, R28. [CrossRef] [PubMed]

56. Rangel, N.; Villegas, V.E.; Rondon-Lagos, M. Profiling of gene expression regulated by 17β-estradiol and tamoxifen in estrogen receptor-positive and estrogen receptor-negative human breast cancer cell lines. *Breast Cancer* **2017**, *9*, 537–550. [PubMed]

57. Carroll, J.S.; Meyer, C.A.; Song, J.; Li, W.; Geistlinger, T.R.; Eeckhoute, J.; Brodsky, A.S.; Keeton, E.K.; Fertuck, K.C.; Hall, G.F.; et al. Genome-wide analysis of estrogen receptor binding sites. *Nat. Genet.* **2006**, *38*, 1289–1297. [CrossRef] [PubMed]

58. Joseph, R.; Orlov, Y.L.; Huss, M.; Sun, W.; Kong, S.L.; Ukil, L.; Pan, Y.F.; Li, G.; Lim, M.; Thomsen, J.S.; et al. Integrative model of genomic factors for determining binding site selection by estrogen receptor-α. *Mol. Syst. Biol.* **2010**, *6*, 456. [CrossRef] [PubMed]

59. Zaret, K.S.; Carroll, J.S. Pioneer transcription factors: Establishing competence for gene expression. *Genes Dev.* **2011**, *25*, 2227–2241. [CrossRef] [PubMed]

60. Hurtado, A.; Holmes, K.A.; Ross-Innes, C.S.; Schmidt, D.; Carroll, J.S. Foxa1 is a key determinant of estrogen receptor function and endocrine response. *Nat. Genet.* **2011**, *43*, 27–33. [CrossRef] [PubMed]

61. Verde, G.; De Llobet, L.I.; Wright, R.H.G.; Quilez, J.; Peiro, S.; Le Dily, F.; Beato, M. Progesterone receptor maintains estrogen receptor gene expression by regulating DNA methylation in hormone-free breast cancer cells. *bioRxiv* **2017**. [CrossRef]

62. Slattery, M.; Zhou, T.; Yang, L.; Dantas Machado, A.C.; Gordan, R.; Rohs, R. Absence of a simple code: How transcription factors read the genome. *Trends Biochem. Sci.* **2014**, *39*, 381–399. [CrossRef] [PubMed]

63. Berg, O.G.; Winter, R.B.; von Hippel, P.H. Diffusion-driven mechanisms of protein translocation on nucleic acids. 1. Models and theory. *Biochemistry* **1981**, *20*, 6929–6948. [CrossRef] [PubMed]

64. Cortini, R.; Filion, G. Principles of transcription factor traffic on folded chromatin. *bioRxiv* **2017**. [CrossRef]

65. Li, G.; Cai, L.; Chang, H.; Hong, P.; Zhou, Q.; Kulakova, E.V.; Kolchanov, N.A.; Ruan, Y. Chromatin interaction analysis with paired-end tag (chia-pet) sequencing technology and application. *BMC Genom.* **2014**, *15* (Suppl. 12), S11. [CrossRef] [PubMed]

66. Sanyal, A.; Lajoie, B.R.; Jain, G.; Dekker, J. The long-range interaction landscape of gene promoters. *Nature* **2012**, *489*, 109–113. [CrossRef] [PubMed]

67. Shen, Y.; Yue, F.; McCleary, D.F.; Ye, Z.; Edsall, L.; Kuan, S.; Wagner, U.; Dixon, J.; Lee, L.; Lobanenkov, V.V.; et al. A map of the *cis*-regulatory sequences in the mouse genome. *Nature* **2012**, *488*, 116–120. [CrossRef] [PubMed]

68. Lupianez, D.G.; Kraft, K.; Heinrich, V.; Krawitz, P.; Brancati, F.; Klopocki, E.; Horn, D.; Kayserili, H.; Opitz, J.M.; Laxova, R.; et al. Disruptions of topological chromatin domains cause pathogenic rewiring of gene-enhancer interactions. *Cell* **2015**, *161*, 1012–1025. [CrossRef] [PubMed]

69. Hon, G.; Hawkins, R.D.; Caballero, O.L.; Lo, C.; Lister, R.; Pelizzola, M.; Valsesia, A.; Ye, Z.; Kuan, S.; Edsall, L.E.; et al. Global DNA hypomethylation coupled to repressive chromatin domain formation and gene silencing in breast cancer. *Genome Res.* **2012**, *22*, 246–258. [CrossRef] [PubMed]

70. Bert, S.A.; Robinson, M.D.; Strbenac, D.; Statham, A.L.; Song, J.Z.; Hulf, T.; Sutherland, R.L.; Coolen, M.W.; Stirzaker, C.; Clark, S.J. Regional activation of the cancer genome by long-range epigenetic remodeling. *Cancer Cell* **2013**, *23*, 9–22. [CrossRef] [PubMed]

71. McDonald, O.G.; Wu, H.; Timp, W.; Doi, A.; Feinberg, A.P. Genome-scale epigenetic reprogramming during epithelial-to-mesenchymal transition. *Nat. Struct. Mol. Biol.* **2011**, *18*, 867–874. [CrossRef] [PubMed]

72. Antony, J.; Dasgupta, T.; Rhodes, J.M.; McEwan, M.V.; Print, C.G.; O'Sullivan, J.M.; Horsfield, J.A. Cohesin modulates transcription of estrogen-responsive genes. *Biochim. Biophys. Acta* **2015**, *1849*, 257–269. [CrossRef] [PubMed]

73. Ross-Innes, C.S.; Brown, G.D.; Carroll, J.S. A co-ordinated interaction between ctcf and er in breast cancer cells. *BMC Genom.* **2011**, *12*, 593. [CrossRef] [PubMed]

74. Quintin, J.; Le Peron, C.; Palierne, G.; Bizot, M.; Cunha, S.; Serandour, A.A.; Avner, S.; Henry, C.; Percevault, F.; Belaud-Rotureau, M.A.; et al. Dynamic estrogen receptor interactomes control estrogen-responsive trefoil factor (TFF) locus cell-specific activities. *Mol. Cell. Biol.* **2014**, *34*, 2418–2436. [CrossRef] [PubMed]

75. Schmidt, D.; Schwalie, P.C.; Ross-Innes, C.S.; Hurtado, A.; Brown, G.D.; Carroll, J.S.; Flicek, P.; Odom, D.T. A ctcf-independent role for cohesin in tissue-specific transcription. *Genome Res.* **2010**, *20*, 578–588. [CrossRef] [PubMed]

76. Chan, C.S.; Song, J.S. Ccctc-binding factor confines the distal action of estrogen receptor. *Cancer Res.* **2008**, *68*, 9041–9049. [CrossRef] [PubMed]

77. Wu, J.I.; Lessard, J.; Crabtree, G.R. Understanding the words of chromatin regulation. *Cell* **2009**, *136*, 200–206. [CrossRef] [PubMed]

78. Johnson, T.A.; Elbi, C.; Parekh, B.S.; Hager, G.L.; John, S. Chromatin remodeling complexes interact dynamically with a glucocorticoid receptor-regulated promoter. *Mol. Biol. Cell* **2008**, *19*, 3308–3322. [CrossRef] [PubMed]

79. Vicent, G.P.; Nacht, A.S.; Font-Mateu, J.; Castellano, G.; Gaveglia, L.; Ballare, C.; Beato, M. Four enzymes cooperate to displace histone h1 during the first minute of hormonal gene activation. *Genes Dev.* **2011**, *25*, 845–862. [CrossRef] [PubMed]

80. Bretschneider, N.; Kangaspeska, S.; Seifert, M.; Reid, G.; Gannon, F.; Denger, S. E2-mediated cathepsin d (ctsd) activation involves looping of distal enhancer elements. *Mol. Oncol.* **2008**, *2*, 182–190. [CrossRef] [PubMed]

81. Hsu, P.Y.; Hsu, H.K.; Singer, G.A.; Yan, P.S.; Rodriguez, B.A.; Liu, J.C.; Weng, Y.I.; Deatherage, D.E.; Chen, Z.; Pereira, J.S.; et al. Estrogen-mediated epigenetic repression of large chromosomal regions through DNA looping. *Genome Res.* **2010**, *20*, 733–744. [CrossRef] [PubMed]

82. Belakavadi, M.; Fondell, J.D. Role of the mediator complex in nuclear hormone receptor signaling. *Rev. Physiol. Biochem. Pharmacol.* **2006**, *156*, 23–43. [PubMed]

83. Malik, S.; Roeder, R.G. The metazoan mediator co-activator complex as an integrative hub for transcriptional regulation. *Nat. Rev. Genet.* **2010**, *11*, 761–772. [CrossRef] [PubMed]

84. Bojcsuk, D.; Nagy, G.; Balint, B.L. Inducible super-enhancers are organized based on canonical signal-specific transcription factor binding elements. *Nucleic Acids Res.* **2017**, *45*, 3693–3706. [CrossRef] [PubMed]

85. Li, W.; Notani, D.; Ma, Q.; Tanasa, B.; Nunez, E.; Chen, A.Y.; Merkurjev, D.; Zhang, J.; Ohgi, K.; Song, X.; et al. Functional roles of enhancer RNAs for oestrogen-dependent transcriptional activation. *Nature* **2013**, *498*, 516–520. [CrossRef] [PubMed]

86. Hah, N.; Murakami, S.; Nagari, A.; Danko, C.G.; Kraus, W.L. Enhancer transcripts mark active estrogen receptor binding sites. *Genome Res.* **2013**, *23*, 1210–1223. [CrossRef] [PubMed]

87. Lai, F.; Orom, U.A.; Cesaroni, M.; Beringer, M.; Taatjes, D.J.; Blobel, G.A.; Shiekhattar, R. Activating RNAs associate with mediator to enhance chromatin architecture and transcription. *Nature* **2013**, *494*, 497–501. [CrossRef] [PubMed]

88. Le Dily, F.; Vidal, E.; Cuartero, Y.; Quilez, J.; Nacht, S.; Vicent, G.P.; Sharma, P.; Verde, G.; Beato, M. Hormone control regions mediate opposing steroid receptor-dependent genome organizations. *bioRxiv* **2017**. [CrossRef]

89. Hakim, O.; Sung, M.H.; Voss, T.C.; Splinter, E.; John, S.; Sabo, P.J.; Thurman, R.E.; Stamatoyannopoulos, J.A.; de Laat, W.; Hager, G.L. Diverse gene reprogramming events occur in the same spatial clusters of distal regulatory elements. *Genome Res.* **2011**, *21*, 697–706. [CrossRef] [PubMed]

90. Rafique, S.; Thomas, J.S.; Sproul, D.; Bickmore, W.A. Estrogen-induced chromatin decondensation and nuclear re-organization linked to regional epigenetic regulation in breast cancer. *Genome Biol.* **2015**, *16*, 145. [CrossRef] [PubMed]

91. Jubb, A.W.; Boyle, S.; Hume, D.A.; Bickmore, W.A. Glucocorticoid receptor binding induces rapid and prolonged large-scale chromatin decompaction at multiple target loci. *Cell Rep.* **2017**, *21*, 3022–3031. [CrossRef] [PubMed]

92. Nye, A.C.; Rajendran, R.R.; Stenoien, D.L.; Mancini, M.A.; Katzenellenbogen, B.S.; Belmont, A.S. Alteration of large-scale chromatin structure by estrogen receptor. *Mol. Cell. Biol.* **2002**, *22*, 3437–3449. [CrossRef] [PubMed]

93. Mourad, R.; Hsu, P.Y.; Juan, L.; Shen, C.; Koneru, P.; Lin, H.; Liu, Y.; Nephew, K.; Huang, T.H.; Li, L. Estrogen induces global reorganization of chromatin structure in human breast cancer cells. *PLoS ONE* **2014**, *9*, e113354. [CrossRef] [PubMed]

94. Hu, Q.; Kwon, Y.S.; Nunez, E.; Cardamone, M.D.; Hutt, K.R.; Ohgi, K.A.; Garcia-Bassets, I.; Rose, D.W.; Glass, C.K.; Rosenfeld, M.G.; et al. Enhancing nuclear receptor-induced transcription requires nuclear motor and lsd1-dependent gene networking in interchromatin granules. *Proc. Natl. Acad. Sci. USA* **2008**, *105*, 19199–19204. [CrossRef] [PubMed]

95. Kocanova, S.; Kerr, E.A.; Rafique, S.; Boyle, S.; Katz, E.; Caze-Subra, S.; Bickmore, W.A.; Bystricky, K. Activation of estrogen-responsive genes does not require their nuclear co-localization. *PLoS Genet.* **2010**, *6*, e1000922. [CrossRef] [PubMed]

96. Levin, E.R. Integration of the extranuclear and nuclear actions of estrogen. *Mol. Endocrinol.* **2005**, *19*, 1951–1959. [CrossRef] [PubMed]

97. Vicent, G.P.; Ballare, C.; Nacht, A.S.; Clausell, J.; Subtil-Rodriguez, A.; Quiles, I.; Jordan, A.; Beato, M. Induction of progesterone target genes requires activation of erk and msk kinases and phosphorylation of histone h3. *Mol. Cell* **2006**, *24*, 367–381. [CrossRef] [PubMed]

98. Migliaccio, A.; Piccolo, D.; Castoria, G.; Di Domenico, M.; Bilancio, A.; Lombardi, M.; Gong, W.; Beato, M.; Auricchio, F. Activation of the src/p21ras/erk pathway by progesterone receptor via cross-talk with estrogen receptor. *EMBO J.* **1998**, *17*, 2008–2018. [CrossRef] [PubMed]

99. Revankar, C.M.; Cimino, D.F.; Sklar, L.A.; Arterburn, J.B.; Prossnitz, E.R. A transmembrane intracellular estrogen receptor mediates rapid cell signaling. *Science* **2005**, *307*, 1625–1630. [CrossRef] [PubMed]

100. Wright, R.H.; Castellano, G.; Bonet, J.; Le Dily, F.; Font-Mateu, J.; Ballare, C.; Nacht, A.S.; Soronellas, D.; Oliva, B.; Beato, M. CDK2-dependent activation of PARP-1 is required for hormonal gene regulation in breast cancer cells. *Genes Dev.* **2012**, *26*, 1972–1983. [CrossRef] [PubMed]

101. Wright, R.H.; Lioutas, A.; Le Dily, F.; Soronellas, D.; Pohl, A.; Bonet, J.; Nacht, A.S.; Samino, S.; Font-Mateu, J.; Vicent, G.P.; et al. ADP-ribose-derived nuclear ATP synthesis by NUDIX5 is required for chromatin remodeling. *Science* **2016**, *352*, 1221–1225. [CrossRef] [PubMed]

102. Altmeyer, M.; Neelsen, K.J.; Teloni, F.; Pozdnyakova, I.; Pellegrino, S.; Grofte, M.; Rask, M.B.; Streicher, W.; Jungmichel, S.; Nielsen, M.L.; et al. Liquid demixing of intrinsically disordered proteins is seeded by poly(adp-ribose). *Nat. Commun.* **2015**, *6*, 8088. [CrossRef] [PubMed]

103. Patel, A.; Malinovska, L.; Saha, S.; Wang, J.; Alberti, S.; Krishnan, Y.; Hyman, A.A. ATP as a biological hydrotrope. *Science* **2017**, *356*, 753–756. [CrossRef] [PubMed]

104. Franke, M.; Ibrahim, D.M.; Andrey, G.; Schwarzer, W.; Heinrich, V.; Schopflin, R.; Kraft, K.; Kempfer, R.; Jerkovic, I.; Chan, W.L.; et al. Formation of new chromatin domains determines pathogenicity of genomic duplications. *Nature* **2016**, *538*, 265–269. [CrossRef] [PubMed]

105. Zhang, Y.; Gong, M.; Yuan, H.; Park, H.G.; Frierson, H.F.; Li, H. Chimeric transcript generated by *cis*-splicing of adjacent genes regulates prostate cancer cell proliferation. *Cancer Discov.* **2012**, *2*, 598–607. [CrossRef] [PubMed]

106. Lin, C.; Yang, L.; Tanasa, B.; Hutt, K.; Ju, B.G.; Ohgi, K.; Zhang, J.; Rose, D.W.; Fu, X.D.; Glass, C.K.; et al. Nuclear receptor-induced chromosomal proximity and DNA breaks underlie specific translocations in cancer. *Cell* **2009**, *139*, 1069–1083. [CrossRef] [PubMed]

International Journal of
Molecular Sciences

MDPI

Review

Hypoxia and Hormone-Mediated Pathways Converge at the Histone Demethylase KDM4B in Cancer

Jun Yang [1,*], Adrian L. Harris [2] and Andrew M. Davidoff [1]

[1] Department of Surgery, St. Jude Children's Research Hospital, 262 Danny Thomas Place,
     Memphis, TN 38105, USA; Andrew.Davidoff@stjude.org
[2] Molecular Oncology Laboratories, Department of Oncology, Weatherall Institute of Molecular Medicine,
     University of Oxford, Oxford OX3 9DS, UK; aharris.lab@imm.ox.ac.uk
*   Correspondence: Jun.Yang2@stjude.org

Received: 19 December 2017; Accepted: 9 January 2018; Published: 13 January 2018

Abstract: Hormones play an important role in pathophysiology. The hormone receptors, such as estrogen receptor alpha and androgen receptor in breast cancer and prostate cancer, are critical to cancer cell proliferation and tumor growth. In this review we focused on the cross-talk between hormone and hypoxia pathways, particularly in breast cancer. We delineated a novel signaling pathway from estrogen receptor to hypoxia-inducible factor 1, and discussed the role of this pathway in endocrine therapy resistance. Further, we discussed the estrogen and hypoxia pathways converging at histone demethylase KDM4B, an important epigenetic modifier in cancer.

Keywords: estrogen receptor alpha; hypoxia-inducible factor 1; KDM4B; endocrine therapy resistance

1. Introduction

A solid tumor is a heterogeneous mass that is comprised of not only genetically and epigenetically distinct clones, but also of areas with varying degree of hypoxia that result from rapid cancer cell proliferation that outgrows its blood supply. To survive in hostile hypoxic environments, cancer cells decelerate their proliferation rate, alter metabolism and cellular pH, and induce angiogenesis [1]. These responses of cells to hypoxia are largely coordinated by hypoxia-inducible factors HIF-1 and HIF-2 [2], which drive the expression of a plethora of target genes, such as vascular endothelial growth factor (*VEGF*), carbonic anhydrase IX (*CA9*), and glucose transporter 1 (*Glut-1*) [2]. HIF is a heterodimer composed of an alpha subunit (HIF-1α or HIF-2α) and a β subunit (HIF-1β) [3]. HIF-1β is constitutively expressed, whereas HIF-α is regulated by oxygen availability. Under conditions of normoxia, HIF-α is hydroxylated at conserved prolyl residues by oxygen-dependent prolyl hydroxylases (PHDs), resulting in binding of the von Hippel-Lindau protein (VHL), an E3 ubiquitin ligase, for ubiquitin-mediated degradation [4–6]. Under conditions of hypoxia, however, HIFα is stabilized through inhibition of hydroxylation, leading to transactivation of its target genes. Preclinical and clinical studies show that the hypoxia/HIF-1 pathway plays an important role in promoting local tumor invasion and distal metastasis, as well as negatively influencing the responses to radiotherapy and chemotherapy [2,7–11]. The hypoxia/HIF-1 pathway is also involved in immunosuppression and resistance to immunotherapy [12]. In addition to oxygen-mediated degradation of HIFα through VHL, the hypoxia signaling pathway is regulated by a variety of oncogenes (e.g., ERK [13], HER2 [14], mTOR [15], Ras [16,17]) tumor suppressors (e.g., LKB1 [18], PML [19], PTEN [20,21], p53 [22], SDHB [23]), and metabolites (e.g., 2-hydroxyglutarate [24], succinate [25], and fumarate [26]). Here, we focus on the cross-talk between hypoxia and estrogen-mediated pathways, which converge to regulate epigenetic modulators in breast cancer, one of the most common cancers with 450,000 deaths each year worldwide.

Int. J. Mol. Sci. **2018**, *19*, 240

2. The Cross-Talk between Hypoxia and Estrogen

Fifty years ago, Mirand et al. reported that in rodents, estradiol cyclopentylpropionate (ECP) was able to inhibit erythropoiesis by suppressing the production of erythropoiesis stimulating factor (ESF), which is now known as erythropoietin (EPO) [27], a direct target of hypoxia inducible factor. A subsequent study in 1973 by Gordon et al. further confirmed that estrogen inhibits the production of EPO in female rats exposed to various degrees of hypoxia [28]. Paradoxically, estrogen increases splenic erythropoiesis that is accompanied with elevated plasma EPO levels [29]. Interestingly, the expression of *EPO* mRNA is stimulated by both estradiol and hypoxia in the mouse uterus [30], and the hypoxic induction of *EPO* requires the presence of estradiol [30]. In 1999, Ruohola et al. reported that estradiol caused an increase of the HIF target *VEGF* mRNA in MCF-7 breast cancer cells, which was blocked by antiestrogen ICI 182780, suggesting that the effect was mediated by the estrogen receptor [31]. Subsequent studies further demonstrated the dual regulation of *VEGF* by hypoxia and estrogen [32–34]. These data indicate that estrogen and hypoxia pathways are connected. A later study showed that 17-β estradiol attenuates the hypoxic induction of HIF-1α and EPO in Hep3B cells [35]. However, in estrogen receptor-positive breast cancer cells, estrogen induces activation of HIF-1α [34] and co-operates with hypoxia to regulate the expression of a subgroup of genes [36]. Estrogen receptor antagonists (e.g., tamoxifen, raloxifene, or bazedoxifene) all suppress HIF-1α protein accumulation in osteoclast precursor cells [37]. Therefore, estrogen-mediated signaling can either negatively or positively affect the hypoxia pathway in different cellular contexts.

Estrogen receptor alpha (ERα) is an estrogen-dependent nuclear transcription factor that is not only critical for mammary epithelial cell division, but also breast cancer progression [38,39]. Despite the multiple molecular subtypes that have been classified based on transcriptomic and genetic features [40], ERα is one of the most important biomarkers directing breast cancer treatment. It is recommended that all patients with ERα positivity should have adjuvant endocrine therapy. ERα is expressed in approximately 70% of breast tumors [41], the majority of which depend on estrogen signaling, thereby providing the rationale for using anti-estrogens as adjuvant therapy to treat breast cancer [42]. Endocrine therapy drugs for breast cancer include selective ER modulators, such as tamoxifen, antagonists such as fulvestrant, and aromatase inhibitors such as anastrozole. Tamoxifen is a first-generation selective ER modulator (SERM) and has been widely used in breast cancer prevention and treatment [42]. It antagonizes ERα function in breast cancer cells by competing with estrogen for ERα binding while preserving its activating and estrogen-like functions in the bone [43]. Although now replaced by aromatase inhibitors (AI) as first-line treatment in post-menopausal women, tamoxifen still remains important in premenopausal breast cancer and after failure of AIs. The antagonist fulvestrant leads to ERα protein degradation [44], while aromatase inhibitors block the conversion of androgens to estrogens thereby reducing overall estrogen levels [45]. The application of endocrine therapies has led to a significant reduction in breast cancer mortality [46]. However, not all ER-positive patients respond to endocrine therapies and nearly all women with advanced cancer will eventually die from metastatic disease [47,48], as resistance often develops [49]. Many mechanisms have been proposed to account for endocrine therapy resistance [50,51], including loss of ERα expression or expression of truncated ER isoforms, posttranslational modification of ERα, deregulation of ERα co-activators, and increased receptor tyrosine kinase signaling. Recent studies further indicate that somatic ERα mutation [52,53], as well as genomic amplification of distant ER response elements [54] could contribute to hormone therapy resistance. Hypoxia is also involved in endocrine therapy resistance. Clinical studies have shown that HIF-1α expression is associated with an aggressive phenotype of breast cancer, i.e., large tumor size, high grade, high proliferation rate, and lymph node metastasis [55]. Increased HIF-1α is also associated with ERα positivity [55], whilst HIF-1β, the partner of HIF-1α, has been shown to function as a potent co-activator of ER-dependent transcription [56]. Importantly, HIF-1α protein expression was associated with tamoxifen resistance in neoadjuvant, primary therapy of ERα-positive breast cancers [57], as well as resistance to chemoendocrine therapy [58].

The exact nature of the relationship between hypoxia and estrogen pathways was a puzzle until our recent findings showing that the *HIF-1α* gene is a direct target of ERα [59]. In this study, we analyzed the global gene expression profile in response to hypoxia and the ERα antagonist fulvestrant and found a subgroup of genes that were dually responsive to the hormone and to oxygen. These genes were upregulated by hypoxia but the ERα antagonist fulvestrant significantly reduced their expression. These data were consistent with previous studies that showed some genes, such as *KDM4B*, *STC2*, and *VEGF*, bear both a hypoxia response element and estrogen response element [60–66]. Most interestingly, we found that ERα signaling directly regulates HIF-1α expression. When MCF7 cells were grown without estrogen and then placed in hypoxia or treated with the hypoxia mimetic deferoximine, estradiol greatly enhanced HIF-1α expression and this was reversed by fulvestrant and ERα depletion. By analyzing the *HIF-1α* genomic sequence that bears 15 exons and 14 introns, we identified a canonical estrogen response element (ERE) located in the first intron (Figure 1A). Interestingly, there is also a FOXA1 binding site that is 64 nucleotides downstream of ERE, further supporting it as a bona fide ERα binding element, because FOXA1 is a pioneer factor that facilitates ERα recruitment [67]. Actually, one study has shown that overexpression of FOXA1 in ER-positive breast cancer cell lines promotes resistance to tamoxifen and to estrogen deprivation [68]. We further validated our findings by chromatin immunoprecipitation-PCR and a luciferase reporter assay, showing that ERα directly binds to this locus, driving *HIF-1α* gene expression. This finding not only explains the early findings that estrogen and hypoxia pathways crosstalk, but also indicates that overactive HIF-1α function may partially compensate for estrogen signaling when ERα function is compromised, such as in the circumstance of hormone therapy, leading to hormone therapy resistance (Figure 1B,C).

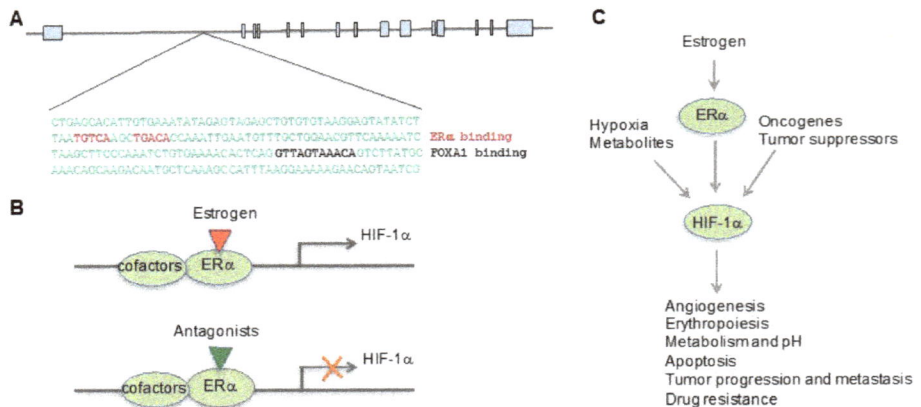

Figure 1. Estrogen pathway directly drives HIF-1α expression. (**A**) *HIF-1α* gene bears a canonical estrogen receptor binding element (ERE), with a FOXA1 binding site downstream of ERE; (**B**) When ERα is bound by its ligand it drives the expression of HIF-1α. However, ERα antagonists block the expression of HIF-1α; (**C**) The pathways mediated by hypoxia, estrogen, metabolites, and cancer genes converge on HIF-1α, which drives a plethora of genes that are involved in multiple biological processes, cancer progression, and therapeutic resistance.

3. The Hypoxia and Estrogen Signaling Pathways Converge on Histone Demethylases

Genetic abnormalities that drive tumorigenesis are usually coupled with epigenetic alterations that engage multiple important biological processes such as DNA replication, DNA repair, and gene expression [69–73]. One such aberrant chromatin modification is histone lysine methylation [69,70], which was believed to be irreversible until the discovery of lysine-specific demethylase 1 (LSD1) [74].

Subsequent studies identified another family of histone demethylases, the Jumonji C (JmjC) domain–containing demethylases [75], which require iron and 2-oxoglutarate (2-OG) for their activities. The JmjC histone lysine demethylase family (KDMs) is composed of 17 members and is responsible for reversing most of the histone methyl marks in the human genome. Dysregulated histone lysine methylation is commonly seen in various cancers [76], which is consistent with observed genetic alterations and/or dysregulation of histone methyltransferases and KDMs [75,77–81]. Interestingly, we and others have shown that many JmjC histone demethylases are hypoxia-inducible [61,82], some of which, including KDM3A [60,61,83], KDM4B [60,61,63], and KDM4C [84], are direct targets of HIFs.

The KDM4 subfamily of histone demethylases consists of four members. KDM4A, KDM4B, and KDM4C share high sequence homology in their catalytic domains, and they remove methyl groups from H3K9me2/me3 and H3K36me2/me3 [75]. KDM4A-4C members also bear other similar functional domains that include two PHD domains and two Tudor domains. However, KDM4D is less conserved and removes methyl groups only from H3K9me2/me3. KDM4B plays important roles in the self-renewal of embryonic stem cells and the conversion of induced pluripotent stem cells [85,86], and is linked to many forms of cancer [87]. KDM4B is amplified in medulloblastoma [88] and malignant peripheral nerve sheath tumors [89], and is overexpressed in many other cancers [90–92]. KDM4B regulates the expression of key oncogenes, such as *C-MYC* [93–96] and *CDK6* [97], and is involved in cancer invasiveness, metastasis, and therapeutic resistance [98–100]. Interestingly, KDM4B is a direct target of p53, exerting its DNA repair function in response DNA damage [101–103]. Recently, we showed that KDM4B is involved in neuroblastoma growth and tumor maintenance [104]. The expression of KDM4B was highly correlated with that of the *MYCN* oncogene in neuroblastoma, and it formed a complex with N-Myc protein, thereby facilitating its function by maintaining low levels of repressive H3K9me2/me3 marks at Myc-binding sites. In breast cancer we have shown that HIF-1α and ERα can coordinate expression of genes, such as KDM4B, whose expression is driven by both ERα and HIF-1α and epigenetically regulates the G2/M phase of cell cycle progression in breast cancer cells [63] and other cancer cell lines (unpublished data), as the expression of several key cell cycle genes is correlated with changes in the KDM4B substrate, H3K9me3 [63,104]. Similar to other dual responsive genes such as *VEGF* and *STC2* that are regulated by both estrogens and hypoxia, the genomic locus of KDM4B bears both HIF-1α and ERα binding elements [60,63] (Figure 2A). The cross-talk between HIF-1α and ERα converges at KDM4B, which is important for cell cycle progression and tumor growth in ER positive breast cancer [63,93]. Importantly, in endocrine therapy-resistant breast cancer cells, the regulation of KDM4B by HIF-1α and ERα is intact and KDM4B is still required for G2/M phase progression [59] (Figure 2B). In addition, KDM4B is not only required for enhancing androgen receptor (AR) transcriptional activity through histone modification, but it also enhances AR protein stability via inhibition of AR ubiquitination [105], demonstrating the functional connection between AR and KDM4B in prostate cancer. Therefore, HIF-1α plays an important role in modulating anti-androgen responses via KDM4B in prostate cancer.

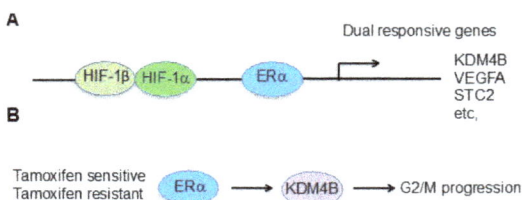

Figure 2. Hypoxia and estrogen pathways converge at KDM4B for cancer cell proliferation in ERα positive breast cancer. (**A**) KDM4B is one of the genes responsive to both estrogen and hypoxia-mediated pathways; (**B**) Regardless of endocrine therapy resistance, ERα drives KDM4B expression, which is required for G2/M phase progression.

4. Future Prospects

Many transcriptional factors, such as Myc, ERα, and AR, exert oncogenic functions to drive cancer cell proliferation. Directly targeting these oncogenic transcription factors is either technically challenging or leads to therapeutic resistance. Therefore, new approaches need to be developed to overcome these obstacles. Transcription factors need to complex with other cofactors to drive gene expression and many of these cofactors are histone modifiers. Thus, development of small molecules to target the histone modifiers, such as KDM4B, may provide an opportunity to enhance the efficacy of standard chemotherapeutics or to overcome drug resistance. Recently, efforts have been made by us and other groups to identify and develop KDM4B inhibitors for cancer treatment [106–109]. By using a chemoinformatics in combination with high-content imaging approach we identified ciclopirox as a novel histone demethylase inhibitor. Ciclopirox targeted KDM4B, inhibited Myc signaling, resulting in suppression of neuroblastoma cell viability and tumor growth associated with an induction of differentiation [107]. We also found that MCF7 cells (ERα-positive) were much more sensitive to MDA-MB-231 cells (ERα-negative) (Jun Yang, St Jude Children's Research Hospital, Memphis, TN, USA. unpublished data), suggesting ERα-positive breast cancer cells are more addicted to KDM4B. Chu et al. identified a KDM4B inhibitor that significantly blocked the viability of cultured prostate cancer cells, which was accompanied by transcriptional silencing of growth-related genes, a substantial portion of which were AR-responsive [106]. Recently, a more potent and selective KDM4 inhibitor was developed by Cellgene [108,109], which was efficacious in breast and colon cancer models. Although whether these KDM4B inhibitors are able to reverse endocrine therapy resistance needs to be tested, we believe specific and potent KDM4B inhibitors hold a promise for overcoming endocrine therapy resistance to breast cancer and prostate cancer.

Acknowledgments: Andrew M. Davidoff and Jun Yang was supported by the Assisi Foundation of Memphis, the American Lebanese Syrian Associated Charities (ALSAC), the US Public Health Service Childhood Solid Tumor Program Project grant no. CA23099, and the Cancer Center Support grant no. 21766 from the National Cancer Institute. Jun Yang was also supported by American Cancer Society-Research Scholar 130421-RSG-17-071-01-TBG, the National Cancer Institute of the National Institutes of Health under award number R03CA212802.

Conflicts of Interest: The authors declare no conflict of interest.

References

1. Semenza, G.L. HIF-1 mediates metabolic responses to intratumoral hypoxia and oncogenic mutations. *J. Clin. Investig.* **2013**, *123*, 3664–3671. [CrossRef] [PubMed]
2. Bertout, J.A.; Patel, S.A.; Simon, M.C. The impact of O_2 availability on human cancer. *Nat. Rev. Cancer* **2008**, *8*, 967–975. [CrossRef] [PubMed]
3. Wang, G.L.; Jiang, B.H.; Rue, E.A.; Semenza, G.L. Hypoxia-inducible factor 1 is a basic-helix-loop-helix-PAS heterodimer regulated by cellular O_2 tension. *Proc. Natl. Acad. Sci. USA* **1995**, *92*, 5510–5514. [CrossRef] [PubMed]
4. Maxwell, P.H.; Wiesener, M.S.; Chang, G.W.; Clifford, S.C.; Vaux, E.C.; Cockman, M.E.; Wykoff, C.C.; Pugh, C.W.; Maher, E.R.; Ratcliffe, P.J. The tumour suppressor protein VHL targets hypoxia-inducible factors for oxygen-dependent proteolysis. *Nature* **1999**, *399*, 271–275. [CrossRef] [PubMed]
5. Jaakkola, P.; Mole, D.R.; Tian, Y.M.; Wilson, M.I.; Gielbert, J.; Gaskell, S.J.; von Kriegsheim, A.; Hebestreit, H.F.; Mukherji, M.; Schofield, C.J.; et al. Targeting of HIF-alpha to the von Hippel-Lindau ubiquitylation complex by O_2-regulated prolyl hydroxylation. *Science* **2001**, *292*, 468–472. [CrossRef] [PubMed]
6. Ivan, M.; Kondo, K.; Yang, H.; Kim, W.; Valiando, J.; Ohh, M.; Salic, A.; Asara, J.M.; Lane, W.S.; Kaelin, W.G., Jr. HIFalpha targeted for VHL-mediated destruction by proline hydroxylation: Implications for O_2 sensing. *Science* **2001**, *292*, 464–468. [CrossRef] [PubMed]
7. Keith, B.; Johnson, R.S.; Simon, M.C. HIF1alpha and HIF2alpha: Sibling rivalry in hypoxic tumour growth and progression. *Nat. Rev. Cancer* **2012**, *12*, 9–22.
8. Brown, J.M.; Wilson, W.R. Exploiting tumour hypoxia in cancer treatment. *Nat. Rev. Cancer* **2004**, *4*, 437–447. [CrossRef] [PubMed]

9.   Warfel, N.A.; El-Deiry, W.S. HIF-1 signaling in drug resistance to chemotherapy. *Curr. Med. Chem.* **2014**, *21*, 3021–3028. [CrossRef] [PubMed]

10. Brown, J.M. Tumor hypoxia, drug resistance, and metastases. *J. Natl. Cancer Inst.* **1990**, *82*, 338–339. [CrossRef] [PubMed]

11. Erler, J.T.; Giaccia, A.J. Lysyl oxidase mediates hypoxic control of metastasis. *Cancer Res.* **2006**, *66*, 10238–10241. [CrossRef] [PubMed]

12. Barsoum, I.B.; Koti, M.; Siemens, D.R.; Graham, C.H. Mechanisms of hypoxia-mediated immune escape in cancer. *Cancer Res.* **2014**, *74*, 7185–7190. [CrossRef] [PubMed]

13. Richard, D.E.; Berra, E.; Gothie, E.; Roux, D.; Pouyssegur, J. p42/p44 mitogen-activated protein kinases phosphorylate hypoxia-inducible factor 1alpha (HIF-1α) and enhance the transcriptional activity of HIF-1. *J. Biol. Chem.* **1999**, *274*, 32631–32637. [CrossRef] [PubMed]

14. Laughner, E.; Taghavi, P.; Chiles, K.; Mahon, P.C.; Semenza, G.L. HER2 (neu) signaling increases the rate of hypoxia-inducible factor 1alpha (HIF-1α) synthesis: Novel mechanism for HIF-1-mediated vascular endothelial growth factor expression. *Mol. Cell. Biol.* **2001**, *21*, 3995–4004. [CrossRef] [PubMed]

15. Majumder, P.K.; Febbo, P.G.; Bikoff, R.; Berger, R.; Xue, Q.; McMahon, L.M.; Manola, J.; Brugarolas, J.; McDonnell, T.J.; Golub, T.R.; et al. mTOR inhibition reverses Akt-dependent prostate intraepithelial neoplasia through regulation of apoptotic and HIF-1-dependent pathways. *Nat. Med.* **2004**, *10*, 594–601. [CrossRef] [PubMed]

16. Kikuchi, H.; Pino, M.S.; Zeng, M.; Shirasawa, S.; Chung, D.C. Oncogenic KRAS and BRAF differentially regulate hypoxia-inducible factor-1α and -2α in colon cancer. *Cancer Res.* **2009**, *69*, 8499–8506. [CrossRef] [PubMed]

17. Gerald, D.; Berra, E.; Frapart, Y.M.; Chan, D.A.; Giaccia, A.J.; Mansuy, D.; Pouyssegur, J.; Yaniv, M.; Mechta-Grigoriou, F. JunD reduces tumor angiogenesis by protecting cells from oxidative stress. *Cell* **2004**, *118*, 781–794. [CrossRef] [PubMed]

18. Shackelford, D.B.; Vasquez, D.S.; Corbeil, J.; Wu, S.; Leblanc, M.; Wu, C.L.; Vera, D.R.; Shaw, R.J. mTOR and HIF-1α-mediated tumor metabolism in an LKB1 mouse model of Peutz-Jeghers syndrome. *Proc. Natl. Acad. Sci. USA* **2009**, *106*, 11137–11142. [CrossRef] [PubMed]

19. Bernardi, R.; Guernah, I.; Jin, D.; Grisendi, S.; Alimonti, A.; Teruya-Feldstein, J.; Cordon-Cardo, C.; Simon, M.C.; Rafii, S.; Pandolfi, P.P. PML inhibits HIF-1α translation and neoangiogenesis through repression of mTOR. *Nature* **2006**, *442*, 779–785. [CrossRef] [PubMed]

20. Zhong, H.; Chiles, K.; Feldser, D.; Laughner, E.; Hanrahan, C.; Georgescu, M.M.; Simons, J.W.; Semenza, G.L. Modulation of hypoxia-inducible factor 1α expression by the epidermal growth factor/phosphatidylinositol 3-kinase/PTEN/AKT/FRAP pathway in human prostate cancer cells: Implications for tumor angiogenesis and therapeutics. *Cancer Res.* **2000**, *60*, 1541–1545. [PubMed]

21. Zundel, W.; Schindler, C.; Haas-Kogan, D.; Koong, A.; Kaper, F.; Chen, E.; Gottschalk, A.R.; Ryan, H.E.; Johnson, R.S.; Jefferson, A.B.; et al. Loss of PTEN facilitates HIF-1-mediated gene expression. *Genes Dev.* **2000**, *14*, 391–396. [PubMed]

22. Yang, J.; Ahmed, A.; Poon, E.; Perusinghe, N.; de Haven Brandon, A.; Box, G.; Valenti, M.; Eccles, S.; Rouschop, K.; Wouters, B.; et al. Small-molecule activation of p53 blocks hypoxia-inducible factor 1α and vascular endothelial growth factor expression in vivo and leads to tumor cell apoptosis in normoxia and hypoxia. *Mol. Cell. Biol.* **2009**, *29*, 2243–2253. [CrossRef] [PubMed]

23. Pollard, P.J.; Briere, J.J.; Alam, N.A.; Barwell, J.; Barclay, E.; Wortham, N.C.; Hunt, T.; Mitchell, M.; Olpin, S.; Moat, S.J.; et al. Accumulation of Krebs cycle intermediates and over-expression of HIF1α in tumours which result from germline FH and SDH mutations. *Hum. Mol. Genet.* **2005**, *14*, 2231–2239. [CrossRef] [PubMed]

24. Koivunen, P.; Lee, S.; Duncan, C.G.; Lopez, G.; Lu, G.; Ramkissoon, S.; Losman, J.A.; Joensuu, P.; Bergmann, U.; Gross, S.; et al. Transformation by the (R)-enantiomer of 2-hydroxyglutarate linked to EGLN activation. *Nature* **2012**, *483*, 484–488. [CrossRef] [PubMed]

25. Selak, M.A.; Armour, S.M.; MacKenzie, E.D.; Boulahbel, H.; Watson, D.G.; Mansfield, K.D.; Pan, Y.; Simon, M.C.; Thompson, C.B.; Gottlieb, E. Succinate links TCA cycle dysfunction to oncogenesis by inhibiting HIF-alpha prolyl hydroxylase. *Cancer Cell* **2005**, *7*, 77–85. [CrossRef] [PubMed]

26. Isaacs, J.S.; Jung, Y.J.; Mole, D.R.; Lee, S.; Torres-Cabala, C.; Chung, Y.L.; Merino, M.; Trepel, J.; Zbar, B.; Toro, J.; et al. HIF overexpression correlates with biallelic loss of fumarate hydratase in renal cancer: Novel role of fumarate in regulation of HIF stability. *Cancer Cell* **2005**, *8*, 143–153. [CrossRef] [PubMed]

27. Mirand, E.A.; Gordon, A.S. Mechanism of estrogen action in erythropoiesis. *Endocrinology* **1966**, *78*, 325–332. [CrossRef] [PubMed]

28. Peschle, C.; Rappaport, I.A.; Sasso, G.F.; Condorelli, M.; Gordon, A.S. The role of estrogen in the regulation of erythropoietin production. *Endocrinology* **1973**, *92*, 358–362. [PubMed]

29. Anagnostou, A.; Zander, A.; Barone, J.; Fried, W. Mechanism of the increased splenic erythropoiesis in mice treated with estradiol benzoate. *J. Lab. Clin. Med.* **1976**, *88*, 700–706. [PubMed]

30. Yasuda, Y.; Masuda, S.; Chikuma, M.; Inoue, K.; Nagao, M.; Sasaki, R. Estrogen-dependent production of erythropoietin in uterus and its implication in uterine angiogenesis. *J. Biol. Chem.* **1998**, *273*, 25381–25387. [CrossRef] [PubMed]

31. Ruohola, J.K.; Valve, E.M.; Karkkainen, M.J.; Joukov, V.; Alitalo, K.; Harkonen, P.L. Vascular endothelial growth factors are differentially regulated by steroid hormones and antiestrogens in breast cancer cells. *Mol. Cell. Endocrinol.* **1999**, *149*, 29–40. [CrossRef]

32. Bausero, P.; Ben-Mahdi, M.; Mazucatelli, J.; Bloy, C.; Perrot-Applanat, M. Vascular endothelial growth factor is modulated in vascular muscle cells by estradiol, tamoxifen, and hypoxia. *Am. J. Physiol. Heart Circ. Physiol.* **2000**, *279*, H2033–H2042. [CrossRef] [PubMed]

33. Maity, A.; Sall, W.; Koch, C.J.; Oprysko, P.R.; Evans, S.M. Low pO2 and β-estradiol induce VEGF in MCF-7 and MCF-7-5C cells: Relationship to in vivo hypoxia. *Breast Cancer Res. Treat.* **2001**, *67*, 51–60. [CrossRef] [PubMed]

34. Kazi, A.A.; Koos, R.D. Estrogen-induced activation of hypoxia-inducible factor-1α, vascular endothelial growth factor expression, and edema in the uterus are mediated by the phosphatidylinositol 3-kinase/Akt pathway. *Endocrinology* **2007**, *148*, 2363–2374. [CrossRef] [PubMed]

35. Mukundan, H.; Kanagy, N.L.; Resta, T.C. 17-beta estradiol attenuates hypoxic induction of HIF-1α and erythropoietin in Hep3B cells. *J. Cardiovasc. Pharmacol.* **2004**, *44*, 93–100. [CrossRef] [PubMed]

36. Seifeddine, R.; Dreiem, A.; Tomkiewicz, C.; Fulchignoni-Lataud, M.C.; Brito, I.; Danan, J.L.; Favaudon, V.; Barouki, R.; Massaad-Massade, L. Hypoxia and estrogen co-operate to regulate gene expression in T-47D human breast cancer cells. *J. Steroid Biochem. Mol. Biol.* **2007**, *104*, 169–179. [CrossRef] [PubMed]

37. Morita, M.; Sato, Y.; Iwasaki, R.; Kobayashi, T.; Watanabe, R.; Oike, T.; Miyamoto, K.; Toyama, Y.; Matsumoto, M.; Nakamura, M.; et al. Selective estrogen receptor modulators suppress Hif1α protein accumulation in mouse osteoclasts. *PLoS ONE* **2016**, *11*, e0165922. [CrossRef] [PubMed]

38. Carroll, J.S.; Brown, M. Estrogen receptor target gene: An evolving concept. *Mol. Endocrinol.* **2006**, *20*, 1707–1714. [CrossRef] [PubMed]

39. McDonnell, D.P.; Norris, J.D. Connections and regulation of the human estrogen receptor. *Science* **2002**, *296*, 1642–1644. [CrossRef] [PubMed]

40. Cancer Genome Atlas, N. Comprehensive molecular portraits of human breast tumours. *Nature* **2012**, *490*, 61–70.

41. Harvey, J.M.; Clark, G.M.; Osborne, C.K.; Allred, D.C. Estrogen receptor status by immunohistochemistry is superior to the ligand-binding assay for predicting response to adjuvant endocrine therapy in breast cancer. *J. Clin. Oncol.* **1999**, *17*, 1474–1481. [CrossRef] [PubMed]

42. Jordan, V.C. Third annual William L. McGuire Memorial Lecture. "Studies on the estrogen receptor in breast cancer"–20 years as a target for the treatment and prevention of cancer. *Breast Cancer Res. Treat.* **1995**, *36*, 267–285. [CrossRef] [PubMed]

43. Jaiyesimi, I.A.; Buzdar, A.U.; Decker, D.A.; Hortobagyi, G.N. Use of tamoxifen for breast cancer: Twenty-eight years later. *J. Clin. Oncol.* **1995**, *13*, 513–529. [CrossRef] [PubMed]

44. Howell, A.; Osborne, C.K.; Morris, C.; Wakeling, A.E. ICI 182,780 (Faslodex): Development of a novel, "pure" antiestrogen. *Cancer* **2000**, *89*, 817–825.

45. Osborne, C.K. Aromatase inhibitors in relation to other forms of endocrine therapy for breast cancer. *Endocr. Relat. Cancer* **1999**, *6*, 271–276. [CrossRef] [PubMed]

46. Baum, M.; Buzdar, A.; Cuzick, J.; Forbes, J.; Houghton, J.; Howell, A.; Sahmoud, T.; Group, A.T. Anastrozole alone or in combination with tamoxifen versus tamoxifen alone for adjuvant treatment of postmenopausal women with early-stage breast cancer: Results of the ATAC (Arimidex, Tamoxifen Alone or in Combination) trial efficacy and safety update analyses. *Cancer* **2003**, *98*, 1802–1810. [PubMed]

47. Massarweh, S.; Osborne, C.K.; Jiang, S.; Wakeling, A.E.; Rimawi, M.; Mohsin, S.K.; Hilsenbeck, S.; Schiff, R. Mechanisms of tumor regression and resistance to estrogen deprivation and fulvestrant in a model of estrogen receptor-positive, HER-2/neu-positive breast cancer. *Cancer Res.* **2006**, *66*, 8266–8273. [CrossRef] [PubMed]

48. Howell, A.; DeFriend, D.; Robertson, J.; Blamey, R.; Walton, P. Response to a specific antioestrogen (ICI 182780) in tamoxifen-resistant breast cancer. *Lancet* **1995**, *345*, 29–30. [CrossRef]

49. Clarke, R.; Leonessa, F.; Welch, J.N.; Skaar, T.C. Cellular and molecular pharmacology of antiestrogen action and resistance. *Pharmacol. Rev.* **2001**, *53*, 25–71. [PubMed]

50. Groner, A.C.; Brown, M. Role of steroid receptor and coregulator mutations in hormone-dependent cancers. *J. Clin. Investig.* **2017**, *127*, 1126–1135. [CrossRef] [PubMed]

51. Musgrove, E.A.; Sutherland, R.L. Biological determinants of endocrine resistance in breast cancer. *Nat. Rev. Cancer* **2009**, *9*, 631–643. [CrossRef] [PubMed]

52. Toy, W.; Shen, Y.; Won, H.; Green, B.; Sakr, R.A.; Will, M.; Li, Z.; Gala, K.; Fanning, S.; King, T.A.; et al. ESR1 ligand-binding domain mutations in hormone-resistant breast cancer. *Nat. Genet.* **2013**, *45*, 1439–1445. [CrossRef] [PubMed]

53. Robinson, D.R.; Wu, Y.M.; Vats, P.; Su, F.; Lonigro, R.J.; Cao, X.; Kalyana-Sundaram, S.; Wang, R.; Ning, Y.; Hodges, L.; et al. Activating ESR1 mutations in hormone-resistant metastatic breast cancer. *Nat. Genet.* **2013**, *45*, 1446–1451. [CrossRef] [PubMed]

54. Hsu, P.Y.; Hsu, H.K.; Lan, X.; Juan, L.; Yan, P.S.; Labanowska, J.; Heerema, N.; Hsiao, T.H.; Chiu, Y.C.; Chen, Y.; et al. Amplification of distant estrogen response elements deregulates target genes associated with tamoxifen resistance in breast cancer. *Cancer Cell* **2013**, *24*, 197–212. [CrossRef] [PubMed]

55. Bos, R.; Zhong, H.; Hanrahan, C.F.; Mommers, E.C.; Semenza, G.L.; Pinedo, H.M.; Abeloff, M.D.; Simons, J.W.; van Diest, P.J.; van der Wall, E. Levels of hypoxia-inducible factor-1 α during breast carcinogenesis. *J. Natl. Cancer Inst.* **2001**, *93*, 309–314. [CrossRef] [PubMed]

56. Brunnberg, S.; Pettersson, K.; Rydin, E.; Matthews, J.; Hanberg, A.; Pongratz, I. The basic helix-loop-helix-PAS protein ARNT functions as a potent coactivator of estrogen receptor-dependent transcription. *Proc. Natl. Acad. Sci. USA* **2003**, *100*, 6517–6522. [CrossRef] [PubMed]

57. Generali, D.; Buffa, F.M.; Berruti, A.; Brizzi, M.P.; Campo, L.; Bonardi, S.; Bersiga, A.; Allevi, G.; Milani, M.; Aguggini, S.; et al. Phosphorylated ERα, HIF-1α, and MAPK signaling as predictors of primary endocrine treatment response and resistance in patients with breast cancer. *J. Clin. Oncol.* **2009**, *27*, 227–234. [CrossRef] [PubMed]

58. Generali, D.; Berruti, A.; Brizzi, M.P.; Campo, L.; Bonardi, S.; Wigfield, S.; Bersiga, A.; Allevi, G.; Milani, M.; Aguggini, S.; et al. Hypoxia-inducible factor-1alpha expression predicts a poor response to primary chemoendocrine therapy and disease-free survival in primary human breast cancer. *Clin. Cancer Res.* **2006**, *12*, 4562–4568. [CrossRef] [PubMed]

59. Yang, J.; AlTahan, A.; Jones, D.T.; Buffa, F.M.; Bridges, E.; Interiano, R.B.; Qu, C.; Vogt, N.; Li, J.L.; Baban, D.; et al. Estrogen receptor-alpha directly regulates the hypoxia-inducible factor 1 pathway associated with antiestrogen response in breast cancer. *Proc. Natl. Acad. Sci. USA* **2015**, *112*, 15172–15177. [CrossRef] [PubMed]

60. Beyer, S.; Kristensen, M.M.; Jensen, K.S.; Johansen, J.V.; Staller, P. The histone demethylases JMJD1A and JMJD2B are transcriptional targets of hypoxia-inducible factor HIF. *J. Biol. Chem.* **2008**, *283*, 36542–36552. [CrossRef] [PubMed]

61. Yang, J.; Ledaki, I.; Turley, H.; Gatter, K.C.; Montero, J.C.; Li, J.L.; Harris, A.L. Role of hypoxia-inducible factors in epigenetic regulation via histone demethylases. *Ann. N. Y. Acad. Sci.* **2009**, *1177*, 185–197. [CrossRef] [PubMed]

62. Bouras, T.; Southey, M.C.; Chang, A.C.; Reddel, R.R.; Willhite, D.; Glynne, R.; Henderson, M.A.; Armes, J.E.; Venter, D.J. Stanniocalcin 2 is an estrogen-responsive gene coexpressed with the estrogen receptor in human breast cancer. *Cancer Res.* **2002**, *62*, 1289–1295. [PubMed]

63. Yang, J.; Jubb, A.M.; Pike, L.; Buffa, F.M.; Turley, H.; Baban, D.; Leek, R.; Gatter, K.C.; Ragoussis, J.; Harris, A.L. The histone demethylase JMJD2B is regulated by estrogen receptor alpha and hypoxia, and is a key mediator of estrogen induced growth. *Cancer Res.* **2010**, *70*, 6456–6466. [CrossRef] [PubMed]

64. Forsythe, J.A.; Jiang, B.H.; Iyer, N.V.; Agani, F.; Leung, S.W.; Koos, R.D.; Semenza, G.L. Activation of vascular endothelial growth factor gene transcription by hypoxia-inducible factor 1. *Mol. Cell. Biol.* **1996**, *16*, 4604–4613. [CrossRef] [PubMed]

65. Mueller, M.D.; Vigne, J.L.; Minchenko, A.; Lebovic, D.I.; Leitman, D.C.; Taylor, R.N. Regulation of vascular endothelial growth factor (VEGF) gene transcription by estrogen receptors α and β. *Proc. Natl. Acad. Sci. USA* **2000**, *97*, 10972–10977. [CrossRef] [PubMed]

66. Law, A.Y.; Wong, C.K. Stanniocalcin-2 is a HIF-1 target gene that promotes cell proliferation in hypoxia. *Exp. Cell Res.* **2010**, *316*, 466–476. [CrossRef] [PubMed]

67. Hurtado, A.; Holmes, K.A.; Ross-Innes, C.S.; Schmidt, D.; Carroll, J.S. FOXA1 is a key determinant of estrogen receptor function and endocrine response. *Nat. Genet.* **2011**, *43*, 27–33. [CrossRef] [PubMed]

68. Fu, X.; Jeselsohn, R.; Pereira, R.; Hollingsworth, E.F.; Creighton, C.J.; Li, F.; Shea, M.; Nardone, A.; De Angelis, C.; Heiser, L.M.; et al. FOXA1 overexpression mediates endocrine resistance by altering the ER transcriptome and IL-8 expression in ER-positive breast cancer. *Proc. Natl. Acad. Sci. USA* **2016**, *113*, E6600–E6609. [CrossRef] [PubMed]

69. You, J.S.; Jones, P.A. Cancer genetics and epigenetics: Two sides of the same coin? *Cancer Cell* **2012**, *22*, 9–20. [CrossRef] [PubMed]

70. Portela, A.; Esteller, M. Epigenetic modifications and human disease. *Nat. Biotechnol.* **2010**, *28*, 1057–1068. [CrossRef] [PubMed]

71. Greer, E.L.; Shi, Y. Histone methylation: A dynamic mark in health, disease and inheritance. *Nat. Rev. Genet.* **2012**, *13*, 343–357. [CrossRef] [PubMed]

72. Rodriguez-Paredes, M.; Esteller, M. Cancer epigenetics reaches mainstream oncology. *Nat. Med.* **2011**, *17*, 330–339. [CrossRef] [PubMed]

73. Martin, C.; Zhang, Y. The diverse functions of histone lysine methylation. *Nat. Rev. Mol. Cell Biol.* **2005**, *6*, 838–849. [CrossRef] [PubMed]

74. Shi, Y.; Lan, F.; Matson, C.; Mulligan, P.; Whetstine, J.R.; Cole, P.A.; Casero, R.A.; Shi, Y. Histone demethylation mediated by the nuclear amine oxidase homolog LSD1. *Cell* **2004**, *119*, 941–953. [CrossRef] [PubMed]

75. Hojfeldt, J.W.; Agger, K.; Helin, K. Histone lysine demethylases as targets for anticancer therapy. *Nat. Rev. Drug Discov.* **2013**, *12*, 917–930. [CrossRef] [PubMed]

76. Esteller, M. Epigenetics in cancer. *N. Engl. J. Med.* **2008**, *358*, 1148–1159. [CrossRef] [PubMed]

77. Chi, P.; Allis, C.D.; Wang, G.G. Covalent histone modifications–miswritten, misinterpreted and mis-erased in human cancers. *Nat. Rev. Cancer* **2010**, *10*, 457–469. [CrossRef] [PubMed]

78. Kandoth, C.; McLellan, M.D.; Vandin, F.; Ye, K.; Niu, B.; Lu, C.; Xie, M.; Zhang, Q.; McMichael, J.F.; Wyczalkowski, M.A.; et al. Mutational landscape and significance across 12 major cancer types. *Nature* **2013**, *502*, 333–339. [CrossRef] [PubMed]

79. Huether, R.; Dong, L.; Chen, X.; Wu, G.; Parker, M.; Wei, L.; Ma, J.; Edmonson, M.N.; Hedlund, E.K.; Rusch, M.C.; et al. The landscape of somatic mutations in epigenetic regulators across 1000 paediatric cancer genomes. *Nat. Commun.* **2014**, *5*, 3630. [CrossRef] [PubMed]

80. Shi, Y.; Whetstine, J.R. Dynamic regulation of histone lysine methylation by demethylases. *Mol. Cell* **2007**, *25*, 1–14. [CrossRef] [PubMed]

81. Shi, Y. Histone lysine demethylases: Emerging roles in development, physiology and disease. *Nat. Rev. Genet.* **2007**, *8*, 829–833. [CrossRef] [PubMed]

82. Xia, X.; Lemieux, M.E.; Li, W.; Carroll, J.S.; Brown, M.; Liu, X.S.; Kung, A.L. Integrative analysis of HIF binding and transactivation reveals its role in maintaining histone methylation homeostasis. *Proc. Natl. Acad. Sci. USA* **2009**, *106*, 4260–4265. [CrossRef] [PubMed]

83. Wellmann, S.; Bettkober, M.; Zelmer, A.; Seeger, K.; Faigle, M.; Eltzschig, H.K.; Buhrer, C. Hypoxia upregulates the histone demethylase JMJD1A via HIF-1. *Biochem. Biophys. Res. Commun.* **2008**, *372*, 892–897. [CrossRef] [PubMed]

84. Pollard, P.J.; Loenarz, C.; Mole, D.R.; McDonough, M.A.; Gleadle, J.M.; Schofield, C.J.; Ratcliffe, P.J. Regulation of Jumonji-domain-containing histone demethylases by hypoxia-inducible factor (HIF)-1alpha. *Biochem. J.* **2008**, *416*, 387–394. [CrossRef] [PubMed]

85. Das, P.P.; Shao, Z.; Beyaz, S.; Apostolou, E.; Pinello, L.; De Los Angeles, A.; O'Brien, K.; Atsma, J.M.; Fujiwara, Y.; Nguyen, M.; et al. Distinct and combinatorial functions of Jmjd2b/Kdm4b and Jmjd2c/Kdm4c in mouse embryonic stem cell identity. *Mol. Cell* **2014**, *53*, 32–48. [CrossRef] [PubMed]

86. Chen, J.; Liu, H.; Liu, J.; Qi, J.; Wei, B.; Yang, J.; Liang, H.; Chen, Y.; Wu, Y.; Guo, L.; et al. H3K9 methylation is a barrier during somatic cell reprogramming into iPSCs. *Nat. Genet.* **2012**, *45*, 34–42. [CrossRef] [PubMed]

87. Berry, W.L.; Janknecht, R. KDM4/JMJD2 histone demethylases: Epigenetic regulators in cancer cells. *Cancer Res.* **2013**, *73*, 2936–2942. [CrossRef] [PubMed]
88. Northcott, P.A.; Nakahara, Y.; Wu, X.; Feuk, L.; Ellison, D.W.; Croul, S.; Mack, S.; Kongkham, P.N.; Peacock, J.; Dubuc, A.; et al. Multiple recurrent genetic events converge on control of histone lysine methylation in medulloblastoma. *Nat. Genet.* **2009**, *41*, 465–472. [CrossRef] [PubMed]
89. Pryor, J.G.; Brown-Kipphut, B.A.; Iqbal, A.; Scott, G.A. Microarray comparative genomic hybridization detection of copy number changes in desmoplastic melanoma and malignant peripheral nerve sheath tumor. *Am. J. Dermatopathol.* **2011**, *33*, 780–785. [CrossRef] [PubMed]
90. Cloos, P.A.; Christensen, J.; Agger, K.; Helin, K. Erasing the methyl mark: Histone demethylases at the center of cellular differentiation and disease. *Genes Dev.* **2008**, *22*, 1115–1140. [CrossRef] [PubMed]
91. Kooistra, S.M.; Helin, K. Molecular mechanisms and potential functions of histone demethylases. *Nat. Rev. Mol. Cell Biol.* **2012**, *13*, 297–311. [CrossRef] [PubMed]
92. Walters, Z.S.; Villarejo-Balcells, B.; Olmos, D.; Buist, T.W.; Missiaglia, E.; Allen, R.; Al-Lazikani, B.; Garrett, M.D.; Blagg, J.; Shipley, J. JARID2 is a direct target of the PAX3-FOXO1 fusion protein and inhibits myogenic differentiation of rhabdomyosarcoma cells. *Oncogene* **2014**, *33*, 1148–1157. [CrossRef] [PubMed]
93. Kawazu, M.; Saso, K.; Tong, K.I.; McQuire, T.; Goto, K.; Son, D.O.; Wakeham, A.; Miyagishi, M.; Mak, T.W.; Okada, H. Histone demethylase JMJD2B functions as a co-factor of estrogen receptor in breast cancer proliferation and mammary gland development. *PLoS ONE* **2011**, *6*, e17830. [CrossRef] [PubMed]
94. Shi, L.; Sun, L.; Li, Q.; Liang, J.; Yu, W.; Yi, X.; Yang, X.; Li, Y.; Han, X.; Zhang, Y.; et al. Histone demethylase JMJD2B coordinates H3K4/H3K9 methylation and promotes hormonally responsive breast carcinogenesis. *Proc. Natl. Acad. Sci. USA* **2011**, *108*, 7541–7546. [CrossRef] [PubMed]
95. Berry, W.L.; Kim, T.D.; Janknecht, R. Stimulation of beta-catenin and colon cancer cell growth by the KDM4B histone demethylase. *Int. J. Oncol.* **2014**, *44*, 1341–1348. [CrossRef] [PubMed]
96. Qiu, M.T.; Fan, Q.; Zhu, Z.; Kwan, S.Y.; Chen, L.; Chen, J.H.; Ying, Z.L.; Zhou, Y.; Gu, W.; Wang, L.H.; et al. KDM4B and KDM4A promote endometrial cancer progression by regulating androgen receptor, c-myc, and p27kip1. *Oncotarget* **2015**, *6*, 31702–31720. [CrossRef] [PubMed]
97. Toyokawa, G.; Cho, H.S.; Iwai, Y.; Yoshimatsu, M.; Takawa, M.; Hayami, S.; Maejima, K.; Shimizu, N.; Tanaka, H.; Tsunoda, T.; et al. The histone demethylase JMJD2B plays an essential role in human carcinogenesis through positive regulation of cyclin-dependent kinase 6. *Cancer Prev. Res.* **2011**, *4*, 2051–2061. [CrossRef] [PubMed]
98. Bur, H.; Haapasaari, K.M.; Turpeenniemi-Hujanen, T.; Kuittinen, O.; Auvinen, P.; Marin, K.; Soini, Y.; Karihtala, P. Strong KDM4B and KDM4D Expression associates with radioresistance and aggressive phenotype in classical hodgkin lymphoma. *Anticancer Res.* **2016**, *36*, 4677–4683. [CrossRef] [PubMed]
99. Wilson, C.; Qiu, L.; Hong, Y.; Karnik, T.; Tadros, G.; Mau, B.; Ma, T.; Mu, Y.; New, J.; Louie, R.J.; et al. The histone demethylase KDM4B regulates peritoneal seeding of ovarian cancer. *Oncogene* **2016**. [CrossRef] [PubMed]
100. Zhao, L.; Li, W.; Zang, W.; Liu, Z.; Xu, X.; Yu, H.; Yang, Q.; Jia, J. JMJD2B promotes epithelial-mesenchymal transition by cooperating with β-catenin and enhances gastric cancer metastasis. *Clin. Cancer Res.* **2013**, *19*, 6419–6429. [CrossRef] [PubMed]
101. Castellini, L.; Moon, E.J.; Razorenova, O.V.; Krieg, A.J.; von Eyben, R.; Giaccia, A.J. KDM4B/JMJD2B is a p53 target gene that modulates the amplitude of p53 response after DNA damage. *Nucleic Acids Res.* **2017**, *45*, 3674–3692. [CrossRef] [PubMed]
102. Young, L.C.; McDonald, D.W.; Hendzel, M.J. Kdm4b histone demethylase is a DNA damage response protein and confers a survival advantage following gamma-irradiation. *J. Biol. Chem.* **2013**, *288*, 21376–21388. [CrossRef] [PubMed]
103. Zheng, H.; Chen, L.; Pledger, W.J.; Fang, J.; Chen, J. p53 promotes repair of heterochromatin DNA by regulating JMJD2b and SUV39H1 expression. *Oncogene* **2014**, *33*, 734–744. [CrossRef] [PubMed]
104. Yang, J.; AlTahan, A.M.; Hu, D.; Wang, Y.; Cheng, P.H.; Morton, C.L.; Qu, C.; Nathwani, A.C.; Shohet, J.M.; Fotsis, T.; et al. The role of histone demethylase KDM4B in Myc signaling in neuroblastoma. *J. Natl. Cancer Inst.* **2015**, *107*, djv080. [CrossRef] [PubMed]
105. Coffey, K.; Rogerson, L.; Ryan-Munden, C.; Alkharaif, D.; Stockley, J.; Heer, R.; Sahadevan, K.; O'Neill, D.; Jones, D.; Darby, S.; et al. The lysine demethylase, KDM4B, is a key molecule in androgen receptor signalling and turnover. *Nucleic Acids Res.* **2013**, *41*, 4433–4446. [CrossRef] [PubMed]

106. Chu, C.H.; Wang, L.Y.; Hsu, K.C.; Chen, C.C.; Cheng, H.H.; Wang, S.M.; Wu, C.M.; Chen, T.J.; Li, L.T.; Liu, R.; et al. KDM4B as a target for prostate cancer: Structural analysis and selective inhibition by a novel inhibitor. *J. Med. Chem.* **2014**, *57*, 5975–5985. [CrossRef] [PubMed]

107. Yang, J.; Milasta, S.; Hu, D.; AlTahan, A.M.; Interiano, R.B.; Zhou, J.; Davidson, J.; Low, J.; Lin, W.; Bao, J.; et al. Targeting histone demethylases in MYC-driven neuroblastomas with ciclopirox. *Cancer Res.* **2017**, *77*, 4626–4638. [CrossRef] [PubMed]

108. Metzger, E.; Stepputtis, S.S.; Strietz, J.; Preca, B.T.; Urban, S.; Willmann, D.; Allen, A.; Zenk, F.; Iovino, N.; Bronsert, P.; et al. KDM4 inhibition targets breast cancer stem-like cells. *Cancer Res.* **2017**, *77*, 5900–5912. [CrossRef] [PubMed]

109. Chen, Y.K.; Bonaldi, T.; Cuomo, A.; del Rosario, J.R.; Hosfield, D.J.; Kanouni, T.; Kao, S.C.; Lai, C.; Lobo, N.A.; Matuszkiewicz, J.; et al. Design of KDM4 inhibitors with antiproliferative effects in cancer models. *ACS Med. Chem. Lett.* **2017**, *8*, 869–874. [CrossRef] [PubMed]

International Journal of
Molecular Sciences

MDPI

Review

Emerging Roles of Estrogen-Related Receptors in the Brain: Potential Interactions with Estrogen Signaling

Kenji Saito [1] and Huxing Cui [1,2,3,*]

[1] Department of Pharmacology, University of Iowa Carver College of Medicine, Iowa City, IA 52242, USA; kenji-saito@uiowa.edu
[2] F.O.E. Diabetes Research Center, University of Iowa Carver College of Medicine, Iowa City, IA 52242, USA
[3] Obesity Research and Education Initiative, University of Iowa Carver College of Medicine, Iowa City, IA 52242, USA
* Correspondence: huxing-cui@uiowa.edu; Tel.: +1-319-335-6954

Received: 3 February 2018; Accepted: 30 March 2018; Published: 5 April 2018

Abstract: In addition to their well-known role in the female reproductive system, estrogens can act in the brain to regulate a wide range of behaviors and physiological functions in both sexes. Over the past few decades, genetically modified animal models have greatly increased our knowledge about the roles of estrogen receptor (ER) signaling in the brain in behavioral and physiological regulations. However, less attention has been paid to the estrogen-related receptors (ERRs), the members of orphan nuclear receptors whose sequences are homologous to ERs but lack estrogen-binding ability. While endogenous ligands of ERRs remain to be determined, they seemingly share transcriptional targets with ERs and their expression can be directly regulated by ERs through the estrogen-response element embedded within the regulatory region of the genes encoding ERRs. Despite the broad expression of ERRs in the brain, we have just begun to understand the fundamental roles they play at molecular, cellular, and circuit levels. Here, we review recent research advancement in understanding the roles of ERs and ERRs in the brain, with particular emphasis on ERRs, and discuss possible cross-talk between ERs and ERRs in behavioral and physiological regulations.

Keywords: estrogen; estrogen-related receptors; estrogen receptor; brain; central nervous system; mitochondria

---

1. Introduction: Estrogen Receptors (ERs) and Estrogen-Related Receptors (ERRs)

Estrogens are steroid hormones known to regulate a wide range of physiological functions, including but not limited to reproduction, cardiovascular physiology, homeostatic regulation of energy balance, and a variety of social and learning behaviors. Traditionally, the actions of circulating estrogen were believed to be mediated mainly by binding to two specific receptors, estrogen receptors α (ERα) and estrogen receptors β (ERβ), which recognize and activate gene transcription through binding to the genomic element called the estrogen-response element (ERE), either as a homodimer or heterodimer with coactivators [1,2]. Notably, apart from their well-known roles in transcriptional regulation, estrogens were also recently reported to rapidly activate extracellular signal-regulated kinases (ERKs) according to a new mode of action of ERs as well as the expression of an orphan G-protein-coupled receptor 30 (GPR30), that functions as a novel type of ER. As such, even after nearly a century since their discovery, the exact mechanisms by which estrogens regulate different physiological functions are still incompletely understood and remain an active area of research.

The estrogen-related receptors α and β (ERRα and ERRβ) were the two first orphan nuclear receptors identified based on their sequence similarity to the ERα [3]. Together with ERRγ, these three receptors consist of the ERR subfamily of the group III steroid nuclear receptor superfamily. Other group III nuclear receptors include glucocorticoid, mineralocorticoid, progesterone, and androgen receptors as well as ERs. Although ERRs share sequence homologies with ERs, estrogens are not their natural ligands and ERRs exhibit constitutive activity and can work as transcriptional regulators in the absence of ligands [4]. The ERRs contain DNA-binding domains (DBDs) constituting two highly conserved zinc finger motifs that target the receptor to a specific DNA sequence (TCAAGGTCA) called the estrogen-related response element (ERRE). ERRs bind to ERRE as a monomer or a homodimer or as a heterodimer with co-activators [5,6]. In addition to ERRE, ERRs can also bind to ERE and, conversely, ERα, but not ERβ, and can bind to ERRE as well [7], implying shared transcriptional networks driven by both ERRs and ERα. Not surprisingly, in many tissues both ERα and ERRs are highly expressed, including metabolically active skeletal muscle, fat and brain [8,9], but whether and how they are coordinated to control shared and/or distinct transcriptional events remain unclear. Compared to ERs, our knowledge about the tissue- and cell type-specific roles of ERRs are limited. Further studies are needed to uncover transcriptional networks driven by ERRs in different cell types and to investigate how they will affect whole-body physiology either independently or in coordination with estrogen signaling.

2. ERs and Their Modes of Action

As classical nuclear receptors, upon ligand binding ERs translocate to the nucleus and are directly recruited to the EREs on the target genes. This mode of action is called the genomic action of estrogens. However, as mentioned, the estrogen signals can also be mediated through rapid, cytosolic ER-initiated signaling cascades. Mutant female mice in which ERα's ability to bind to the EREs was disrupted, are infertile and display a variety of abnormalities in the reproductive system [10]. However, this mutation in ERα null background restores the obese phenotype of ERα knockout mice [11], indicating that ERα's role in the homeostatic regulation of the energy balance is independent of its genomic action. One likely signaling pathway downstream of ERα to exert its rapid, membrane-initiated action is PI3K/Akt pathway. Estradiol activates the PI3K/Akt pathway in hypothalamic nuclei [11–13]. Genetic inactivation of the PI3K pathway in hypothalamic nuclei blunts the anti-obese effects of estrogens [14,15]. Although the involvement of classical genomic ERα signaling cannot be fully ruled out, these studies suggest a critical role of rapid, membrane-initiated actions of ERα on energy homeostasis. Thus, it is plausible that different modes of action of estrogen can exert different physiological functions. The latest member of estrogen receptors, GPR30 (also known as GPER), is a G protein-coupled estrogen receptor that was cloned by several groups in the 1990s [16–19]. Later, this was followed by numerous reports showing GPR30-dependent 17β-estradiol signaling and actual binding of 17β-estradiol with GPR30 [20–27]. So far, it is unknown whether ERRs can also activate rapid, membrane-initiated signaling pathways as seen in ERs.

3. ERRs and Their Potential Ligands

Although ERRs were identified based on their sequence similarities to ERs, estrogen is not a natural ligand of ERRs and to date, endogenous ligand(s) of ERRs remain unclear. Given the established roles of ERRs in energy metabolism and the development of certain types of cancers [28–32], there have been active and continuous efforts to identify their endogenous ligands, transcriptional co-activators, and the synthetic compounds that can be used to modulate the activity of ERRs. Interestingly, a recent study using affinity chromatography of tissue lipidomes with the ERRα ligand-binding domain identified cholesterol as an endogenous ERRα agonist [33]. While this study represents a first successful screening of potential endogenous ligands for ERRs, the specific transcriptional dynamics as well as physiological functions of cholesterol binding to ERRα remain to be fully determined.

Although endogenous ligands of ERRs remain uncertain, several transcriptional co-activators interacting with ERRs have been identified, which include peroxisome proliferator-activated receptor gamma coactivator 1-alpha (PGC1-α) and PGC1-β [4,28,31]. Both PGC1-α and PGC1-β function as protein ligands for ERRs and the transcriptional activities driven by these interactions have been shown to be essential for mitochondrial biogenesis and, thus, cellular energy metabolism [4,34,35].

Therefore, targeting these transcriptional networks may hold promise for treating the metabolic disorders including diabetes, obesity and cancer [8,34,36–38].

Additionally, multiple inverse agonists or antagonists have been developed. Synthetic estrogen diethylstilbestrol acts as an inverse agonist for all three ERRs [29]. More selective inverse agonists, such as XCT 790 and DY 131, have been developed for ERRα, ERRβ/γ, respectively [39]. Most of these inverse agonists were designed to block the interactions with their protein-binding partners. XCT 790 blocks the interaction between ERRα and PGC1-α and has been shown to inhibit the expression of ERRα target genes [40] and cancer cell proliferation [41]. Other inverse agonists of ERRα are compound 29 and 50 developed by Janssen Pharmaceuticals, which show high selectivity for ERRα and have strong therapeutic potential for treating obesity and diabetes [38]. The estrogenic chemical Bisphenol A has also been reported to be a potential agonist for ERRγ [42]. Additionally, there is some evidence showing that dietary products, such as resveratrol, genistein, rutacarpine, piceatanol and flavone, could function as potential agonists of ERRα [43,44]. More recently, a study screening the Tox21 compound library has identified multiple potential novel ERR agonists, including a potent histone deacetylase inhibitor Suberoylanilide hydroxamic acid (SAHA) and a class of lipid-lowering medication statins, including atorvastatin, fluvastatin, and lovastatin [45]. As such, the use of these natural and synthesized ligands will facilitate the processes of studying the whole-body effects of modulating the transcriptional activities of ERRs and help to develop an effective therapeutic strategy for the diseases associated with ERRs, including cancers and metabolic diseases such as obesity and diabetes.

4. Transcriptional Regulations by Both ERs and ERRs

ER-signaling is involved in the development of breast cancer. Three quarters of breast tumors are considered to express ERα [46,47]. The estrogen signaling through ERα regulates the expression of various genes that play key roles in cell proliferation and cell-cycle progression [48,49]. As a primary treatment for breast cancer, ER-positive breast cancers are preferentially treated with reagents that suppress ERα signaling such as tamoxifen and aromatase inhibitors. Unfortunately, after years of treatment, the recurrence of breast cancer could happen with a resistance to estrogen-signaling inhibitors. Recently, ERRs have been attracting much attention for the prognosis of ER-positive and negative breast cancer. ERRα expression is high in breast cancer, especially in cancer cells lacking ERα, and is considered as a negative prognostic factor for breast-cancer survival [50,51].

While estrogen is not an endogenous ligand of ERRs, there is possible cross-talk between estrogen signaling and ERRs in different ways (Figure 1). Studies showed that the ERRα promoter has multiple steroid hormone response-element half-sites, where ERα could bind, and estrogens stimulate ERRα expression in vivo and in vitro [52,53], suggesting that ERRα is one of the transcriptional targets of ERα. Chromatin immunoprecipitation for ERα and ERRα coupled with microarray revealed that some targeted genes are shared by these two receptors [54]. Both ERα and ERRα stimulate the transcription of Runx2-I, a master regulator of bone development, through a common ERE, and this transcriptional regulation by ERRα is changed dependent on its binding partner; the binding with PGC1-α acts as a transcriptional activator, while the binding with PGC1-β acts as a transcriptional repressor [55]. These studies suggest that ERs and ERRs can cross-talk and mutually regulate the expression of common target genes. However, chromatin immunoprecipitation (ChIP) study also indicates that the occupancy of the shared targets by both ERα and ERRα is relatively modest among each of their transcriptional target genes and, therefore, it is likely that, depending on tissue, both ERα and ERRα maintain a high degree of independence for their transcriptional regulation [54].

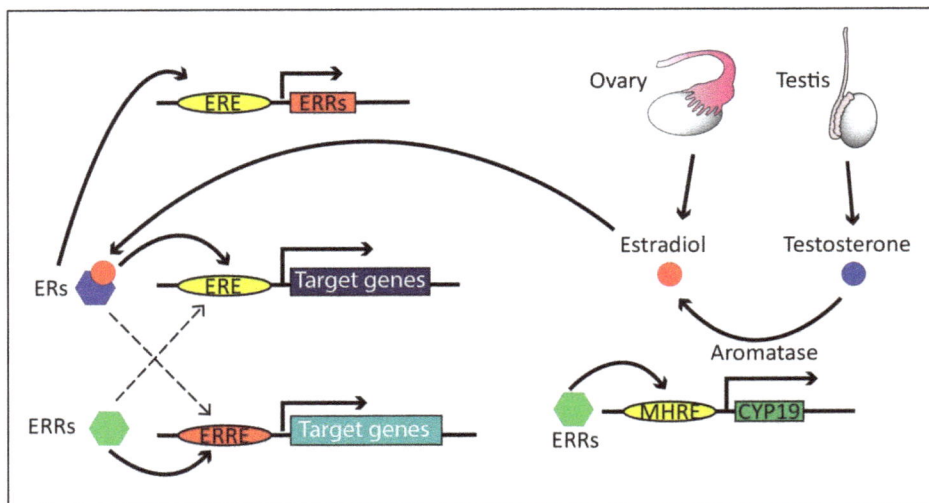

Figure 1. Schematic drawing showing potential cross-talk between estrogen, estrogen receptors (ERs), and estrogen-related receptors (ERRs). Dotted lines indicate relatively weak binding ability. MHRE, multiple-hormone response element; CYP19, aromatase.

Possible interaction between estrogen signaling and ERRs is also supported by a study showing that ERRα regulates the expression of aromatase, an enzyme responsible for the conversion of testosterone to estrogen [56,57]. Aromatase expression is also regulated by ERα [58]. Additionally, it was shown that ERRβ could directly bind to ERα in order to restrain ERα morbidity and suppress estrogen-dependent cellular function [59]. These reciprocal interactions between estrogen, ERs and ERRs warrant future research to investigate (1) whether the levels of ERRα expression changes by ovariectomy or along with different stages of the estrus cycle; (2) whether ERα knockout mice have altered levels of ERRα expression across the tissues; and (3) conversely, do ERRα knockout mice have impaired estrogen signaling and/or reproductive dysfunctions?

5. Functions of ERs and ERRs in the Brain

5.1. Actions in the Central Regulation of Energy Homeostasis

Estrogen signaling has been well known to play an essential role in body-weight regulation [60]. Postmenopausal women experience a remarkable decline in circulating 17β-estradiol (E2), which is often associated with the development or accumulation of body fat, obesity, type II diabetes, hypertension, and the metabolic syndrome [61]. The involvement of estrogens in energy homeostasis is more obvious in experimental animal models. The withdrawal of endogenous estrogens by ovariectomy in female animals leads to hyperadiposity and body-weight gain, and this obese phenotype can be prevented by E2 supplementation [62,63]. Conversely, microinjections of E2 into various brain regions suppress feeding behavior and body-weight gain [64,65]. The importance of central estrogen signaling in energy homeostasis was later confirmed by genetic mouse models. Among three cloned estrogen receptors, the estrogenic effects on energy homeostasis are believed to be primarily mediated by ERα. ERα is widely expressed throughout the brain including, but not limited to, those hypothalamic and brainstem nuclei that are important for the homeostatic regulation of energy balance, such as the ventromedial nucleus of the hypothalamus (VMH), the arcuate nucleus of the hypothalamus (ARC), the medial amygdala (MeA), and the nucleus of the solitary tract (NTS). Humans with a mutation in ERα and mice lacking ERα throughout the body or specifically in the brain are obese due to both hyperphagia and/or reduced physical activity and energy expenditure [66–68]. Furthermore, the anti-obesity effects of E2 replacement in ovariectomized mice are blocked in ERα knockout mice [62].

More specifically, the knockdown of ERα in the VMH by shRNA blunts E2-mediated weight loss and leads to obesity associated with increased visceral fat [69], likely due to decreased physical activity and impaired thermogenesis, but not food intake. Consistent with these findings, the mice with VMH-specific ERα knockout [68] showed modest weight gain due to reduced energy expenditure but not food intake, and were infertile.

ERα is also abundantly expressed in the ARC [69]. The ARC contains two primarily distinct but intermingled neuronal populations that express either anorexigenic pro-opiomelanocortin (POMC) or orexigenic agouti-related peptide (AgRP) and neuropeptide Y (NPY). ERα is expressed in POMC neurons [68,70,71] and POMC levels change in response to estrogens [72]. Genetic mouse study revealed that conditional deletion of ERα in POMC neurons leads to hyperphasia and modest weight gain [68]. AgRP/NPY neurons are also modulated by estrogen signaling. Dhillon and Belsham have shown that estrogens inhibit NPY release in immortalized hypothalamic cells through a ERα-dependent mechanism [73]. AgRP and NPY expressions fluctuate along with the estrus stages and the anorexigenic effect of 17β-estradiol was blunted in mice with the ablation of AgRP neurons [74]. NPY neuronal excitability is also modulated by estrogen via a change in K$^+$ channel expression [75]. However, which type of estrogen receptors are responsible for estrogen effects on AgRP/NPY neurons is controversial. Immunohistochemistry failed to detect ERα in NPY neurons despite clear estrogen effects on this neuronal population [74]. On the other hand, others reported colocalization of NPY and ERα [76,77].

ERα is also highly expressed in the medial amygdala (MeA). Conditional deletion of ERα in the MeA led to weight gain in both male and female mice, mainly due to decreased energy expenditure associated with low physical activity, but not food intake [78]. Interestingly, both male and female aromatase knockout mice develop obesity due to lowered physical activity, and this body weight gain is not associated with hyperphasia [79], resembling phenotypes observed in mice lacking ERα in MeA.

As such, ERα expressed by distinct types of neurons in the brain seemingly plays differential roles in maintaining whole-body energy homeostasis as reviewed elsewhere [60]. By contrast with the positive energy balance observed in ERα knockout mice, however, it has been reported that conventional ERRα knockout mice have reduced body weight and fat mass compared to their control littermates, especially when challenged with high fat diet (HFD) [80]. While food intake seemed comparable between knockout and control groups in an early report [80], we have recently found a significant reduction of palatable HFD intake in ERRα knockout mice that is associated with a significant reduction of body weight [81]. This is not only with general consumption of HFD, and we also found that hungry ERRα knockout mice display less willingness to obtain HFD pellets in an operant-responding behavioral paradigm with a progressive ratio schedule compared to their control littermates, suggesting a reduced motivation to work for palatable food [81]. Interestingly, through a family-based genetic linkage study combined with whole exome sequencing in a family in which multiple members are affected by eating disorders, particularly anorexia nervosa, we have identified a missense mutation in the ERRα gene that co-segregates with illness [82]. A subsequent study in mice revealed that the level of expression of ERRα in the brain is increased by caloric restriction, implying that brain ERRα may sense peripheral energy status and convert it into protective behavioral actions, including motivation to obtain and consume food [81]. It is possible that a genetic deficit of ERRα disrupts this adaptive (protective?) physiological process upon caloric restriction, leading to pathological conditions such as eating disorders [81]. Somewhat consistent with the differential role of ERRα and ERα in body-weight homeostasis, the expression pattern of ERRα in the hypothalamus is different from ERα. Unlike the high expression of ERα in the mediobasal hypothalamus, which is critical for the coordinated control of energy balance [68], ERRα expression is nearly absent in the mediobasal hypothalamus (Figure 2). However, it should also be noted that other than the mediobasal hypothalamus, in many brain regions, including the cortex and hippocampus, both ERα and ERRα are homogeneously expressed (Figure 2). From the viewpoint of brain reward circuits [83,84], however, some of the extra-hypothalamic regions, such as several frontal cortices and the hippocampus, express relatively high levels of ERRα, and the ventral striatum, ventral pallidum, and lateral hypothalamus with moderate levels [81]. Interestingly, the ventral tegmental area (VTA), a key brain region of brain reward function, had minimal expression of ERRα [81]. We have previously shown that specific knockdown of ERRα in the medial prefrontal cortex (mPFC) recapitulates reduced motivation for HFD observed in Esrra-null female mice, indicating that Esrra expression in the mPFC may affects top-down control of food reward behaviors [81]. Nonetheless, it is possible that ERRα expression in the periphery, rather than the brain, is responsible for protected weight gain observed in ERRα knockout mice on HFD feeding. Further studies with conditional a ERRα deletion approach are necessary to prove this possibility.

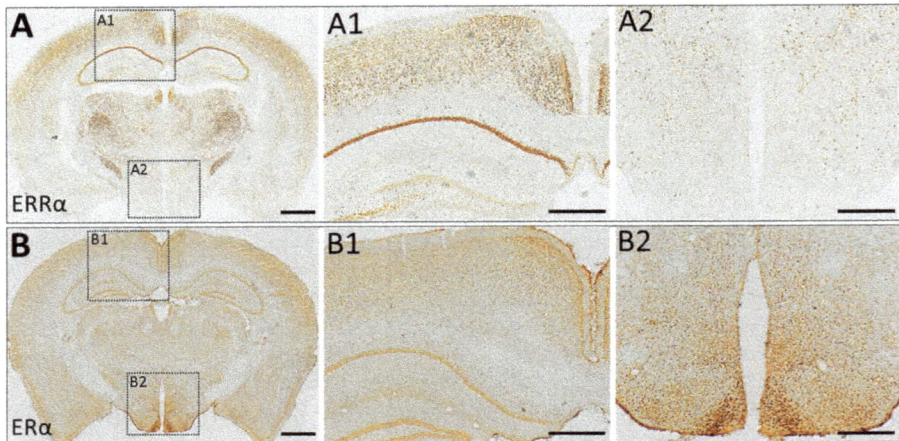

Figure 2. Representative immunohistochemistry (IHC) images showing ERRα (**A**) and ERα (**B**) expression in the mouse brain. Digital zooms of individual boxed regions are shown in A1, A2, B1, and B2. Note that expression pattern of ERRα and ERα is similar in the cortex and the hippocampus (**A1,B1**), but strikingly different expression was observed in the mediobasal hypothalamus (**A2,B2**) that is critical for the homeostatic regulation of energy balance. Note: the IHC image shown in (**A**) was from our previous publication with a zoom-in view [81]. The IHC image shown in (**B**) was from the brain sections of an adult female wild-type mouse stained with validated commercially available ERα antibody (1:1000, Millipore) as reported previously [71]. This ERα antibody was validated in conditional ERα KO mice previously [68]. Scale bar = 1 mm in A and B, and 500 μm in A1, A2, B1, and B2.

ERRγ knockout mice die shortly after birth due to cardiac failure [85] and, therefore, the role of ERRγ in the regulation of body weight homeostasis remains elusive. A recent study with exclusive overexpression of ERRγ in skeletal muscle of obese db/db mice revealed that gain of function of ERRγ in skeletal muscle does not ameliorate obesity or diabetic phenotypes in leptin receptor deficiency [34]. ERRγ is also abundantly expressed throughout the brain [86,87], and colocalization of ERRγ and ERα is confirmed within the same neurons in selected regions [86]. Again, studies with conditional ERRγ deletion models are necessary to determine the effects of ERRγ in the long-term regulation of energy homeostasis.

Like ERRγ knockout, global knockout of ERRβ is also lethal for mice [88]. However, a recent study with conditional deletion of ERRβ using Sox2-Cre revealed that these mice are viable and exhibit significantly decreased body weight compared to their littermate controls mainly due to increased activity and energy expenditure [89]. In fact, food intake was significantly increased, which is likely a compensatory response for extremely increased energy expenditure. Paradoxically, however, when ERRβ is deleted from the central nervous system using nestin-Cre, these mice exhibit significantly increased body weight while maintaining higher activity and energy expenditure. Consistent with these results, it was found that increased body weight was mainly due to increased lean mass, but not fat mass, explaining their increased energy expenditure. Interestingly, a loss of ERRβ caused significant upregulation of ERRγ which, the author concluded, might be responsible for the decreased NPY expression in these mice affecting the satiety response [89]. Thus, the ERRβ is clearly involved in long-term regulation of the energy balance, but underlying mechanisms seem complicated, involving a combination of changes of food intake, meal pattern, activity, and energy expenditure by different mechanisms.

Compared to peripheral tissues, the study of the cross-talk between ERs and ERRs in the brain is limited. Further investigation is needed to clarify their cellular colocalization and to better understand the coordinated actions of these relatives.

5.2. Actions in Learning and Memory

Estrogens are known to affect the hippocampus, a large brain structure critical for learning and memory. Estrogen can acutely modulate the electrophysiological properties of hippocampal neurons in ex vivo slice preparations [90–92]. Since this effect is rapid, estrogen is thought to work through a rapid, membrane-initiated mechanism in this regard. Membrane-initiated estrogen-signaling activates various protein kinase cascades such as PI3K, protein kinase A, protein kinase C, phospho lipase C, and mitogen-activated protein kinase, leading to the modulation of signal transduction, protein phosphorylation, and ion channel activity [76]. Two estrogen receptors, ERα and ERβ, appear to be located predominantly in synapses, axons, dendrites and dendritic spines [93,94], and work differentially in the hippocampal inhibitory and excitatory synapses, respectively. Estradiol acutely enhances excitatory postsynaptic currents, which can be recapitulated by ERβ-specific agonist diaryl-propionitrile (DPN), but not ERα-specific agonist propyl-pyrazole triol (PPT) [95]. On the other hand, estradiol works through ERα to suppress inhibitory neurotransmission in the hippocampal CA1 neurons [96]. Estrogen treatment rapidly increases dendritic spine density in CA1 of the hippocampus associated with improved spatial learning and memory [97]. Rapid estrogen signaling is observed even with E2-BSA, a membrane-impermeable conjugate of estrogen [98,99]. In addition to these classical ERs, a number of studies using different technologies indicate that rapid estrogen signaling mediated by receptors other than ERα/β exists. E2 can potentiate kainate-induced currents, which can be observed even in the hippocampus of ERα KO mice. In addition, this E2-induced potentiation is unaffected even in the presence of ERα/β blocker [100]. Selective ligand to G-protein coupled estrogen receptor STX, that does not bind to ERα and ERβ, and activates G-protein signaling cascade [99].

Despite considerably high expression of ERRs, especially ERRα and ERRγ, in hippocampal formation, their roles in memory and learning are largely unexplored. Some evidence suggests that ERRγ-mediated gene transcriptions may affect hippocampal functions. It was shown that bisphenol A, a potential agonist for ERRγ and ERα/β, modulates spinogenesis in adult hippocampal neurons through ERRγ, but not ERα/β. Pei et al. showed that ERRγ-deficient hippocampal neurons exhibit lower metabolic capacity [101]. In addition, ERRγ-deficient hippocampal slices showed a significant reduction in long-term potentiation (LTP), which could be rescued by pyruvate supplementation, indicating that the impaired LTP is likely caused by the metabolic deficiency. Consistent with these observations, ERRγ knockout mice showed impaired spatial learning and memory [101]. Although ERRα is also widely expressed throughout hippocampal formation (Figure 2), its role in learning and memory remains to be explored. We have previously shown that the ability of learning and memory in ERRα knockout mice was comparable to wild-type littermates in a Barnes maze test, but reversible learning in this behavioral paradigm was significantly impaired in ERRα knockout mice [81]. However, impaired reversible learning in a Barnes maze test could be interpreted as an indication of behavioral rigidity, rather than memory per se. Interestingly, Wrann et al. have recently shown that exercise-induced increases of brain-derived neurotrophic factor (BDNF), a neurotrophic factor essential for synaptic plasticity, hippocampal function and learning, is mediated by elevated expression levels of fibronectin type III domain containing 5 (FNDC5) (precursor of a novel, circulating myokine irisin) that is driven by a ERRα/PGC1α transcriptional network [40]. These findings indicate that, while the ERRα may plays a minimal role in learning under normal physiological conditions, it might be an important molecular mediator of exercise-induced beneficial effects in improving learning and memory through the ERRα/PGC1α→FNDC5→BDNF pathway. Indeed, it has been postulated that, rather than in a baseline condition, the roles ERRα plays become more apparent when animals are subjected to various physiological and environmental challenges requiring them to make adaptive responses [102]. Therefore, the potential role of ERRα in learning and memory needs to be further investigated under different physiological conditions or by different behavioral learning and memory tasks. Additionally, we have previously found that both presynaptic vesicle pool density and the numbers of dendritic spines are significantly decreased in the striatum of female ERRα knockout mice [103]. Although it remains unclear whether similar changes also occur in the hippocampus, these findings are suggestive of impaired synaptic plasticity in ERRα knockout mice by impaired synaptic vesicle trafficking and/or synaptogenesis. It will be interesting to test whether exercise-induced beneficial effects on improving learning and memory are lost or impaired in ERRα-deficient mice.

5.3. Actions in Social Behaviors

Estrogens have been shown to impact a wide range of social behaviors, and ERα and ERβ may play differential roles in this regard in a gender-specific manner [104]. Global ERα knockout male mice are less aggressive [105], while ERβ knockout mice show enhanced aggressive behavior [106], indicating that ERα is essential for expressing aggressive behavior whereas ERβ works on it in an antagonistic manner. Furthermore, ERα in distinct brain nuclei differentially regulates male aggressive and sexual behaviors. Adeno-associated virus (AAV)-mediated ERα knockdown in the medial preoptic area reduced sexual but not aggressive behavior while the knockdown in the VMH suppressed both behaviors. On the other hand, the knockdown in the MeA had no effects on either behavior [107]. MeA also expresses aromatase. Selective ablation of aromatase-expressing cells in the MeA suppresses male aggressive behavior and female maternal aggressive behavior [108].

The involvement of estrogen-signaling in social behaviors should not ignore its function in the perceptions of others [109]. Ovariectomized female animals are less attractive to males than intact female animals, which can be restored by estrogen supplementation [110,111]. Genetic studies suggest that odors produced from either ERα knockout or ERβ knockout female mice are different from those from wild-type female mice, and wild-type male mice show less interest in ERα knockout or ERβ knockout female mice compared to wild-type female mice [112]. The same thing can be said of the perception of ERα and ERβ knockout male mice by wild-type female mice. Wild-type female mice can successfully discriminate the odors of wild-type males from those of knockout males and display significantly higher preference for the WT male odors.

Classic lesion and electrical stimulation studies have identified the brain loci involved in aggression and other social behaviors, which include the anteroventral periventricular nucleus, the medial preoptic area, the bed nucleus of stria terminalis, and VMH. Additionally, the recent development of optogenetic and pharmacogenetic techniques also greatly facilitated the process of delineating brain circuits that might mediate estrogen actions in these behavioral regulations [99]. The VMH is one of the important nuclei that mediate estrogenic actions in energy homeostasis. In addition to energy homeostasis, recent studies using optogenetic and pharmacogenetic tools have begun to unravel the importance of ERα-expressing neurons in the VMH in social behaviors, such as sexual behavior and aggression. Optogenetic and pharmacogenetic activation of ERα-expressing neurons in the VMH trigger aggressive behaviors in both males and females [113,114]. This ERα-expressing subpopulation of VMH neurons seem to not only control aggressive behavior but also involve other social behaviors such as investigation and mounting in males, and increasing both the number of active neurons and the activity level of each neuron can shift the behavioral responses from mounting to attacking [113]. A separate study targeting ERα-expressing neurons with same Cre mouse model also showed that VMH ERα-positive neurons are highly active during attacking and are necessary for female aggressive behavior as well [114]. These studies clearly showed the involvement of ERα-expressing neurons in aggressive behavior, but it remains unclear whether signaling through ERs itself is important for the aggressive behaviors.

ERRs, especially ERRα, also regulate social behaviors. Impaired social function was one of the characteristic behavioral deficits observed in ERRα knockout mice [81]. ERRα knockout mice showed reduced interaction with a novel mouse tested in a social interaction test. Furthermore, in the tube test which measures the dominance tendency, ERRα knockout mice were almost always the losers [81]. The recent sophisticated behavioral and optogenetic studies showed that the activity of the dorsomedial prefrontal cortex (dmPFC), where ERRα is highly expressed, is important for instant winning or losing in the tube test [115,116]. It will be interesting to test if optogenetic activation of dmPFC ERRα-positive neurons can rescue social subordination seen in ERRα knockout mice. It is difficult to associate the VMH-regulated aggressive behavior with dmPFC-regulated social dominance. Social dominance is important for survival in social animals and aggression could be a means to reach the top of the social hierarchy. It is noteworthy that the selective knockdown of ERRα in distinct subnuclei of the PFC can recapitulate some behavioral deficits observed in ERRα knockout mice, such as reduced body weight and food intake, reduced effort responding for food reward, and increased grooming [81]. Recently, synchronized activity among distant brain nuclei (PFC-lateral septum-lateral hypothalamus) is suggested to be involved in food-seeking behavior [117]. Whether such a synchronicity exists between PFC and VMH for the coordinated control of social dominance and/or aggressive behavior remains to be determined. Cortical parvalbumin (PV)-expressing inhibitory interneurons are a potent regulator of local network activities [118] and coherent activity of PV neurons orchestrates synchronous gamma oscillations [119,120]. Widely distributed expression patterns of ERRs (both ERRα [81] and ERRγ [86,87]) throughout the brain are indicative of their likely expression in both excitatory and inhibitory neurons. Future studies of conditional deletion models are required to distinguish their roles in different types of neurons in behavioral regulations. Given an established role of ERRs in mitochondrial biogenesis [121,122] and, thus, cellular energy supply, it would be interesting to investigate whether ERRs are enriched in cortical inhibitory interneurons with high energy demand, particularly fast-spiking PV neurons [123,124], and whether loss of ERRs will cause aberrant firing of these neurons due to insufficient cellular energy production, leading to behavioral abnormalities including social behaviors. In line with the involvement of ERRs in mitochondrial biogenesis in GABAergic interneurons, its protein ligand PGC1-α, a master regulator of mitochondrial biogenesis, is predominantly expressed in these neuronal populations [125] and bidirectionally regulates PV expression [126]. PGC1-α in PV neurons transcriptionally regulates genes relevant to synaptic transmission, as well as metabolism-related genes and conditional deletion of PGC1-α in PV neurons results in asynchronous GABA release and impaired long-term memory [127]. Notably, a recent comprehensive gene expression profiling in the hypothalamus, frontal cortex, and amygdala by RNA-Seq combined with ChIP-Seq revealed that ERRα is in the center of coordinating transcriptional networks for adaptive responses when animals are challenged to agonistic social encounters [128]. Transcriptional regulatory dynamics that take place under social challenges could be essential for animals to learn from this type of social challenge and affect future behaviors critical for survival. The roles of ERRα in adaptive metabolic and behavioral responses to social stressors warrant future investigation. Additionally, a variety of behavioral deficits caused by ERRα deletion are sexually dimorphic [81], yet no sexually dimorphic expression patterns of ERRα in the brain were observed. Further studies are needed to clarify the underlying mechanisms of sexual dimorphic roles of ERRα in these behavioral regulations.

6. Concluding Remark

It has been well known that both ERs and ERRs play important roles in physiological regulations through their abundant expression in peripheral tissues, particularly for metabolic homeostasis and energy metabolism (Figure 3). Mounting evidence indicates that the brain is also one of the primary targets of estrogen (via ERs) to regulate a variety of behaviors and physiological functions including reproduction, energy homeostasis, and learning and memory. ERRs share sequence similarity with ERs, but estrogen is not their endogenous ligand and little attention has been paid to the cross-talk between estrogen-signaling and ERRs. Existing evidence supports the idea that estrogen-signaling and ERRs may cross-talk via transcriptional regulation, or reciprocal binding on each responsive element, or even intercellularly through the regulation of estrogen synthesis by aromatase. However, the roles of ERRs in the brain and functional segregation of the isoforms remain largely unknown. Additionally, the functional overlaps between ERs and ERRs are almost untouched at the behavior level. Gene-expression profiling studies in peripheral tissues and cell lines indicate that shared target genes by both receptor families may be modest, with a high degree of independence. While the expression patterns of ERα and ERRα suggest that these two receptors might colocalize in some brain regions, to what extent, if any, they share transcriptional targets in the brain is unclear. It is obvious that both families are involved in the processes important for brain functions such as synaptic transmission, neuronal firing and mitochondrial biogenesis. A more comprehensive understanding of the target genes and the transcriptional cross-talk between these receptors may provide more insights into the estrogen-dependent and -independent regulation of brain functions.

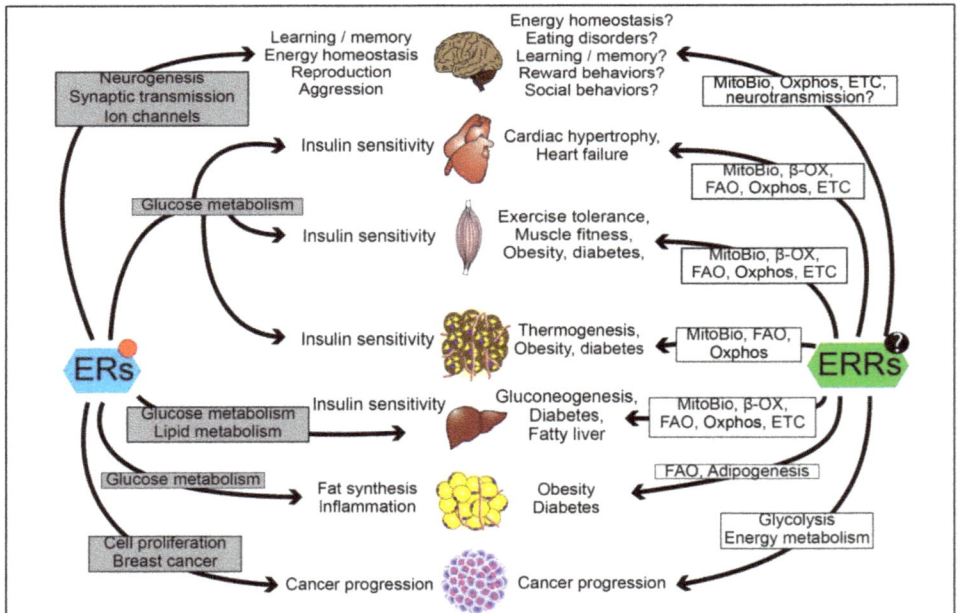

Figure 3. ERs and ERRs regulate a variety of important cellular functions and any disruption in these processes can lead to different pathological conditions. MitoBio, mitochondrial biogenesis; β-OX, beta-oxidation; FAO, fatty acid oxidation; Oxphos, oxidative phosphorylation; ETC, electron transport chain. Red dot indicates estrogens and "?" indicates unknown ligands for ERRs.

Acknowledgments: This work was supported by grants from the National Institutes of Health (HL127673 and MH109920), F.O.E. Diabetes Research Center, and the UIHC Center for Hypertension Research.

Author Contributions: Kenji Saito wrote initial manuscript which was edited and finalized by Huxing Cui.

Conflicts of Interest: The authors declare no conflict of interest.

Abbreviations

AgRP	Agouti-Related Peptide
Akt	Protein Kinase B
ARC	Arcuate nucleus Of Hypothalamus
BDNF	Brain-Derived Neurotrophic Factor
β-OX	β-Oxidation
ChIP	Chromatin Immunoprecipitation
CYP19	Aromatase
DBD	DNA Binding Domain
dmPFC	Dorsomedial Prefrontal Cortex
DPN	Diaryl-Propionitrile
E2	17β-Estradiol
ER	Estrogen Receptor
ERE	Estrogen Response Element
ERR	Estrogen-Related Receptor
ERRE	Estrogen-Related Response Element
ERK	Extracellular Signal-Regulated Kinase
ETC	Electron Transport Chain
FAO	Fatty Acid Oxidation
FNDC5	Fibronectin Type III Domain Containing 5
GPER	G protein-Coupled Estrogen Receptor
HFD	High Fat Diet
IHC	Immunohistochemistry
LTP	Long-Term Potenciation
MeA	Medial Amygdala
MHRE	Multiple-Hormone Response element
MitoBio	Mitochondrial Biogenesis
NPY	Neuropeptide Y
NTS	Nucleus of Solitary Tract
Oxphos	Oxidative Phosphorylation
PV	Parvalbumin
PGC	Proliferator-Activated Receptor γ Coactivator
PI3K	Phosphatidyl Inositide 3-Kinase
POMC	Pro-Opiomelanocortin
PPT	Propyl-Pyrazole Triol
VMH	Ventromedial Nucleus of Hypothalamus
VTA	Ventral Tegmental Area

References

1. Klinge, C.M. Estrogen receptor interaction with co-activators and co-repressors. *Steroids* **2000**, *65*, 227–251. [CrossRef]
2. Pettersson, K.; Svensson, K.; Mattsson, R.; Carlsson, B.; Ohlsson, R.; Berkenstam, A. Expression of a novel member of estrogen response element-binding nuclear receptors is restricted to the early stages of chorion formation during mouse embryogenesis. *Mech. Dev.* **1996**, *54*, 211–223. [CrossRef]
3. Giguère, V.; Yang, N.; Segui, P.; Evans, R.M. Identification of a new class of steroid hormone receptors. *Nature* **1988**, *331*, 91–94. [CrossRef] [PubMed]

4. Huss, J.M.; Garbacz, W.G.; Xie, W. Constitutive activities of estrogen-related receptors: Transcriptional regulation of metabolism by the ERR pathways in health and disease. *Biochim. Biophys. Acta (BBA) Mol. Basis Dis.* **2015**, *1852*, 1912–1927. [CrossRef] [PubMed]

5. Gearhart, M.D.; Holmbeck, S.M.A.; Evans, R.M.; Dyson, H.J.; Wright, P.E. Monomeric complex of human orphan estrogen related receptor-2 with DNA: A pseudo-dimer interface mediates extended half-site recognition. *J. Mol. Biol.* **2003**, *327*, 819–832. [CrossRef]

6. Huppunen, J.; Aarnisalo, P. Dimerization modulates the activity of the orphan nuclear receptor ERRgamma. *Biochem. Biophys. Res. Commun.* **2004**, *314*, 964–970. [CrossRef] [PubMed]

7. Vanacker, J.M.; Pettersson, K.; Gustafsson, J.A.; Laudet, V. Transcriptional targets shared by estrogen receptor-related receptors (ERRs) and estrogen receptor (ER) alpha, but not by ERbeta. *EMBO J.* **1999**, *18*, 4270–4279. [CrossRef] [PubMed]

8. Giguère, V. Transcriptional control of energy homeostasis by the estrogen-related receptors. *Endocr. Rev.* **2008**, *29*, 677–696. [CrossRef] [PubMed]

9. Bookout, A.L.; Jeong, Y.; Downes, M.; Yu, R.T.; Evans, R.M.; Mangelsdorf, D.J. Anatomical profiling of nuclear receptor expression reveals a hierarchical transcriptional network. *Cell* **2006**, *126*, 789–799. [CrossRef] [PubMed]

10. Jakacka, M.; Ito, M.; Martinson, F.; Ishikawa, T.; Lee, E.J.; Jameson, J.L. An Estrogen Receptor (ER)α Deoxyribonucleic Acid-Binding Domain Knock-In Mutation Provides Evidence for Nonclassical ER Pathway Signaling In Vivo. *Mol. Endocrinol.* **2002**, *16*, 2188–2201. [CrossRef] [PubMed]

11. Park, C.J.; Zhao, Z.; Glidewell-Kenney, C.; Lazic, M.; Chambon, P.; Krust, A.; Weiss, J.; Clegg, D.J.; Dunaif, A.; Jameson, J.L.; et al. Genetic rescue of nonclassical ERalpha signaling normalizes energy balance in obese Eralpha-null mutant mice. *J. Clin. Investig.* **2011**, *121*, 604–612. [CrossRef] [PubMed]

12. Malyala, A.; Zhang, C.; Bryant, D.N.; Kelly, M.J.; Rønnekleiv, O.K. PI3K signaling effects in hypothalamic neurons mediated by estrogen. *J. Comp. Neurol.* **2008**, *506*, 895–911. [CrossRef] [PubMed]

13. Micevych, P.E.; Kelly, M.J. Membrane estrogen receptor regulation of hypothalamic function. *Neuroendocrinology* **2012**, *96*, 103–110. [CrossRef] [PubMed]

14. Saito, K.; He, Y.; Yang, Y.; Zhu, L.; Wang, C.; Xu, P.; Hinton, A.O.; Yan, X.; Zhao, J.; Fukuda, M.; et al. PI3K in the ventromedial hypothalamic nucleus mediates estrogenic actions on energy expenditure in female mice. *Sci. Rep.* **2016**, *6*, 1–10. [CrossRef] [PubMed]

15. Zhu, L.; Xu, P.; Cao, X.; Yang, Y.; Hinton, A.O.; Xia, Y.; Saito, K.; Yan, X.; Zou, F.; Ding, H.; et al. The ERα-PI3K cascade in proopiomelanocortin progenitor neurons regulates feeding and glucose balance in female mice. *Endocrinology* **2015**, *156*, 4474–4491. [CrossRef] [PubMed]

16. Carmeci, C.; Thompson, D.A.; Ring, H.Z.; Francke, U.; Weigel, R.J. Identification of a gene (GPR30) with homology to the G-protein-coupled receptor superfamily associated with estrogen receptor expression in breast cancer. *Genomics* **1997**, *45*, 607–617. [CrossRef] [PubMed]

17. O'Dowd, B.F.; Nguyen, T.; Marchese, A.; Cheng, R.; Lynch, K.R.; Heng, H.H.; Kolakowski, L.F.; George, S.R. Discovery of three novel G-protein-coupled receptor genes. *Genomics* **1998**, *47*, 310–313. [CrossRef] [PubMed]

18. Owman, C.; Nilsson, C.; Lolait, S.J. Cloning of cDNA encoding a putative chemoattractant receptor. *Genomics* **1996**, *37*, 187–194. [CrossRef] [PubMed]

19. Takada, Y.; Kato, C.; Kondo, S.; Korenaga, R.; Ando, J. Cloning of cDNAs encoding G protein-coupled receptor expressed in human endothelial cells exposed to fluid shear stress. *Biochem. Biophys. Res. Commun.* **1997**, *240*, 737–741. [CrossRef] [PubMed]

20. Davis, K.E.; Carstens, E.J.; Irani, B.G.; Gent, L.M.; Hahner, L.M.; Clegg, D.J. Sexually dimorphic role of G protein-coupled estrogen receptor (GPER) in modulating energy homeostasis. *Horm. Behav.* **2014**, *66*, 196–207. [CrossRef] [PubMed]

21. Isensee, J.; Meoli, L.; Zazzu, V.; Nabzdyk, C.; Witt, H.; Soewarto, D.; Effertz, K.; Fuchs, H.; Gailus-Durner, V.; Busch, D.; et al. Expression pattern of G protein-coupled receptor 30 in LacZ reporter mice. *Endocrinology* **2009**, *150*, 1722–1730. [CrossRef] [PubMed]

22. Kwon, O.; Kang, E.S.; Kim, I.; Shin, S.; Kim, M.; Kwon, S.; Oh, S.R.; Ahn, Y.S.; Kim, C.H. GPR30 mediates anorectic estrogen-induced STAT3 signaling in the hypothalamus. *Metab. Clin. Exp.* **2014**, *63*, 1455–1461. [CrossRef] [PubMed]

23. Martensson, U.E.A.; Salehi, S.A.; Windahl, S.; Gomez, M.F.; Swärd, K.; Daszkiewicz-Nilsson, J.; Wendt, A.; Andersson, N.; Hellstrand, P.; Gründe, P.O.; et al. Deletion of the G protein-coupled receptor 30 impairs glucose tolerance, reduces bone growth, increases blood pressure, and eliminates estradiol-stimulated insulin release in female mice. *Endocrinology* **2009**, *150*, 687–698. [CrossRef] [PubMed]

24. Revankar, C.M. A Transmembrane Intracellular Estrogen Receptor Mediates Rapid Cell Signaling. *Science* **2005**, *307*, 1625–1630. [CrossRef] [PubMed]

25. Sharma, G.; Hu, C.; Brigman, J.L.; Zhu, G.; Hathaway, H.J.; Prossnitz, E.R. GPER deficiency in male mice results in insulin resistance, dyslipidemia, and a proinflammatory state. *Endocrinology* **2013**, *154*, 4136–4145. [CrossRef] [PubMed]

26. Wang, C.; Dehghani, B.; Magrisso, I.J.; Rick, E.A.; Bonhomme, E.; Cody, D.B.; Elenich, L.A.; Subramanian, S.; Murphy, S.J.; Kelly, M.J.; et al. GPR30 contributes to estrogen-induced thymic atrophy. *Mol. Endocrinol. (Baltimore, MD)* **2008**, *22*, 636–648. [CrossRef] [PubMed]

27. Waters, E.M.; Thompson, L.I.; Patel, P.; Gonzales, A.D.; Ye, H.; Filardo, E.J.; Clegg, D.J.; Gorecka, J.; Akama, K.T.; McEwen, B.S.; et al. G-Protein-Coupled Estrogen Receptor 1 Is Anatomically Positioned to Modulate Synaptic Plasticity in the Mouse Hippocampus. *J. Neurosci.* **2015**, *35*, 2384–2397. [CrossRef] [PubMed]

28. Misawa, A.; Inoue, S. Estrogen-related receptors in breast cancer and prostate cancer. *Front. Endocrinol.* **2015**, *6*, 1–7. [CrossRef] [PubMed]

29. Giguère, V. To ERR in the estrogen pathway. *Trends Endocrinol. Metab. TEM* **2002**, *13*, 220–225. [CrossRef]

30. Hu, P.; Kinyamu, H.K.; Wang, L.; Martin, J.; Archer, T.K.; Teng, C. Estrogen induces estrogen-related receptor α gene expression and chromatin structural changes in estrogen receptor (ER)-positive and ER-negative breast cancer cells. *J. Biol. Chem.* **2008**, *283*, 6752–6763. [CrossRef] [PubMed]

31. Deblois, G.; Giguère, V. Oestrogen-related receptors in breast cancer: Control of cellular metabolism and beyond. *Nat. Rev. Cancer* **2013**, *13*, 27–36. [CrossRef] [PubMed]

32. Michalek, R.D.; Gerriets, V.A.; Nichols, A.G.; Inoue, M.; Kazmin, D.; Chang, C.-Y.; Dwyer, M.A.; Nelson, E.R.; Pollizzi, K.N.; Ilkayeva, O.; et al. Estrogen-related receptor-α is a metabolic regulator of effector T-cell activation and differentiation. *Proc. Natl. Acad. Sci. USA* **2011**, *108*, 18348–18353. [CrossRef] [PubMed]

33. Wei, W.; Schwaid, A.G.; Wang, X.; Wang, X.; Chen, S.; Chu, Q.; Saghatelian, A.; Wan, Y. Ligand activation of ERRα by cholesterol mediates statin and bisphosphonate effects. *Cell Metab.* **2016**, *23*, 479–491. [CrossRef] [PubMed]

34. Badin, P.-M.; Vila, I.K.; Sopariwala, D.H.; Yadav, V.; Lorca, S.; Louche, K.; Kim, E.R.; Tong, Q.; Song, M.S.; Moro, C.; et al. Exercise-like effects by Estrogen-related receptor-gamma in muscle do not prevent insulin resistance in db/db mice. *Sci. Rep.* **2016**, *6*, 26442. [CrossRef] [PubMed]

35. Gantner, M.L.; Hazen, B.C.; Eury, E.; Brown, E.L.; Kralli, A. Complementary roles of estrogen-related receptors in brown adipocyte thermogenic function. *Endocrinology* **2016**, *157*, 4770–4781. [CrossRef] [PubMed]

36. Handschin, C.; Mootha, V.K. Estrogen-related receptor α (ERRα): A novel target in type 2 diabetes. *Drug Discov. Today Ther. Strateg.* **2005**, *2*, 151–156. [CrossRef]

37. Powelka, A.M.; Seth, A.; Virbasius, J.V.; Kiskinis, E.; Nicoloro, S.M.; Guilherme, A.; Tang, X.; Straubhaar, J.; Cherniack, A.D.; Parker, M.G.; et al. Suppression of oxidative metabolism and mitochondrial biogenesis by the transcriptional corepressor RIP140 in mouse adipocytes. *J. Clin. Investig.* **2006**, *116*, 125–136. [CrossRef] [PubMed]

38. Patch, R.J.; Huang, H.; Patel, S.; Cheung, W.; Xu, G.; Zhao, B.P.; Beauchamp, D.A.; Rentzeperis, D.; Geisler, J.G.; Askari, H.B.; et al. Indazole-based ligands for estrogen-related receptor α as potential anti-diabetic agents. *Eur. J. Med. Chem.* **2017**, *138*, 830–853. [CrossRef] [PubMed]

39. Gibson, D.A.; Saunders, P.T.K. Estrogen dependent signaling in reproductive tissues—A role for estrogen receptors and estrogen related receptors. *Mol. Cell. Endocrinol.* **2012**, *348*, 361–372. [CrossRef] [PubMed]

40. Wrann, C.D.; White, J.P.; Salogiannnis, J.; Laznik-Bogoslavski, D.; Wu, J.; Ma, D.; Lin, J.D.; Greenberg, M.E.; Spiegelman, B.M. Exercise induces hippocampal BDNF through a PGC-1α/FNDC5 pathway. *Cell Metab.* **2013**, *18*, 649–659. [CrossRef] [PubMed]

41. Bianco, S.; Lanvin, O.; Tribollet, V.; Macari, C.; North, S.; Vanacker, J.-M. Modulating estrogen receptor-related receptor-alpha activity inhibits cell proliferation. *J. Biol. Chem.* **2009**, *284*, 23286–23292. [CrossRef] [PubMed]

42. Takayanagi, S.; Tokunaga, T.; Liu, X.; Okada, H.; Matsushima, A.; Shimohigashi, Y. Endocrine disruptor bisphenol A strongly binds to human estrogen-related receptor gamma (ERRgamma) with high constitutive activity. *Toxicol. Lett.* **2006**, *167*, 95–105. [CrossRef] [PubMed]

43. Teng, C.T.; Beames, B.; Alex Merrick, B.; Martin, N.; Romeo, C.; Jetten, A.M. Development of a stable cell line with an intact PGC-1α/ERRα axis for screening environmental chemicals. *Biochem. Biophys. Res. Commun.* **2014**, *444*, 177–181. [CrossRef] [PubMed]

44. Teng, C.T.; Hsieh, J.-H.; Zhao, J.; Huang, R.; Xia, M.; Martin, N.; Gao, X.; Dixon, D.; Auerbach, S.S.; Witt, K.L.; et al. Development of Novel Cell Lines for High-Throughput Screening to Detect Estrogen-Related Receptor Alpha Modulators. *SLAS Discov. Adv. Life Sci. R D* **2017**, *22*, 720–731. [CrossRef] [PubMed]

45. Lynch, C.; Zhao, J.; Huang, R.; Kanaya, N.; Bernal, L.; Hsieh, J.-H.; Auerbach, S.S.; Witt, K.L.; Merrick, B.A.; Chen, S.; et al. Identification of Estrogen-Related Receptor Alpha Agonists in the Tox21 Compound Library. *Endocrinology* **2017**, *159*, 744–753. [CrossRef] [PubMed]

46. Masood, S. Estrogen and progesterone receptors in cytology: A comprehensive review. *Diagn. Cytopathol.* **1992**, *8*, 475–491. [CrossRef] [PubMed]

47. Mohibi, S.; Mirza, S.; Band, H.; Band, V. Mouse models of estrogen receptor-positive breast cancer. *J. Carcinog.* **2011**, *10*, 35. [PubMed]

48. Caldon, C.E.; Sutherland, R.L.; Musgrove, E. Cell cycle proteins in epithelial cell differentiation: Implications for breast cancer. *Cell Cycle (Georgetown, TX)* **2010**, *9*, 1918–1928. [CrossRef] [PubMed]

49. Musgrove, E.A.; Caldon, C.E.; Barraclough, J.; Stone, A.; Sutherland, R.L. Cyclin D as a therapeutic target in cancer. *Nat. Rev. Cancer* **2011**, *11*, 558–572. [CrossRef] [PubMed]

50. Ariazi, E.A.; Clark, G.M.; Mertz, J.E. Estrogen-related receptor alpha and estrogen-related receptor gamma associate with unfavorable and favorable biomarkers, respectively, in human breast cancer. *Cancer Res.* **2002**, *62*, 6510–6518. [PubMed]

51. Suzuki, T.; Miki, Y.; Moriya, T.; Shimada, N.; Ishida, T.; Hirakawa, H.; Ohuchi, N.; Sasano, H. Estrogen-related receptor α in human breast carcinoma as a potent prognostic factor. *Cancer Res.* **2004**, *64*, 4670–4676. [CrossRef] [PubMed]

52. Liu, D.; Zhang, Z.; Gladwell, W.; Teng, C.T. Estrogen stimulates estrogen-related receptor alpha gene expression through conserved hormone response elements. *Endocrinology* **2003**, *144*, 4894–4904. [CrossRef] [PubMed]

53. Shigeta, H.; Zuo, W.; Yang, N.; DiAugustine, R.; Teng, C.T. The mouse estrogen receptor-related orphan receptor alpha 1: Molecular cloning and estrogen responsiveness. *J. Mol. Endocrinol.* **1997**, *19*, 299–309. [CrossRef] [PubMed]

54. Deblois, G.; Hall, J.A.; Perry, M.C.; Laganière, J.; Ghahremani, M.; Park, M.; Hallett, M.; Giguère, V. Genome-wide identification of direct target genes implicates estrogen-related receptor α as a determinant of breast cancer heterogeneity. *Cancer Res.* **2009**, *69*, 6149–6157. [CrossRef] [PubMed]

55. Kammerer, M.; Gutzwiller, S.; Stauffer, D.; Delhon, I.; Seltenmeyer, Y.; Fournier, B. Estrogen Receptor α (ERα) and Estrogen Related Receptor α (ERRα) are both transcriptional regulators of the Runx2-I isoform. *Mol. Cell. Endocrinol.* **2013**, *369*, 150–160. [CrossRef] [PubMed]

56. Yang, C.; Zhou, D.; Chen, S. Modulation of aromatase expression in the breast tissue by ERR alpha-1 orphan receptor. *Cancer Res.* **1998**, *58*, 5695–5700. [PubMed]

57. Miao, L.; Shi, J.; Wang, C.-Y.; Zhu, Y.; Du, X.; Jiao, H.; Mo, Z.; Klocker, H.; Lee, C.; Zhang, J. Estrogen receptor-related receptor alpha mediates up-regulation of aromatase expression by prostaglandin E2 in prostate stromal cells. *Mol. Endocrinol.* **2010**, *24*, 1175–1186. [CrossRef] [PubMed]

58. Kumar, P.; Kamat, A.; Mendelson, C.R. Estrogen receptor alpha (ERα) mediates stimulatory effects of estrogen on aromatase (CYP19) gene expression in human placenta. *Mol. Endocrinol.* **2009**, *23*, 784–793. [CrossRef] [PubMed]

59. Tanida, T.; Matsuda, K.I.; Yamada, S.; Hashimoto, T.; Kawata, M. Estrogen-related Receptor β Reduces the Subnuclear Mobility of Estrogen Receptor α and Suppresses Estrogen-dependent Cellular Function. *J. Biol. Chem.* **2015**, *290*, 12332–12345. [CrossRef] [PubMed]

60. Saito, K.; Cao, X.; He, Y.; Xu, Y. Progress in the molecular understanding of central regulation of body weight by estrogens. *Obesity* **2015**, *23*, 919–926. [CrossRef] [PubMed]

61. Allende-Vigo, M.Z. Women and the metabolic syndrome: An overview of its peculiarities. *P. R. Health Sci. J.* **2008**, *27*, 190–195. [PubMed]

62. Geary, N.; Asarian, L.; Korach, K.S.; Pfaff, D.W.; Ogawa, S. Deficits in E2-dependent control of feeding, weight gain, and cholecystokinin satiation in ER-α null mice. *Endocrinology* **2001**, *142*, 4751–4757. [CrossRef] [PubMed]
63. Roesch, D.M. Effects of selective estrogen receptor agonists on food intake and body weight gain in rats. *Physiol. Behav.* **2005**, *87*, 39–44. [CrossRef] [PubMed]
64. Butera, P.C.; Beikirch, R.J. Central implants of diluted estradiol: Independent effects on ingestive and reproductive behaviors of ovariectomized rats. *Brain Res.* **1989**, *491*, 266–273. [CrossRef]
65. Palmer, K.; Gray, J.M. Central vs peripheral effects of estrogen on food intake and lipoprotein lipase activity in ovariectomized rats. *Physiol. Behav.* **1986**, *37*, 187–189. [CrossRef]
66. Heine, P.A.; Taylor, J.A.; Iwamoto, G.A.; Lubahn, D.B.; Cooke, P.S. Increased adipose tissue in male and female estrogen receptor-alpha knockout mice. *Proc. Natl. Acad. Sci. USA* **2000**, *97*, 12729–12734. [CrossRef] [PubMed]
67. Okura, T.; Koda, M.; Ando, F.; Niino, N.; Ohta, S.; Shimokata, H. Association of polymorphisms in the estrogen receptor alpha gene with body fat distribution. *Int. J. Obes. Relat. Metab. Disord.* **2003**, *27*, 1020–1027. [CrossRef] [PubMed]
68. Xu, Y.; Nedungadi, T.P.; Zhu, L.; Sobhani, N.; Irani, B.G.; Davis, K.E.; Zhang, X.; Zou, F.; Gent, L.M.; Hahner, L.D.; et al. Distinct hypothalamic neurons mediate estrogenic effects on energy homeostasis and reproduction. *Cell Metab.* **2011**, *14*, 453–465. [CrossRef] [PubMed]
69. Merchenthaler, I.; Lane, M.V.; Numan, S.; Dellovade, T.L. Distribution of estrogen receptor alpha and beta in the mouse central nervous system: In vivo autoradiographic and immunocytochemical analyses. *J. Comp. Neurol.* **2004**, *473*, 270–291. [CrossRef] [PubMed]
70. De Souza, F.S.J.; Nasif, S.; Leal, R.; Levi, D.H.; Low, M.J.; Rubinsten, M. López- The estrogen receptor α colocalizes with proopiomelanocortin in hypothalamic neurons and binds to a conserved motif present in the neuron-specific enhancer nPE2. *Eur. J. Pharmacol.* **2010**, *660*, 181–187. [CrossRef] [PubMed]
71. Saito, K.; He, Y.; Yan, X.; Yang, Y.; Wang, C.; Xu, P.; Hinton, A.O.; Shu, G.; Yu, L.; Tong, Q.; et al. Visualizing estrogen receptor-α-expressing neurons using a new ERα-ZsGreen reporter mouse line. *Metabolism* **2015**, *65*, 522–532. [CrossRef] [PubMed]
72. Slamberova, R.; Hnatczuk, O.C.; Vathy, I. Expression of proopiomelanocortin and proenkephalin mRNA in sexually dimorphic brain regions are altered in adult male and female rats treated prenatally with morphine. *J. Pept. Res.* **2004**, *63*, 399–408. [CrossRef] [PubMed]
73. Dhillon, S.S.; Belsham, D.D. Estrogen inhibits NPY secretion through membrane-associated estrogen receptor (ER)-α in clonal, immortalized hypothalamic neurons. *Int. J. Obes.* **2010**, *35*, 198–207. [CrossRef] [PubMed]
74. Olofsson, L.E.; Pierce, A.A.; Xu, A.W. Functional requirement of AgRP and NPY neurons in ovarian cycle-dependent regulation of food intake. *Proc. Natl. Acad. Sci. USA* **2009**, *106*, 15932–15937. [CrossRef] [PubMed]
75. Roepke, T.A.; Qiu, J.; Smith, A.W.; Ronnekleiv, O.K.; Kelly, M.J. Fasting and 17beta-estradiol differentially modulate the M-current in neuropeptide Y neurons. *J. Neurosci.* **2011**, *31*, 11825–11835. [CrossRef] [PubMed]
76. Roepke, T.A.; Ronnekleiv, O.K.; Kelly, M.J. Physiological consequences of membrane-initiated estrogen signaling in the brain. *Front. Biosci.* **2011**, *16*, 1560–1573. [CrossRef]
77. Sun, F.; Yu, J. The effect of a special herbal tea on obesity and anovulation in androgen-sterilized rats. *Proc. Soc. Exp. Biol. Med.* **2000**, *223*, 295–301. [CrossRef] [PubMed]
78. Xu, P.; Cao, X.; He, Y.; Zhu, L.; Yang, Y.; Saito, K.; Wang, C.; Yan, X.; Hinton, A.O., Jr.; Zou, F.; et al. Estrogen receptor-alpha in medial amygdala neurons regulates body weight. *J. Clin. Investig.* **2015**, *125*, 2861–2876. [CrossRef] [PubMed]
79. Jones, M.E.; Thorburn, A.W.; Britt, K.L.; Hewitt, K.N.; Wreford, N.G.; Proietto, J.; Oz, O.K.; Leury, B.J.; Robertson, K.M.; Yao, S.; et al. Aromatase-deficient (ArKO) mice have a phenotype of increased adiposity. *Proc. Natl. Acad. Sci. USA* **2000**, *97*, 12735–12740. [CrossRef] [PubMed]
80. Luo, J.; Sladek, R.; Carrier, J.; Bader, J.A.; Richard, D.; Giguere, V. Reduced fat mass in mice lacking orphan nuclear receptor estrogen-related receptor alpha. *Mol. Cell. Biol.* **2003**, *23*, 7947–7956. [CrossRef] [PubMed]
81. Cui, H.; Lu, Y.; Khan, M.Z.; Anderson, R.M.; McDaniel, L.; Wilson, H.E.; Yin, T.C.; Radley, J.J.; Pieper, A.A.; Lutter, M. Behavioral disturbances in estrogen-related receptor alpha-null mice. *Cell Rep.* **2015**, *11*, 344–350. [CrossRef] [PubMed]

82. Cui, H.; Moore, J.; Ashimi, S.S.; Mason, B.L.; Drawbridge, J.N.; Han, S.; Hing, B.; Matthews, A.; McAdams, C.J.; Darbro, B.W.; et al. Eating disorder predisposition is associated with ESRRA and HDAC4 mutations. *J. Clin. Investig.* **2013**, *123*, 4706–4713. [CrossRef] [PubMed]

83. Haber, S.N.; Knutson, B. The reward circuit: Linking primate anatomy and human imaging. *Neuropsychopharmacology* **2010**, *35*, 4–26. [CrossRef] [PubMed]

84. Stice, E.; Figlewicz, D.P.; Gosnell, B.A.; Levine, A.S.; Pratt, W.E. The contribution of brain reward circuits to the obesity epidemic. *Neurosci. Biobehav. Rev.* **2013**, *37*, 2047–2058. [CrossRef] [PubMed]

85. Alaynick, W.A.; Kondo, R.P.; Xie, W.; He, W.; Dufour, C.R.; Downes, M.; Jonker, J.W.; Giles, W.; Naviaux, R.K.; Giguere, V.; et al. ERRgamma directs and maintains the transition to oxidative metabolism in the postnatal heart. *Cell Metab.* **2007**, *6*, 13–24. [CrossRef] [PubMed]

86. Tanida, T.; Matsuda, K.I.; Yamada, S.; Kawata, M.; Tanaka, M. Immunohistochemical profiling of estrogen-related receptor gamma in rat brain and colocalization with estrogen receptor α in the preoptic area. *Brain Res.* **2017**, *1659*, 71–80. [CrossRef] [PubMed]

87. Lorke, D.E.; Susens, U.; Borgmeyer, U.; Hermans-Borgmeyer, I. Differential expression of the estrogen receptor-related receptor gamma in the mouse brain. *Brain Res. Mol. Brain Res.* **2000**, *77*, 277–280. [CrossRef]

88. Luo, J.; Sladek, R.; Bader, J.A.; Matthyssen, A.; Rossant, J.; Giguere, V. Placental abnormalities in mouse embryos lacking the orphan nuclear receptor ERR-beta. *Nature* **1997**, *388*, 778–782. [CrossRef] [PubMed]

89. Byerly, M.S.; Swanson, R.D.; Wong, G.W.; Blackshaw, S. Estrogen-related receptor beta deficiency alters body composition and response to restraint stress. *BMC Physiol.* **2013**, *13*, 10. [CrossRef] [PubMed]

90. Wong, M.; Moss, R.L. Long-term and short-term electrophysiological effects of estrogen on the synaptic properties of hippocampal CA1 neurons. *J. Neurosci.* **1992**, *12*, 3217–3225. [PubMed]

91. Rudick, C.N.; Woolley, C.S. Selective estrogen receptor modulators regulate phasic activation of hippocampal CA1 pyramidal cells by estrogen. *Endocrinology* **2003**, *144*, 179–187. [CrossRef] [PubMed]

92. Kramar, E.A.; Chen, L.Y.; Brandon, N.J.; Rex, C.S.; Liu, F.; Gall, C.M.; Lynch, G. Cytoskeletal changes underlie estrogen's acute effects on synaptic transmission and plasticity. *J. Neurosci.* **2009**, *29*, 12982–12993. [CrossRef] [PubMed]

93. Milner, T.A.; Ayoola, K.; Drake, C.T.; Herrick, S.P.; Tabori, N.E.; McEwen, B.S.; Warrier, S.; Alves, S.E. Ultrastructural localization of estrogen receptor beta immunoreactivity in the rat hippocampal formation. *J. Comp. Neurol.* **2005**, *491*, 81–95. [CrossRef] [PubMed]

94. Milner, T.A.; McEwen, B.S.; Hayashi, S.; Li, C.J.; Reagan, L.P.; Alves, S.E. Ultrastructural evidence that hippocampal alpha estrogen receptors are located at extranuclear sites. *J. Comp. Neurol.* **2001**, *429*, 355–371. [CrossRef]

95. Smejkalova, T.; Woolley, C.S. Estradiol acutely potentiates hippocampal excitatory synaptic transmission through a presynaptic mechanism. *J. Neurosci.* **2010**, *30*, 16137–16148. [CrossRef] [PubMed]

96. Huang, G.Z.; Woolley, C.S. Estradiol Acutely Suppresses Inhibition in the Hippocampus through a Sex-Specific Endocannabinoid and mGluR-Dependent Mechanism. *Neuron* **2012**, *74*, 801–808. [CrossRef] [PubMed]

97. Phan, A.; Gabor, C.S.; Favaro, K.J.; Kaschack, S.; Armstrong, J.N.; MacLusky, N.J.; Choleris, E. Low doses of 17β-estradiol rapidly improve learning and increase hippocampal dendritic spines. *Neuropsychopharmacology* **2012**, *37*, 2299–2309. [CrossRef] [PubMed]

98. Woolley, C.S. Acute effects of estrogen on neuronal physiology. *Annu. Rev. Pharmacol. Toxicol.* **2007**, *47*, 657–680. [CrossRef] [PubMed]

99. Kelly, M.J.; Ronnekleiv, O.K. A selective membrane estrogen receptor agonist maintains autonomic functions in hypoestrogenic states. *Brain Res.* **2013**, *1514*, 75–82. [CrossRef] [PubMed]

100. Gu, Q.; Korach, K.S.; Moss, R.L. Rapid action of 17beta-estradiol on kainate-induced currents in hippocampal neurons lacking intracellular estrogen receptors. *Endocrinology* **1999**, *140*, 660–666. [CrossRef] [PubMed]

101. Pei, L.; Mu, Y.; Leblanc, M.; Alaynick, W.; Barish, G.D.; Pankratz, M.; Tseng, T.W.; Kaufman, S.; Liddle, C.; Yu, R.T.; et al. Dependence of hippocampal function on ERRγ-regulated mitochondrial metabolism. *Cell Metab.* **2015**, *21*, 628–636. [CrossRef] [PubMed]

102. Villena, J.A.; Kralli, A. ERRα: A metabolic function for the oldest orphan. *Trends Endocrinol. Metab.* **2008**, *19*, 269–276. [CrossRef] [PubMed]

103. De Jesus-Cortes, H.; Lu, Y.; Anderson, R.M.; Khan, M.Z.; Nath, V.; McDaniel, L.; Lutter, M.; Radley, J.J.; Pieper, A.A.; Cui, H. Loss of estrogen-related receptor alpha disrupts ventral-striatal synaptic function in female mice. *Neuroscience* **2016**, *329*, 66–73. [CrossRef] [PubMed]

104. Tetel, M.J.; Pfaff, D.W. Contributions of estrogen receptor-alpha and estrogen receptor-ss to the regulation of behavior. *Biochim. Biophys. Acta* **2010**, *1800*, 1084–1089. [CrossRef] [PubMed]

105. Ogawa, S.; Washburn, T.F.; Taylor, J.; Lubahn, D.B.; Korach, K.S.; Pfaff, D.W. Modifications of testosterone-dependent behaviors by estrogen receptor-alpha gene disruption in male mice. *Endocrinology* **1998**, *139*, 5058–5069. [CrossRef] [PubMed]

106. Ogawa, S.; Lubahn, D.B.; Korach, K.S.; Pfaff, D.W. Behavioral effects of estrogen receptor gene disruption in male mice. *Proc. Natl. Acad. Sci. USA* **1997**, *94*, 1476–1481. [CrossRef] [PubMed]

107. Sano, K.; Tsuda, M.C.; Musatov, S.; Sakamoto, T.; Ogawa, S. Differential effects of site-specific knockdown of estrogen receptor alpha in the medial amygdala, medial pre-optic area, and ventromedial nucleus of the hypothalamus on sexual and aggressive behavior of male mice. *Eur. J. Neurosci.* **2013**, *37*, 1308–1319. [CrossRef] [PubMed]

108. Unger, E.K.; Burke, K.J., Jr.; Yang, C.F.; Bender, K.J.; Fuller, P.M.; Shah, N.M. Medial amygdalar aromatase neurons regulate aggression in both sexes. *Cell Rep.* **2015**, *10*, 453–462. [CrossRef] [PubMed]

109. Ervin, K.S.; Lymer, J.M.; Matta, R.; Clipperton-Allen, A.E.; Kavaliers, M.; Choleris, E. Estrogen involvement in social behavior in rodents: Rapid and long-term actions. *Horm. Behav.* **2015**, *74*, 53–76. [CrossRef] [PubMed]

110. Beach, F.A. Sexual attractivity, proceptivity, and receptivity in female mammals. *Horm. Behav.* **1976**, *7*, 105–138. [CrossRef]

111. Tennent, B.J.; Smith, E.R.; Davidson, J.M. The effects of estrogen and progesterone on female rat proceptive behavior. *Horm. Behav.* **1980**, *14*, 65–75. [CrossRef]

112. Kavaliers, M.; Agmo, A.; Choleris, E.; Gustafsson, J.A.; Korach, K.S.; Muglia, L.J.; Pfaff, D.W.; Ogawa, S. Oxytocin and estrogen receptor alpha and beta knockout mice provide discriminably different odor cues in behavioral assays. *Genes Brain Behav.* **2004**, *3*, 189–195. [CrossRef] [PubMed]

113. Lee, H.; Kim, D.W.; Remedios, R.; Anthony, T.E.; Chang, A.; Madisen, L.; Zeng, H.; Anderson, D.J. Scalable control of mounting and attack by Esr1+ neurons in the ventromedial hypothalamus. *Nature* **2014**, *509*, 627–632. [CrossRef] [PubMed]

114. Hashikawa, K.; Hashikawa, Y.; Tremblay, R.; Zhang, J.; Feng, J.E.; Sabol, A.; Piper, W.T.; Lee, H.; Rudy, B.; Lin, D. Esr1(+) cells in the ventromedial hypothalamus control female aggression. *Nat. Neurosci.* **2017**, *20*, 1580–1590. [CrossRef] [PubMed]

115. Zhou, T.; Zhu, H.; Fan, Z.; Wang, F.; Chen, Y.; Liang, H.; Yang, Z.; Zhang, L.; Lin, L.; Zhan, Y.; et al. History of winning remodels thalamo-PFC circuit to reinforce social dominance. *Science* **2017**, *357*, 162–168. [CrossRef] [PubMed]

116. Wang, F.; Zhu, J.; Zhu, H.; Zhang, Q.; Lin, Z.; Hu, H. Bidirectional control of social hierarchy by synaptic efficacy in medial prefrontal cortex. *Science* **2011**, *334*, 693–697. [CrossRef] [PubMed]

117. Carus-Cadavieco, M.; Gorbati, M.; Ye, L.; Bender, F.; van der Veldt, S.; Kosse, C.; Borgers, C.; Lee, S.Y.; Ramakrishnan, C.; Hu, Y.; et al. Gamma oscillations organize top-down signalling to hypothalamus and enable food seeking. *Nature* **2017**, *542*, 232–236. [CrossRef] [PubMed]

118. Hu, H.; Gan, J.; Jonas, P. Fast-spiking, parvalbumin(+) GABAergic interneurons: From cellular design to microcircuit function. *Science* **2014**, *345*, 1255263. [CrossRef] [PubMed]

119. Cardin, J.A.; Carlen, M.; Meletis, K.; Knoblich, U.; Zhang, F.; Deisseroth, K.; Tsai, L.H.; Moore, C.I. Driving fast-spiking cells induces gamma rhythm and controls sensory responses. *Nature* **2009**, *459*, 663–667. [CrossRef] [PubMed]

120. Sohal, V.S.; Zhang, F.; Yizhar, O.; Deisseroth, K. Parvalbumin neurons and gamma rhythms enhance cortical circuit performance. *Nature* **2009**, *459*, 698–702. [CrossRef] [PubMed]

121. Scarpulla, R.C. Metabolic control of mitochondrial biogenesis through the PGC-1 family regulatory network. *Biochim. Biophys. Acta* **2011**, *1813*, 1269–1278. [CrossRef] [PubMed]

122. Audet-Walsh, E.; Giguere, V. The multiple universes of estrogen-related receptor alpha and gamma in metabolic control and related diseases. *Acta Pharmacol. Sin.* **2015**, *36*, 51–61. [CrossRef] [PubMed]

123. Carter, B.C.; Bean, B.P. Sodium entry during action potentials of mammalian neurons: Incomplete inactivation and reduced metabolic efficiency in fast-spiking neurons. *Neuron* **2009**, *64*, 898–909. [CrossRef] [PubMed]

124. Carter, B.C.; Bean, B.P. Incomplete inactivation and rapid recovery of voltage-dependent sodium channels during high-frequency firing in cerebellar Purkinje neurons. *J. Neurophysiol.* **2010**, *105*, 860–871. [CrossRef] [PubMed]

125. Cowell, R.M.; Blake, K.R.; Russell, J.W. Localization of the transcriptional coactivator PGC-1alpha to GABAergic neurons during maturation of the rat brain. *J. Comp. Neurol.* **2007**, *502*, 1–18. [CrossRef] [PubMed]

126. Lucas, E.K.; Markwardt, S.J.; Gupta, S.; Meador-Woodruff, J.H.; Lin, J.D.; Overstreet-Wadiche, L.; Cowell, R.M. Parvalbumin deficiency and GABAergic dysfunction in mice lacking PGC-1alpha. *J. Neurosci.* **2010**, *30*, 7227–7235. [CrossRef] [PubMed]

127. Lucas, E.K.; Dougherty, S.E.; McMeekin, L.J.; Reid, C.S.; Dobrunz, L.E.; West, A.B.; Hablitz, J.J.; Cowell, R.M. PGC-1alpha provides a transcriptional framework for synchronous neurotransmitter release from parvalbumin-positive interneurons. *J. Neurosci.* **2014**, *34*, 14375–14387. [CrossRef] [PubMed]

128. Saul, M.C.; Seward, C.H.; Troy, J.M.; Zhang, H.; Sloofman, L.G.; Lu, X.; Weisner, P.A.; Caetano-Anolles, D.; Sun, H.; Zhao, S.D.; et al. Transcriptional regulatory dynamics drive coordinated metabolic and neural response to social challenge in mice. *Genome Res.* **2017**, *27*, 959–972. [CrossRef] [PubMed]

International Journal of
Molecular Sciences

MDPI

Review

Estrogen, Estrogen Receptor and Lung Cancer

Li-Han Hsu [1,2,3], Nei-Min Chu [4] and Shu-Huei Kao [1,5,*]

1   Ph.D. Program in Medical Biotechnology, College of Medical Science and Technology,
    Taipei Medical University, Taipei 110, Taiwan; lhhsu@kfsyscc.org
2   Division of Pulmonary and Critical Care Medicine, Sun Yat-Sen Cancer Center, Taipei 112, Taiwan
3   Department of Medicine, National Yang-Ming University Medical School, Taipei 112, Taiwan
4   Department of Medical Oncology, Sun Yat-Sen Cancer Center, Taipei 112, Taiwan; nmchu@kfsyscc.org
5   School of Medical Laboratory Science and Biotechnology, College of Medical Science and Technology,
    Taipei Medical University, Taipei 110, Taiwan
*   Correspondence: kaosh@tmu.edu.tw; Tel.: +886-2-2736-1661 (ext.3317); Fax: +886-2-2732-4510

Received: 25 June 2017; Accepted: 3 August 2017; Published: 5 August 2017

Abstract: Estrogen has been postulated as a contributor for lung cancer development and progression. We reviewed the current knowledge about the expression and prognostic implications of the estrogen receptors (ER) in lung cancer, the effect and signaling pathway of estrogen on lung cancer, the hormone replacement therapy and lung cancer risk and survival, the mechanistic relationship between the ER and the epidermal growth factor receptor (EGFR), and the relevant clinical trials combining the ER antagonist and the EGFR antagonist, to investigate the role of estrogen in lung cancer. Estrogen and its receptor have the potential to become a prognosticator and a therapeutic target in lung cancer. On the other hand, tobacco smoking aggravates the effect of estrogen and endocrine disruptive chemicals from the environment targeting ER may well contribute to the lung carcinogenesis. They have gradually become important issues in the course of preventive medicine.

Keywords: epidermal growth factor receptor; estrogen; estrogen receptor; hormone; lung cancer; lung adenocarcinoma

---

1. Introduction

Estrogens are steroid hormones. 17-β-Estradiol (E2) is the primary reproductive hormone synthesized in the ovary under the stimulation of the follicular stimulating hormone and the luteinizing hormone [1–4]. Estrone and estriol are mostly synthesized in the liver from E2. The functions of estrogen and its receptors in reproductive organs, especially in a female, have been known for several decades. The importance of the estrogen signaling pathway in various physiologic, pathologic functions and carcinogenesis has also been extensively investigated, especially in the context of breast cancer.

There are two types of classical estrogen receptor (ER). ER alpha (ERα, also known as ESR1), product of genes on chromosome 6, first cloned in 1986 and distributed in breast, ovary, and endometrium. ER beta (ERβ, also known as ESR2), product of genes on chromosome 14, discovered in 1996 [5], with a wider distribution including the bone, brain, colon, endothelium, kidney, lung, ovary, prostate, and testes. They share similar structures and are composed of five domains. The A/B domain is the site of the transcriptional activation, with the coactivator, AF-1. The C-domain is the DNA-binding site. The D-domain hinge contains a nuclear localization signal. The E-domain is the ligand binding domain and the site of the transcriptional activation, with the coactivator, AF-2. The F-domain may play a complex regulatory role. The 55% homology between the ERα and ERβ in the ligand binding domain results in the variable affinities. While both exhibit similar affinities to E2, ERα has a higher affinity to estrone and ERβ has a higher affinity to estriol. In addition to the wide-type ERs, several splicing variants or isoforms of the ERs have been described with variable DNA- or ligand-binding properties.

Upon binding with estrogen, the ERs form either homo- or heterodimers and bind to the estrogen responsive element, the ERE within the promoter of a target gene, and then regulate its transcription in the case of the genomic pathway. ERs may also regulate gene expression via the binding to other transcription factors such as the activator protein 1, AP-1 or the stimulating protein 1, Sp1. On the other hand, ERs may translocate to the membrane, where they may mediate a non-genomic pathway that results in more rapid responses, such as the activation of protein kinase, the production of second messengers, or the regulation of ion channels.

Lung cancer is a leading cause of cancer-related mortality worldwide, including Taiwan. Studies conducted in Western countries estimated that 85–90% of lung cancer cases were attributed to smoking [6,7]. Although 80% of female lung cancer patients worldwide have smoked, less than 10% of Taiwanese women are smokers [8]. In our previous lung cancer study, only 6.4% of the female patients had smoked cigarettes at some time in their lives [9]. There is a lung adenocarcinoma epidemic with an equal occurrence and prognosis in both genders who have never smoked in Taiwan [10]. Smoking history appeared to be a poor prognostic factor for patients with lung adenocarcinoma, rather than as a risk factor. The low smoking prevalence and high incidence rate of adenocarcinoma constituted distinctive characteristics of lung cancer in Asian countries, and leads to the suggested existence of non-tobacco related risk factors in the pathogenesis of lung adenocarcinoma.

Another study showed a more significant survival advantage for elderly women with lung adenocarcinoma, as compared with their male counterparts [11]. In addition to the inferior survival of elderly male patients attributed to the accumulated adverse effect of a higher prevalent smoking habit, the superior survival of the postmenopausal female patients was possibly due to the less estrogen cancer promoting effect. The premenopausal women, who comprised one-fifth of the non-smoking female patients with lung adenocarcinoma, were found to have had the more advanced disease and a shorter survival rate than the postmenopausal women. The epidemiology results suggest that estrogen adversely affects the prognosis of patients with lung adenocarcinoma.

Estrogen is speculated as playing an important role in lung carcinogenesis [12,13]. In healthy lung tissue, ERβ is highly expressed in pneumocytes and in the bronchial epithelial cells, and is required for the maintenance of the extracellular matrix of the lung [14,15]. The lung tissues of ERβ null mice were found to have a decreased number of alveoli and a lesser amount of surfactant [16]. Studies of ER deficient mice have shown that ERα mediates the determination of the alveolar number and the surface area while ERβ affects the lung tissue elastic recoil [17]. Estrogen receptors (ER) are consistently found in lung cancer tissues and cell lines, especially adenocarcinoma, and mostly in the form of the ERβ [18–21]. Estrogen has been reported to adversely affect the prognosis of lung cancer patients [22–30]. However, there are several studies with conflicting results about the effect of estrogen on the risk and/or survival of lung cancer [31–36].

We demonstrated ERβ was the predominant ER in the A549 and PE089 lung cancer cell lines, and malignant pleural effusions from the patients with lung adenocarcinoma. Osteopontin (OPN) is a small integrin-binding ligand *N*-linked glycoprotein regulating signaling pathways involved in tumor progression and metastasis [37–39]. Enhanced OPN expression has been noted in the plasma of advanced lung cancer patients, and OPN has also been speculated to be involved in the formation of malignant pleural effusion [40,41]. Estrogen up-regulated the OPN expression and promoted lung cancer cell migration via the ERβ activation of the MEK/ERK signaling pathway. An additive effect of the ER antagonist and the epidermal growth factor receptor (EGFR) antagonist on the inhibition of lung cancer cell migration was also observed. Osteopontin supposedly contributes to the cross-talk between the ER and EGFR signaling pathways [11]. In current clinical practice, breast cancer survivors offer a unique patient cohort to evaluate the effect of anti-estrogen on the survival of lung cancer patients [42,43]. We have evaluated the outcome of 26 women who have had second primary lung cancer among 6361 breast cancer patients diagnosed and treated between January 2000, and December

2009, at Sun Yat-Sen Cancer Center and found that the patients who were treated with anti-estrogens for breast cancer had a longer cancer-specific survival rate than those without anti-estrogens [44]. Multivariate analysis confirmed that the anti-estrogen treatment was an independent prognostic factor. These findings reinforced the evidence that estrogen had, in fact, contributed to the lung cancer progression.

This review aims to summarize the current knowledge with regard to the expression and prognostic implications of the ERs in lung cancer, the effect and signaling pathway of estrogen on lung cancer, the hormone replacement therapy and lung cancer risk and survival, the mechanistic relationship between the ER and the EGFR, and the relevant clinical trials combining the ER antagonist and the EGFR antagonist, to investigate the role of estrogen in lung cancer. Interaction between tobacco smoking and estrogen, and the role of endocrine disruptive chemicals targeting ER from the environment in the lung carcinogenesis were also discussed from the viewpoint of preventive medicine. As small cell lung cancer (SCLC) is a distinct neuroendocrine tumor, composed of about 10% to 15% of lung cancer, and the association between estrogen and SCLC was scarcely studied and mostly obsolete [45,46]. The following issues will focus on the non-small cell lung cancer (NSCLC), being mostly adenocarcinoma.

2. Estrogen Receptor in Lung Cancer

Baik et al. have systemically reviewed the detection rates of the ERα and ERβ in lung cancer [47,48]. For the ERα, the detection rate using 1D5 with the epitope in N-terminus is 0% to 55%, in contrast with 36% to 84% using HC-20 with the epitope in C-terminus, and 0% to 78% using 6F11 for the full length. For the ERβ, the detection rate using H-150 or 14C8 with the epitope in N-terminus is 49% to 98%, and 16% to 86%, respectively. The detection rate is 9% to 84% using PPG5/10 with the epitope in the C-terminus. The results were variable. Such an inconsistency may be due to the differences in the methodology, i.e., which antibody is used, heterogeneous definitions of positivity, and various patient populations, i.e., pathology, stage, gender, and smoking history [47–49]. The ERα antibody with epitope in the C-terminus reported a higher detection rate than that with epitope in the N-terminus in the NSCLC, and was mostly cytoplasm-located. The ERα probably occurs as the N-terminal deleted mutants in the NSCLC and lacks the nuclear localization [47–49]. Unlike the ERα, both of the full-length and splicing variants of the ERβ exist in the NSCLC cells. A strong expression of the ERβ was observed in the cytoplasm as well as the nucleus. Standardized measurement, i.e., which antibody was used, or a different approach from immunohistochemistry, e.g., western blot, mRNA expression by real time quantitative PCR, is necessary to make the ERs as useful biomarkers in the future.

Estrogen receptor β appears to be the predominant form in lung cancer from the literature [18–21]. Five splicing variants had been identified with ERβ1 being the only full-length receptor able to bind ligand and form homodimers in human. The rest of the isoforms are inactive, but they can form heterodimers with ERβ1 to regulate its transcriptional activity [50].

The expressions of ERα and ERβ as a prognosticator for NSCLC have been reported in several studies [20,21,51–64] (Table 1). Contrary to that in breast cancer, ERα in lung cancer was mainly observed in the cytoplasm and associated with a poor prognosis. Most reports found that the nuclear ERβ was predictive of a better prognosis, and the cytoplasmic ERβ was associated with a poor prognosis [47,48]. Nonetheless, opposing results have also been reported [54,60,62,63,65,66]. Co-expression of the cytoplasmic ERβ and the nuclear ERβ that had been reported correlated with a poor survival rate when compared to those without co-expression [67]. The nuclear and cytoplasmic ERs may have a distinct function and affect the prognosis differentially via the genomic or non-genomic pathway. ERβ has also been shown to localize with the mitochondria in a ligand-dependent or -independent manner and can affect the bioenergetics and anti-apoptotic signaling. Mitochondrial ERβ sequesters Bad and inhibit Bad-Bcl-XL, and Bad-Bcl-2 interactions, to protect against apoptosis, thereby suggesting its value as a new therapeutic target [27,68,69]. Further study is warranted to analyze the function of different ERβ isoforms and their cellular localization, which is essential to completely understand the role of the ERβ in lung cancer. According to the study of Kadota, although nuclear ERα expression was observed in only 17% of the patients with pT1a lung adenocarcinomas, it was an independent predictor of recurrence [61]. The nuclear ERα expression positively correlated with the tumoral FoxP3+ lymphocytes, and poor prognostic immune microenvironments.

Table 1. Estrogen receptor (ER) detected by immune-histochemical stain as prognosticators in NSCLC.

References	ER Subtype	Location	Prognosis
Kawai 2005 [20]	α	Cytoplasm	Worse
	β	Nucleus	Better
Schwartz 2005 [21]	β	Non-specified	Better (male)
			Worse (female) *
Wu 2005 [51]	β	Nucleus	Better
Skov 2005 [52]	β	Nucleus	Better (male)
			Worse (female)
Nose 2009 [53]	α	Cytoplasm	Worse
	β	Nucleus	Better
Raso 2009 [54]	β	Nucleus	Worse
Stabile 2011 [55]	β	Cytoplasm	Worse
Rouquette 2012 [56]	α	Nucleus	Better
Rades 2012 [57]	α	Non-specified	Worse
Karlsson 2012 [58]	β	Nucleus	Better
Navaratnam 2012 [59]	β1	Nucleus	Better in earlier stage
			Worse in later stage
Liu 2013 [60]	β2,5	Cytoplasm	Better
Kadota 2015 [61]	α	Nucleus	Worse
Liu 2015 [62]	β	Cytoplasm	Better
Skjefstad 2016 [63]	β	Nucleus	Worse (female)
Tanaka 2016 [64]	β	Non-specified	Worse (male)

* Not significant but with a trend.

The G-protein-coupled estrogen receptor (GPER), discovered in 2005, was proposed to be involved in the cancer cell proliferation, migration and invasion, and acts as a modulator of the neoplastic transformation [4,70,71]. It is not only located in the cell membrane, it has also been detected in the Golgi apparatus and endoplasmic reticulum [72,73]. Increased expression of the GPER was observed in the lung cancer cell lines as well as the human and mice lung cancer tissue, and more was located in the cytoplasm [74,75]. Paradoxically, the antagonists/modulators of the classical estrogen receptors such as tamoxifen, raloxifen and fulvestrant, were found to be the GPER agonists [70].

In contrast with GPER, the classical ERs do not contain a hydrophobic part that may serve as a transmembrane domain. However, the presence of ERs in the membrane of somatic and cancer cells have been reported. The membrane translocation of the ERs is mediated by the SRC family of tyrosine kinase [76]. Specific motifs and modifications are required. The knowledge on how the classical ERs translocate to the membrane together with the knowledge on the GPER action, and the interactions between the GPER and the classical ERs is of the greatest importance to understand the membrane-associated non-genomic pathways of estrogen.

In premenopausal women, estrogens produced by their ovaries play a major role in the female reproductive organs through the ERα. In postmenopausal women, however, estrogens produced/activated by peripherally localized estrogen-metabolizing enzymes, such as aromatase, which converts androgen into estrogens, are thought to play physiologically and pathologically important roles in various organs through the ERβ, distributing systemically [77]. Estrogen can be synthesized in situ in lung cancer. Ikeda et al. measured the estrogen concentrations in the noncancerous peripheral lung tissue using liquid chromatography/electrospray tandem mass spectrometry in the postmenopausal female patients with synchronous multiple lung adenocarcinomas, and found a significantly higher level than the control cases with a single lung adenocarcinoma [78] (Figure 1). Our study of the malignant pleural effusion of lung adenocarcinoma revealed that some postmenopausal women had extraordinarily high pleural fluid estradiol concentrations, and there was no correlation between the pleural fluid concentrations of estradiol and the vascular endothelial growth factor, a marker of pleural vascular hyperpermeability [79]. In addition, the EGFR wild-type lung adenocarcinoma is probably an estrogen-dependent carcinoma, as a higher expression and potent poor prognosticator of aromatase and the ERβ in the group [62].

Thyroid transcription factor 1 (TTF-1) expression, as a lineage marker of the terminal respiratory unit, is helpful to distinguish the primary (TTF-1 positive) from the metastatic (usually TTF1 negative) lung adenocarcinoma, the pleural lung carcinoma (TTF-1 positive) from the mesothelioma [80]. The TTF-1-positive adenocarcinomas had a statistically significant prevalence of the female, non-smoker, and associated with the EGFR mutation [81,82]. The ER and TTF-1 immunoreactivity is commonly used as a means of distinguishing breast carcinomas from the adenocarcinomas of other primary sites, including the lung, but mostly using the antibody of the ERα. The TTF-1 positivity may be associated with the ERβ expression in lung adenocarcinoma with clinical significance, which therefore deserves further study [83].

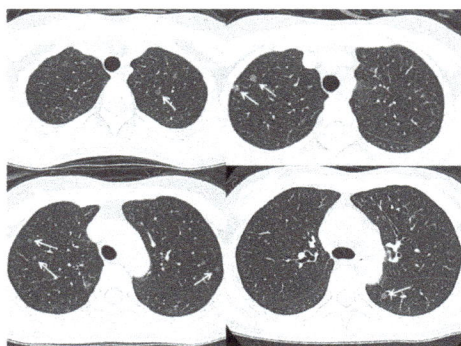

Figure 1. A 48 year-old non-smoking woman was found to have multiple subcentimetre ground glass opacities (arrows) in her bilateral lungs on a low-dose CT screening. Video-assisted thoracoscopic surgery with a right upper lobe wedge resection confirmed the diagnosis of synchronous multiple lung adenocarcinomas harboring the EGFR wild-type.

3. Hormone Replacement Therapy and Lung Cancer Risk and Survival

There were also controversies in the relationship between the hormone replacement therapy (HRT) and lung cancer. Although most studies reported estrogen or HRT adversely affected the prognosis of lung cancer patients [22–30], some reported HRT decreased the risk and favorably affected the prognosis [31–36]. In the Women's Health Initiative Trial, HRT using estrogen plus progestin in postmenopausal women did not increase the incidence of lung cancer, but increased the risk (60%) of dying from NSCLC [29]. Unlike the use of estrogen plus progestin, the usage of conjugated equine estrogen alone did not increase the incidence or death from lung cancer [36]. In another Vitamins and Lifestyle Study, postmenopausal women taking estrogen plus progestin were reported to have a 50% increased risk of incident lung cancer for usage of 10 years or longer and an advanced stage at diagnosis [30]. Greiser et al. made a systemic review and meta-analysis from 18 studies for the risk of lung cancer after HRT [84]. Ever use of HRT in non-smoking women may well increase the risk of lung adenocarcinoma. Data from the randomized controlled trials suggested that estrogen/progestin therapy increased the lung cancer mortality. The increased risk of death from lung cancer during the estrogen plus progestin usage in the Women's Health Initiative Trial was recently reported attenuated after the discontinuation of the medication in a 14-year cumulative follow-up [85].

Siegfried and Stabile provided explanations for the discrepancy of the HRT effect [49]. Different influences of estrogen on the balance of differentiation induction and proliferation in normal lung epithelium and malignant epithelium have been reported [55]. Compared with the matched normal lung tissues, ERβ is overexpressed in lung cancer, which could lead to an abnormal response to estrogen. The ability of the immune system to reject the malignant lung tissues during the early process could be enhanced by HRT [49] and related to a different level of ER expression [61,86,87]. Exogenous hormone usage reduces the local estrogen production by inhibiting the pulmonary aromatase expression. The exact HRT used, i.e., type, duration, timing, and adjusted covariates may modulate the effects of HRT. More specifically designed studies to address the HRT type, smoking, and histology, are therefore warranted to arrive at the more definitive conclusions. However, since the HRT is now recommended to be used for a limited duration, its effects on lung cancer risk or survival may be less pronounced in the future.

4. ER as Targets for Lung Cancer Therapy and Relationship with EGFR

Estradiol is locally produced in the NSCLC mainly by aromatase, which is localized in both the epithelial cell components of lung tumors as well as in the infiltrating macrophages; even exclusively confined to the inflammatory cells infiltrated in the pre-neoplastic and neoplastic areas in some of the animal models [88,89]. Patients whose tumors harbored a higher expression of aromatase and ERβ have a lower survival rate, especially in postmenopausal women [13,90]. The use of selective ER modulators and/or aromatase inhibitors have been reported to be clinically effective in the NSCLC that are positive for both the ER and aromatase [91,92]. Recently, Hamilton et al. utilized a quantitative high-throughput screening of approved drugs, and identified the ER antagonist, fulvestrant, as being capable of reducing the mesenchymal features of lung cancer cells and sensitize to the cytotoxic effect of the chemotherapy [93].

As aforementioned, the activities of the ERβ could be genomic or non-genomic [94] (Figure 2). The estrogen-ERβ complex binds to the nuclear estrogen response elements directly or through the transcription factor, to promote the gene expression. Estrogen also combines with membrane-bound ERβ to activate the cytoplasmic signaling pathway and interacts with the EGFR signaling pathways [95,96] (Figure 3). EGFR has been reported to directly phosphorylate ER at specific serine residues (a ligand-independent signaling) in 87.5% of the ER-positive lung tumors [96,97]. In addition to the MEK/ERK signaling pathway, estrogen also activates the PI3K/AKT signaling pathway, another downstream pathway of the EGFR activation, to promote lung cancer cell metastasis through epithelial mesenchymal transition [98]. Other non-genomic activities have also been explored. Fan et al. found a higher ERβ expression in the lymph node as compared to the primary tumor tissues, and estrogen promotes the lung cancer cell metastasis via the ERβ-mediated up-regulation of the matrix-metalloproteinase-2 [99]. In the mRNA analyses, when comparing the high versus low ERβ expressing tumors by the group of Siegfried and Stabile, the top differentially expressed genes in the high ERβ tumors involved the fibroblast growth factor signaling and the human embryonic stem cell pluripotency [100].

ER and EGFR, as targets for dual lung cancer therapy, have been studied. A combination of the ER antagonist and the EGFR tyrosine kinase inhibitor has been shown to decrease cell proliferation and tumor growth more than one individual treatment in both in vitro and in vivo studies [53,97,101,102]. In the NSCLC cell lines, the EGFR protein expression was down-regulated in response to estrogen and up-regulated in response to anti-estrogens in vitro. Conversely, the ERβ expression is decreased in response to the epidermal growth factor and increased in response to gefitinib [101]. A strong association has been reported between the expression of the ERβ and EGFR mutations in lung adenocarcinoma [53,54,103,104]. These studies have provided evidence of a functional interaction between the ER and EGFR pathways in lung cancer and have supported a rationale to use the combined therapy [95,105].

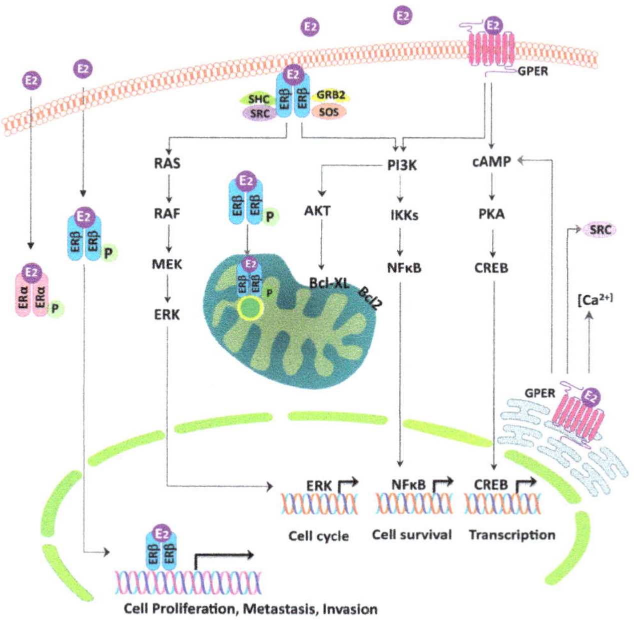

Figure 2. The putative role of the estrogen receptor in regulating the lung cancer cells growth. The estrogen receptor β (ERβ) appears to be the predominant form in lung cancer and is present in the cytoplasm, nucleus, mitochondria and plasma membrane. The ERβ has been found to activate the PI3K/IKK/NFκB, PI3K/AKT/Bcl-XL and the RAS/RAF/MEK/ERK signaling pathways to regulate the cell proliferation, invasion, metastasis, mitochondrial biogenesis and anti-apoptosis. The G-protein-coupled estrogen receptor (GPER) activates the cAMP/PKA/CREB and the PI3K/IKK/NFκB signaling pathways and acts as a modulator of the neoplastic transformation.

The available strategies to target the estrogen signaling pathway include the aromatase inhibitors, the reversible nonsteroidal agents (e.g., letrozole, anastrozole) or the irreversible steroidal inactivator (e.g., exemestane), the nonsteroidal elective ER modulator (e.g., tamoxifen, raloxifene), and the ER antagonists (e.g., fulvestrant) [47]. Giovannini et al. reported that the additional effect of letrozole in a patient with lung adenocarcinoma and scalp metastasis persisted on gefitinib [106]. A pilot study revealed that treatment combining the gefitinib and fulvestrant for postmenopausal women with advanced NSCLC was well-tolerated and demonstrated a result [107]. Several phase II clinical trials are currently ongoing to investigate their effects on advanced NSCLC, mostly in a second-line setting and combined with the EGFR tyrosine kinase inhibitor [108] (Table 2). Some studies have included correlative tissue analysis of the ER and the progesterone receptor status to evaluate their role as a predictor of response. Besides the treatment strategies through inhibiting estrogen synthesis or blocking its effect, dexamethasone has been demonstrated to induce the estrogen sulfotransferase to decreases the estradiol levels in tumor tissues and suppress the A549 xenograft tumor growth [109,110].

No significant difference in the clinicopathological characteristic between the ERβ-positive and ERβ-negative lung adenocarcinoma has been mentioned, except some reported that the ER expression correlated with the tumor differentiation [111]. Detailed pathologic examination of the ERβ-positive adenocarcinoma may be necessary to show the genotype-phenotype correlations, similar to those found in the ALK-rearranged or the EGFR-mutated adenocarcinoma [112,113]. Patients with these characteristic histologic features might be good candidates for, and could benefit from, therapy targeting the ER signaling pathways.

Novel technologies, e.g., next generation DNA sequencing, epigenetics, transcriptomics, proteomics, and metabolomics, can make an abundant contribution in the understanding of lung cancer [114–116]. The Genetic Epidemiological Study of Lung Adenocarcinoma (GELAC) in Taiwan had found that the gene polymorphisms related to the estrogen biosynthesis and metabolism was associated with an increased occurrence of L858R mutation of the EGFR in non-smoking female lung adenocarcinoma patients [117]. The use of HRT may modify the association of protective EGFR single nucleotide polymorphisms (SNPs) with lung adenocarcinoma risk [118]. The EGFR SNPs have a cumulative effect on decreasing the lung adenocarcinoma risk in non-smoking women with HRT. The ER gene SNPs are associated with a lung adenocarcinoma risk in non-smoking women [119]. The joint effects of the ER and EGFR gene SNPs and HRT usage on lung adenocarcinoma risk highlight the gene-environment interaction in lung carcinogenesis.

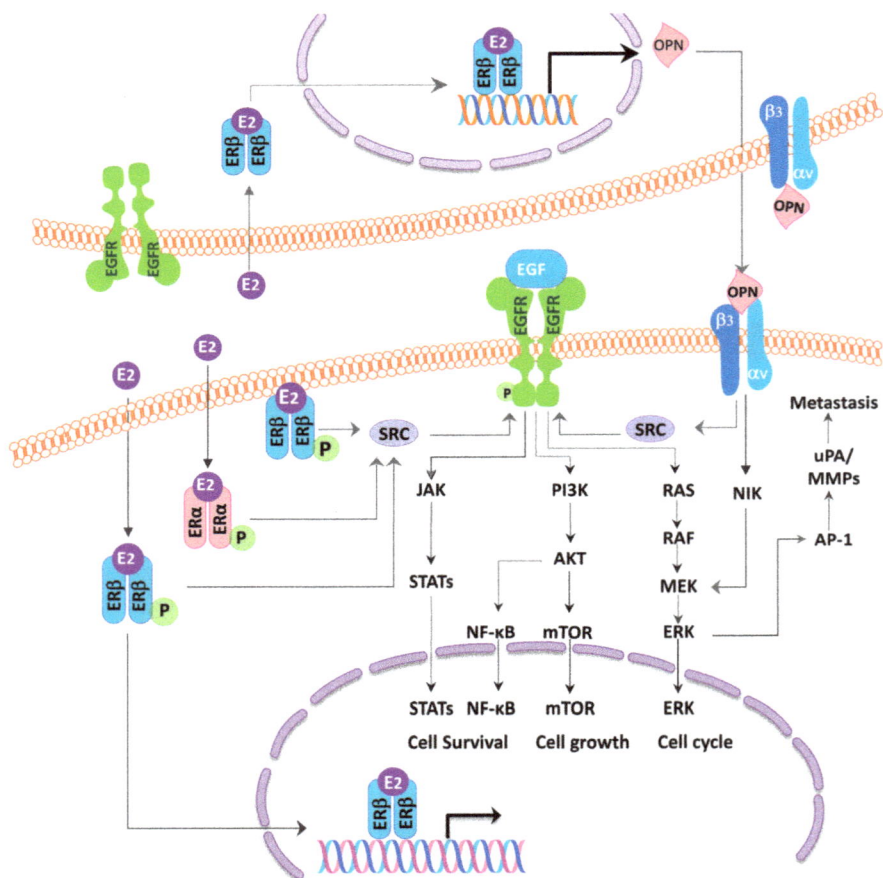

Figure 3. The schematic diagram illustrating the mechanisms of how the estrogen receptor (ER) coordinates with the epidermal growth factor receptor (EGFR) to affect the cell growth in the lung adenocarcinoma. Estrogen stimulates the steroid receptor coactivator (SRC) protein, which in turn, activates the EGFR signaling pathways. In addition, estrogen upregulates the osteopontin (OPN) expression and promotes the lung cancer cell migration via the MEK/ERK signaling pathway. The SRC and OPN contribute to the cross-talk between the ER and the EGFR.

Table 2. Clinical trials of hormone therapy in advanced NSCLC (http://www.clinicaltrial.gov/, accessed on 11 June 2017).

Patient Population	Allowed Prior Therapy	Treatment	Correlate Response with Receptors Expression	ClinicalTrials.Gov Identifier & Status
Stage IIIB or IV NSCLC, both gender	≥1 prior chemotherapy	Erlotinib + fulvestrant vs. Erlotinib	Yes	NCT00100854 Active, not recruiting (2004~)
Stage IIIB or IV NSCLC, both gender, ER or PR positive	Stable disease on erlotinib >2 months, prior chemotherapy not defined	Erlotinib + fulvestrant (single arm)	Before trial entry	NCT00592007 Terminated with results (2007~)
Stage IIIB or IV, postmenopausal women	Completed 4 cycles of induction platinum-based chemotherapy	Arm B-1. Best supportive care (BSC); Arm B-2. BSC + bevacizumab; Arm A-1. Fulvestrant + anastrozole; Arm A-2. Fulvestrant + anastrazole + bevacizumab	Yes	NCT00932152 Terminated with results (2010~)
Stage III or IV NSCLC, postmenopausal women	Chemotherapy, 0–1 line for EGFR mutations and 1–2 lines for EGFR wild type	Gefitinib + fulvestrant vs. Gefitinib for EGFR mutations; Erlotinib + fulvestrant vs. Erlotinib for EGFR wild type	No	NCT01556191 Recruiting (2012~)
Stage IV NSCLC, postmenopausal women	Phase I dose escalating study	Exemestane + premetrexed, carboplatin	No	NCT01664754 Active, not recruiting (2012~)
Stage III or IV NSCLC, postmenopausal women	Chemotherapy 1–3 line	Exemestane (single arm)	No	NCT02666105 Recruiting (2016~)

* In the order of study start date.

5. Smoking Aggravates the Effect of Estrogen and Endocrine Disruptive Chemical Targeting ERβ from the Environment May Contribute to the Lung Carcinogenesis

Tobacco smoking is a common source of complex environmental chemical exposure. More than 3000 chemicals have been identified in tobacco smoke, and many of them are both mutagenic and carcinogenic. There exists a phenomena associated estrogenic metabolism with tobacco combustion. Higher levels of polycyclic aromatic hydrocarbon-derived DNA adducts have been reported in female smokers than in male smokers. Estrogen synergize with the tobacco compounds through the induction of CYP1B1, an enzyme responsible for estrogenic metabolism, which leads to enhanced reactive oxygen species formation and carcinogenesis [4,120–122].

On the other hand, there have been constant concerns about the endocrine disruptive chemical (EDC) in the environment. EDC that have estrogenic properties are known as xenoestrogens. Although their estrogenic activity is weaker than that of estradiol, newer types of EDC and inadvertent forms of exposure continue to be discovered. There is increasing concern about their cumulative effects in carcinogenesis [123]. Endocrine disruptive chemical, e.g., polychlorinated dibenzo-*p*-dioxins, bisphenol A, polychlorinated biphenyls, polybrominated flame retardants, and methoxychlor, were supposed to be a factor in the environment leading to an increased incidence of lung adenocarcinoma [124,125]. They target ERβ with highly variable effects. Their combination with the aryl hydrocarbon receptor and its nuclear translocator could also modulate the ER activity.

Air pollution containing a mixture of particulate matters (PM) and gas contaminants is generally considered to play a role in the development of lung cancer. According to the particles' size, they are categorized into coarse particles (<10 and >2.5 μm in aerodynamic diameter, PM10), fine particles ($\leq$2.5 and >0.1 μm in aerodynamic diameter, PM2.5), and ultrafine particles ($\leq$0.1 μm). In addition to the concern of particle size, the combustion of fossil fuels, road traffic, industries, and waste dumps, are known to emit a number of different mutagens and carcinogens, many of which possess xenoestrogenic activity [126,127]. Multifactorial risk assessment incorporating personal exposure history, genetic polymorphisms related to estrogen biosynthesis and metabolism, ER polymorphisms, and biomonitoring data collected from the environment may well identify the population at risk. Collaborations between oncology, system biology, and environmental science will provide an important step to elucidate the etiology of lung cancer and help to make the relevant legislation in the future.

6. Conclusions

In addition to the well-known drivers of lung cancer, EGFR (55.7%), KRAS (5.2%), BRAF (2.0%), HER2 (0.7%) mutations, and EML4-ALK translocation (9.8%) [128], a body of epidemiological evidence, preclinical in vitro and in vivo studies, and recent data from the clinical trials, support estrogen as an important factor that contributes to lung carcinogenesis, lung cancer growth, metastasis, and affecting the prognosis. Different pathways of the ER activation and interactions with EGFR were proposed. Estrogen, with its receptor, has the potential to be a prognosticator and a therapeutic target in lung cancer. The ER antagonist may well become a new and effective treatment modality for patients with lung adenocarcinoma and an alternative treatment for patients with acquired resistance to the EGFR antagonists [101,105,129]. However, there were many conflicting results in the literature that need to be addressed [31–36,130], of which include the standardized measurements of the ER expression before adopting them as a useful biomarker, the mechanisms that underlie the controversy in the effect of hormone replacement therapy, the role of different estrogen and various ER in lung cancer cell proliferation, migration, and invasion, and the pathways involved in their interactions with other mediators. The risk of EDC exposure also raises the concern of genetic and environmental interaction in lung carcinogenesis.

Acknowledgments: The authors would like to offer their sincere thanks to Michael Wise for his help with the English language editing; Yun-Ying Chen and Shiao-Chiu Huang for their assistance with the figures, data and references preparation.

Author Contributions: Li-Han Hsu and Shu-Huei Kao conceived the paper; Li-Han Hsu and Shu-Huei Kao wrote the paper; Li-Han Hsu, Nei-Min Chu, and Shu-Huei Kao revised the paper.

Conflicts of Interest: The authors declare no conflict of interest.

Abbreviations

EDC	Endocrine disruptive chemical
EGFR	Epidermal growth factor receptor
ER	Estrogen receptor
ERα	Estrogen receptor α
ERβ	Estrogen receptor β
GPER	G-protein-coupled estrogen receptor
HRT	Hormone replacement therapy
NSCLC	Non-small cell lung cancer
OPN	Osteopontin
PM	Particulate matters
SCLC	Small cell lung cancer
SNPs	Single nucleotide polymorphisms

References

1. Nilsson, S.; Mäkelä, S.; Treuter, E.; Tujague, M.; Thomsen, J.; Andersson, G.; Enmark, E.; Pettersson, K.; Warner, M.; Gustafsson, J.A. Mechanisms of estrogen action. *Physiol. Rev.* **2001**, *81*, 1535–1565. [PubMed]
2. Paterni, I.; Granchi, C.; Katzenellenbogen, J.A.; Minutolo, F. Estrogen receptors alpha (ERα) and beta (ERβ): Subtype-selective ligands and clinical potential. *Steroids* **2014**, *90*, 13–29. [CrossRef] [PubMed]
3. Dostalova, P.; Zatecka, E.; Dvorakova-Hortova, K. Of oestrogens and sperm: A review of the roles of oestrogens and oestrogen receptors in male reproduction. *Int. J. Mol. Sci.* **2017**, *18*, 904. [CrossRef] [PubMed]
4. Slowikowski, B.K.; Lianeri, M.; Jagodzinski, P.P. Exploring estrogenic activity in lung cancer. *Mol. Biol. Rep.* **2017**, *44*, 35–50. [CrossRef] [PubMed]
5. Mosselman, S.; Polman, J.; Dijkema, R. ER beta: Identification and characterization of a novel human estrogen receptor. *FEBS Lett.* **1996**, *392*, 49–53. [CrossRef]
6. Landis, S.H.; Murray, T.; Bolden, S.; Wingo, P.A. Cancer statistics, 1999. *CA Cancer J. Clin.* **1999**, *49*, 8–31. [CrossRef] [PubMed]
7. Centers for Disease Control (CDC); National Center for Chronic Disease Prevention and Health Promotion; Office on Smoking and Health. *Women and Smoking: A Report of the Surgeon General*; U.S. Public Health Service, Office of the Surgeon General: Washington, DC, USA, 2001. Available online: http://www.cdc.gov/tobacco (accessed on 20 May 2017).
8. Health Promotion Administration, Ministry of Health and Welfare, The Executive Yuan. *Adult Smoking Behavior Survey*. Available online: http://www.hpa.gov.tw (accessed on 20 May 2017).
9. Hsu, L.H.; Chu, N.M.; Liu, C.C.; Tsai, S.Y.; You, D.L.; Ko, J.S.; Lu, M.C.; Feng, A.C. Sex-associated differences in non-small cell lung cancer in the new era: Is gender an independent prognostic factor? *Lung Cancer* **2009**, *66*, 262–267. [CrossRef] [PubMed]
10. Taiwan Cancer Registry. Cancer incidence and mortality rates in Taiwan. Available online: http://tcr.cph.ntu.edu.tw (accessed on 20 May 2017).
11. Hsu, L.H.; Liu, K.J.; Tsai, M.F.; Wu, C.R.; Feng, A.C.; Chu, N.M.; Kao, S.H. Estrogen adversely affects the prognosis of patients with lung adenocarcinoma. *Cancer Sci.* **2015**, *106*, 51–59. [CrossRef] [PubMed]
12. Zang, E.A.; Wynder, E.L. Differences in lung cancer risk between men and women: Examination of the evidence. *J. Natl. Cancer Inst.* **1996**, *88*, 183–192. [CrossRef] [PubMed]
13. Siegfried, J.M. Women and lung cancer: Does oestrogen play a role? *Lancet Oncol.* **2001**, *2*, 506–513. [CrossRef]
14. Carey, M.A.; Card, J.W.; Voltz, J.W.; Germolec, D.R.; Korach, K.S.; Zeldin, D.C. The impact of sex and sex hormones on lung physiology and disease: Lessons from animal studies. *Am. J. Physiol. Lung Cell. Mol. Physiol.* **2007**, *293*, L272–L278. [CrossRef] [PubMed]
15. Brandenberger, A.W.; Tee, M.K.; Lee, J.Y.; Chao, V.; Jaffe, R.B. Tissue distribution of estrogen receptors alpha (ER-alpha) and beta (ER-beta) mRNA in the midgestational human fetus. *J. Clin. Endocrinol. Metab.* **1997**, *82*, 3509–3512. [PubMed]

16. Morani, A.; Barros, R.P.; Imamov, O.; Hultenby, K.; Arner, A.; Warner, M.; Gustafsson, J.A. Lung dysfunction causes systemic hypoxia in estrogen receptor beta knockout (ERbeta$^{-/-}$) mice. *Proc. Natl. Acad. Sci. USA* **2006**, *103*, 7165–7169. [CrossRef] [PubMed]

17. Patrone, C.; Cassel, T.N.; Pettersson, K.; Piao, Y.S.; Cheng, G.; Ciana, P.; Maggi, A.; Warner, M.; Gustafsson, J.A.; Nord, M. Regulation of postnatal lung development and homeostasis by estrogen receptor beta. *Mol. Cell. Biol.* **2003**, *23*, 8542–8552. [CrossRef] [PubMed]

18. Zhang, G.; Liu, X.; Farkas, A.M.; Parwani, A.V.; Lathrop, K.L.; Lenzner, D.; Land, S.R.; Srinivas, H. Estrogen receptor beta functions through nongenomic mechanisms in lung cancer cells. *Mol. Endocrinol.* **2009**, *23*, 146–156. [CrossRef] [PubMed]

19. Stabile, L.P.; Siegfried, J.M. Estrogen receptor pathways in lung cancer. *Curr. Oncol. Rep.* **2004**, *6*, 259–267. [CrossRef] [PubMed]

20. Kawai, H.; Ishii, A.; Washiya, K.; Konno, T.; Kon, H.; Yamaya, C.; Ono, I.; Minamiya, Y.; Ogawa, J. Estrogen receptor alpha and beta are prognostic factors in non-small cell lung cancer. *Clin. Cancer Res.* **2005**, *11*, 5084–5089. [CrossRef] [PubMed]

21. Schwartz, A.G.; Prysak, G.M.; Murphy, V.; Lonardo, F.; Pass, H.; Schwartz, J.; Brooks, S. Nuclear estrogen receptor beta in lung cancer: Expression and survival differences by sex. *Clin. Cancer Res.* **2005**, *11*, 7280–7287. [CrossRef] [PubMed]

22. Omoto, Y.; Kobayashi, Y.; Nishida, K.; Tsuchiya, E.; Eguchi, H.; Nakagawa, K.; Ishikawa, Y.; Yamori, T.; Iwase, H.; Fujii, Y.; et al. Expression, function, and clinical implications of the estrogen receptor beta in human lung cancers. *Biochem. Biophys. Res. Commun.* **2001**, *285*, 340–347. [CrossRef] [PubMed]

23. Stabile, L.P.; Davis, A.L.; Gubish, C.T.; Hopkins, T.M.; Luketich, J.D.; Christie, N.; Finkelstein, S.; Siegfried, J.M. Human non-small cell lung tumors and cells derived from normal lung express both estrogen receptor alpha and beta and show biological responses to estrogen. *Cancer Res.* **2002**, *62*, 2141–2150. [PubMed]

24. Ganti, A.K.; Sahmoun, A.E.; Panwalkar, A.W.; Tendulkar, K.K.; Potti, A. Hormone replacement therapy is associated with decreased survival in women with lung cancer. *J. Clin. Oncol.* **2006**, *24*, 59–63. [CrossRef] [PubMed]

25. Liu, Y.; Inoue, M.; Sobue, T.; Tsugane, S. Reproductive factors, hormone use and the risk of lung cancer among middle-aged never-smoking Japanese women: A large-scale population-based cohort study. *Int. J. Cancer* **2005**, *117*, 662–666. [CrossRef] [PubMed]

26. Niikawa, H.; Suzuki, T.; Miki, Y.; Suzuki, S.; Nagasaki, S.; Akahira, J.; Honma, S.; Evans, D.B.; Hayashi, S.; Kondo, T.; et al. Intratumoral estrogens and estrogen receptors in human non-small cell lung carcinoma. *Clin. Cancer Res.* **2008**, *14*, 4417–4426. [CrossRef] [PubMed]

27. Zhang, G.; Yanamala, N.; Lathrop, K.L.; Zhang, L.; Klein-Seetharaman, J.; Srinivas, H. Ligand-independent antiapoptotic function of estrogen receptor-beta in lung cancer cells. *Mol. Endocrinol.* **2010**, *24*, 1737–1747. [CrossRef] [PubMed]

28. Mah, V.; Marquez, D.; Alavi, M.; Maresh, E.L.; Zhang, L.; Yoon, N.; Horvath, S.; Bagryanova, L.; Fishbein, M.C.; Chia, D.; et al. Expression levels of estrogen receptor beta in conjunction with aromatase predict survival in non-small cell lung cancer. *Lung Cancer* **2011**, *74*, 318–325. [CrossRef] [PubMed]

29. Chlebowski, R.T.; Schwartz, A.G.; Wakelee, H.; Anderson, G.L.; Stefanick, M.L.; Manson, J.E.; Rodabough, R.J.; Chien, J.W.; Wactawski-Wende, J.; Gass, M.; et al. Women's Health Initiative, I. Oestrogen plus progestin and lung cancer in postmenopausal women (Women's Health Initiative trial): A post-hoc analysis of a randomised controlled trial. *Lancet* **2009**, *374*, 1243–1251. [CrossRef]

30. Slatore, C.G.; Chien, J.W.; Au, D.H.; Satia, J.A.; White, E. Lung cancer and hormone replacement therapy: Association in the vitamins and lifestyle study. *J. Clin. Oncol.* **2010**, *28*, 1540–1546. [CrossRef] [PubMed]

31. Schabath, M.B.; Wu, X.; Vassilopoulou-Sellin, R.; Vaporciyan, A.A.; Spitz, M.R. Hormone replacement therapy and lung cancer risk: A case-control analysis. *Clin. Cancer Res.* **2004**, *10*, 113–123. [CrossRef] [PubMed]

32. Schwartz, A.G.; Wenzlaff, A.S.; Prysak, G.M.; Murphy, V.; Cote, M.L.; Brooks, S.C.; Skafar, D.F.; Lonardo, F. Reproductive factors, hormone use, estrogen receptor expression and risk of non small-cell lung cancer in women. *J. Clin. Oncol.* **2007**, *25*, 5785–5792. [CrossRef] [PubMed]

33. Chen, K.Y.; Hsiao, C.F.; Chang, G.C.; Tsai, Y.H.; Su, W.C.; Perng, R.P.; Huang, M.S.; Hsiung, C.A.; Chen, C.J.; Yang, P.C.; et al. Hormone replacement therapy and lung cancer risk in Chinese. *Cancer* **2007**, *110*, 1768–1775. [CrossRef] [PubMed]

34. Huang, B.; Carloss, H.; Wyatt, S.W.; Riley, E. Hormone replacement therapy and survival in lung cancer in postmenopausal women in a rural population. *Cancer* **2009**, *115*, 4167–4175. [CrossRef] [PubMed]

35. Ayeni, O.; Robinson, A. Hormone replacement therapy and outcomes for women with non-small-cell lung cancer: Can an association be confirmed? *Curr. Oncol.* **2009**, *16*, 21–25. [PubMed]

36. Chlebowski, R.T.; Anderson, G.L.; Manson, J.E.; Schwartz, A.G.; Wakelee, H.; Gass, M.; Rodabough, R.J.; Johnson, K.C.; Wactawski-Wende, J.; Kotchen, J.M.; et al. Lung cancer among postmenopausal women treated with estrogen alone in the Women's Health Initiative randomized trial. *J. Natl. Cancer Inst.* **2010**, *102*, 1413–1421. [CrossRef] [PubMed]

37. Bellahcene, A.; Castronovo, V.; Ogbureke, K.U.; Fisher, L.W.; Fedarko, N.S. Small integrin-binding ligand *N*-linked glycoproteins (SIBLINGs): Multifunctional proteins in cancer. *Nat. Rev. Cancer* **2008**, *8*, 212–226. [CrossRef] [PubMed]

38. Zirngibl, R.A.; Chan, J.S.; Aubin, J.E. Divergent regulation of the Osteopontin promoter by the estrogen receptor-related receptors is isoform- and cell context dependent. *J. Cell Biochem.* **2013**, *114*, 2356–2362. [CrossRef] [PubMed]

39. De Silva Rudland, S.; Martin, L.; Roshanlall, C.; Winstanley, J.; Leinster, S.; Platt-Higgins, A.; Carroll, J.; West, C.; Barraclough, R.; et al. Association of S100A4 and osteopontin with specific prognostic factors and survival of patients with minimally invasive breast cancer. *Clin. Cancer Res.* **2006**, *12*, 1192–1200. [CrossRef] [PubMed]

40. Chang, Y.S.; Kim, H.J.; Chang, J.; Ahn, C.M.; Kim, S.K.; Kim, S.K. Elevated circulating level of osteopontin is associated with advanced disease state of non-small cell lung cancer. *Lung Cancer* **2007**, *57*, 373–380. [CrossRef] [PubMed]

41. Cui, R.; Takahashi, F.; Ohashi, R.; Takahashi, F.; Ohashi, R.; Yoshioka, M.; Gu, T.; Tajima, K.; Unnoura, T.; Iwakami, S.; et al. Osteopontin is involved in the formation of malignant pleural effusion in lung cancer. *Lung Cancer* **2009**, *63*, 368–374. [CrossRef] [PubMed]

42. Bouchardy, C.; Benhamou, S.; Schaffar, R.; Verkooijen, H.M.; Fioretta, G.; Schubert, H.; Vinh-Hung, V.; Soria, J.C.; Vlastos, G.; Rapiti, E. Lung cancer mortality risk among breast cancer patients treated with anti-estrogens. *Cancer* **2011**, *117*, 1288–1295. [CrossRef] [PubMed]

43. Lother, S.A.; Harding, G.A.; Musto, G.; Navaratnam, S.; Pitz, M.W. Antiestrogen use and survival of women with non-small cell lung cancer in Manitoba, Canada. *Horm. Cancer* **2013**, *4*, 270–276. [CrossRef] [PubMed]

44. Hsu, L.H.; Feng, A.C.; Kao, S.H.; Liu, C.C.; Tsai, S.Y.; Shih, L.S.; Chu, N.M. Second primary lung cancers among breast cancer patients treated with anti-estrogens have a longer cancer-specific survival. *Anticancer Res.* **2015**, *35*, 1121–1127. [PubMed]

45. Del Prete, S.A.; Maurer, L.H.; Brinck-Johnsen, T.; Sorenson, G.D. 17-Beta-estradiol levels in patients with small cell carcinoma of the lung. *J. Steroid Biochem.* **1983**, *18*, 195–196. [CrossRef]

46. Sorenson, G.D.; Pettengill, O.S.; Brinck-Johnsen, T.; Cate, C.C.; Maurer, L.H. Hormone production by cultures of small-cell carcinoma of the lung. *Cancer* **1981**, *47*, 1289–1296. [CrossRef]

47. Baik, C.S.; Eaton, K.D. Estrogen signaling in lung cancer: An opportunity for novel therapy. *Cancers* **2012**, *4*, 969–988. [CrossRef] [PubMed]

48. Kawai, H. Estrogen receptors as the novel therapeutic biomarker in non-small cell lung cancer. *World J. Clin. Oncol.* **2014**, *5*, 1020–1027. [CrossRef] [PubMed]

49. Siegfried, J.M.; Stabile, L.P. Estrongenic steroid hormones in lung cancer. *Semin. Oncol.* **2014**, *41*, 5–16. [CrossRef] [PubMed]

50. Leung, Y.K.; Mak, P.; Hassan, S.; Ho, S.M. Estrogen receptor (ER)-beta isoforms: A key to understanding ER-beta signaling. *Proc. Natl. Acad. Sci. USA* **2006**, *103*, 13162–13167. [CrossRef] [PubMed]

51. Wu, C.T.; Chang, Y.L.; Shih, J.Y.; Lee, Y.C. The significance of estrogen receptor beta in 301 surgically treated non-small cell lung cancers. *J. Thorac. Cardiovasc. Surg.* **2005**, *130*, 979–986. [CrossRef] [PubMed]

52. Skov, B.G.; Fischer, B.M.; Pappot, H. Oestrogen receptor beta over expression in males with non-small cell lung cancer is associated with better survival. *Lung Cancer* **2008**, *59*, 88–94. [CrossRef] [PubMed]

53. Nose, N.; Sugio, K.; Oyama, T.; Nozoe, T.; Uramoto, H.; Iwata, T.; Onitsuka, T.; Yasumoto, K. Association between estrogen receptor-beta expression and epidermal growth factor receptor mutation in the postoperative prognosis of adenocarcinoma of the lung. *J. Clin. Oncol.* **2009**, *27*, 411–417. [CrossRef] [PubMed]

54. Raso, M.G.; Behrens, C.; Herynk, M.H.; Liu, S.; Prudkin, L.; Ozburn, N.C.; Woods, D.M.; Tang, X.; Mehran, R.J.; Moran, C.; et al. Immunohistochemical expression of estrogen and progesterone receptors identifies a subset of NSCLCs and correlates with EGFR mutation. *Clin. Cancer Res.* **2009**, *15*, 5359–5368. [CrossRef] [PubMed]

55. Stabile, L.P.; Dacic, S.; Land, S.R.; Lenzner, D.E.; Dhir, R.; Acquafondata, M.; Landreneau, R.J.; Grandis, J.R.; Siegfried, J.M. Combined analysis of estrogen receptor beta-1 and progesterone receptor expression identifies lung cancer patients with poor outcome. *Clin. Cancer Res.* **2011**, *17*, 154–164. [CrossRef] [PubMed]

56. Rouquette, I.; Lauwers-Cances, V.; Allera, C.; Brouchet, L.; Milia, J.; Nicaise, Y.; Laurent, J.; Delisle, M.B.; Favre, G.; Didier, A.; et al. Characteristics of lung cancer in women: Importance of hormonal and growth factors. *Lung Cancer* **2012**, *76*, 280–285. [CrossRef] [PubMed]

57. Rades, D.; Setter, C.; Dahl, O.; Schild, S.E.; Noack, F. The prognostic impact of tumor cell expression of estrogen receptor-alpha, progesterone receptor, and androgen receptor in patients irradiated for nonsmall cell lung cancer. *Cancer* **2012**, *118*, 157–163. [CrossRef] [PubMed]

58. Karlsson, C.; Helenius, G.; Fernandes, O.; Karlsson, M.G. Oestrogen receptor beta in NSCLC—Prevalence, proliferative influence, prognostic impact and smoking. *Acta Pathol. Microbiol. Immunol. Scand.* **2012**, *120*, 451–458. [CrossRef] [PubMed]

59. Navaratnam, S.; Skliris, G.; Qing, G.; Banerji, S.; Badiani, K.; Tu, D.; Bradbury, P.A.; Leighl, N.B.; Shepherd, F.A.; Nowatzki, J.; et al. Differential role of estrogen receptor beta in early versus metastatic non-small cell lung cancer. *Horm. Cancer* **2012**, *3*, 93–100. [CrossRef] [PubMed]

60. Liu, Z.; Liao, Y.; Tang, H.; Chen, G. The expression of estrogen receptors beta2, 5 identifies and is associated with prognosis in non-small cell lung cancer. *Endocrine* **2013**, *44*, 517–524. [CrossRef] [PubMed]

61. Kadota, K.; Eguchi, T.; Villena-Vargas, J.; Woo, K.M.; Sima, C.S.; Jones, D.R.; Travis, W.D.; Adusumilli, P.S. Nuclear estrogen receptor-alpha expression is an independent predictor of recurrence in male patients with pT1aN0 lung adenocarcinomas, and correlates with regulatory T-cell infiltration. *Oncotarget* **2015**, *6*, 27505–27518. [CrossRef] [PubMed]

62. Liu, C.M.; Chiu, K.L.; Chen, T.S.; Chang, S.M.; Yang, S.Y.; Chen, L.H.; Ni, Y.L.; Sher, Y.P.; Yu, S.L.; Ma, W.L. Potential therapeutic benefit of combining gefitinib and tamoxifen for treating advanced lung adenocarcinoma. *Biomed. Res. Int.* **2015**, *2015*, 642041. [CrossRef] [PubMed]

63. Skjefstad, K.; Grindstad, T.; Khanehkenari, M.R.; Richardsen, E.; Donnem, T.; Kilvaer, T.; Andersen, S.; Bremnes, R.M.; Busund, L.T.; Al-Saad, S. Prognostic relevance of estrogen receptor alpha, beta and aromatase expression in non-small cell lung cancer. *Steroids* **2016**, *113*, 5–13. [CrossRef] [PubMed]

64. Tanaka, K.; Shimizu, K.; Kakegawa, S.; Ohtaki, Y.; Nagashima, T.; Kaira, K.; Horiguchi, J.; Oyama, T.; Takeyoshi, I. Prognostic significance of aromatase and estrogen receptor beta expression in EGFR wild-type lung adenocarcinoma. *Am. J. Transl. Res.* **2016**, *8*, 81–97. [PubMed]

65. Li, W.; Tse, L.A.; Wang, F. Prognostic value of estrogen receptors mRNA expression in non-small cell lung cancer: A systematic review and meta-analysis. *Steroids* **2015**, *104*, 129–136. [CrossRef] [PubMed]

66. Ma, L.; Zhan, P.; Liu, Y.; Zhou, Z.; Zhu, Q.; Miu, Y.; Wang, X.; Jin, J.; Li, Q.; Lv, T.; et al. Prognostic value of the expression of estrogen receptor beta in patients with non-small cell lung cancer: A meta-analysis. *Transl. Lung Cancer Res.* **2016**, *5*, 202–207. [CrossRef] [PubMed]

67. Wang, Z.; Li, Z.; Ding, X.; Shen, Z.; Liu, Z.; An, T.; Duan, J.; Zhong, J.; Wu, M.; Zhao, J.; et al. ERbeta localization influenced outcomes of EGFR-TKI treatment in NSCLC patients with EGFR mutations. *Sci. Rep.* **2015**, *5*, 11392. [CrossRef] [PubMed]

68. Liao, T.L.; Tzeng, C.R.; Yu, C.L.; Wang, Y.P.; Kao, S.H. Estrogen receptor-beta in mitochondria: Implications for mitochondrial bioenergetics and tumorigenesis. *Ann. N. Y. Acad. Sci.* **2015**, *1350*, 52–60. [CrossRef] [PubMed]

69. Xie, Q.; Huang, Z.; Liu, Y.; Liu, X.; Huang, L. Mitochondrial estrogen receptor beta inhibits non-small cell lung cancer cell apoptosis via interaction with Bad. *Nan Fang Yi Ke Da Xue Xue Bao* **2015**, *35*, 98–102. [PubMed]

70. Jacenik, D.; Cygankiewicz, A.I.; Krajewska, W.M. The G protein-coupled estrogen receptor as a modulator of neoplastic transformation. *Mol. Cell. Endocrinol.* **2016**, *429*, 10–18. [CrossRef] [PubMed]

71. Konings, G.F.; Reynaert, N.L.; Delvoux, B.; Verhamme, F.M.; Bracke, K.R.; Brusselle, G.G.; Romano, A.; Vernooy, J.H. Increased levels of enzymes involved in local estradiol synthesis in chronic obstructive pulmonary disease. *Mol. Cell. Endocrinol.* **2017**, *443*, 23–31. [CrossRef] [PubMed]

72. Revankar, C.M.; Cimino, D.F.; Sklar, L.A.; Arterburn, J.B.; Prossnitz, E.R. A transmembrane intracellular estrogen receptor mediates rapid cell signaling. *Science* **2005**, *307*, 1625–1630. [CrossRef] [PubMed]

73. Sakamoto, H.; Matsuda, K.; Hosokawa, K.; Nishi, M.; Morris, J.F.; Prossnitz, E.R.; Kawata, M. Expression of G protein-coupled receptor-30, a G protein-coupled membrane estrogen receptor, in oxytocin neurons of the rat paraventricular and supraoptic nuclei. *Endocrinology* **2007**, *148*, 5842–5850. [CrossRef] [PubMed]

74. Jala, V.R.; Radde, B.N.; Haribabu, B.; Klinge, C.M. Enhanced expression of G-protein coupled estrogen receptor (GPER/GPR30) in lung cancer. *BMC Cancer* **2012**, *12*, 624. [CrossRef] [PubMed]

75. Liu, C.; Liao, Y.; Fan, S.; Tang, H.; Jiang, Z.; Zhou, B.; Xiong, J.; Zhou, S.; Zou, M.; Wang, J. G protein-coupled estrogen receptor (GPER) mediates NSCLC progression induced by 17beta-estradiol (E2) and selective agonist G1. *Med. Oncol.* **2015**, *32*, 104. [CrossRef] [PubMed]

76. Lucas, T.F.; Siu, E.R.; Esteves, C.A.; Monteiro, H.P.; Oliveira, C.A.; Porto, C.S.; Lazari, M.F. 17 beta-estradiol induces the translocation of the estrogen receptors ESR1 and ESR2 to the cell membrane, MAPK3/1 phosphorylation and proliferation of cultured immature rat Sertoli cells. *Biol. Reprod.* **2008**, *78*, 101–114. [CrossRef] [PubMed]

77. Honma, N.; Hosoi, T.; Arai, T.; Takubo, K. Estrogen and cancers of the colorectum, breast, and lung in postmenopausal women. *Pathol. Int.* **2015**, *65*, 451–459. [CrossRef] [PubMed]

78. Ikeda, K.; Shiraishi, K.; Yoshida, A.; Shinchi, Y.; Sanada, M.; Motooka, Y.; Fujino, K.; Mori, T.; Suzuki, M. Synchronous multiple lung adenocarcinomas: Estrogen concentration in peripheral lung. *PLoS ONE* **2016**, *11*, e0160910. [CrossRef] [PubMed]

79. Hsu, L.H.; Hsu, P.C.; Liao, T.L.; Feng, A.C.; Chu, N.M.; Kao, S.H. Pleural fluid osteopontin, vascular endothelial growth factor, and urokinase-type plasminogen activator levels as predictors of pleurodesis outcome and prognosticators in patients with malignant pleural effusion: A prospective cohort study. *BMC Cancer* **2016**, *16*, 463. [CrossRef] [PubMed]

80. Yatabe, Y.; Mitsudomi, T.; Takahashi, T. TTF-1 expression in pulmonary adenocarcinomas. *Am. J. Surg. Pathol.* **2002**, *26*, 767–773. [CrossRef] [PubMed]

81. Shanzhi, W.; Yiping, H.; Ling, H.; Jianming, Z.; Qiang, L. The relationship between TTF-1 expression and EGFR mutations in lung adenocarcinomas. *PLoS ONE* **2014**, *9*, e95479. [CrossRef] [PubMed]

82. Chung, K.P.; Huang, Y.T.; Chang, Y.L.; Yu, C.J.; Yang, C.H.; Chang, Y.C.; Shih, J.Y.; Yang, P.C. Clinical significance of thyroid transcription factor-1 in advanced lung adenocarcinoma under epidermal growth factor receptor tyrosine kinase inhibitor treatment. *Chest* **2012**, *141*, 420–428. [CrossRef] [PubMed]

83. Lau, S.K.; Chu, P.G.; Weiss, L.M. Immunohistochemical expression of estrogen receptor in pulmonary adenocarcinoma. *Appl. Immunohistochem. Mol. Morphol.* **2006**, *14*, 83–87. [CrossRef] [PubMed]

84. Greiser, C.M.; Greiser, E.M.; Doren, M. Menopausal hormone therapy and risk of lung cancer—Systematic review and meta-analysis. *Maturitas* **2010**, *65*, 198–204. [CrossRef] [PubMed]

85. Chlebowski, R.T.; Wakelee, H.; Pettinger, M.; Rohan, T.; Liu, J.; Simon, M.; Tindle, H.; Messina, C.; Johnson, K.; Schwartz, A.; et al. Estrogen plus progestin and lung cancer: Follow-up of the Women's Health Initiative randomized trial. *Clin. Lung Cancer* **2016**, *17*, 10–17. [CrossRef] [PubMed]

86. Jiang, X.; Shapiro, D.J. The immune system and inflammation in breast cancer. *Mol. Cell. Endocrinol.* **2014**, *382*, 673–682. [CrossRef] [PubMed]

87. Grivennikov, S.I.; Greten, F.R.; Karin, M. Immunity, inflammation, and cancer. *Cell* **2010**, *140*, 883–899. [CrossRef] [PubMed]

88. Miki, Y.; Suzuki, T.; Abe, K.; Suzuki, S.; Niikawa, H.; Iida, S.; Hata, S.; Akahira, J.; Mori, K.; Evans, D.B.; et al. Intratumoral localization of aromatase and interaction between stromal and parenchymal cells in the non-small cell lung carcinoma microenvironment. *Cancer Res.* **2010**, *70*, 6659–6669. [CrossRef] [PubMed]

89. Stabile, L.P.; Rothstein, M.E.; Cunningham, D.E.; Land, S.R.; Dacic, S.; Keohavong, P.; Siegfried, J.M. Prevention of tobacco carcinogen-induced lung cancer in female mice using antiestrogens. *Carcinogenesis* **2012**, *33*, 2181–2189. [CrossRef] [PubMed]

90. Mah, V.; Seligson, D.B.; Li, A.; Marquez, D.C.; Wistuba, I.I.; Elshimali, Y.; Fishbein, M.C.; Chia, D.; Pietras, R.J.; Goodglick, L. Aromatase expression predicts survival in women with early-stage non small cell lung cancer. *Cancer Res.* **2007**, *67*, 10484–10490. [CrossRef] [PubMed]

91. Weinberg, O.K.; Marquez-Garban, D.C.; Fishbein, M.C.; Goodglick, L.; Garban, H.J.; Dubinett, S.M.; Pietras, R.J. Aromatase inhibitors in human lung cancer therapy. *Cancer Res.* **2005**, *65*, 11287–11291. [CrossRef] [PubMed]

92. Marquez-Garban, D.C.; Chen, H.W.; Goodglick, L.; Fishbein, M.C.; Pietras, R.J. Targeting aromatase and estrogen signaling in human non-small cell lung cancer. *Ann. N. Y. Acad. Sci.* **2009**, *1155*, 194–205. [CrossRef] [PubMed]

93. Hamilton, D.H.; Griner, L.M.; Keller, J.M.; Hu, X.; Southall, N.; Marugan, J.; David, J.M.; Ferrer, M.; Palena, C. Targeting estrogen receptor signaling with fulvestrant enhances immune and chemotherapy-mediated cytotoxicity of human lung cancer. *Clin. Cancer Res.* **2016**, *22*, 6204–6216. [CrossRef] [PubMed]

94. Swedenborg, E.; Power, K.A.; Cai, W.; Pongratz, I.; Ruegg, J. Regulation of estrogen receptor beta activity and implications in health and disease. *Cell. Mol. Life Sci.* **2009**, *66*, 3873–3894. [CrossRef] [PubMed]

95. Levin, E.R. Bidirectional signaling between the estrogen receptor and the epidermal growth factor receptor. *Mol. Endocrinol.* **2003**, *17*, 309–317. [CrossRef] [PubMed]

96. Kato, S.; Endoh, H.; Masuhiro, Y.; Kitamoto, T.; Uchiyama, S.; Sasaki, H.; Masushige, S.; Gotoh, Y.; Nishida, E.; Kawashima, H.; et al. Activation of the estrogen receptor through phosphorylation by mitogen-activated protein kinase. *Science* **1995**, *270*, 1491–1494. [CrossRef] [PubMed]

97. Marquez-Garban, D.C.; Chen, H.W.; Fishbein, M.C.; Goodglick, L.; Pietras, R.J. Estrogen receptor signaling pathways in human non-small cell lung cancer. *Steroids* **2007**, *72*, 135–143. [CrossRef] [PubMed]

98. Zhao, X.Z.; Liu, Y.; Zhou, L.J.; Wang, Z.Q.; Wu, Z.H.; Yang, X.Y. Role of estrogen in lung cancer based on the estrogen receptor-epithelial mesenchymal transduction signaling pathways. *OncoTargets Ther.* **2015**, *8*, 2849–2863. [CrossRef] [PubMed]

99. Fan, S.; Liao, Y.; Liu, C.; Huang, Q.; Liang, H.; Ai, B.; Fu, S.; Zhou, S. Estrogen promotes tumor metastasis via estrogen receptor beta-mediated regulation of matrix-metalloproteinase-2 in non-small cell lung cancer. *Oncotarget* **2017**. [CrossRef] [PubMed]

100. Siegfried, J.M.; Farooqui, M.; Rothenberger, N.J.; Dacic, S.; Stabile, L.P. Interaction between the estrogen receptor and fibroblast growth factor receptor pathways in non-small cell lung cancer. *Oncotarget* **2017**, *8*, 24063–24076. [CrossRef] [PubMed]

101. Stabile, L.P.; Lyker, J.S.; Gubish, C.T.; Zhang, W.; Grandis, J.R.; Siegfried, J.M. Combined targeting of the estrogen receptor and the epidermal growth factor receptor in non-small cell lung cancer shows enhanced antiproliferative effects. *Cancer Res.* **2005**, *65*, 1459–1470. [CrossRef] [PubMed]

102. Pietras, R.J.; Marquez, D.C.; Chen, H.W.; Tsai, E.; Weinberg, O.; Fishbein, M. Estrogen and growth factor receptor interactions in human breast and non-small cell lung cancer cells. *Steroids* **2005**, *70*, 372–381. [CrossRef] [PubMed]

103. Deng, F.; Li, M.; Shan, W.L.; Qian, L.T.; Meng, S.P.; Zhang, X.L.; Wang, B.L. Correlation between epidermal growth factor receptor mutations and the expression of estrogen receptor-beta in advanced non-small cell lung cancer. *Oncol. Lett.* **2017**, *13*, 2359–2365. [PubMed]

104. Kawaguchi, T.; Koh, Y.; Ando, M.; Ito, N.; Takeo, S.; Adachi, H.; Tagawa, T.; Kakegawa, S.; Yamashita, M.; Kataoka, K.; et al. Prospective analysis of oncogenic driver mutations and environmental factors: Japan Molecular Epidemiology For Lung Cancer Study. *J. Clin. Oncol.* **2016**, *34*, 2247–2257. [CrossRef] [PubMed]

105. Dubey, S.; Siegfried, J.M.; Traynor, A.M. Non-small-cell lung cancer and breast carcinoma: Chemotherapy and beyond. *Lancet Oncol.* **2006**, *7*, 416–424. [CrossRef]

106. Giovannini, M.; Belli, C.; Villa, E.; Gregorc, V. Estrogen receptor and epidermal growth factor receptor as targets for dual lung cancer therapy: Not just a case? *J. Thorac. Oncol.* **2008**, *3*, 684–685. [CrossRef] [PubMed]

107. Traynor, A.M.; Schiller, J.H.; Stabile, L.P.; Kolesar, J.M.; Eickhoff, J.C.; Dacic, S.; Hoang, T.; Dubey, S.; Marcotte, S.M.; Siegfried, J.M. Pilot study of gefitinib and fulvestrant in the treatment of post-menopausal women with advanced non-small cell lung cancer. *Lung Cancer* **2009**, *64*, 51–59. [CrossRef] [PubMed]

108. ClinicalTrials.gov, U.S. National Institutes of Health. Available online: http://www.clinicaltrial.gov/ (accessed on 11 June 2017).

109. Iida, S.; Kakinuma, H.; Miki, Y.; Abe, K.; Sakurai, M.; Suzuki, S.; Niikawa, H.; Akahira, J.; Suzuki, T.; Sasano, H. Steroid sulphatase and oestrogen sulphotransferase in human non-small-cell lung carcinoma. *Br. J. Cancer* **2013**, *108*, 1415–1424. [CrossRef] [PubMed]

110. Wang, L.J.; Li, J.; Hao, F.R.; Yuan, Y.; Li, J.Y.; Lu, W.; Zhou, T.Y. Dexamethasone suppresses the growth of human non-small cell lung cancer via inducing estrogen sulfotransferase and inactivating estrogen. *Acta Pharmacol. Sin.* **2016**, *37*, 845–856. [CrossRef] [PubMed]

111. Chen, X.Q.; Zheng, L.X.; Li, Z.Y.; Lin, T.Y. Clinicopathological significance of oestrogen receptor expression in non-small cell lung cancer. *J. Int. Med. Res.* **2017**, *45*, 51–58. [CrossRef] [PubMed]

112. Inamura, K.; Takeuchi, K.; Togashi, Y.; Hatano, S.; Ninomiya, H.; Motoi, N.; Mun, M.Y.; Sakao, Y.; Okumura, S.; Nakagawa, K.; et al. EML4-ALK lung cancers are characterized by rare other mutations, a TTF-1 cell lineage, an acinar histology, and young onset. *Mod. Pathol.* **2009**, *22*, 508–515. [CrossRef] [PubMed]

113. Inamura, K.; Ninomiya, H.; Ishikawa, Y.; Matsubara, O. Is the epidermal growth factor receptor status in lung cancers reflected in clinicopathologic features? *Arch. Pathol. Lab. Med.* **2010**, *134*, 66–72. [PubMed]

114. Lehtio, J.; De Petris, L. Lung cancer proteomics, clinical and technological considerations. *J. Proteom.* **2010**, *73*, 1851–1863. [CrossRef] [PubMed]

115. Indovina, P.; Marcelli, E.; Pentimalli, F.; Tanganelli, P.; Tarro, G.; Giordano, A. Mass spectrometry-based proteomics: The road to lung cancer biomarker discovery. *Mass Spectrom. Rev.* **2013**, *32*, 129–142. [CrossRef] [PubMed]

116. Hagemann, I.S.; Devarakonda, S.; Lockwood, C.M.; Spencer, D.H.; Guebert, K.; Bredemeyer, A.J.; Al-Kateb, H.; Nguyen, T.T.; Duncavage, E.J.; Cottrell, C.E.; et al. Clinical next-generation sequencing in patients with non-small cell lung cancer. *Cancer* **2015**, *121*, 631–639. [CrossRef] [PubMed]

117. Yang, S.Y.; Yang, T.Y.; Chen, K.C.; Li, Y.J.; Hsu, K.H.; Tsai, C.R.; Chen, C.Y.; Hsu, C.P.; Hsia, J.Y.; Chuang, C.Y.; et al. EGFR L858R mutation and polymorphisms of genes related to estrogen biosynthesis and metabolism in never-smoking female lung adenocarcinoma patients. *Clin. Cancer Res.* **2011**, *17*, 2149–2158. [CrossRef] [PubMed]

118. Chen, K.Y.; Hsiao, C.F.; Chang, G.C.; Tsai, Y.H.; Su, W.C.; Chen, Y.M.; Huang, M.S.; Hsiung, C.A.; Chen, C.J.; Yang, P.C. EGFR polymorphisms, hormone replacement therapy and lung adenocarcinoma risk: Analysis from a genome-wide association study in never-smoking women. *Carcinogenesis* **2013**, *34*, 612–619. [CrossRef] [PubMed]

119. Chen, K.Y.; Hsiao, C.F.; Chang, G.C.; Tsai, Y.H.; Su, W.C.; Chen, Y.M.; Huang, M.S.; Tsai, F.Y.; Jiang, S.S.; Chang, I.S.; et al. Estrogen receptor gene polymorphisms and lung adenocarcinoma risk in never-smoking women. *J. Thorac. Oncol.* **2015**, *10*, 1413–1420. [CrossRef] [PubMed]

120. Belous, A.R.; Hachey, D.L.; Dawling, S.; Roodi, N.; Parl, F.F. Cytochrome P450 1B1-mediated estrogen metabolism results in estrogen-deoxyribonucleoside adduct formation. *Cancer Res.* **2007**, *67*, 812–817. [CrossRef] [PubMed]

121. Meireles, S.I.; Esteves, G.H.; Hirata, R., Jr.; Peri, S.; Devarajan, K.; Slifker, M.; Mosier, S.L.; Peng, J.; Vadhanam, M.V.; Hurst, H.E.; et al. Early changes in gene expression induced by tobacco smoke: Evidence for the importance of estrogen within lung tissue. *Cancer Prev. Res.* **2010**, *3*, 707–717. [CrossRef] [PubMed]

122. Peng, J.; Xu, X.; Mace, B.E.; Vanderveer, L.A.; Workman, L.R.; Slifker, M.J.; Sullivan, P.M.; Veenstra, T.D.; Clapper, M.L. Estrogen metabolism within the lung and its modulation by tobacco smoke. *Carcinogenesis* **2013**, *34*, 909–915. [CrossRef] [PubMed]

123. Olea, N.; Pazos, P.; Exposito, J. Inadvertent exposure to xenoestrogens. *Eur. J. Cancer Prev.* **1998**, *7*, S17–S23. [CrossRef] [PubMed]

124. Swedenborg, E.; Ruegg, J.; Makela, S.; Pongratz, I. Endocrine disruptive chemicals: Mechanisms of action and involvement in metabolic disorders. *J. Mol. Endocrinol.* **2009**, *43*, 1–10. [CrossRef] [PubMed]

125. Swedenborg, E.; Pongratz, I.; Gustafsson, J.A. Endocrine disruptors targeting ERbeta function. *Int. J. Androl.* **2010**, *33*, 288–297. [CrossRef] [PubMed]

126. Fucic, A.; Gamulin, M.; Ferencic, Z.; Rokotov, D.S.; Katic, J.; Bartonova, A.; Lovasic, I.B.; Merlo, D.F. Lung cancer and environmental chemical exposure: A review of our current state of knowledge with reference to the role of hormones and hormone receptors as an increased risk factor for developing lung cancer in man. *Toxicol. Pathol.* **2010**, *38*, 849–855. [CrossRef] [PubMed]

127. Fucic, A.; Gamulin, M.; Ferencic, Z.; Katic, J.; von Krauss, M.K.; Bartonova, A.; Merlo, D.F. Environmental exposure to xenoestrogens and oestrogen related cancers: Reproductive system, breast, lung, kidney, pancreas, and brain. *Environ. Health* **2012**, *11*, S8. [CrossRef] [PubMed]

128. Hsu, K.H.; Ho, C.C.; Hsai, T.C.; Tseng, J.S.; Su, K.Y.; Wu, M.F.; Chiu, K.L.; Yang, T.Y.; Chen, K.C.; Ooi, H.; et al. Identification of five driver gene mutations in patients with treatment-naive lung adenocarcinoma in Taiwan. *PLoS ONE* **2015**, *10*, e0120852. [CrossRef] [PubMed]

129. Garon, E.B.; Pietras, R.J.; Finn, R.S.; Kamranpour, N.; Pitts, S.; Marquez-Garban, D.C.; Desai, A.J.; Dering, J.; Hosmer, W.; von Euw, E.M.; et al. Antiestrogen fulvestrant enhances the antiproliferative effects of epidermal growth factor receptor inhibitors in human non-small-cell lung cancer. *J. Thorac. Oncol.* **2013**, *8*, 270–278. [CrossRef] [PubMed]

130. Patel, J.D.; Gray, R.G.; Stewart, J.A.; Skinner, H.G.; Schiller, J.H. Tamoxifen does not reduce the risk of lung cancer in women. *J. Clin. Oncol.* **2005**, *23*, s7212. [CrossRef]

International Journal of
Molecular Sciences

MDPI

Review

Natural Anti-Estrogen Receptor Alpha Antibodies Able to Induce Estrogenic Responses in Breast Cancer Cells: Hypotheses Concerning Their Mechanisms of Action and Emergence †

Guy Leclercq

Laboratoire de Cancérologie Mammaire, Institut J. Bordet, Centre des Tumeurs de l'Université Libre de Bruxelles, 1, rue Héger-Bordet, 1000 Brussels, Belgium; guy.leclercq@ulb.ac.be

† This paper may be considered as a tribute to my colleague Albert Borkowski, Internist and Director of the Endocrinology Laboratory of the Institute J. Bordet, who passed away in 2007 and who first described certain biological functions of these antibodies.

Received: 25 December 2017; Accepted: 25 January 2018; Published: 30 January 2018

Abstract: The detection of human anti-estrogen receptor α antibodies (ERαABs) inducing estrogenic responses in MCF-7 mammary tumor cells suggests their implication in breast cancer emergence and/or evolution. A recent report revealing a correlation between the titer of such antibodies in sera from patients suffering from this disease and the percentage of proliferative cells in samples taken from their tumors supports this concept. Complementary evidence of the ability of ERαABs to interact with an epitope localized within the estradiol-binding core of ERα also argues in its favor. This epitope is indeed inserted in a regulatory platform implicated in ERα-initiated signal transduction pathways and transcriptions. According to some experimental observations, two auto-immune reactions may already be advocated to explain the emergence of ERαABs: one involving probably the idiotypic network to produce antibodies acting as estrogenic secretions and the other based on antibodies able to abrogate the action of a natural ERα inhibitor or to prevent the competitive inhibitory potency of released receptor degradation products able to entrap circulating estrogens and co-activators. All of this information, the aspect of which is mainly fundamental, may open new ways in the current tendency to combine immunological and endocrine approaches for the management of breast cancer.

Keywords: estrogen receptor α; natural antibodies; estrogenic responses; mechanism of action; auto-immune diseases

1. Introduction

Among modulators of steroid hormone receptors, natural anti-estrogen receptor antibodies (ERABs) are of peculiar interest in view of their implication in the emergence and/or evolution of autoimmune diseases and cancers [1]. The present paper focuses on the potential biological relevance of these antibodies in the context of the hormone-dependence of breast cancer, a topic on which I have been working for more than four decades.

The recent finding by the group of Pierdominici and Ortona of a correlation between the titer of ERABs raised against the alpha form of the receptor (ERαABs) in sera from a series of women with breast cancer and the percentage of Ki67-positive cells (a known marker of proliferation) in samples taken from their tumors [2] offered to me an opportunity to discuss here the possible implication of these antibodies in the development of breast cancers. In fact, this concept had already been proposed in the late 1980s by my colleague Borkowski, who detected a sub-population of IgGs able to interact with the estradiol (E_2) binding site of ERα in sera from healthy women [3,4]. This work, in which I collaborated, revealed moreover the ability of these IgGs to induce estrogenic (or estrogenic-like) responses in ERα-positive MCF-7 breast cancer cells, suggesting that they act on these cells as the hormone [4]. Further studies revealed that this view was only partly true: the major estrogenic activity of the IgGs seemed to derive from the neutralization of ERα-related peptides able to inhibit its activation [5]. Skepticism concerning the biological significance of these various observations, as well as their potential insertion in therapeutic programs forced us to stop our investigations. We hope that the recent investigations of Pierdominici and Ortona, which also concern the prominent role of estrogens in autoimmune diseases [6], may encourage the scientific community to assess again questions relevant to the suspected role of such natural anti-ERα antibodies in breast cancer.

The present paper devoted to this hope mainly concerns the mechanism(s) by which ERαABs may operate; processes implicated in their emergence will be also evoked. Available data being quite tenuous, my proposals are largely speculative. Nevertheless, I anticipate that they may open avenues for new experimentations not necessarily restricted to ERα, since the existence of natural antibodies raised against other steroids hormone receptors has been reported, as will be recalled briefly in the next section.

2. Natural Antibodies against Steroid Hormone Receptors, the Existence of Which Had Been Reported about Three Decades Ago

To my knowledge, the first evocation of such antibodies must be attributed to the group of O'Malley that reported in 1981 the existence of "spontaneous" antibodies raised against the progesterone receptor in two thirds of sheep sera [7]. Surprisingly, these authors limited their investigation to the assessment of the binding properties of these antibodies for the α and β isoforms of this receptor without raising any questions relevant to their biological role. This topic was addressed in the following year by Liao and Witte who reported a high titer of anti-androgen receptors in patients with prostate disease, when compared with normal subjects [8]. These authors logically proposed some relevance to this detection in terms of disease management. The discovery of the existence of anti-ERα may be ascribed to Borkowski [3], as well as to Muddaris and Peck Jr. [9], who detected them at the same time. While Borkowski focused his studies on the biological function of these antibodies, Muddaris and Peck reported striking sex and age-related differences in the level of the latter: young females displayed a higher titer than corresponding males. This level also declined in middle age, before increasing in old age, in contrast to males in which it continuously decreased. Although these various observations were quite provocative, they failed to generate a significant interest for about two decades, as previously mentioned.

3. Major Properties of ERαABs

3.1. Ability to Induce Estrogenic (or Estrogenic-Like) Responses

As reported below, ERαABs act as ERα agonists through both non-genomic and genomic procedures, which operate sequentially, the non-genomic preceding largely the genomic procedures [10,11]. This suspected co-operative mechanism [11–14], detected with MCF-7 breast cancer cells, seems to be initiated at the plasma membrane (Section 3.2).

3.1.1. Signal Transduction Activation and Subsequent Cell Proliferation Enhancement

Highly purified ERαABs almost immediately activate the phosphorylation of ERK (Extracellular regulated kinase) in MCF-7 cells without producing any similar effect on Akt (Protein kinase B) [2]. The maximal effect of the antibodies occurs after 5 min and subsequently declines, returning to the original level after 30 min. As expected, a significant increase in proliferation is recorded after one day of treatment.

3.1.2. Transcriptions and Related ERα Level Changes

Over-night exposure of MCF-7 cells to highly purified ERαABs (IgGs) enhances their level of progesterone receptors in a dose-dependent manner, as observed with E_2 used as the control; this increase is progressively inhibited by pure antiestrogens [4,5]. The same behavior is recorded for cathepsin D secretion. A loss of the capacity of the cells to specifically incorporate [^{3}H]E_2 (ERα whole cell assay) occurs in parallel, which may be ascribed to a decrease of the ERα level, detected by Western blotting. IgGs also partially abrogate the capacity of the cells to incorporate [^{3}H]E_2 in the presence of an analog of hydroxy-tamoxifen, which stabilizes the receptor within the nucleus [15], as does E_2.

3.2. Selective Ability to Associate with the E_2-Binding Site of the Native Form of ERα Localized at the Plasma Membrane

When submitted to low-salt sucrose gradient sedimentations, ERα from cytosolic extracts is known to migrate within two distinct oligomeric structures, i.e., of 4 and 8S (note that these velocities may slightly differ according to the nature of the tissues from which ERα is extracted, the experimental conditions, as well as the choice of the sedimentation markers used for their assessment) [16,17]. The 4S entity contains proteolytic products of the receptor, while the latter is maintained within the 8S entity in its native form (67 kDa) by a protective action of chaperones with which it associates. Interactions between highly purified ERαABs (IgGs) and ERα occur in the region of its E_2-binding site since an increase of sedimentation velocity of the 8S oligomer is detected when the [^{3}H]E_2 labeling of the receptor is performed after sedimentation on the fractions collected from the gradient. With pre-labeled cytosols, this sedimentation shift is replaced by a partial displacement of bound [^{3}H]E_2 by the IgGs [3]. Complementary experiments including an assessment of the binding parameters of [^{3}H]E_2 to ERα, in the absence and the presence of increasing amounts of these IgGs, respectively, confirmed the implication of the E_2 binding site of the receptor in this complex. Accordingly, these IgGs behaved as competitive inhibitors (increase of K_d values) [3], a finding in agreement with the recent identification of an epitope able to recognize ERαABs (Y^{459}TFLSSTLKSLEE^{471}; Figure 1) within the E_2-binding core of ERα (Asn309-Lys529; MW: 26 kDa [18]) [2].

Interestingly, this Tyr459-Glu471 epitope contains a small motif (Thr465-Ser468), which is cleaved under proteolytic attack without any loss of E_2 binding ability [18], a property resulting from a cutting of the estrogen-binding core of ERα in two distinct entities (7 and 17 kDa) that stick together through hydrophobic contacts [19]. According to our sedimentation data, such a complex would logically be sufficient for ERαABs recruitment by the "pseudo" native ERα when it is stabilized in a peculiar oligomeric quaternary structure. Hence, one may understand that the known dissociation of such a structure at the time of ERα activation under the action of an appropriate modulator affects the topology of the ERα-binding core, giving rise therefore to a loss of its recruitment potency for ERαABs, a property that manifestly does not hold for E_2 and most probably other conventional estrogenic ligands [20–23].

This suspected binding selectivity, as well as the large size of ERαABs may explain their association in living cells with the plasma membrane-bound receptor form (mERα) [2], principally localized within caveolae [10,11]. This peculiar localization, which results, at least in part, from the palmitoylation of the native (newly-synthetized) receptor [13], appears especially appropriate for this association contributing to rapid, non-genomic, responses (in the present context, ERK phosphorylation; Section 3.1.1). It does not indeed imply any navigation of ERαABs across the plasma membrane to reach oligomeric complexes in which they would moreover not easily internalize to interact with the native and non-markedly altered receptor forms.

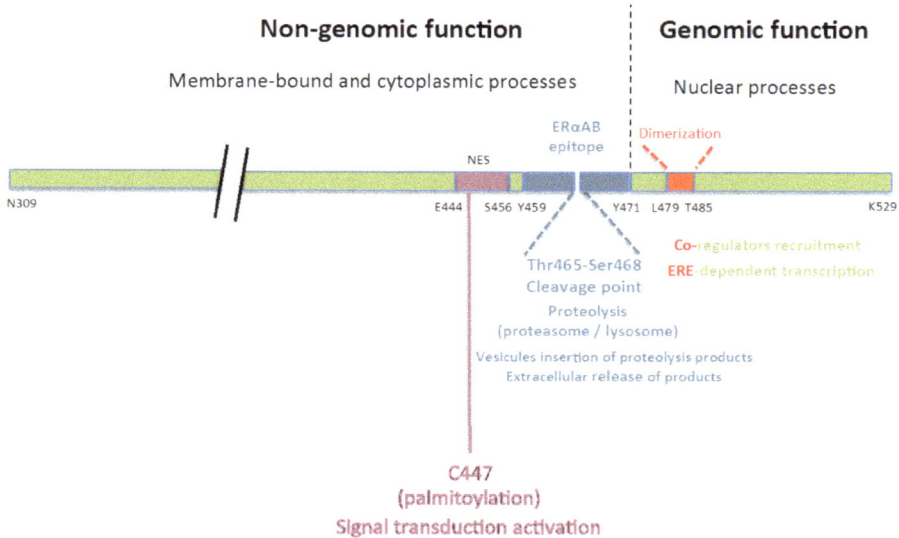

Figure 1. Schematic representation of the regulatory platform of the E_2-binding core of ERα (N309-K529), postulated to mainly contribute to the onset of non-genomic and genomic responses induced by E_2 and ERαABs. The ERαABs' epitope (Y459-E471) occupies a central, pivotal position localized between two motifs, each of them being implicated in one of these two types of responses (non-genomic, E444-S456; genomic, L479-T485). Functions of these three amino-acids sequences, as well as biological consequences resulting from E_2/ERαABs binding and consecutive activation of related inter-relationships between motifs of the platform are defined below (for details, see Section 3.3).

3.3. Potent Regulatory Functions of the ERαAB-Binding Epitope

The sensitivity to proteolytic attacks of the Thr464-Ser468 amino acid sequence of the ERαAB-binding epitope of ERα suggests its inclusion within a surface-exposed region, a property usually recorded with "regulatory platforms" subjected to recruitment and exchange of co-regulators [24]. The identification within this epitope of two functional motifs localized respectively on the left and right sides of the ERαAB-binding epitope supports such a view.

The left-side sequence (E^{444}FVCLKSIILLNS^{456}; Figure 1) corresponds indeed to an identified nuclear exclusion signal that contributes to the return of the activated ERα within the cytoplasm, where it is subjected to proteasomal degradation [24–27]. This step is key for the pursuit of previously initiated transcriptional processes. Hence, this motif would play a role in ERα intracellular trafficking as well as in its resulting turnover rate and related biological activity [11,13,24]. The presence within this motif of Cys-447, the palmitoylation of which favors the anchorage of the receptor with the plasma membrane, validates this proposal [28]. In contrast, the right-side motif (L^{479}DKTITDT^{485}) seems mainly to contribute to (Estrogen response element) ERE-dependent transcription since it corresponds to one of the three amino-acids sequences of the ERα homo-dimerization interface required for such transcription [29].

Hence, the pivotal position of the Thr465-Ser468 sequence within the E_2-binding core of ERα (which contains the ERαABs binding epitope) confers to this sequence a primordial role in the onset of quasi-immediate non-genomic responses, as well as subsequent genomic responses. Such a dual capacity of action is reminiscent of a model proposed to explain how a ligand of the so-called nuclear receptor family may activate rapid signal transduction pathways issued from the cellular membrane, as well as genes' expression, either individually or sequentially [30,31].

According to this model, all ligands' binding sites of the nuclear receptor family are composed of two adjacent cavities in which potent agonists and antagonists may penetrate [30,31]; for ERα and β, see [32]. One of these cavities corresponds to a channel conducting to the other cavity in which selected ligands may be engulfed; the capacity of the ligands to open a protective barrier localized at the entrance of this second cavity might regulate this selection. Molecular interactions between receptors, chaperones and co-regulators are also implicated in this access-regulatory process. The entrance channel, in which access is less restrictive, is directly implicated in quasi-immediate activation of signal transduction pathways, while the cavity in which ligands are engulfed corresponds to the pocket contributing to receptor-mediated transcription, the topology of which has been established by X-ray diffraction crystallography. Cellular localization of the receptor is logically a complementary factor involved in this dual regulation.

Logically, the rapid ERαAB-induced ERK phosphorylation implicated in the enhancement of MCF-7 cells' proliferation (Section 3.1.1) may derive from a relatively low specific interaction of these antibodies with the entrance cavity, the structure of which may be related to the left-side motif implicated in non-genomic responses. Such a hypothesis may also hold to some extent for subsequent indirect induction of ERE-dependent transcription, since this left-side motif seems to play a role in the intracellular trafficking of the receptor, which regulates such transcription. ERαAB-mediated enhancement of the progesterone receptor level may obviously not result from an engulfment of these antibodies within the putative adjacent cavity implicated in gene expression. Access to this adjacent cavity being under the control of a barrier, one may propose that interactions between ERαABs and specific residues of the entrance cavity in which they may penetrate would suppress the repressive function of the barrier, favoring thereby ERα-mediated transcriptions. Receptor-related binding motifs of the plasma membrane may contribute to this property.

4. ERα-Related Sites Potentially Able to Contribute to the Mechanism of Action of ERαABs

Several sites identified on the plasma membrane may legitimately be proposed as potential alternative targets for ERαAB-induced responses, some of them acting cooperatively with mERs [33]. Some of these sites are devoid of any E_2-binding ability (i.e., HER2, EGFR), while others attract the hormone as demonstrated with synthetized E_2-conjugates unable to penetrate the cells [34]. Among such E_2-binding targets, two splice receptor variants (ERα36 and ERα46; see [35,36] and the references therein), as well as a G protein-coupled receptor (GPR30 [37–39]) have been especially well studied. The capacity of GPR30 to interact with calmodulin, as well as with the calmodulin-binding site of ERα, implicating its dimerization for the enhancement of ERE-dependent transcription [40–43], would confer to this peculiar receptor a potent role in ERαAB-induced genomic functions. In fact, the capacity of GPR30 to move between the plasma membrane, the endoplasmic reticulum and the nucleus advocates in favor of its contribution to other ERα-mediated processes under the control of the antibodies [37,44]: GPR30 appears indeed to be an actor involved in the intracellular trafficking of the receptor governing its various biological functions.

Of course, the implication of such receptor-related sites in the onset of ERαAB-induced responses needs to be validated or rejected. Measurement of markers (Ca^{2+} fluxes or secondary messengers such as c-AMP or IP3) may be helpful in this regard, especially for the evaluation of complementary ERα-independent processes [33]. In this context, specific antagonists with a special emphasis on compounds abrogating the action of HER2, EGFR or GPR30 need also to be tested. This approach being at the present time quite marginal [2,5], one may consider that any use of such antagonists in the clinical perspective is out of scope, even if humanized versions of monoclonal antibodies raised against HER2 (trastuzumab, pertuzumab) seem appropriate for a first-line experimental assessment [45,46]. Induction by such drugs of a decrease of efficiency of signal transductions initiated by the putative action of ERαABs at the level of HER2 might alter growth of breast cancer cells, which in connection with the known antibody-dependent cellular cytotoxicity (ADCC) of these compounds related to their ability to recruit and activate natural killer cells (NK) would generate a major curative effect, even in the absence of ERα. Note in this context that pertuzumab abrogates the hetero-dimerization of HER2 with other members of the HER family, while trastuzumab mainly affects its homo-dimerization. Since such dimerizations are implicated in the activation of signal transductions enhancing cell growth, one may consider that pertuzumab might be more efficient for blocking a putative ERαABs association with membrane ERα-related receptors, promoting proliferation.

Finally, it should be stressed that the estrogen activity of ERαABs should not necessarily be derived from a direct interaction with the plasma membrane-bound E_2-binding site. This assumption results from experiments conducted with anti-E_2~BSA antibodies and highly purified ERαBAs (IgGs), which displayed an estrogenic activity [5]. Anti-E_2~BSA antibodies sharing most likely some structural similarities with the hormone binding site of ERα, the authors of this observation concluded that these two classes of antibodies may act as "soluble ERα forms" present in the blood to liberate, by a competitive process, the receptor from the repressive action exerted by a peptide inhibitor looking structurally like E_2~BSA. Hence, the estrogenic activity of a subpopulation of ERαABs may result, at least in part, from the ability to abrogate the effect of ERα co-repressors. If confirmed, this concept would logically also hold for other possible ERαABs targets, as described below.

All hypotheses evoked in Sections 3 and 4 to explain the mechanisms by which ERαABs may generate estrogenic responses are schematically summarized in Figure 2.

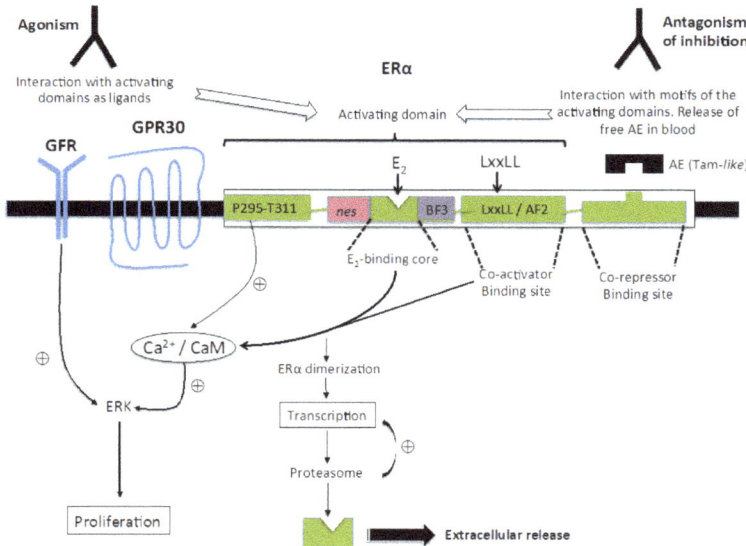

Figure 2. Schematic representation of ERαABs-induced mechanisms initiated at the plasma membrane to promote enhanced proliferation and ERE-dependent transcription. Reported ERαAB activities (agonism, antagonism of inhibition) were integrated in a classical model explaining co-operation between binding sites for growth factors and steroid hormones in the onset of non-genomic and genomic responses [11,12,14,27,33,37,38]. Note the pivotal role of the Ca^{2+}/calmodulin complex in the inter-relationships between GPR30 and recruitment sites of the receptor for ERαABs and adjacent co-activators. AE: antiestrogen; Tam-like: Tamoxifen-like.

5. Mechanisms Implicated in the Emergence of ERαABs

The present section will solely refer to the emergence of ERαABs for which ERα binding properties have been overviewed. For any topics concerning E_2-related changes in immune functions or auto-immunity, I invite the reader to consult [1,6], which are extensive in this regard.

For me, insufficient experimental data have been reported to propose mechanisms giving rise to the production of ERαABs. A priori, two auto-immune reactions involving eventually a contribution of the idiotypic network may theoretically be advocated, as suggested in the previous sections: one giving rise to antibodies acting as endogenous estrogenic secretions or any expositions to environmental estrogens, the other to antibodies abrogating the repressive action of a natural antagonist. In this context, one may wonder about the participation of recently identified circulating ER (α and β) forms in human sera in the emergence of ERABs (see Section 5.3). The next sections will analyze the relevance of these possibilities, schematically presented in Figure 3.

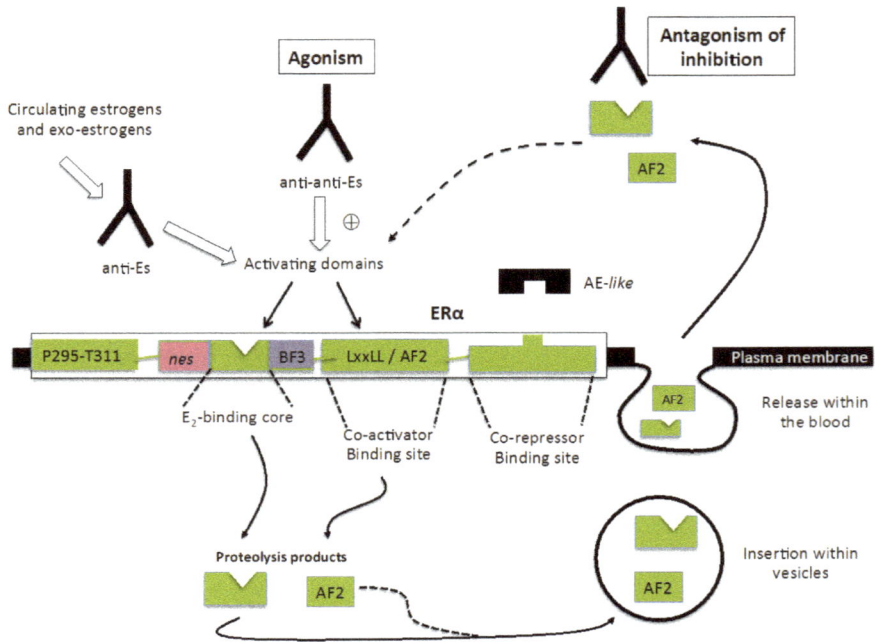

Figure 3. Schematic representation of suspected mechanisms able to contribute to ERαABs' emergence. Agonists: antibodies able to mimic the action of circulating estrogens (natural, synthetic and xenoestrogens). Anti-antagonists: antibodies against a natural extra-cellular repressor recognizing a specific inhibitory site of ERα or preventing the access of activating modulators to the receptor by a competitive binding process. ERα degradation products including binding sites for estrogens or LXXLL motifs of co-activators [24], released within the blood, may generate this last class of antibodies. Es: Estrogens, BF3: Binding function 3, AF2: Activation function 2.

5.1. Anti-Idiotypic Antibodies Acting as Physiological Estrogens

Similarities between E_2 and ERαABs, in terms of interactions with the native ERα form, reflect most probably structural identities between the hormone and the active site of natural anti-idiotypic antibodies raised against anti-E_2 IgGs. Since circulating E_2 is mainly conjugated to serum proteins, one may postulate that such a cross-reaction may also hold for such conjugates, especially anti-E_2~BSA IgGs, the level of which would largely dominate. Since such a concept is not restrictive to this hormone, it should be extended to all other physiological estrogens, as well as so-called "xenoestrogens" (natural phytoestrogens and synthetic "endocrine disrupting chemicals" [22,23]) able to interact with the ligand-binding site of ERα to induce estrogenic (or estrogenic-like) responses. The implication of these molecules in autoimmune response has been, indeed, evoked [47].

The production in the early 1990s of a monoclonal antibody (clone 1D5) directed against the binding site of an anti-E_2 monoclonal antibody lends credence to this concept: 1D5 was found to interact with the hormone-binding domain of the receptor to mimic some estrogenic actions (creatine kinase induction, rapid Ca^{2+} flux enhancement), both in vivo and in vitro [48,49]. These experimental data were proposed to be mainly dependent on an interaction with a membrane-bound form of ERα. Nevertheless, they were also postulated to result to some extent from an intra-cellular penetration of 1D5. A lack of knowledge concerning the relationships between the plasma membrane and intra-cellular ERα forms at the time of such studies may explain this statement, which may appear now quite obsolete, even if this could not be excluded.

5.2. Antibodies with Anti-Repressive Activities

In theory, all receptor-mediated agonistic activities may result from an ability to abrogate the antagonism of a modulator acting at the level of the ligand-binding site or an adjacent site implicated in the recruitment of co-repressors. Such a view has been proposed to explain, at least in part, the estrogenic activity of ERαABs (Section 4, last paragraph). While a potency to liberate the E_2-binding site of the plasma-bound receptor from the antagonism of a specific inhibitor has been solely addressed [5], it seems that other sites of the hormone-binding domain involved in the recruitment of co-activators (LXXLL/AF2, BF3, etc. [24,50]) must also be taken into account. However, the presence of such sites in the whole family of steroid hormone receptors would largely limit the specificity of action of antibodies raised against them, giving rise to inappropriate adverse effects. Hence, their importance seems quite dubious.

In this context, a possible interaction of ERαABs with an identified ERα-binding site implicated in the recruitment of tamoxifen and other mixed antagonists/agonists [51,52] may also be advocated. Experiments revealed that that treatment of cytosolic ERα preparations with tamoxifen enhances the immuno-reactivity of this site for a monoclonal antibody (H222) raised against an epitope of the receptor ligand-binding domain [53], revealing that this compound may expose an occult antigenic determinant accessible to a subpopulation of ERαABs. Whatever could be the finality of such an interaction with a site contributing to the activity of tamoxifen, either agonist or antagonist, one may consider that it may modulate the SERM character of this compound.

5.3. Implication of ERα or ERα Fragments Released within the Blood in the Onset of ERαABs

Could ERα and β recently detected within human sera [54] be implicated in the emergence of ERαABs against these two receptors? This important question has some justification in the finding that the latter display anti-inflammatory properties, the net action of which depends on their relative proportions (β > α) and localization; ERα is moreover associated with auto-immune processes [55].

These circulating ERα and β forms (detected in patients with Crohn's disease with a commercial ELISA) most probably correspond to various receptor fragments issued from their intracellular proteasomal and lysosomal degradation, released within the blood as small vesicles (exosomes) implicated in immune responses or processed for MHC (major histocompatibility complex) presentation after autophagy [56,57]. Hence, one may logically assume that ERα fragments may be implicated in the emergence of ERαABs with a repressive activity, some of them abrogating the effect of natural inhibitors present in the blood, others abrogating the potent competitive inhibitory potency of ERα degradation products able to recruit circulating activators (mainly E_2, co-activators), liberating thereby these agents for the accomplishment of their function.

The detection in media from E_2-stimulated cells of a 44-amino-acid peptide including a repressive motif of ERα (Pro295-Thr311) [58], able to interact with the Pro365-Asp369 type II β turn element of its BF3 motif that regulates the dimerization of the receptor ([59], and see Section 6), may appear as a stone in the edification of this concept. A synthetic peptide corresponding to the Pro295-Thr311 motif (ERα17p) induces indeed estrogenic responses, as well as some receptor-independent actions in various breast cancer cell lines [60], the lack of specificity of these actions resulting most probably from distinct interactions with the type II β turn/BF3 motifs of the various steroid hormone receptors expressed in these cells. Antibodies raised against the P295-T311 sequence (Gentaur: 04-rb-ERα17p) would logically generate a similar absence of specificity of action in contrast to antibodies raised against the E2-binding core of the receptor (Section 5.2, first paragraph). Such a lack of specificity would not be necessarily detrimental for therapeutic purposes, especially in the case of antiestrogen resistance, as proposed for antagonists aimed to antagonize the recruitment of co-activators [24,61].

6. Conclusions and Perspectives

Structural studies of the estrogenic core of ERα, reported here, reveal that the ERαABs epitope localizes at a place of prominent importance for the successive onset of non-genomic and genomic responses. The finding that this epitope is adjacent to regulatory motifs governing these responses argues in favor of such a statement. Complementary inclusion of these data into a model established from X-ray crystallographic investigations relevant to the activated intracellular ERα form indicated that the Leu479-Thr485 motif, which contributes to the dimerization of the receptor, corresponds to a part of its BF3 motif implicated in ERE-dependent transcription [61] (Figure 4; analysis performed by my colleague Yves Jacquot, Sorbonne Universités, Université Pierre et Marie Curie, Ecole Normale Supérieure, Paris, France). This information strongly suggests that ligand-induced conformational changes relevant to the intracellular receptor may also hold for its plasma membrane-bound form, justifying the dimerization ability of the latter. Hence, the biochemical assessment of the interactions between ligands aimed at targeting the "insoluble" ERα entrapped within the plasma membrane and the conventional "soluble" cytoplasmic and nuclear receptor forms would be a valuable approach to the decryption of the mechanism by which ERα operates. Hence, interest in ERαABs would not be restricted to physio/pathological purposes.

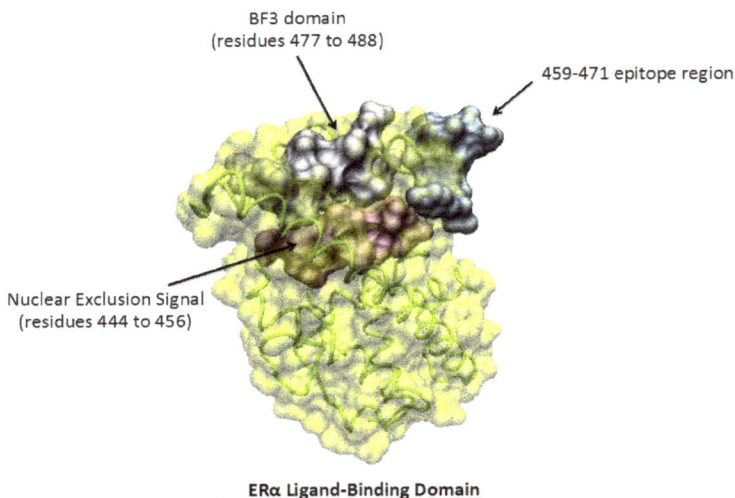

Figure 4. Surface structure of the BF3- (in grey) and ERαAB epitope/E2 (in blue)-binding domains of the human estrogen receptor α (ERα in yellow, Connolly surface). The BF3 domain is composed of two regions, i.e., the 365–369 type II β-turn region and the 477–488 helix 10 (H10) region, the latter overlapping the 479–485 sequence implicated in the dimerization of the receptor. The ERαABs epitope is in close contact with this BF3 domain, as well as the 444–456 nuclear exclusion site (nes, for nuclear exclusion signal, in pink). Interaction between the 301–311 region with the 365–269 type II β turn seems to repress the dimerization potency of the 477–488 helix. Transparency allows the visualization of the helices (in green) that comprise the receptor.

In this regard, experimental data critically reviewed here leave no doubt about the importance of ERαABs in breast cancer emergence and/or evolution, even if these biological aspects have only been marginally addressed at the present time [2,5]. The capacity of ERαABs to stimulate MCF-7 cell growth suggests some potential implication in the resistance to endocrine treatments. Such a topic needs to be rapidly assessed with tamoxifen-resistant cell lines. On the other hand, since the mammary gland is under the control of both ERα and β, which respectively promote or repress its neoplasia [62], the search for natural antibodies raised against ERβ seems of major interest. Such a task may open new pathways in the current tendency to combine immunological and endocrine approaches in the management of cancer. The present review being mainly devoted to fundamental aspects of ERαABs, I encourage immunologists and endocrinologists to extend my work to reported clinical observations, especially those that, by ignorance, I failed to refer. Such an issue will be extremely helpful to confirm or reject a tendency to see a strong autoimmune ER function in breast cancer, which, in the affirmative, would be taken into account in the design of future therapeutic programs.

Acknowledgments: I warmly thank Yves Jacquot (Sorbonne Universités, Université Pierre et Marie Curie, Ecole Normale Supérieure, Paris, France), my friend, for his help in the design of the figures. Our long-term cooperation in the investigation of the processes by which ERα operates must be stressed. In the present context, this obviously contributes to the concepts that I propose.

Conflicts of Interest: The author declares no conflict of interest. As Honorary (retired) Professor, I failed to have access to any kind of financial support.

Abbreviations

Akt	Protein kinase B
AF2	Activation function 2
BF3	Binding function 3
ERK	Extracellular regulated kinase
EGFR	Epidermal growth factor receptor
ER	Estrogen receptor
ERAB	Estrogen receptor antibody
ERE	Estrogen response element
HER 2	Human epidermal growth factor 2
IP3	Inositol triphosphate

References

1. Ortona, E.; Pierdominici, M.; Berstein, L. Autoantibodies to estrogen receptors and their involvement in autoimmune diseases and cancer. *J. Steroids Biochem. Biophys. Mol. Biol.* **2014**, *144*, 260–267. [CrossRef] [PubMed]
2. Maselli, A.; Capoccia, S.; Pugliese, P.; Raggi, C.; Cirulli, F.; Fabi, A.; Maloni, W.; Pierdominici, M.; Ortona, E. Autoantibodies specific to estrogen receptor alpha act as estrogen agonists and their levels correlate with breast cancer cell proliferation. *Oncoimmunology* **2016**, *5*, e1074375. [CrossRef] [PubMed]
3. Borkowski, A.; Gyling, M.; Muquardt, C.; Body, J.J.; Leclercq, G. A subpopulation of immunoglobin G in man selectively interacts with the hormone–binding sites of estrogen receptors. *J. Clin. Endocrinol. Metab.* **1987**, *64*, 356–363. [CrossRef] [PubMed]
4. Borkowski, A.; Gyling, M.; Muquardt, C.; Body, J.J.; Leclercq, G. Estrogen-like activity of a subpopulation of natural antiestrogen receptor autoantibodies in man. *Endocrinology* **1991**, *128*, 3283–3292. [CrossRef] [PubMed]
5. Tassignon, J.; Haeseleer, F.; Borkowski, A. Natural antiestrogen receptor autoantibodies in man with estrogenic activity in mammary carcinoma cell culture: Study of their mechanism of action; evidence for involvement of estrogen-like epitopes. *J. Clin. Endocrinol. Metab.* **1997**, *82*, 3464–3470. [CrossRef] [PubMed]
6. Ortona, E.; Pierdominici, M.; Maselli, A.; Veronesi, C.; Aloisi, F.; Shoenfeld, Y. Sex-based differences in autoimmune diseases. *Annali Dell'Istituto Superiore Di Sanita* **2016**, *52*, 205–212. [CrossRef] [PubMed]

7. Weigel, N.L.; Pousette, A.; Schrader, W.T.; O'Malley, B.W. Analysis of chicken progesterone receptor structure using a spontaneous sheep antibody. *Biochemistry* **1981**, *20*, 6798–6802. [CrossRef] [PubMed]
8. Liao, S.; Witte, D. Auto immune anti-androgen-receptor antibodies in human serum. *Proc. Natl. Acad. Sci. USA* **1985**, *82*, 8345–8348. [CrossRef] [PubMed]
9. Muddaris, A.; Peck, E.J., Jr. Human anti-estrogen receptor antibodies: Assay, characterization and age- and sex-related differences. *J. Clin. Endocrinol. Metab.* **1987**, *64*, 246–254. [CrossRef] [PubMed]
10. Hammes, S.R.; Levin, E.R. Extranuclear steroid receptors: Nature and actions. *Endocr. Rev.* **2007**, *28*, 726–741. [CrossRef] [PubMed]
11. Schwartz, N.; Verma, A.; Bivens, C.B.; Schwartz, Z.; Boyan, B.D. Rapid steroid hormone actions via membrane receptors. *Biochem. Biophys. Acta* **2016**, *1863*, 2289–2298. [CrossRef] [PubMed]
12. Leclercq, G.; Lacroix, M.; Laios, J.; Laurent, G. Estrogen receptor alpha: Impact of ligands on intracellular shuttling and turnover rate in breast cancer cells. *Curr. Cancer Drug Targets* **2006**, *6*, 39–64. [CrossRef] [PubMed]
13. La Rosa, P.; Perisi, V.; Leclercq, G.; Marino, M.; Acconcia, F. Palmitoylation regulates 17β-estradiol-induced estrogen receptor-α degradation and transcriptional activity. *Mol. Endocrinol.* **2012**, *26*, 762–764. [CrossRef] [PubMed]
14. Tecalco-Cruz, A.C.; Pérez-Alvarado, I.A.; Ramirez-Jarquin, J.O. Nucleo-cytoplasmic transport of estrogen receptor alpha in breast cancer cells. *Cell. Signal.* **2017**, *34*, 121–132. [CrossRef] [PubMed]
15. Laïos, I.; Journe, F.; Laurent, G.; Nonclercq, D.; Toillon, R.A.; Seo, H.S.; Leclercq, G. Mechanisms governing the accumulation of estrogen receptor alpha in MCF7 cells treated with hydroxytamoxifen and reated antiestrogens. *J. Steroid Biochem. Mol. Biol.* **2003**, *87*, 207–211. [CrossRef] [PubMed]
16. Jensen, E.V.; Block, G.E.; Smith, S.; Kyser, K.; De Sombre, E.R. Estrogen receptors and breast cancer response to adrenalectomy. Prediction of response in cancer therapy. *NCI Monogr.* **1971**, *34*, 55–70.
17. Leclercq, G.; Heuson, J.C. Specific estrogen receptor of the DMBA-induced mammary tumor of the rat and its requiring molecular transformation. *Eur. J. Cancer* **1973**, *9*, 675–680. [CrossRef]
18. Seielstad, D.A.; Carson, K.; Kushner, P.J.; Greene, G.L.; Katzenellengogen, J.A. Analysis of the structural core of the human estrogen ligand binding domain by selective proteolysis/mass spectrometry analysis. *Biochemistry* **1995**, *34*, 12605–12615. [CrossRef] [PubMed]
19. Thole, H.H.; Jungblut, P.W. The ligand-binding site of the estradiol receptor resides in a non-covalent complex of two consecutive peptides of 17 and 7 kDa. *Biochem. Biophys. Res. Commun.* **1994**, *199*, 826–833. [CrossRef] [PubMed]
20. Anstead, G.M.; Carlson, K.E.; Katzenellenbogen, J.A. The estradiol pharmacophore: Ligand-estrogen-receptor binding affinity relationship and a model for the receptor binding site. *Steroids* **1997**, *62*, 268–303. [CrossRef]
21. Fink, B.E.; Mortensen, D.S.; Staufer, S.R.; Aron, Z.; Katzenellenbogen, J.A. Novel structural templates for estrogen-receptor ligands and prospects for combinatorial synthesis of estrogens. *Chem. Biol.* **1999**, *6*, 205–219. [CrossRef]
22. Jacquot, Y.; Rojas, C.; Refouvelet, B.; Robert, J.F.; Leclercq, G.; Xicluna, A. Recent advances in the development of phytoestrogens and derivatives: An update of the promising perspectives in the prevention of postmenopausal diseases. *Mini-Rev. Med. Chem.* **2003**, *2*, 387–400. [CrossRef]
23. Lóránd, T.; Vigh, E.; Garai, J. Hormonal action of plant derived and antropogenic non-steroidal estrogenic compounds; phytoestrogens and xenoestrogens. *Curr. Med. Chem.* **2010**, *17*, 3542–3574. [CrossRef] [PubMed]
24. Leclercq, G.; Gallo, D.; Cossy, J.; Laïos, I.; Larsimont, D.; Laurent, G.; Jacquot, Y. Peptides targeting estrogen receptor alpha-potential applications for breast cancer treatment. *Curr. Pharm. Des.* **2011**, *17*, 2632–2653. [CrossRef] [PubMed]
25. Lombardi, M.; Castoria, G.; Migliaccio, A.; Barone, M.V.; Di Stasio, R.; Ciociola, A.; Bottero, D.; Yagamuchi, H.; Appela, E.; Auricchio, F. Hormone-dependent nuclear export of estradiol receptor and DNA synthesis in breast cancer cells. *J. Cell Biol.* **2008**, *182*, 327–340. [CrossRef] [PubMed]
26. Nonclercq, D.; Journe, F.; Laïos, I.; Chaboteaux, C.; Toillon, R.A.; Leclercq, G.; Laurent, G. Effect of nuclear export inhibition on estrogen receptor regulation in breast cancer cells. *Mol. Endocrinol.* **2007**, *39*, 105–118. [CrossRef] [PubMed]
27. Tecalco-Cruz, A.C. Molecular pathways involved in the transport of nuclear receptors from the nucleus to the cytoplasm. *J. Steroid Biochem. Mol. Biol.* **2018**, in press. [CrossRef] [PubMed]

28. Acconcia, F.; Ascenzi, P.; Fabozzi, G.; Visca, P.; Marino, M. S-palmitoylation modulates human estrogen receptor–alpha functions. *Biochem. Biophys. Res. Commun.* **2004**, *316*, 878–883. [CrossRef] [PubMed]
29. Chalkaraborty, S.; Biswas, B.K.; Asare, B.K.; Rajnarayanan, R.V. Designer interface peptide grafts target estrogen receptor alpha dimerization. *Biochem. Biophys. Res. Commun.* **2016**, *47*, 116–122. [CrossRef] [PubMed]
30. Mizwicki, M.T.; Keidel, D.; Bula, C.M.; Bishop, C.M.; Zanello, L.; Wurtz, J.M.; Moras, D.; Norman, A.W. Identification of an alternative ligand-binding pocket in the nuclear vitamine D receptor and its functional importance in 1α,25(OH)$_2$-vitamin D$_3$ signaling. *Proc. Natl. Acad. Sci. USA* **2004**, *101*, 12876–12881. [CrossRef] [PubMed]
31. Norman, A.W.; Mizwicki, M.T.; Norman, D.P. Steroid-hormone rapid actions. Membrane receptors and conformational ensemble model. *Nat. Rev. Drug Discov.* **2004**, *3*, 27–41. [CrossRef] [PubMed]
32. Van Hoorn, W.P. Identification of a second binding site in estrogen receptor. *J. Med. Chem.* **2002**, *45*, 584–589. [CrossRef] [PubMed]
33. Bennesch, M.A.; Picard, D. Minireview. Tipping the balance: Ligand-independent activation of steroids receptors. *Mol. Endocrinol.* **2015**, *29*, 349–363. [CrossRef] [PubMed]
34. Shearer, K.E.; Rickter, E.L.; Peterson, A.C.; Weatherman, R.V. Dissecting rapid estrogen signaling with conjugates. *Steroids* **2012**, *77*, 968–973. [CrossRef] [PubMed]
35. Zhang, X.; Wang, Z.Y. Estrogen receptor-α variant, ER-α36, is involved in tamoxifen resistance and estrogen hypersensitivity. *Endocrinology* **2013**, *154*, 1990–1998. [CrossRef] [PubMed]
36. Lin, A.H.; Li, R.W.; Ho, E.Y.; Leung, S.W.; VanHoute, P.M.; Man, R.Y. Differential ligand binding affinities of human estrogen receptor-α isoforms. *PLoS ONE* **2013**, *8*, e63199. [CrossRef] [PubMed]
37. Prossnitz, E.R.; Maggliolini, M. Mechanism of estrogen signaling and gene expression via GPR30. *Mol. Cell. Endocrinol.* **2009**, *308*, 32–38. [CrossRef] [PubMed]
38. Levin, E.R. G protein-coupled receptor 30: Estrogen receptor or collaborator? *Endocrinology* **2009**, *150*, 1563–1565. [CrossRef] [PubMed]
39. Notas, G.; Kampa, M.; Pelekanou, V.; Castanas, E. Interplay of estrogen receptors and GPR30 for the regulation of early initiated transcriptional effects: Pharmacological approach. *Steroids* **2012**, *77*, 943–950. [CrossRef] [PubMed]
40. Tran, Q.K.; Vermeer, M. Biosensor-based approach identifies four distinct calmodulin–binding domains in the G-coupled estrogen receptor 1. *PLoS ONE* **2014**, *9*, e89669. [CrossRef] [PubMed]
41. Leiber, D.; Burlina, F.; Byrne, C.; Robin, P.; Piesse, C.; Gonzalez, L.; Leclercq, G.; Tanfin, Z.; Jacquot, Y. The sequence Pro[295]-Ther[311] of the hinge of oestrogen receptor α is involved in ERK1/2 activation via GPR30 in leiomyoma cells. *Biochem. J.* **2015**, *472*, 97–109. [CrossRef] [PubMed]
42. Gallo, D.; Jacquot, Y.; Laurent, G.; Leclercq, G. Calmodulin, a regulatory partner of the estrogen receptor alpha in breast cancer cells. *Mol. Cell. Endocrinol.* **2008**, *268*, 20–26. [CrossRef] [PubMed]
43. Li, Z.; Zhang, Y.; Hedman, A.C.; Ames, J.B.; Sacks, D.B. Calmodulin lobes facilitate dimerization and activation of estrogen receptor-α. *J. Biol. Chem.* **2017**, *292*, 4614–4622. [CrossRef] [PubMed]
44. Cheng, S.B.; Graeber, C.T.; Quin, J.A.; Filardo, E.J. Retrograde transport of the transmembrane estrogen receptor, G-protein-coupled-receptor-30 (GPR30/GPER) from the plasma membrane towards thenucleus. *Steroids* **2011**, *76*, 892–896. [CrossRef] [PubMed]
45. Vu, T.; Claret, F.X. Trastuzumab: Updated mechanisms of action and resistance in breast cancer. *Front. Oncol.* **2012**, *2*, 62. [CrossRef] [PubMed]
46. Harbeck, N.; Beckmann, M.W.; Rody, A.; Schneeweiss, A.; Müller, V.; Fehm, T.; Marschner, N.; Gluz, O.; Schader, I.; Heinrich, G.; et al. HER2 dimerization inhibitor pertuzumab—Mode of action and clinical data in brest cancer. *Brest Cancer* **2013**, *8*, 49–55. [CrossRef]
47. Chighizola, C.; Meroni, P.L. The role of environmental estrogens and autoimmunity. *Autoimmun. Rev.* **2012**, *11*, A493–A501. [CrossRef] [PubMed]
48. Mor, G.; Amir-Zaltsman, Y.; Barnard, G.; Kohen, F. Characterization of an antiidiotypic antibody mimicking the actions of estradiol and its interaction with estrogen receptors. *Endocrinology* **1992**, *130*, 3633–3640. [CrossRef] [PubMed]
49. Sömjen, D.; Kohen, F.; Lieberherr, M. Nongenomic effects of an anti-idiotypic antibody as an estrogen mimetic in female human and rat osteoblasts. *J. Cell. Biochem.* **2007**, *65*, 53–66. [CrossRef]

50. Buzón, V.; Carbo, L.R.; Estruch, S.B.; Fletterick, R.J.; Estébanez-Perpina, E. A conserved surface on the ligand binding domain of nuclear receptors for allosteric control. *Mol. Cell. Endocrinol.* **2012**, *348*, 394–402. [CrossRef] [PubMed]

51. Berthois, Y.; Pons, M.; Dussert, C.; Crastes de Paulet, A.; Martin, P.M. Agonist-antagonist activity of anti-estrogens in the human breast cancer cell line MCF-7: An hypothesis for the interaction with a site distinct from the estrogen binding site. *Mol. Cell. Endocrinol.* **1994**, *99*, 259–268. [CrossRef]

52. Kojetin, D.J.; Burris, T.P.; Jensen, E.V.; Khan, S.A. Implications of the binding of tamoxifen to the coactivator recognition site of the estrogen receptor. *Endocr. Relat. Cancer* **2008**, *15*, 851–870. [CrossRef] [PubMed]

53. Martin, P.M.; Berthois, Y.; Jensen, E.V. Binding of antiestrogens exposes an occult antigenic determinant in the human estrogen receptor. *Proc. Natl. Acad. Sci. USA* **1988**, *85*, 2533–2537. [CrossRef] [PubMed]

54. Linares, P.M.; Algaba, A.; Urzainqui, A.; Guijarro-Rojas, M.; Gonzalez-Tajuelo, R.; Garrido, J.; Chaparro, M.; Gisbert, J.P.; Bermejo, F.; Guerra, I.; et al. Ratio of circulating estrogen receptors beta and alpha (ERβ/ERα) indicates endoscopic activity in patients with Crohn's disease. *Dig. Dis. Sci.* **2017**, *62*, 2744–2754. [CrossRef] [PubMed]

55. Hill, L.; Jeganathan, V.; Chinnasamy, P.; Grimaldi, C.; Diamond, B. Differential roles of estrogen receptors α and β in control of B-cell maturation and selection. *Mol. Med.* **2011**, *17*, 211–220. [CrossRef] [PubMed]

56. Beninson, L.A.; Fleshner, M. Exosomes: An emerging factor in stress-induced immunomodulation. *Semin. Immunol.* **2014**, *26*, 394–401. [CrossRef] [PubMed]

57. Schmid, D.; Münz, C. Innate and adaptive immunity through autophagy. *Immunity* **2007**, *27*, 11–21. [CrossRef] [PubMed]

58. Gallo, D.; Leclercq, G.; Haddad, J.; Vinh, J.; Castanas, E.; Kampa, M.; Pelekanou, V.; Jacquot, Y. Estrogen Receptor Alpha Polypeptide Sequence, Diagnostic and Therapeutic Applications Thereof. U.S. Patent WO 20120449229 A1, 19 April 2012.

59. Jacquot, Y.; Gallo, D.; Leclercq, Y. Estrogen receptor alpha—Identification by a modelling approach of a potential polyproline II recognizing domain within the AF-2 region of the receptor that would play a role of prime importance in its mechanism of action. *J. Steroid. Biochem. Mol. Biol.* **2007**, *104*, 1–10. [CrossRef] [PubMed]

60. Notas, G.; Kampa, M.; Pelekanou, V.; Troullinki, M.; Jacquot, Y.; Leclercq, G.; Castanas, E. Whole transcriptome analysis of the ERα synthetic fragment P295-T311 (ERα17p) identifies specific ERα isoform (ERα, ERα36)-dependent and-independent actions in breast cancer cells. *Mol. Oncol.* **2013**, *7*, 595–610. [CrossRef] [PubMed]

61. Gallo, D.; Leclercq, G.; Jacquot, Y. The N-terminal part of the ligand-binding domain of the human estrogen receptor α: A new target for estrogen disruptors. In *Medicinal Chemistry Research Progress*; Colombo, G.P., Ricci, S., Eds.; Nova: Hauppauge, NY, USA, 2009; pp. 207–224.

62. Treeck, O.; Lattrich, C.; Springwald, A.; Ortmann, O. Estrogen receptor beta exerts growth–inhibitory effects on human mammary epithelial cells. *Breast Cancer Res. Treat.* **2010**, 557–565. [CrossRef] [PubMed]

International Journal of
Molecular Sciences

MDPI

Review

The Role of the Estrogen Pathway in the Tumor Microenvironment

Natalie J Rothenberger [1], Ashwin Somasundaram [1,2] and Laura P. Stabile [3,4,*

[1]   Department of Medicine, Division of Hematology/Oncology, University of Pittsburgh,
       Pittsburgh, PA 15232, USA; njr31@pitt.edu (N.J.R.); somasundarama@upmc.edu (A.S.)
[2]   Department of Immunology, University of Pittsburgh, Pittsburgh, PA 15213, USA
[3]   Department of Pharmacology & Chemical Biology, University of Pittsburgh, Pittsburgh, PA 15213, USA
[4]   UPMC Hillman Cancer Center, Pittsburgh, PA 15213, USA
*   Correspondence: stabilela@upmc.edu; Tel.: +1-412-623-2015

Received: 11 January 2018; Accepted: 16 February 2018; Published: 19 February 2018

Abstract: Estrogen receptors are broadly expressed in many cell types involved in the innate and adaptive immune responses, and differentially regulate the production of cytokines. While both genomic and non-genomic tumor cell promoting mechanisms of estrogen signaling are well characterized in multiple carcinomas including breast, ovarian, and lung, recent investigations have identified a potential immune regulatory role of estrogens in the tumor microenvironment. Tumor immune tolerance is a well-established mediator of oncogenesis, with increasing evidence indicating the importance of the immune response in tumor progression. Immune-based therapies such as antibodies that block checkpoint signals have emerged as exciting therapeutic approaches for cancer treatment, offering durable remissions and prolonged survival. However, only a subset of patients demonstrate clinical response to these agents, prompting efforts to elucidate additional immunosuppressive mechanisms within the tumor microenvironment. Evidence drawn from multiple cancer types, including carcinomas traditionally classified as non-immunogenic, implicate estrogen as a potential mediator of immunosuppression through modulation of protumor responses independent of direct activity on tumor cells. Herein, we review the interplay between estrogen and the tumor microenvironment and the clinical implications of endocrine therapy as a novel treatment strategy within immuno-oncology.

Keywords: estrogen; cancer; tumor microenvironment; immunotherapy; immunosuppression

---

1. Introduction

Estrogens are pleiotropic steroids that play a regulatory role in a myriad of physiological processes from reproduction to lipid metabolism [1]. Biosynthetically converted from precursor androgens by the enzyme aromatase (CYP19A1), estrogens exert both genomic and non-genomic biological effects mediated by interactions with one of two cognate receptors, estrogen receptor α (ERα) or estrogen receptor β (ERβ). Albeit encoded by separate genes, both ER isoforms exhibit similar functional and structural organization [1]. Displaying high sequence homology within the DNA and ligand binding domains, both receptors interact similarly with endogenous estrogens, mainly 17β-estradiol (E2) [2,3]. In addition to mediating biological mechanisms involved in homeostasis, E2 also plays a role in the development and malignant progression of multiple cancers. The oncogenic role of estrogens is well characterized in both classical and nonclassical hormone-sensitive carcinomas including breast, prostate, endometrial, ovarian, colon, and lung [4]. ERs are located in both the nucleus and the cytoplasm of tumor cells enabling tumor-promoting transcriptional regulation of genes involved in cell survival and proliferation [5,6], and non-genomic crosstalk with growth factor pathways, including epidermal growth factor (EGF), insulin growth factor (IGF), and fibroblast growth factor

Int. J. Mol. Sci. **2018**, *19*, 611; doi:10.3390/ijms19020611   79   www.mdpi.com/journal/ijms

(FGF) [7–9]. Due to these tumorigenic mechanisms, therapies that interfere with E2 signaling, such as selective estrogen receptor modulators or degraders (SERMs or SERDs) and aromatase inhibitors (AIs), have been developed and clinically implemented for the treatment of ER-positive breast cancer. While agents that target the estrogen pathway have been seminal in reducing breast cancer mortality over the past three decades [10], most studies in breast cancer and other cancer types have focused strictly on tumoral ER expression and signaling.

Along with tumor cells, non-cancerous cells comprising the tumor microenvironment (TME) are now recognized as critical mediators of tumor progression. Mounting evidence suggests that in addition to intracellular mechanisms such as mutational load and neoantigen presentation, interplay between cancer cells, stromal cells, immune cells, and extracellular molecules within the TME profoundly influence anti-tumor immunity and immunotherapeutic response [11–14]. The notion that enhancing tumor immunogenicity and inhibiting immunosuppressive mediators can functionally suppress progression of malignant tumors has led to the development of promising immunotherapeutic strategies. However, the clinical utility of current immunotherapies remains limited due to marginal response rates and acquired resistance mechanisms [15–17]. Therefore, greater elucidation of targetable cellular machinery involved in tumor immune evasion is necessary to improve the clinical benefit of immunotherapies.

The numerous biological effects of the E2 pathway are facilitated by distinct ER isoform expression found not only on tumor cells, but also on most immune cell types [18–21]. The impact of E2 in autoimmune pathogenesis remains heavily investigated, with reports of paradoxical and disease-dependent effects. The influence of E2 in autoimmunity is potentially concentration-dependent, and immune cell-specific. Several reviews detail E2-mediated immune responses, including transcriptional regulation of immune mediating genes possessing ERE sequences and regulation of lymphopoiesis and immune cell differentiation [22–25]. Given the prevalence of E2 modulation in both innate and adaptive immune responses, along with its evident role in tumor progression, there exist several implications for immunomodulatory effects of E2 within the TME. Herein, we will discuss findings within current literature evaluating the protumoral impact of E2 on the TME and the implications of targeting the E2 pathway in cancer to promote an anti-tumor immune response.

2. Estrogen Receptor and Aromatase Expression in Tumor Cells: Correlations with Clinical Outcome

Tumoral ER expression is reported in nearly 30 different types of cancer, predominately in hormone-sensitive tumors such as breast, ovarian, endometrial, and prostate [26,27]. Studies comparing clinicopathological characteristics with ER protein expression (typically evaluated by immunohistochemistry (IHC)) in tumor tissue show differential relation to disease prognosis based on cellular localization and cancer type. In breast cancer, while predominately expressed in the nucleus, ERα protein expression in either the nucleus and/or cytoplasm correlates with features of advanced disease, including larger tumor size and lymph node metastasis [28]. However, ERα-positive breast cancer patients exhibit improved overall survival (OS) compared to ERα-negative patients, likely owing to the clinical benefit of adjuvant endocrine therapies for ERα-positive patients [18,29]. The clinical relevance of ERβ expression in breast cancer remains controversial largely due to challenges associated with ERβ splice variants and post-translational modifications, as well as the lack of a clinically standardized ERβ antibody [19,30,31]. As an integral enzyme in estrogen production, intratumoral aromatase has also been evaluated in breast cancer. While one study reported an association between aromatase activity and poor prognosis, others have failed to correlate aromatase activity or protein expression with clinical outcomes, suggesting that paracrine sources of estrogen may be of greater significance in hormone-dependent breast cancers [32–35]. In contrast to breast cancer, non-small cell lung cancer (NSCLC) ERα protein expression is more commonly expressed in the cytoplasm and is a negative prognostic marker [36,37]. Similarly, elevated cytoplasmic ERβ protein expression in NSCLC is associated with poorer OS [38], potentially indicative of the predominance of non-genomic mechanisms in NSCLC. Alternatively, nuclear ERβ expression in NSCLC correlates favorably with OS in some studies and negatively in others (reviewed in [39]). Tumoral aromatase protein expression and activity is also reported in NSCLC, with elevated expression identified as a predictor of poorer survival in women with early stage disease [40]. In advanced ovarian cancer tumors, while aromatase activity and ERβ mRNA expression do not correlate with any clinical outcomes [41,42], a recent meta-analysis revealed ERα protein expression was associated with improved OS [43]. Finally, while clinical correlations with aromatase have yet to be evaluated, both ERα and ERβ expression are associated with improved OS in endometrial cancer [44]. These clinical correlations, combined with mounting preclinical studies, indicate an intricate and pervasive protumoral role for hormonal signaling in multiple cancers, providing rationale for further investigation of ER expression and oncogenic cellular modulation.

3. Estrogen Receptor and Aromatase Expression and Estrogen-Mediated Effects in the Tumor Microenvironment

In addition to neoplastic cells, ERs and aromatase are also expressed on stromal and immune cells within the TME (Table 1). Numerous studies over the past decade have demonstrated that interactions between tumor cells and surrounding recruited stromal cells are integral in disrupting homeostasis and potentiating tumorigenesis (reviewed in [14,45]). Albeit highly heterogeneous within and across tumor types, regularly observed cellular components of the TME include: cancer associated fibroblasts (CAFs), tumor associated macrophages (TAMs), myeloid derived suppressor cells (MDSCs), immune T and B cells, natural killer (NK) cells, and endothelial cells [14]. ER and aromatase expression in TME stromal and immune cells suggest a potential immunomodulatory role of ER signaling in cancer biology as detailed by cell type below.

Table 1. Estrogen receptor (ER) and aromatase expression in stromal and immune cells in the tumor microenvironment.

TME Cell Type	Cancer Type	Human Expression	Murine Expression	Method of Evaluation	Reference
Stromal	Breast	Aromatase	ERα	PCR, IHC	[46,47]
	Melanoma		ERα	IHC	[47]
	Lung		ERα	IHC	[47]
	Endometrial	Aromatase		IHC	[48]
CAF	Breast	ERα		PCR	[49]
	Prostate	ERα, ERβ		IHC	[50,51]
	Endometrial	ERα, ERβ		PCR	[52]
	Ovarian	ERα		IHC	[53]
TAM	Ovarian	ERα, ERβ		IF, IHC	[54]
	Breast	Aromatase		IHC, PCR	[55]
	Lung	Aromatase	Aromatase	IHC	[56,57]
MDSC	Ovarian	ERα	ERα	PCR, Western	[53]

Studies were identified by PubMed searches using keywords: ERα, ERβ, aromatase, stromal, CAF, TAM, MDSC, expression, cancer. CAF: cancer associated fibroblast; TAM: tumor associated macrophage; MDSC: myeloid derived suppressor cell; IHC: immunohistochemistry; PCR: polymerase chain reaction; IF: immunofluorescence; Western: western blotting analysis.

3.1. Stromal Cells

It has become increasingly evident that tumor progression is reliant not only on tumor cells present in malignant tissue, but also the distinctive stromal cells recruited to the TME that signal among the tumor cells and each other. An in vivo murine model evaluating tumor cell-independent mechanisms of ER signaling within the TME has identified ERα expression and modulation in stromal cell types. In ovariectomized syngeneic mice transplanted with ER-negative melanoma, breast, or lung cancer cells,

E2 treatment significantly enhanced tumor growth of each cell type compared to untreated controls via interactions with stromal ERα [47]. Further, E2-stimulated tumor growth was increased when evaluated in immunocompromised mice, suggesting this effect may be more reliant on the innate immune response [47]. In addition to tumor growth, E2 also enhanced angiogenesis by increasing blood vessel density 2.1-fold in E2-treated mice compared to controls, an effect reliant on host ERα expression [47]. Peritumoral aromatase expression is also reported in endometrial cancer stromal cells, correlating with advanced disease and poor OS [48,58]. Aromatase is also observed in breast cancer stromal adipocytes of obese postmenopausal women, and several studies have identified mechanistic associations between obesity, inflammation, elevated aromatase, and breast cancer development [46,59,60].

3.2. Cancer Associated Fibroblasts

CAFs are among the most prevalent stromal cell type within the TME and act as a paracrine source of chemokines and soluble growth factors that activate signaling pathways involved in tumor cell survival, invasion, and metastasis [61]. A study using nuclear receptor arrays to compare gene expression profiles between normal human breast adipose fibroblasts and primary CAFs from malignant human breast tissue, observed ERα expression in fibroblasts from primary breast cancer tissue [49]. Despite similar levels of ERα expression observed in both cancerous and normal fibroblasts, the E2 responsive gene, liver receptor homolog-1 (*LRH-1*) was upregulated in CAFs compared to normal fibroblasts [49]. *LRH-1* is also an estrogen response gene and a direct transcriptional regulator of the aromatase encoding gene *CYP19A1* [62–64]. Aromatase is found to be co-expressed in breast cancers with LRH-1, suggesting a paracrine mechanism of E2 synthesis and ER-mediated oncogenesis in the breast cancer TME [65]. Endometrial CAFs also express both ERs and can promote tumor cell proliferation when co-cultured with human endometrial tumor cells [52]. Endometrial CAFs induce in vitro tumor cell proliferation in part through activation of the phosphatidylinositide 3-kinase (PI3K) and mitogen-activated protein kinase (MAPK) signaling networks, which are well-known ER-mediated pathways in breast and lung cancer [52,66–68].

ERα is also expressed in prostate CAFs, however, clinical implications remain unclear with some reports identifying CAF ERα and ERβ expression as a marker of clinically advanced disease [50], while other reports suggest ERα expressing CAFs provide a protective effect against tumor cell invasion and macrophage infiltration [69,70]. In the latter studies, stromal ERα reduced both murine and human prostate cancer cell invasion using an in vitro co-culture system, and reduced lymph node metastasis of orthotopically implanted human prostate cancer cells in mice [70]. Mechanistically, ERα-positive CAFs abated migratory behavior of adjacent prostate tumor cells through reduced expression of C–C motif chemokine ligand 5 (CCL5) and IL-6 chemokines, both of which have identified roles in tumor immune recruitment, inflammation, and activation of growth factor signaling [71,72].

3.3. Tumor Associated Macrophages

Macrophages critically regulate innate immune responses under normal physiological conditions; however, several studies have shown that TAMs can promote tumor cell proliferation, an inflammatory microenvironment, and metastasis [73,74]. Macrophage immune responses are tissue-specific and dependent on polarization by different cytokines within the local microenvironment [75]. Fully polarized M1 macrophages produce proinflammatory cytokines including IFNγ, interleukin 12 (IL-12), and TNFα, that contribute to tumor rejection and antigen presentation [75]. Alternatively, macrophages exhibiting an M2 phenotype produce type-2 cytokines including interleukins 4,5,6, and 10 [75], all of which are identified promoters of tumor progression through enhanced tumor cell growth and immune evasion [76]. Infiltrating TAMs observed in malignant tumors display an M2 phenotype, representing another potential protumoral therapeutic target within the TME. TAM infiltration is observed in a wide-range of cancer types and correlates with poor prognosis [77]. For example, TAM infiltration is an independent poor prognostic predictor for ovarian cancer, with higher infiltration observed in cancerous specimens compared to benign lesions, and density-dependent associations with five-year survival rates [78].

Co-localized expression of both ERα and ERβ is reported in human high grade serous ovarian cancer (HGSOC) TAMs, and premenopausal patients show elevated TAM infiltration compared to postmenopausal women, with highest overall TAM density observed in ERα-positive tumors [54]. Conversely, while TAM infiltration has been associated with poor prognosis in both hormone receptor positive and negative breast cancers, TAM enrichment and proliferation is more commonly observed in hormone receptor negative breast tumors [79,80]. However, M1 versus M2 polarization was not evaluated in these studies. Furthermore, a separate IHC analysis of breast cancer specimens revealed aromatase expression in TAMs, enabling local E2 production within the TME and enhanced ER-positive breast tumor cell proliferation [55]. Aromatase is also expressed in TAMs from NSCLC patient tumors [56], and both aromatase and ERβ are observed in infiltrating macrophages of preneoplasias in tobacco carcinogen-induced murine lung tumors [57].

While a paucity of data exists regarding ER expression in TAMs of several cancer types, there is evidence that E2 can induce M2 polarization and tumor infiltration. Using a polyomavirus middle T (PyMT) ER-positive breast cancer murine model, E2 increased tumoral M2 TAM infiltration, while untreated controls alternatively exhibited M1 TAM infiltration [81]. Furthermore, E2 enhanced M2 macrophage secretion of vascular endothelial growth factor (VEGF), an identified mediator of M2 macrophage recruitment [81,82]. E2 has been shown to also upregulate VEGF expression and pulmonary macrophage content in the lungs of mice exposed to a tobacco carcinogen [83]. Evaluation of E2-mediated tumor growth in a HGSOC murine model showed that E2 not only enhanced the growth of ER-negative xenografts, but also increased M2 TAM infiltration compared to untreated ovariectomized mice [54]. In addition to reports of E2-mediated TAM infiltration, a tissue microarray of patient samples coupled with in vitro analysis revealed endometrial M2 TAMs mediate ER activation through epigenetic upregulation of ERα by secreted interleukin-17A (IL-17A), increasing E2-driven malignant endometrial cell proliferation [84]. Taken together, these studies suggest a potential positive feedback mechanism between the estrogen pathway and M2 TAM infiltration in certain cancers. Targeting this interaction may therefore provide therapeutic benefit as recently demonstrated in a lung cancer xenograft model using the phytoestrogen SERM resveratrol [85]. The study showed resveratrol treatment significantly suppressed tumor growth by inhibiting M2 polarization of TAMs and decreasing activation of signal transducer and activator of transcription 3 (STAT3) signaling [85].

3.4. Myeloid Derived Suppressor Cells

MDSCs are another myeloid cell present in the TME known to disrupt immune surveillance and promote tumor development [86]. ERα expression was also recently identified by IHC and confirmed by PCR and immunoblotting in MDSCs isolated from the tumor, bone marrow, and peripheral blood of human ovarian cancer patients [53]. Using an E2-insensitive syngeneic ovarian cancer model, ovariectomized mice exhibited improved survival compared to non-ovariectomized mice following tumor challenge, while E2 supplementation accelerated tumor progression and reversed the protective effect found in estrogen-depleted mice [53]. Notably, this effect was only observed in immunocompetent mice with no survival benefit of ovariectomy observed in tumor-bearing T-cell deficient immunocompromised mice, suggesting the antitumor effects of E2 deficiency is reliant on functional adaptive immunity [53]. E2-treated mice also exhibited significantly fewer helper and cytotoxic T cells, but also exhibited significantly elevated recruitment of MDSCs in both the spleen and tumor beds [53]. Specifically, the immunosuppressive activity of granulocytic MDSCs was increased in this model. ER-dependence of MDSC expansion was demonstrated using the ERα antagonist methylpiperidino pyrazole (MPP) to inhibit MDSC proliferation in vitro [53]. In the peritoneal cavity of ovarian tumor-bearing mice, E2 treatment increased activation of STAT3 signaling, a regulator of myeloid differentiation and development [87], through transcriptional upregulation of JAK2 and SRC activity [53]. Similar findings were also observed in syngeneic lung and breast cancer murine models and the E2-stimulated tumor growth was abrogated by MDSC depletion using anti-Gr1 antibodies [53].

3.5. Tumor Infiltrating Lymphocytes (TIL)

Lymphocyte composition of the TME vastly differs based on cancer type and immune infiltrates exhibit opposing properties promoting tumor progression and antitumor immunity depending on the primary tumor [88]. For example, CD4+ T cell polarization has been identified as a mediator of tumor immune surveillance. T helper 1 (Th1) T cell responses are associated with tumor suppression and upregulation of IFNγ and IL-12, while T helper 2 (Th2) responses are reliant of IL-4 production and exhibit protumor activity [89,90]. Interestingly, several murine and human studies report elevated E2 induces increased Th2 responses and upregulate IL-4 production [22,25]. A recent study utilizing an in silico machine learning based approach, identified increased immune infiltrate including Th1 T cells, B cells, and cytotoxic T lymphocytes (CTLs) in ER-negative breast tumors relative to ER-positive breast tumors [91]. This study observed an inverse correlation between ER activity and immune infiltration of each of these cells in breast cancer tissues, confirming previous reports that increased TIL, specifically CD8+ T cells, in ER-negative tumors significantly correlates with improved OS [91,92]. Furthermore, a post-hoc analysis of gene expression in ER-positive breast cancer patients showed that treatment with the AI letrozole increased the infiltration of B cell and T helper lymphocyte subsets at early and late time points following treatment initiation [91].

3.5.1. Cytotoxic T Cells and Natural Killer Cells

Granule-mediated exocytosis is one pathway by which CTLs and NK cells initiate apoptosis to eliminate pathogenic and tumor cells [93]. Serine proteases such as granzyme B are deposited into the target cells to initiate caspase-dependent apoptosis [94]. Jiang et.al. cultured ERα expressing human liver carcinoma cells with E2 and showed E2 treatment upregulated expression of the granzyme B inhibitor, proteinase inhibitor-9 (PI-9), and protected the cells against NK and CTL-induced apoptosis in DNA fragmentation assays [95]. E2-induced PI-9 expression was also observed in ERα-positive MCF7 breast cancer cells, again protecting cells against NK elimination, while PI-9 knockdown blocked E2's protective effect against NK granule-mediated apoptosis [96]. These studies suggest that E2 enhances immunosuppression through inhibition of NK and CTL-mediated tumor cell elimination.

3.5.2. Regulatory T Cells

T cell activation and effector differentiation is an essential part of the adaptive immune response. FoxP3 expressing Tregs are integral in coordinating suppression of anti-tumor immune responses, secreting immunosuppressive cytokines and inhibiting responder T cell expansion [97]. Physiological doses of E2 administered to immunocompetent ovariectomized female mice have been shown to enhance CD4+CD25+ Treg expansion and upregulate Foxp3 expression in multiple tissues [98]. Furthermore, fluorescence-activated cell sorting (FACs) assays revealed ERα expressing CD4+CD25- cells incubated with E2 acquire CD25 expression [98]. E2 transformed CD4+CD25+ T cells exhibited an immunosuppressive Treg phenotype, significantly inhibiting T cell proliferation in an in vitro mixed lymphocyte reaction [98]. Additional studies have reported E2-stimulated Foxp3 expression in murine Tregs, which is of importance considering that Foxp3 is essential for Treg functionality, and tumoral aggregation of FoxP3+ Tregs in patients is a predictor of poor prognosis in multiple cancers [99–101]. For example, in early-stage NSCLC patients, nuclear ERα expression was found independently associated with increased risk of recurrence and FoxP3+ lymphocyte infiltrate [102]. Further, a recent meta-analysis reported FoxP3+ Treg infiltration significantly correlated with poorer OS in ER-positive breast cancer patients, but improved survival rates in ER-negative patients [103]. In addition, evaluation of ERα-positive breast tumors from patients treated with letrozole showed a significant reduction of FoxP3+ Tregs post-treatment [104].

Moreover, Tregs isolated from mice treated with E2 displayed enhanced suppression and increased intracellular expression of the immune checkpoint protein programmed death-1 (PD-1), while ERα and ERβ knockout reduced Treg suppression and PD-1 expression [105]. Of note, E2 treatment also stimulates in vitro expression of the PD-1 ligand (PD-L1) on ERα-positive endometrial and breast cancer cells through activation of PI3K signaling [106]. Interactions between PD-L1 expressing tumor cells and PD-1 positive T cells induces cytotoxic T cell exhaustion, resulting in tumor immune evasion [107]. Evidence that E2 upregulates both PD-L1 and PD-1, suggests E2 signaling may critically influence the PD-1/PD-L1 pathway.

3.6. Inflammatory Cytokines and Eicosanoids

Chronic inflammation is widely recognized as an ancillary mechanism promoting tumor progression. The TME releases cytokines that activate protumoral pathways mediating proliferation, immune evasion, and metastasis [108]. IL-6, a proinflammatory cytokine, has been shown to enhance ERα-positive breast cancer cell growth and invasion [109]. Local TAFs in breast cancers act as a paracrine source of the elevated IL-6, driving STAT3 activation and ERα-positive tumor cell proliferation both in vitro and in vivo [110]. TNFα, another ubiquitous TME cytokine, regulates expression of genes associated with metastatic phenotypes in ERα-positive breast cancer cells [111]. TNFα has also been shown to upregulate aromatase expression in cultured human adipose stromal cells [112]. Transcriptional linear correlations between aromatase and the cytokines TNFα and IL-6 have been reported in patient breast cancer tissue, but not in adjacent non-cancerous tissue [113]. A similar correlation has also been seen between aromatase and the eicosanoid cyclooxygenase-2 (COX-2) [113]. COX-2 is responsible for the synthesis of inflammatory promoting eicosanoids such as prostaglandin E2 (PGE2) [114]. It is well established that PGE2 promotes upregulated transcription of aromatase through elevated cyclic adenosine monophosphate (cAMP) in breast tumors [115]. Despite conflicting reports, some epidemiological studies show that regular use of COX-2 inhibiting nonsteroidal anti-inflammatory drugs (NSAIDs) can reduce the risk of developing ERα-positive breast cancers, but not ERα-negative cancers [116].

Significant correlations between ERα, TNFα, and NF-κB protein expression have also been reported in breast cancer tissues [117]. NF-κB signaling is well recognized for its role in tumor initiation and inflammation [118]. Constitutive activation of NF-κB is observed in several cancers, and is associated with the cytokines IL-6 and TNFα [118]. Increased DNA binding of NF-κB and activator protein-1 (AP-1) has been observed in SERM-resistant, ERα-positive breast cancer cell line models and patient specimens [119,120]. Furthermore, E2 exposure in a murine model evaluating tobacco-induced lung cancer enhanced pulmonary inflammation through increased activation of NF-κB signaling and expression of VEGF and IL-17A [83]. Alternatively, targeting E2 and inflammatory pathways with combined AI and NSAID treatment maximally prevented carcinogen-induced lung tumor development in mice, significantly reducing STAT3 and MAPK signaling, circulating IL-6, and IL-17A expression [83]. Taken together, these reports indicate potential interactions between the E2 pathway and regulators of tumor-promoting inflammation, representing another beneficial target of E2 inhibition.

4. Clinical Implications of Targeting the Estrogen Pathway in the Tumor Microenvironment

Immunotherapy is a powerful therapeutic strategy for cancer; however, the immunosuppressive TME poses major obstacles for this approach. Currently, immune checkpoint inhibitors of cytotoxic T-lymphocyte-associated antigen 4 (CTLA4) and PD-1/PD-L1 are among the most clinically evaluated immune therapies [121]. These agentshave remarkably advanced cancer treatment, significantly improving response rates and survival compared with standard-of-care chemotherapies [122–125]. However, typical response rates to these therapies remain limited to only around 20–35% of patients, with variable responses depending on stage, tumor type, and PD-L1 staining positivity [126]. Furthermore, while some patients have durable responses, mechanisms of acquired and adaptive resistance are becoming apparent, with 25 to 33% of melanoma patients exhibiting delayed relapse on these therapies [15,16].

Recent efforts to identify molecular events underlying immune evasion and failed therapeutic response report that damaged DNA repair mechanisms, increased non-synonymous somatic mutational load, and neoantigen presentation correlate with tumor immunogenicity and improved clinical outcomes [12,13,127]. Alternatively, mechanisms facilitating immune evasion involve damage to antigen presenting capacity and recurrence of non-antigenic mutations poorly presented by MHC class 1 molecules [128,129]. While these findings provide a greater understanding of tumor immunoediting and potential biomarkers predictive of response, novel therapeutic combinations are still needed to improve the efficacy of current immunotherapeutic agents. The identification of E2 modulation of the tumor immune phenotype justifies investigation of endocrine agents to reverse tumor immune tolerance. As depicted in Figure 1, E2 signaling can modulate the immune TME through enhanced protumoral responses. Therefore, anti-estrogen therapy has the potential to not only reverse an immunosuppressive TME, but also to augment response in E2-sensitive tumors.

Recently, a high-throughput screening assay in lung cancer cells identified the anti-estrogen fulvestrant as the top compound that increased tumor sensitivity to immune-mediated lysis [130]. Fulvestrant is an ideal candidate to combine with anti-PD-1/PD-L1 agents, due to its proven safety profile and non-overlapping toxicities. These new findings of E2 action on immune cells could create a paradigm shift towards utilizing anti-estrogen therapy to target the immunosuppressive TME, thereby increasing the efficacy and duration of response of current immunotherapies [131].

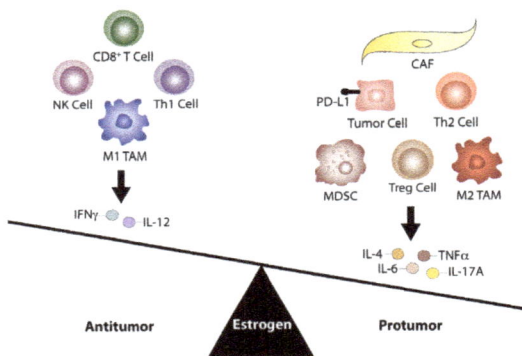

Figure 1. The E2 pathway promotes a protumor TME. The E2 pathway contributes to aberrant regulation of antitumor immunity, enhancing a greater number of protumoral responses within the TME. Current literature suggests E2 may facilitate an immunosuppressive TME by shifting the balance in favor of Th2 responses, production of tumor-promoting cytokines (IL-6, IL-4, TNFα, and IL-17A), and M2 TAM infiltration compared to Th1 responses, associated Th1 cytokines (IL-12 and IFNγ), and M1 TAM infiltration. E2 may further promote tumor immune evasion through proliferation of Treg and MDSC populations, increased tumor cell PD-L1 expression, and inhibition of CD8[+] T cell and NK cell induced apoptosis. CAFs may additionally support a protumor environment by supplying paracrine sources of E2 and IL-6. Therefore, targeted inhibition of the E2 pathway may act as a novel strategy to enhance the effects of immunotherapies and reverse this immune imbalance within the TME.

5. Conclusions and Perspective

The E2 pathway is an identified promoter of tumorigenesis in several cancers, largely for its direct genomic and non-genomic effects on tumor cells. However, evidence of ER and aromatase expression on stromal and immune cells within the TME indicates that additional mechanisms exist by which estrogens enhance malignant progression. It is becoming increasingly evident that cells comprising the TME can impact tumor immunity, either beneficially through enhanced antitumoral immune responses, or detrimentally through increased protumoral responses. Evidence thus far suggests that E2 facilitates a primarily tumor-promoting and immunosuppressive TME in multiple tumor types. While checkpoint blockade immunotherapies have exhibited significant clinical success for the treatment of certain cancers, partial response rates and acquired resistance to these therapies necessitate the development of strategies to boost immunotherapeutic responses. The data summarized here points to the E2 pathway as a regulator of tumor immune responses, suggesting that clinical benefit may be derived from combining estrogen blocking agents with immune checkpoint inhibitors. Prior to clinical analysis of this combination, a more comprehensive characterization of E2-related proteins in the TME of various tumor types is necessary. There is also a need for standardized methods and CLIA-approved assays for the detection of ERβ and aromatase expression. Future studies evaluating response to current immunotherapies based on sex-differences, patient demographics including menopausal status, and obesity are warranted, given the pervasive involvement of the E2 pathway in tumor immunity.

Acknowledgments: A portion of this work was supported by SPORE in Lung Cancer Grant P50 CA090440 from the National Cancer Institute.

Author Contributions: Natalie J Rothenberger prepared the manuscript, table, and figure. Laura P. Stabile provided essential input on content and organization and critically edited the manuscript. Ashwin Somasundaram provided vital immunological information and review of content.

Conflicts of Interest: The authors declare no conflict of interest.

Abbreviations

ERβ	Estrogen receptor β
ERα	Estrogen receptor α
ERE	Estrogen response element
E2	17β-Estradiol
DC	Dendritic cell
Treg	Regulatory T cell
SERM	Selective estrogen receptor modulator
SERD	Selective estrogen receptor degrader
IHC	Immunohistochemistry
IF	Immunofluorescence
Th2	T helper 2
Th1	T helper 1
IL-4	Interleukin-4
IFNγ	Interferon Gamma
IL-6	Interleukin-6
TNFα	Tumor necrosis factor alpha
TME	Tumor microenvironment
EGF	Epidermal growth factor
IGF	Insulin growth factor
FGF	Fibroblast growth factor
OS	Overall survival
NSCLC	Non-small cell lung cancer
CAF	Cancer associated fibroblast
TAM	Tumor associated macrophage
MDSC	Myeloid derived suppressor cell
MPP	Methylpiperidino pyrazole
NK	Natural killer
LRH-1	Liver receptor homolog-1
PCR	Polymerase chain reaction
PI3K	Phosphatidylinositide 3-kinase

MAPK	Mitogen-activated protein kinase
CCL5	C–C motif chemokine ligand 5
IL-12	Interleukin-12
VEGF	Vascular endothelial growth factor
HGSOC	High grade serous ovarian cancer
IL-17A	Interleukin-17A
STAT3	Signal transducer and activator of transcription 3
TIL	Tumor infiltrating lymphocyte
AI	Aromatase inhibitor
CTL	Cytotoxic T lymphocyte
PI-9	Proteinase inhibitor-9
PD-1	Programmed death-1
PD-L1	PD-1 ligand
COX-2	Cyclooxygenase-2
PGE2	Prostaglandin E2
NSAID	Nonsteroidal anti-inflammatory drug
AP-1	Activator protein-1
CTLA4	Cytotoxic T-lymphocyte–associated antigen 4

References

1. Nilsson, S.; Gustafsson, J. Estrogen receptors: Their actions and functional roles in health and human disease. In *Nuclear Receptors: Current Concepts and Future Challenges*; Bunce, C., Campbell, M.J., Eds.; Springer: Dordrecht, The Netherlands, 2010; pp. 91–141.
2. Delaunay, F.; Pettersson, K.; Tujague, M.; Gustafsson, J.A. Functional differences between the amino-terminal domains of estrogen receptors alpha and beta. *Mol. Pharmacol.* **2000**, *58*, 584–590. [CrossRef] [PubMed]
3. Zhu, B.T.; Han, G.Z.; Shim, J.Y.; Wen, Y.; Jiang, X.R. Quantitative structure-activity relationship of various endogenous estrogen metabolites for human estrogen receptor alpha and beta subtypes: Insights into the structural determinants favoring a differential subtype binding. *Endocrinology* **2006**, *147*, 4132–4150. [CrossRef] [PubMed]
4. Folkerd, E.J.; Dowsett, M. Influence of sex hormones on cancer progression. *J. Clin. Oncol.* **2010**, *28*, 4038–4044. [CrossRef] [PubMed]
5. Frasor, J.; Danes, J.M.; Komm, B.; Chang, K.C.; Lyttle, C.R.; Katzenellenbogen, B.S. Profiling of estrogen up- and down-regulated gene expression in human breast cancer cells: Insights into gene networks and pathways underlying estrogenic control of proliferation and cell phenotype. *Endocrinology* **2003**, *144*, 4562–4574. [CrossRef] [PubMed]
6. Hershberger, P.A.; Vasquez, A.C.; Kanterewicz, B.; Land, S.; Siegfried, J.M.; Nichols, M. Regulation of endogenous gene expression in human non-small cell lung cancer cells by estrogen receptor ligands. *Cancer Res.* **2005**, *65*, 1598–1605. [CrossRef] [PubMed]
7. Egloff, A.M.; Rothstein, M.E.; Seethala, R.; Siegfried, J.M.; Grandis, J.R.; Stabile, L.P. Cross-talk between estrogen receptor and epidermal growth factor receptor in head and neck squamous cell carcinoma. *Clin. Cancer Res.* **2009**, *15*, 6529–6540. [CrossRef] [PubMed]
8. Lanzino, M.; Morelli, C.; Garofalo, C.; Panno, M.L.; Mauro, L.; Ando, S.; Sisci, D. Interaction between estrogen receptor alpha and insulin/igf signaling in breast cancer. *Curr. Cancer Drug Targets* **2008**, *8*, 597–610. [CrossRef] [PubMed]
9. Siegfried, J.M.; Farooqui, M.; Rothenberger, N.J.; Dacic, S.; Stabile, L.P. Interaction between the estrogen receptor and fibroblast growth factor receptor pathways in non-small cell lung cancer. *Oncotarget* **2017**, *8*, 24063–24076. [CrossRef] [PubMed]
10. *Cancer Facts & Figures 2017*; American Cancer Society: Atlanta, GA, USA, 2017.
11. Alexandrov, L.B.; Nik-Zainal, S.; Wedge, D.C.; Aparicio, S.A.; Behjati, S.; Biankin, A.V.; Bignell, G.R.; Bolli, N.; Borg, A.; Borresen-Dale, A.L.; et al. Signatures of mutational processes in human cancer. *Nature* **2013**, *500*, 415–421. [CrossRef] [PubMed]
12. Rooney, M.S.; Shukla, S.A.; Wu, C.J.; Getz, G.; Hacohen, N. Molecular and genetic properties of tumors associated with local immune cytolytic activity. *Cell* **2015**, *160*, 48–61. [CrossRef] [PubMed]

13. Rizvi, N.A.; Hellmann, M.D.; Snyder, A.; Kvistborg, P.; Makarov, V.; Havel, J.J.; Lee, W.; Yuan, J.; Wong, P.; Ho, T.S.; et al. Mutational landscape determines sensitivity to pd-1 blockade in non-small cell lung cancer. *Science* **2015**, *348*, 124–128. [CrossRef] [PubMed]

14. Quail, D.F.; Joyce, J.A. Microenvironmental regulation of tumor progression and metastasis. *Nat. Med.* **2013**, *19*, 1423–1437. [CrossRef] [PubMed]

15. Sharma, P.; Hu-Lieskovan, S.; Wargo, J.A.; Ribas, A. Primary, adaptive, and acquired resistance to cancer immunotherapy. *Cell* **2017**, *168*, 707–723. [CrossRef] [PubMed]

16. Schachter, J.; Ribas, A.; Long, G.V.; Arance, A.; Grob, J.J.; Mortier, L.; Daud, A.; Carlino, M.S.; McNeil, C.; Lotem, M.; et al. Pembrolizumab versus ipilimumab for advanced melanoma: Final overall survival results of a multicentre, randomised, open-label phase 3 study (keynote-006). *Lancet* **2017**, *390*, 1853–1862. [CrossRef]

17. Somasundaram, A.; Burns, T.F. The next generation of immunotherapy: Keeping lung cancer in check. *J. Hematol. Oncol.* **2017**, *10*, 87. [CrossRef] [PubMed]

18. Dunnwald, L.K.; Rossing, M.A.; Li, C.I. Hormone receptor status, tumor characteristics, and prognosis: A prospective cohort of breast cancer patients. *Breast Cancer Res.* **2007**, *9*, R6. [CrossRef] [PubMed]

19. Leung, Y.K.; Lee, M.T.; Lam, H.M.; Tarapore, P.; Ho, S.M. Estrogen receptor-beta and breast cancer: Translating biology into clinical practice. *Steroids* **2012**, *77*, 727–737. [CrossRef] [PubMed]

20. Phiel, K.L.; Henderson, R.A.; Adelman, S.J.; Elloso, M.M. Differential estrogen receptor gene expression in human peripheral blood mononuclear cell populations. *Immunol. Lett.* **2005**, *97*, 107–113. [CrossRef] [PubMed]

21. Laffont, S.; Rouquie, N.; Azar, P.; Seillet, C.; Plumas, J.; Aspord, C.; Guery, J.C. X-chromosome complement and estrogen receptor signaling independently contribute to the enhanced tlr7-mediated ifn-alpha production of plasmacytoid dendritic cells from women. *J. Immunol.* **2014**, *193*, 5444–5452. [CrossRef] [PubMed]

22. Fish, E.N. The x-files in immunity: Sex-based differences predispose immune responses. *Nat. Rev. Immunol.* **2008**, *8*, 737–744. [CrossRef] [PubMed]

23. Kovats, S. Estrogen receptors regulate innate immune cells and signaling pathways. *Cell. Immunol.* **2015**, *294*, 63–69. [CrossRef] [PubMed]

24. Kovats, S. Estrogen receptors regulate an inflammatory pathway of dendritic cell differentiation: Mechanisms and implications for immunity. *Horm. Behav.* **2012**, *62*, 254–262. [CrossRef] [PubMed]

25. Khan, D.; Ansar Ahmed, S. The immune system is a natural target for estrogen action: Opposing effects of estrogen in two prototypical autoimmune diseases. *Front. Immunol.* **2015**, *6*, 635. [CrossRef] [PubMed]

26. Cerami, E.; Gao, J.; Dogrusoz, U.; Gross, B.E.; Sumer, S.O.; Aksoy, B.A.; Jacobsen, A.; Byrne, C.J.; Heuer, M.L.; Larsson, E.; et al. The cbio cancer genomics portal: An open platform for exploring multidimensional cancer genomics data. *Cancer Discov.* **2012**, *2*, 401–404. [CrossRef] [PubMed]

27. Gao, J.; Aksoy, B.A.; Dogrusoz, U.; Dresdner, G.; Gross, B.; Sumer, S.O.; Sun, Y.; Jacobsen, A.; Sinha, R.; Larsson, E.; et al. Integrative analysis of complex cancer genomics and clinical profiles using the cbioportal. *Sci. Signal.* **2013**, *6*, pl1. [CrossRef] [PubMed]

28. Li, L.; Wang, Q.; Lv, X.; Sha, L.; Qin, H.; Wang, L.; Li, L. Expression and localization of estrogen receptor in human breast cancer and its clinical significance. *Cell Biochem. Biophys.* **2015**, *71*, 63–68. [CrossRef] [PubMed]

29. Grann, V.R.; Troxel, A.B.; Zojwalla, N.J.; Jacobson, J.S.; Hershman, D.; Neugut, A.I. Hormone receptor status and survival in a population-based cohort of patients with breast carcinoma. *Cancer* **2005**, *103*, 2241–2251. [CrossRef] [PubMed]

30. Haldosen, L.A.; Zhao, C.; Dahlman-Wright, K. Estrogen receptor beta in breast cancer. *Mol. Cell. Endocrinol.* **2014**, *382*, 665–672. [CrossRef] [PubMed]

31. Leygue, E.; Murphy, L.C. A bi-faceted role of estrogen receptor beta in breast cancer. *Endocr. Relat. Cancer* **2013**, *20*, R127–R139. [CrossRef] [PubMed]

32. Miller, W.R.; Anderson, T.J.; Jack, W.J. Relationship between tumour aromatase activity, tumour characteristics and response to therapy. *J. Steroid Biochem. Mol. Biol.* **1990**, *37*, 1055–1059. [CrossRef]

33. Lipton, A.; Santen, R.J.; Santner, S.J.; Harvey, H.A.; Sanders, S.I.; Matthews, Y.L. Prognostic value of breast cancer aromatase. *Cancer* **1992**, *70*, 1951–1955. [CrossRef]

34. Esteban, J.M.; Warsi, Z.; Haniu, M.; Hall, P.; Shively, J.E.; Chen, S. Detection of intratumoral aromatase in breast carcinomas. An immunohistochemical study with clinicopathologic correlation. *Am. J. Pathol.* **1992**, *140*, 337–343. [PubMed]

35. Miki, Y.; Suzuki, T.; Sasano, H. Controversies of aromatase localization in human breast cancer–stromal versus parenchymal cells. *J. Steroid Biochem. Mol. Biol.* **2007**, *106*, 97–101. [CrossRef] [PubMed]

36. Kawai, H.; Ishii, A.; Washiya, K.; Konno, T.; Kon, H.; Yamaya, C.; Ono, I.; Minamiya, Y.; Ogawa, J. Estrogen receptor alpha and beta are prognostic factors in non-small cell lung cancer. *Clin. Cancer Res.* **2005**, *11*, 5084–5089. [CrossRef] [PubMed]

37. Nose, N.; Sugio, K.; Oyama, T.; Nozoe, T.; Uramoto, H.; Iwata, T.; Onitsuka, T.; Yasumoto, K. Association between estrogen receptor-beta expression and epidermal growth factor receptor mutation in the postoperative prognosis of adenocarcinoma of the lung. *J. Clin. Oncol.* **2009**, *27*, 411–417. [CrossRef] [PubMed]

38. Stabile, L.P.; Dacic, S.; Land, S.R.; Lenzner, D.E.; Dhir, R.; Acquafondata, M.; Landreneau, R.J.; Grandis, J.R.; Siegfried, J.M. Combined analysis of estrogen receptor beta-1 and progesterone receptor expression identifies lung cancer patients with poor outcome. *Clin. Cancer Res.* **2011**, *17*, 154–164. [CrossRef] [PubMed]

39. Hsu, L.H.; Chu, N.M.; Kao, S.H. Estrogen, estrogen receptor and lung cancer. *Int. J. Mol. Sci.* **2017**, *18*. [CrossRef] [PubMed]

40. Mah, V.; Seligson, D.B.; Li, A.; Marquez, D.C.; Wistuba, I.I.; Elshimali, Y.; Fishbein, M.C.; Chia, D.; Pietras, R.J.; Goodglick, L. Aromatase expression predicts survival in women with early-stage non small cell lung cancer. *Cancer Res.* **2007**, *67*, 10484–10490. [CrossRef] [PubMed]

41. Slotman, B.J.; Kuhnel, R.; Rao, B.R.; Dijkhuizen, G.H.; de Graaff, J.; Stolk, J.G. Importance of steroid receptors and aromatase activity in the prognosis of ovarian cancer: High tumor progesterone receptor levels correlate with longer survival. *Gynecol. Oncol.* **1989**, *33*, 76–81. [CrossRef]

42. Cunat, S.; Rabenoelina, F.; Daures, J.P.; Katsaros, D.; Sasano, H.; Miller, W.R.; Maudelonde, T.; Pujol, P. Aromatase expression in ovarian epithelial cancers. *J. Steroid Biochem. Mol. Biol.* **2005**, *93*, 15–24. [CrossRef] [PubMed]

43. Shen, Z.; Luo, H.; Li, S.; Sheng, B.; Zhao, M.; Zhu, H.; Zhu, X. Correlation between estrogen receptor expression and prognosis in epithelial ovarian cancer: A meta-analysis. *Oncotarget* **2017**, *8*, 62400–62413. [CrossRef] [PubMed]

44. Zhang, Y.; Zhao, D.; Gong, C.; Zhang, F.; He, J.; Zhang, W.; Zhao, Y.; Sun, J. Prognostic role of hormone receptors in endometrial cancer: A systematic review and meta-analysis. *World J. Surg. Oncol.* **2015**, *13*, 208. [CrossRef] [PubMed]

45. Hanahan, D.; Coussens, L.M. Accessories to the crime: Functions of cells recruited to the tumor microenvironment. *Cancer Cell* **2012**, *21*, 309–322. [CrossRef] [PubMed]

46. Morris, P.G.; Hudis, C.A.; Giri, D.; Morrow, M.; Falcone, D.J.; Zhou, X.K.; Du, B.; Brogi, E.; Crawford, C.B.; Kopelovich, L.; et al. Inflammation and increased aromatase expression occur in the breast tissue of obese women with breast cancer. *Cancer Prev. Res. (Phila.)* **2011**, *4*, 1021–1029. [CrossRef] [PubMed]

47. Pequeux, C.; Raymond-Letron, I.; Blacher, S.; Boudou, F.; Adlanmerini, M.; Fouque, M.J.; Rochaix, P.; Noel, A.; Foidart, J.M.; Krust, A.; et al. Stromal estrogen receptor-alpha promotes tumor growth by normalizing an increased angiogenesis. *Cancer Res.* **2012**, *72*, 3010–3019. [CrossRef] [PubMed]

48. Segawa, T.; Shozu, M.; Murakami, K.; Kasai, T.; Shinohara, K.; Nomura, K.; Ohno, S.; Inoue, M. Aromatase expression in stromal cells of endometrioid endometrial cancer correlates with poor survival. *Clin. Cancer Res.* **2005**, *11*, 2188–2194. [CrossRef] [PubMed]

49. Knower, K.C.; Chand, A.L.; Eriksson, N.; Takagi, K.; Miki, Y.; Sasano, H.; Visvader, J.E.; Lindeman, G.J.; Funder, J.W.; Fuller, P.J.; et al. Distinct nuclear receptor expression in stroma adjacent to breast tumors. *Breast Cancer Res. Treat.* **2013**, *142*, 211–223. [CrossRef] [PubMed]

50. Daniels, G.; Gellert, L.L.; Melamed, J.; Hatcher, D.; Li, Y.; Wei, J.; Wang, J.; Lee, P. Decreased expression of stromal estrogen receptor alpha and beta in prostate cancer. *Am. J. Transl. Res.* **2014**, *6*, 140–146. [PubMed]

51. Leav, I.; Lau, K.M.; Adams, J.Y.; McNeal, J.E.; Taplin, M.E.; Wang, J.; Singh, H.; Ho, S.M. Comparative studies of the estrogen receptors beta and alpha and the androgen receptor in normal human prostate glands, dysplasia, and in primary and metastatic carcinoma. *Am. J. Pathol.* **2001**, *159*, 79–92. [CrossRef]

52. Subramaniam, K.S.; Tham, S.T.; Mohamed, Z.; Woo, Y.L.; Mat Adenan, N.A.; Chung, I. Cancer-associated fibroblasts promote proliferation of endometrial cancer cells. *PLoS ONE* **2013**, *8*, e68923. [CrossRef] [PubMed]

53. Svoronos, N.; Perales-Puchalt, A.; Allegrezza, M.J.; Rutkowski, M.R.; Payne, K.K.; Tesone, A.J.; Nguyen, J.M.; Curiel, T.J.; Cadungog, M.G.; Singhal, S.; et al. Tumor cell-independent estrogen signaling drives disease progression through mobilization of myeloid-derived suppressor cells. *Cancer Discov.* **2017**, *7*, 72–85. [CrossRef] [PubMed]

54. Ciucci, A.; Zannoni, G.F.; Buttarelli, M.; Lisi, L.; Travaglia, D.; Martinelli, E.; Scambia, G.; Gallo, D. Multiple direct and indirect mechanisms drive estrogen-induced tumor growth in high grade serous ovarian cancers. *Oncotarget* **2016**, *7*, 8155–8171. [CrossRef] [PubMed]

55. Mor, G.; Yue, W.; Santen, R.J.; Gutierrez, L.; Eliza, M.; Berstein, L.M.; Harada, N.; Wang, J.; Lysiak, J.; Diano, S.; et al. Macrophages, estrogen and the microenvironment of breast cancer. *J. Steroid Biochem. Mol. Biol.* **1998**, *67*, 403–411. [CrossRef]

56. Siegfried, J.M.; Stabile, L.P. Estrongenic steroid hormones in lung cancer. *Semin. Oncol.* **2014**, *41*, 5–16. [CrossRef] [PubMed]

57. Stabile, L.P.; Rothstein, M.E.; Cunningham, D.E.; Land, S.R.; Dacic, S.; Keohavong, P.; Siegfried, J.M. Prevention of tobacco carcinogen-induced lung cancer in female mice using antiestrogens. *Carcinogenesis* **2012**, *33*, 2181–2189. [CrossRef] [PubMed]

58. Matsumoto, M.; Yamaguchi, Y.; Seino, Y.; Hatakeyama, A.; Takei, H.; Niikura, H.; Ito, K.; Suzuki, T.; Sasano, H.; Yaegashi, N.; et al. Estrogen signaling ability in human endometrial cancer through the cancer-stromal interaction. *Endoc. Relat. Cancer* **2008**, *15*, 451–463. [CrossRef] [PubMed]

59. Subbaramaiah, K.; Morris, P.G.; Zhou, X.K.; Morrow, M.; Du, B.; Giri, D.; Kopelovich, L.; Hudis, C.A.; Dannenberg, A.J. Increased levels of cox-2 and prostaglandin e2 contribute to elevated aromatase expression in inflamed breast tissue of obese women. *Cancer Discov.* **2012**, *2*, 356–365. [CrossRef] [PubMed]

60. Subbaramaiah, K.; Howe, L.R.; Bhardwaj, P.; Du, B.; Gravaghi, C.; Yantiss, R.K.; Zhou, X.K.; Blaho, V.A.; Hla, T.; Yang, P.; et al. Obesity is associated with inflammation and elevated aromatase expression in the mouse mammary gland. *Cancer Prev. Res. (Phila.)* **2011**, *4*, 329–346. [CrossRef] [PubMed]

61. Xing, F.; Saidou, J.; Watabe, K. Cancer associated fibroblasts (cafs) in tumor microenvironment. *Front. Biosci. (Landmark Ed.)* **2010**, *15*, 166–179. [CrossRef] [PubMed]

62. Annicotte, J.S.; Chavey, C.; Servant, N.; Teyssier, J.; Bardin, A.; Licznar, A.; Badia, E.; Pujol, P.; Vignon, F.; Maudelonde, T.; et al. The nuclear receptor liver receptor homolog-1 is an estrogen receptor target gene. *Oncogene* **2005**, *24*, 8167–8175. [CrossRef] [PubMed]

63. Clyne, C.D.; Kovacic, A.; Speed, C.J.; Zhou, J.; Pezzi, V.; Simpson, E.R. Regulation of aromatase expression by the nuclear receptor lrh-1 in adipose tissue. *Mol. Cell. Endocrinol.* **2004**, *215*, 39–44. [CrossRef] [PubMed]

64. Chand, A.L.; Herridge, K.A.; Howard, T.L.; Simpson, E.R.; Clyne, C.D. Tissue-specific regulation of aromatase promoter ii by the orphan nuclear receptor lrh-1 in breast adipose stromal fibroblasts. *Steroids* **2011**, *76*, 741–744. [CrossRef] [PubMed]

65. Miki, Y.; Clyne, C.D.; Suzuki, T.; Moriya, T.; Shibuya, R.; Nakamura, Y.; Ishida, T.; Yabuki, N.; Kitada, K.; Hayashi, S.; et al. Immunolocalization of liver receptor homologue-1 (lrh-1) in human breast carcinoma: Possible regulator of insitu steroidogenesis. *Cancer Lett.* **2006**, *244*, 24–33. [CrossRef] [PubMed]

66. Guo, R.X.; Wei, L.H.; Tu, Z.; Sun, P.M.; Wang, J.L.; Zhao, D.; Li, X.P.; Tang, J.M. 17 beta-estradiol activates pi3k/akt signaling pathway by estrogen receptor (er)-dependent and er-independent mechanisms in endometrial cancer cells. *J. Steroid Biochem. Mol. Biol.* **2006**, *99*, 9–18. [CrossRef] [PubMed]

67. Stabile, L.P.; Lyker, J.S.; Gubish, C.T.; Zhang, W.; Grandis, J.R.; Siegfried, J.M. Combined targeting of the estrogen receptor and the epidermal growth factor receptor in non-small cell lung cancer shows enhanced antiproliferative effects. *Cancer Res.* **2005**, *65*, 1459–1470. [CrossRef] [PubMed]

68. Keshamouni, V.G.; Mattingly, R.R.; Reddy, K.B. Mechanism of 17-beta-estradiol-induced erk1/2 activation in breast cancer cells. A role for her2 and pkc-delta. *J. Biol. Chem.* **2002**, *277*, 22558–22565. [CrossRef] [PubMed]

69. Yeh, C.R.; Slavin, S.; Da, J.; Hsu, I.; Luo, J.; Xiao, G.Q.; Ding, J.; Chou, F.J.; Yeh, S. Estrogen receptor alpha in cancer associated fibroblasts suppresses prostate cancer invasion via reducing ccl5, il6 and macrophage infiltration in the tumor microenvironment. *Mol. Cancer* **2016**, *15*, 7. [CrossRef] [PubMed]

70. Slavin, S.; Yeh, C.R.; Da, J.; Yu, S.; Miyamoto, H.; Messing, E.M.; Guancial, E.; Yeh, S. Estrogen receptor alpha in cancer-associated fibroblasts suppresses prostate cancer invasion via modulation of thrombospondin 2 and matrix metalloproteinase 3. *Carcinogenesis* **2014**, *35*, 1301–1309. [CrossRef] [PubMed]

71. Aldinucci, D.; Colombatti, A. The inflammatory chemokine ccl5 and cancer progression. *Mediat. Inflamm.* **2014**, *2014*, 292376. [CrossRef] [PubMed]

72. Kumari, N.; Dwarakanath, B.S.; Das, A.; Bhatt, A.N. Role of interleukin-6 in cancer progression and therapeutic resistance. *Tumour Biol.* **2016**, *37*, 11553–11572. [CrossRef] [PubMed]

73. Qian, B.Z.; Pollard, J.W. Macrophage diversity enhances tumor progression and metastasis. *Cell* **2010**, *141*, 39–51. [CrossRef] [PubMed]

74. Liu, Y.; Cao, X. The origin and function of tumor-associated macrophages. *Cell. Mol. Immunol.* **2015**, *12*, 1–4. [CrossRef] [PubMed]

75. Mantovani, A.; Sozzani, S.; Locati, M.; Allavena, P.; Sica, A. Macrophage polarization: Tumor-associated macrophages as a paradigm for polarized m2 mononuclear phagocytes. *Trends Immunol.* **2002**, *23*, 549–555. [CrossRef]

76. Lee, S.; Margolin, K. Cytokines in cancer immunotherapy. *Cancers* **2011**, *3*, 3856–3893. [CrossRef] [PubMed]

77. Bingle, L.; Brown, N.J.; Lewis, C.E. The role of tumour-associated macrophages in tumour progression: Implications for new anticancer therapies. *J. Pathol.* **2002**, *196*, 254–265. [CrossRef] [PubMed]

78. Wan, T.; Liu, J.H.; Zheng, L.M.; Cai, M.Y.; Ding, T. Prognostic significance of tumor-associated macrophage infiltration in advanced epithelial ovarian carcinoma. *Chin. J. Cancer* **2009**, *28*, 268–271.

79. Gwak, J.M.; Jang, M.H.; Kim, D.I.; Seo, A.N.; Park, S.Y. Prognostic value of tumor-associated macrophages according to histologic locations and hormone receptor status in breast cancer. *PLoS ONE* **2015**, *10*, e0125728. [CrossRef] [PubMed]

80. Campbell, M.J.; Tonlaar, N.Y.; Garwood, E.R.; Huo, D.; Moore, D.H.; Khramtsov, A.I.; Au, A.; Baehner, F.; Chen, Y.; Malaka, D.O.; et al. Proliferating macrophages associated with high grade, hormone receptor negative breast cancer and poor clinical outcome. *Breast Cancer Res. Treat.* **2011**, *128*, 703–711. [CrossRef] [PubMed]

81. Svensson, S.; Abrahamsson, A.; Rodriguez, G.V.; Olsson, A.K.; Jensen, L.; Cao, Y.; Dabrosin, C. Ccl2 and ccl5 are novel therapeutic targets for estrogen-dependent breast cancer. *Clin. Cancer Res.* **2015**, *21*, 3794–3805. [CrossRef] [PubMed]

82. Okizaki, S.; Ito, Y.; Hosono, K.; Oba, K.; Ohkubo, H.; Kojo, K.; Nishizawa, N.; Shibuya, M.; Shichiri, M.; Majima, M. Vascular endothelial growth factor receptor type 1 signaling prevents delayed wound healing in diabetes by attenuating the production of il-1beta by recruited macrophages. *J. Pathol.* **2016**, *186*, 1481–1498. [CrossRef] [PubMed]

83. Stabile, L.P.; Farooqui, M.; Kanterewicz, B.; Abberbock, S.; Kurland, B.F.; Diergaarde, B.; Siegfried, J.M. Preclinical evidence for combined use of aromatase inhibitors and nsaids as preventive agents of tobacco-induced lung cancer. *J. Thorac. Oncol.* **2017**. [CrossRef] [PubMed]

84. Ning, C.; Xie, B.; Zhang, L.; Li, C.; Shan, W.; Yang, B.; Luo, X.; Gu, C.; He, Q.; Jin, H.; et al. Infiltrating macrophages induce eralpha expression through an il17a-mediated epigenetic mechanism to sensitize endometrial cancer cells to estrogen. *Cancer Res.* **2016**, *76*, 1354–1366. [CrossRef] [PubMed]

85. Sun, L.; Chen, B.; Jiang, R.; Li, J.; Wang, B. Resveratrol inhibits lung cancer growth by suppressing m2-like polarization of tumor associated macrophages. *Cell. Immunol.* **2017**, *311*, 86–93. [CrossRef] [PubMed]

86. Umansky, V.; Blattner, C.; Gebhardt, C.; Utikal, J. The role of myeloid-derived suppressor cells (mdsc) in cancer progression. *Vaccines (Basel)* **2016**, *4*, 36. [CrossRef] [PubMed]

87. Gabrilovich, D.I.; Ostrand-Rosenberg, S.; Bronte, V. Coordinated regulation of myeloid cells by tumours. *Nat. Rev. Immunol.* **2012**, *12*, 253–268. [CrossRef] [PubMed]

88. Fridman, W.H.; Pages, F.; Sautes-Fridman, C.; Galon, J. The immune contexture in human tumours: Impact on clinical outcome. *Nat. Rev. Cancer* **2012**, *12*, 298–306. [CrossRef] [PubMed]

89. Haabeth, O.A.; Lorvik, K.B.; Hammarstrom, C.; Donaldson, I.M.; Haraldsen, G.; Bogen, B.; Corthay, A. Inflammation driven by tumour-specific th1 cells protects against b-cell cancer. *Nat. Commun.* **2011**, *2*, 240. [CrossRef] [PubMed]

90. DeNardo, D.G.; Barreto, J.B.; Andreu, P.; Vasquez, L.; Tawfik, D.; Kolhatkar, N.; Coussens, L.M. Cd4(+) t cells regulate pulmonary metastasis of mammary carcinomas by enhancing protumor properties of macrophages. *Cancer Cell* **2009**, *16*, 91–102. [CrossRef] [PubMed]

91. Dannenfelser, R.; Nome, M.; Tahiri, A.; Ursini-Siegel, J.; Vollan, H.K.M.; Haakensen, V.D.; Helland, A.; Naume, B.; Caldas, C.; Borresen-Dale, A.L.; et al. Data-driven analysis of immune infiltrate in a large cohort of breast cancer and its association with disease progression, er activity, and genomic complexity. *Oncotarget* **2017**, *8*, 57121–57133. [CrossRef] [PubMed]

92. Ali, H.R.; Provenzano, E.; Dawson, S.J.; Blows, F.M.; Liu, B.; Shah, M.; Earl, H.M.; Poole, C.J.; Hiller, L.; Dunn, J.A.; et al. Association between cd8+ t-cell infiltration and breast cancer survival in 12,439 patients. *Ann. Oncol.* **2014**, *25*, 1536–1543. [CrossRef] [PubMed]

93. Cullen, S.P.; Martin, S.J. Mechanisms of granule-dependent killing. *Cell Death Differ.* **2008**, *15*, 251–262. [CrossRef] [PubMed]

94. Lieberman, J. The abcs of granule-mediated cytotoxicity: New weapons in the arsenal. *Nat. Rev. Immunol.* **2003**, *3*, 361–370. [CrossRef] [PubMed]

95. Jiang, X.; Orr, B.A.; Kranz, D.M.; Shapiro, D.J. Estrogen induction of the granzyme b inhibitor, proteinase inhibitor 9, protects cells against apoptosis mediated by cytotoxic t lymphocytes and natural killer cells. *Endocrinology* **2006**, *147*, 1419–1426. [CrossRef] [PubMed]

96. Jiang, X.; Ellison, S.J.; Alarid, E.T.; Shapiro, D.J. Interplay between the levels of estrogen and estrogen receptor controls the level of the granzyme inhibitor, proteinase inhibitor 9 and susceptibility to immune surveillance by natural killer cells. *Oncogene* **2007**, *26*, 4106–4114. [CrossRef] [PubMed]

97. Tanaka, A.; Sakaguchi, S. Regulatory t cells in cancer immunotherapy. *Cell Res.* **2017**, *27*, 109–118. [CrossRef] [PubMed]

98. Tai, P.; Wang, J.; Jin, H.; Song, X.; Yan, J.; Kang, Y.; Zhao, L.; An, X.; Du, X.; Chen, X.; et al. Induction of regulatory t cells by physiological level estrogen. *J. Cell. Physiol.* **2008**, *214*, 456–464. [CrossRef] [PubMed]

99. Polanczyk, M.J.; Carson, B.D.; Subramanian, S.; Afentoulis, M.; Vandenbark, A.A.; Ziegler, S.F.; Offner, H. Cutting edge: Estrogen drives expansion of the cd4+cd25+ regulatory t cell compartment. *J. Immunol.* **2004**, *173*, 2227–2230. [CrossRef] [PubMed]

100. Fontenot, J.D.; Gavin, M.A.; Rudensky, A.Y. Foxp3 programs the development and function of cd4+cd25+ regulatory t cells. *Nat. Immunol.* **2003**, *4*, 330–336. [CrossRef] [PubMed]

101. Chaudhary, B.; Elkord, E. Regulatory t cells in the tumor microenvironment and cancer progression: Role and therapeutic targeting. *Vaccines (Basel)* **2016**, *4*. [CrossRef] [PubMed]

102. Kadota, K.; Eguchi, T.; Villena-Vargas, J.; Woo, K.M.; Sima, C.S.; Jones, D.R.; Travis, W.D.; Adusumilli, P.S. Nuclear estrogen receptor-alpha expression is an independent predictor of recurrence in male patients with pt1an0 lung adenocarcinomas, and correlates with regulatory t-cell infiltration. *Oncotarget* **2015**, *6*, 27505–27518. [CrossRef] [PubMed]

103. Shang, B.; Liu, Y.; Jiang, S.J.; Liu, Y. Prognostic value of tumor-infiltrating foxp3+ regulatory t cells in cancers: A systematic review and meta-analysis. *Sci. Rep.* **2015**, *5*, 15179. [CrossRef] [PubMed]

104. Generali, D.; Bates, G.; Berruti, A.; Brizzi, M.P.; Campo, L.; Bonardi, S.; Bersiga, A.; Allevi, G.; Milani, M.; Aguggini, S.; et al. Immunomodulation of foxp3+ regulatory t cells by the aromatase inhibitor letrozole in breast cancer patients. *Clin. Cancer Res.* **2009**, *15*, 1046–1051. [CrossRef] [PubMed]

105. Polanczyk, M.J.; Hopke, C.; Vandenbark, A.A.; Offner, H. Treg suppressive activity involves estrogen-dependent expression of programmed death-1 (pd-1). *Int. Immunol.* **2007**, *19*, 337–343. [CrossRef] [PubMed]

106. Yang, L.; Huang, F.; Mei, J.; Wang, X.; Zhang, Q.; Wang, H.; Xi, M.; You, Z. Posttranscriptional control of pd-l1 expression by 17beta-estradiol via pi3k/akt signaling pathway in eralpha-positive cancer cell lines. *Int. J. Gynecol. Cancer* **2017**, *27*, 196–205. [CrossRef] [PubMed]

107. Jiang, Y.; Li, Y.; Zhu, B. T-cell exhaustion in the tumor microenvironment. *Cell Death Dis.* **2015**, *6*, e1792. [CrossRef] [PubMed]

108. Yoshimura, A. Signal transduction of inflammatory cytokines and tumor development. *Cancer Sci.* **2006**, *97*, 439–447. [CrossRef] [PubMed]

109. Sasser, A.K.; Sullivan, N.J.; Studebaker, A.W.; Hendey, L.F.; Axel, A.E.; Hall, B.M. Interleukin-6 is a potent growth factor for er-alpha-positive human breast cancer. *FASEB J.* **2007**, *21*, 3763–3770. [CrossRef] [PubMed]

110. Studebaker, A.W.; Storci, G.; Werbeck, J.L.; Sansone, P.; Sasser, A.K.; Tavolari, S.; Huang, T.; Chan, M.W.; Marini, F.C.; Rosol, T.J.; et al. Fibroblasts isolated from common sites of breast cancer metastasis enhance cancer cell growth rates and invasiveness in an interleukin-6-dependent manner. *Cancer Res.* **2008**, *68*, 9087–9095. [CrossRef] [PubMed]

111. Yin, Y.; Chen, X.; Shu, Y. Gene expression of the invasive phenotype of tnf-alpha-treated mcf-7 cells. *Biomed. Pharmacother.* **2009**, *63*, 421–428. [CrossRef] [PubMed]

112. Zhao, Y.; Nichols, J.E.; Valdez, R.; Mendelson, C.R.; Simpson, E.R. Tumor necrosis factor-alpha stimulates aromatase gene expression in human adipose stromal cells through use of an activating protein-1 binding site upstream of promoter 1.4. *Mol. Endocrinol.* **1996**, *10*, 1350–1357. [PubMed]

113. Irahara, N.; Miyoshi, Y.; Taguchi, T.; Tamaki, Y.; Noguchi, S. Quantitative analysis of aromatase mrna expression derived from various promoters (i.4, i.3, pii and i.7) and its association with expression of tnf-alpha, il-6 and cox-2 mrnas in human breast cancer. *Int. J. Cancer* **2006**, *118*, 1915–1921. [CrossRef] [PubMed]

114. Ricciotti, E.; FitzGerald, G.A. Prostaglandins and inflammation. *Arterioscler. Thromb. Vasc. Biol.* **2011**, *31*, 986–1000. [CrossRef] [PubMed]

115. Zhao, Y.; Agarwal, V.R.; Mendelson, C.R.; Simpson, E.R. Estrogen biosynthesis proximal to a breast tumor is stimulated by pge2 via cyclic amp, leading to activation of promoter ii of the cyp19 (aromatase) gene. *Endocrinology* **1996**, *137*, 5739–5742. [CrossRef] [PubMed]

116. Terry, M.B.; Gammon, M.D.; Zhang, F.F.; Tawfik, H.; Teitelbaum, S.L.; Britton, J.A.; Subbaramaiah, K.; Dannenberg, A.J.; Neugut, A.I. Association of frequency and duration of aspirin use and hormone receptor status with breast cancer risk. *JAMA* **2004**, *291*, 2433–2440. [CrossRef] [PubMed]

117. Zhou, X.L.; Fan, W.; Yang, G.; Yu, M.X. The clinical significance of pr, er, nf- kappa b, and tnf- alpha in breast cancer. *Dis. Markers* **2014**, *2014*, 494581. [CrossRef] [PubMed]

118. Hoesel, B.; Schmid, J.A. The complexity of nf-kappab signaling in inflammation and cancer. *Mol. Cancer* **2013**, *12*, 86. [CrossRef] [PubMed]

119. Johnston, S.R.; Lu, B.; Scott, G.K.; Kushner, P.J.; Smith, I.E.; Dowsett, M.; Benz, C.C. Increased activator protein-1 DNA binding and c-jun nh2-terminal kinase activity in human breast tumors with acquired tamoxifen resistance. *Clin. Cancer Res.* **1999**, *5*, 251–256. [PubMed]

120. Zhou, Y.; Yau, C.; Gray, J.W.; Chew, K.; Dairkee, S.H.; Moore, D.H.; Eppenberger, U.; Eppenberger-Castori, S.; Benz, C.C. Enhanced nf kappa b and ap-1 transcriptional activity associated with antiestrogen resistant breast cancer. *BMC Cancer* **2007**, *7*, 59. [CrossRef] [PubMed]

121. Pardoll, D.M. The blockade of immune checkpoints in cancer immunotherapy. *Nat. Rev. Cancer* **2012**, *12*, 252–264. [CrossRef] [PubMed]

122. Wolchok, J.D.; Chiarion-Sileni, V.; Gonzalez, R.; Rutkowski, P.; Grob, J.J.; Cowey, C.L.; Lao, C.D.; Wagstaff, J.; Schadendorf, D.; Ferrucci, P.F.; et al. Overall survival with combined nivolumab and ipilimumab in advanced melanoma. *N. Engl. J. Med.* **2017**, *377*, 1345–1356. [CrossRef] [PubMed]

123. Reck, M.; Rodriguez-Abreu, D.; Robinson, A.G.; Hui, R.; Csoszi, T.; Fulop, A.; Gottfried, M.; Peled, N.; Tafreshi, A.; Cuffe, S.; et al. Pembrolizumab versus chemotherapy for pd-l1-positive non-small-cell lung cancer. *N. Engl. J. Med.* **2016**, *375*, 1823–1833. [CrossRef] [PubMed]

124. Brahmer, J.; Reckamp, K.L.; Baas, P.; Crino, L.; Eberhardt, W.E.; Poddubskaya, E.; Antonia, S.; Pluzanski, A.; Vokes, E.E.; Holgado, E.; et al. Nivolumab versus docetaxel in advanced squamous-cell non-small-cell lung cancer. *N. Engl. J. Med.* **2015**, *373*, 123–135. [CrossRef] [PubMed]

125. Wang, X.; Bao, Z.; Zhang, X.; Li, F.; Lai, T.; Cao, C.; Chen, Z.; Li, W.; Shen, H.; Ying, S. Effectiveness and safety of pd-1/pd-l1 inhibitors in the treatment of solid tumors: A systematic review and meta-analysis. *Oncotarget* **2017**, *8*, 59901–59914. [CrossRef] [PubMed]

126. Patel, S.P.; Kurzrock, R. Pd-l1 expression as a predictive biomarker in cancer immunotherapy. *Mol. Cancer Ther.* **2015**, *14*, 847–856. [CrossRef] [PubMed]

127. Green, A.R.; Aleskandarany, M.A.; Ali, R.; Hodgson, E.G.; Atabani, S.; De Souza, K.; Rakha, E.A.; Ellis, I.O.; Madhusudan, S. Clinical impact of tumor DNA repair expression and t-cell infiltration in breast cancers. *Cancer Immunol. Res.* **2017**, *5*, 292–299. [CrossRef] [PubMed]

128. McGranahan, N.; Rosenthal, R.; Hiley, C.T.; Rowan, A.J.; Watkins, T.B.K.; Wilson, G.A.; Birkbak, N.J.; Veeriah, S.; Van Loo, P.; Herrero, J.; et al. Allele-specific hla loss and immune escape in lung cancer evolution. *Cell* **2017**, *171*, 1259.e11–1271.e11. [CrossRef] [PubMed]

129. Marty, R.; Kaabinejadian, S.; Rossell, D.; Slifker, M.J.; van de Haar, J.; Engin, H.B.; de Prisco, N.; Ideker, T.; Hildebrand, W.H.; Font-Burgada, J.; et al. Mhc-i genotype restricts the oncogenic mutational landscape. *Cell* **2017**, *171*, 1272.e15–1283.e15. [CrossRef] [PubMed]

130. Hamilton, D.H.; Griner, L.M.; Keller, J.M.; Hu, X.; Southall, N.; Marugan, J.; David, J.M.; Ferrer, M.; Palena, C. Targeting estrogen receptor signaling with fulvestrant enhances immune and chemotherapy-mediated cytotoxicity of human lung cancer. *Clin. Cancer Res.* **2016**, *22*, 6204–6216. [CrossRef] [PubMed]

131. Welte, T.; Zhang, X.H.; Rosen, J.M. Repurposing antiestrogens for tumor immunotherapy. *Cancer Discov.* **2017**, *7*, 17–19. [CrossRef] [PubMed]

International Journal of
Molecular Sciences

MDPI

Review

Estrogen Receptor Signaling in Radiotherapy: From Molecular Mechanisms to Clinical Studies

Chao Rong [1,*], Étienne Fasolt Richard Corvin Meinert [1,2] and Jochen Hess [1,2]

[1] Section Experimental and Translational Head and Neck Oncology, Department of Otolaryngology, Head and Neck Surgery, University Hospital Heidelberg, 69120 Heidelberg, Germany; fasolt.meinert@med.uni-heidelberg.de (É.F.R.C.M.); Jochen.Hess@med.uni-heidelberg.de (J.H.)

[2] Research Group Molecular Mechanisms of Head and Neck Tumors, German Cancer Research Center (DKFZ), 69120 Heidelberg, Germany

* Correspondence: Chao.Rong@med.uni-heidelberg.de; Tel.: +49-6221-56-7278; Fax: +49-6221-56-4604

Received: 28 December 2017; Accepted: 26 February 2018; Published: 2 March 2018

Abstract: Numerous studies have established a proof of concept that abnormal expression and function of estrogen receptors (ER) are crucial processes in initiation and development of hormone-related cancers and also affect the efficacy of anti-cancer therapy. Radiotherapy has been applied as one of the most common and potent therapeutic strategies, which is synergistic with surgical excision, chemotherapy and targeted therapy for treating malignant tumors. However, the impact of ionizing radiation on ER expression and ER-related signaling in cancer tissue, as well as the interaction between endocrine and irradiation therapy remains largely elusive. This review will discuss recent findings on ER and ER-related signaling, which are relevant for cancer radiotherapy. In addition, we will summarize pre-clinical and clinical studies that evaluate the consequences of anti-estrogen and irradiation therapy in cancer, including emerging studies on head and neck cancer, which might improve the understanding and development of novel therapeutic strategies for estrogen-related cancers.

Keywords: estrogen; estrogen receptor; radiotherapy; radioresistance; breast cancer; head and neck cancer

1. Introduction

Estrogens exert many physiological functions in target tissues mainly via two members of the nuclear receptor superfamily: Estrogen receptor-α (ERα) and ERβ. They are encoded by separate genes, ESR1 and ESR2, respectively, transcribed from various chromosomal locations, and multiple mRNA splice variants exist for both receptors in normal and disease states [1,2]. On the structural level, both receptors possess five distinct structural and functional domains, harboring a DNA-binding domain (DBD), a ligand-binding domain (LBD), hinge domain and two transcriptional activation functions (AF-1, AF-2). As members of the hormone nuclear receptor superfamily, they share over 50% similarity in their hormone-binding domains and a predicted 96% similarity within the DBDs. However, there is a lower degree of sequence similarity within their hormone-independent AF-1 domains (Figure 1A) [3,4].

ERs and their variants mediate distinct effects as transcription factors in the nucleus when they are bound to their specific ligands through various mechanisms, which could be explained by genomic or non-genomic signaling pathways [5]. In the genomic mode of ER action, the ligands (e.g., estrogen hormones) diffuse into the cell and bind to the LBDs of the receptors, which result in homo- or heterodimer formation and subsequent binding to DNA at estrogen responsive element (ERE) sequences. Once bound to EREs the ligand-ER complex can modify gene expression by recruitment of distinct co-regulatory proteins, known as co-activators and co-repressors, or by interaction with other transcription factors, such as activator protein 1 (AP1), specificity protein 1 (SP1), and others [6]. In contrast, estrogen can elicit rapid response via non-genomic signaling pathways, which depend on the presence of a secondary messenger such as cyclic adenosine monophosphate (cAMP) and calcium, or the activation of protein kinases (Figure 1B) [5].

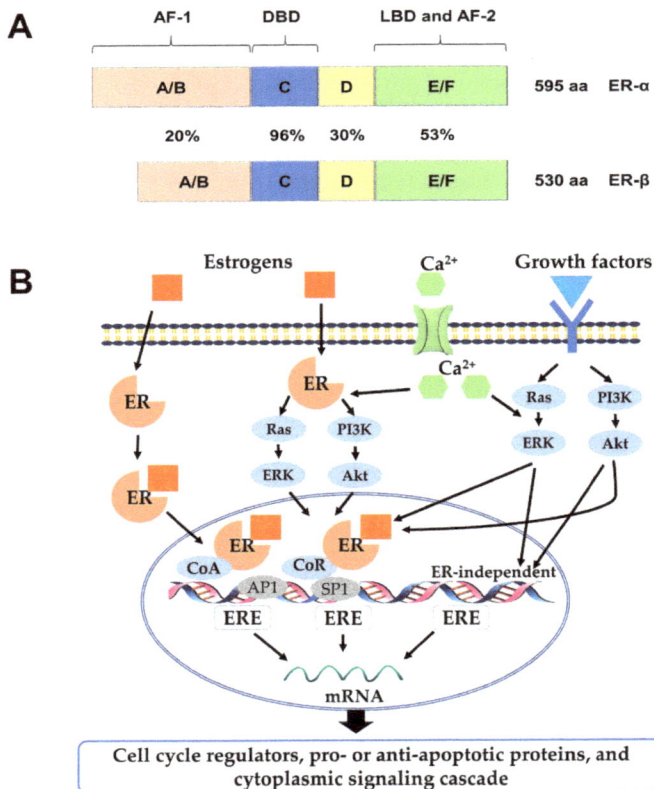

Figure 1. (**A**) Structural and functional domains of the ERα and ERβ. Structural domains of estrogen receptor α (ERα) (595aa) and ERβ (530aa) are labeled A-F. Both receptors have five distinct structural and functional domains: DNA-binding domain (DBD; C), hinge domain (D), ligand-binding domain (LBD; E/F), and two transcriptional activation function domains AF-1 (A/B) and AF-2 (F). The percentage of amino acid homologies between ERα and ERβ domains is also indicated; (**B**) Schematic illustration of ER-mediated signaling pathways. In the classical mechanism of ER action, estrogens (E2) bind to ERs and the E2-ER complex binds directly to estrogen response elements (EREs). Once bound to EREs the E2-ER complex can modify gene expression by the recruitment of distinct co-regulatory proteins, known as co-activators and co-repressors. In the ERE-independent genomic action, nuclear E2-ERs complexes interact with other transcription factors, such as activator protein 1 (AP1) or specificity protein 1 (SP1). In the ligand-independent genomic action, growth factors activate protein kinase cascades, such as Ras-ERK or PI3K-Akt, causing activation of nuclear transcription factors. In the non-genomic action, the E2-ERs complex activates protein-kinase cascades or cyclin adenosine monophosphate (cAMP) and calcium, leading to altered functions of proteins in the cytoplasm. ERK: extracellular signal–regulated kinase; PI3K: phosphatidylinositide 3-kinase; CoR: co-repressor; CoA: co-activator.

Over the last few decades, a growing number of studies have established a proof of concept that abnormal expression and regulation of ERs are crucial events in initiation and development of hormone-related cancers and are related to the outcome of cancer therapy. Many lines of evidence indicate that ERα and ERβ might perform different functions during carcinogenesis and anti-cancer therapy [2,7]. Currently, radiotherapy is used as one of the most common and potent cancer therapeutic strategies. It acts synergistically with surgical excision, chemotherapy and targeted therapy for treating malignant tumors in human. Encouragingly, clinical studies revealed that ER-positive breast cancer can be targeted by radiotherapy in combination with the modulation of ER activity, namely, endocrine therapy [8]. Tamoxifen belongs to the most frequently prescribed selective ER modulators (SERMs), which have been an effective and safe adjuvant endocrine therapy for several decades. However, the molecular mechanisms through which ionizing radiation (IR) regulates ER activity in cancer tissue and whether ER signaling has an impact on the efficacy of radiotherapy in various types of malignancies, remain largely elusive. Furthermore, the variability of ERα and ERβ expression, diverse response of ER and ER-related signaling to irradiation both contribute to the risk of safety and efficacy of cancer therapy.

In this review, we will discuss distinct functions of ER and ER-related signaling that are relevant to cancer radiotherapy. In addition, we will summarize pre-clinical and clinical studies that evaluate the consequences of anti-estrogen and irradiation therapy in cancer, including emerging studies on head and neck cancer, which might improve the understanding and development of novel therapeutic strategies for estrogen-related cancers.

2. Estrogen Receptor Signaling and Ionizing Radiation

2.1. Molecular and Cellular Responses to Ionizing Radiation

Radiotherapy is mainly based on the principle that normal tissue cells exhibit greater DNA repair capacity than carcinoma cells upon damage due to ionizing radiation [9]. Nowadays, ionizing radiation has become a widely applied treatment strategy for the majority of solid cancers. A series of biological effects on genomic DNA, which is considered as the most important target molecule, can be induced by photons, electrons, or heavy ions, which are generated by linear accelerators [10,11]. The biochemical lesions in genomic DNA of cancer cells can be achieved in a direct and indirect manner. A therapeutic dose of linear energy transfer (LET), such as particles or neutrons, can directly cause DNA damage, including single-strand breaks (SSB), modified bases, damage of the sugar backbone, double strand breaks (DSB) as well as effects on DNA repair [12,13]. Indirect effects are enforced by the generation of reactive oxygen species (ROS) that target and damage genomic DNA (Figure 2) [14]. It is worth noting that estrogens can also induce ROS in breast cancer cells, resulting in elevated genomic instability and a higher degree of clonal heterogeneity. The efficacy of radiotherapy might be modulated by estrogen-reduced ROS. The function of estrogen-induced ROS production in breast cancer has been reviewed previously by Okoh and coworkers [15]. Most types of human cells dispose of DNA damage with complicated response mechanisms, collectively named DNA-damage response (DDR), regardless of whether the damage is induced in a direct or indirect mode of action. Mechanisms of DDR can be activated and arrest the cell cycle at specific checkpoints, executing either DNA repair or induce programmed cell death (namely apoptosis) and cellular senescence, which are critical for maintaining cellular genomic integrity and for preventing neoplastic transformation [16].

DSB damage is the most lethal type of DNA damage induced by ionizing radiation [17]. DNA repair is the frontline response to cellular DNA damage, which also contributes to irradiation resistance in tumor cells. Efficient DNA repair enables tumor cells to replicate and survive. Generally, DSB damage repair is carried out by two major pathways: non-homologous end jointing (NHEJ) and conservative homologous recombination (HR), which have been extensively reviewed previously [18–20]. NHEJ is considered as the primary DSB repair pathway, which is activated throughout the cell cycle and relies on rejoining free DNA ends without the requirement for sequence homology. During this repair process, DNA strands of DSB sites are cut or modified, and the ligation of DNA ends are achieved directly and quickly regardless of homology, deletions or insertions. Although this makes NHEJ possibly error-prone, this mechanism can repair the DNA damage rapidly to eliminate potential genetic instability [21]. It is worth noting that estrogens have been shown to induce components of NHEJ in breast cancer cells and that therapeutic targeting of ERs result in irreparable DSB [22]. HR is widely known as a more precise mode of repair, which uses an undamaged template to retrieve the chromatid sequence content missing at the DSB sites. During HR, the damaged chromatid physically contacts with an undamaged sister chromatid with a homologous sequence for genetic information restoration in the late S/G2 phase of the cell cycle [23]. Compelling experimental evidence implicates that estrogens mediate both positive and negative regulation of HR. In melanoma, the tumor suppressor gene *MEN1* and ERα stimulate the transcription of *BRCA1*, *RAD51* and *RAD51AP1*, which encode key players in HR-directed DNA repair. Fulvestrant inhibits *BRCA1*, *RAD51* and *RAD51AP1* expression, resulting in decreased HR activity [24]. However, in medulloblastoma, an enhanced ERβ activity has been associated with nuclear translocation of insulin receptor substrate 1 (IRS-1), which interacts with *RAD51* at the sites of damaged DNA and reduces the HR function [25]. Pharmacological inhibition of ERβ induces medulloblastoma cells resistance to cisplatin by elevated formation of *RAD51* and increased levels of HR [26].

Cell cycle progression can be arrested at distinct cell cycle checkpoints temporarily, which are the G1 checkpoint during transition from G1 to S phase, and the G2 checkpoint of G2/M phase boundary. After perception of DNA lesions induced by IR, various biochemical signals are activated by well-defined cascades of protein kinases. Ataxia telangiectasia mutated (ATM) and ATM- and Rad3-related (ATR) kinases are upstream activators of IR-induced checkpoint arrest. ATM and ATR fulfill their physiological functions via phosphorylation of numerous substrates, such as Chk (checkpoint kinase) 2 and Chk 1, respectively, which are essential for cell cycle arrest at G1/S or G2/M in response to DNA damage [27,28]. The G1/S checkpoint pathway is mainly operated by two key effectors, namely the p53 transcription factor and the cell division cycle 25 A (Cdc25A) phosphatase, which regulate two distinct branches. The key effector for the G2/M checkpoint is the Cyclin B/Cdk1 protein complex, whose activation after IR-induced DNA damage is regulated by ATM/Chk2 and ATR/Chk1 (see Figure 2 for details) [29].

2.2. Interaction between Estrogen Receptor Signaling and Ionizing Irradiation

It is well established that estrogens regulate cell cycle progression in hormone-related carcinomas [30]. Therefore, the influence of estrogens or estrogen modulators on cell cycle progression is a critical factor for the interaction between ER signaling and IR. In MCF-7 cells, a major effect of estrogen is the activation of cell cycle progression by induction of G1 phase entry and shortening of the G1/S transition [31]. This effect is at least in part due to induced transcription of *c-Myc* and *Cyclin D1*, two key regulators of cell cycle progression [31]. Genomic approaches were applied to demonstrate that *c-Myc* regulates radioresistance through transcriptional activation of *Chk1* and *Chk2* by direct binding to their gene promoters in nasopharyngeal carcinoma cells, revealing a potential therapeutic strategy in reduction of radioresistance through blockade of the c-Myc-Chk1/Chk2 pathway [32]. In breast cancer, several studies provided a functional link between estrogen-related signaling and *c-Myc* regulated transcription [33]. Induction of *c-Myc* by estrogen is achieved via binding of ER to an atypical estrogen-responsive cis-acting element (ERE) in the promoter sequence [34]. Antisense c-Myc phosphorothioate oligonucleotides restrained proliferation of estrogen-stimulated cancer cells [35]. Moreover, induction of *c-Myc* in estrogen deprivation-arrested cells simulated the function of estrogen by restarting the cell cycle progression [36]. It is also worth noting that *Cyclin D1* has been involved in estrogen/anti-estrogen regulation of cell cycle progression by binding and activating *Cdk2* and *Cdk4*. Elevated mRNA levels of *Cyclin D1* precede modifications at the protein level, suggesting that the function of estrogen in Cyclin D1 protein expression is mediated at the transcript level [37,38]. Estrogen triggers transcription of *Cyclin D1* by a cAMP response element (CRE) in the promoter region [39,40]. Induction of Cyclin D1 leads to the formation of Cyclin E-Cdk2 complexes, which results in increased phosphorylation of pRb and S phase progression [41]. These complexes also contribute to decreased Cdk inhibitor p21^{Cip1} and p27^{Kip1} protein levels [31]. In the same study, estrogen-activated Cdk2 and DNA synthesis was restrained by antisense Cdc25A oligonucleotides, while inactive Cyclin E-Cdk2 complexes were reactivated by Cdc25A in vitro and in vivo, identifying Cdc25A as another grow-promoting effector of estrogen action [31].

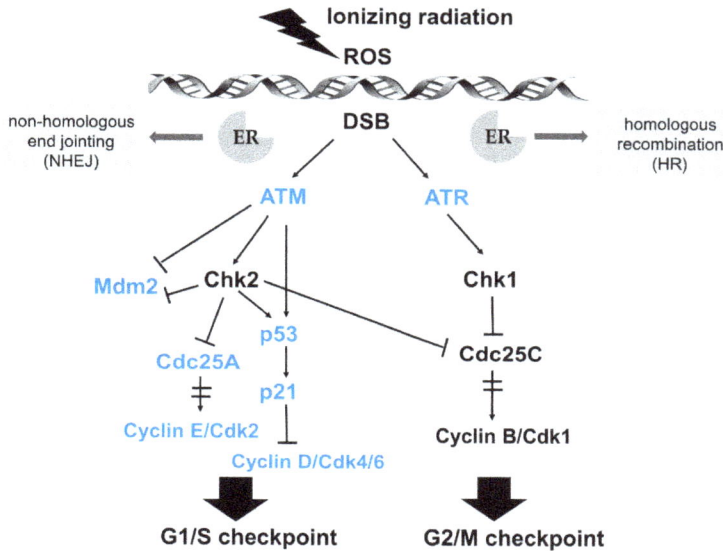

Figure 2. Schematic illustration of the signaling pathways in response to DNA damage and key effectors that interact with ER signaling. After perception of DNA lesions induced by IR directly or indirectly (by ROS generation), various biochemical signals are activated by cascades of protein kinases. Ataxia telangiectasia mutated (ATM) and ATM- and Rad3-related (ATR) kinases are upstream activators of IR-induced G1/S and G2/M checkpoint arrest. The G1/S checkpoint pathway is operated by p53 and Cdc25A in distinct branches. Firstly, ATM or Chk2 directly phosphorylates the p53 transcription factor and targets mouse double minute 2 homolog (Mdm2), achieving the stabilization and accumulation of the p53 protein. The critical effector of p53-dependent transcription is p21, which is a Cdk inhibitor and binds the complexes of Cyclin E/Cdk2 and Cyclin D/Cdk4/6. Another branch of the G1/S checkpoint pathway is activated rapidly via ATM-dependent phosphorylation of Chk2. Subsequently, Cdc25A, an activator of the Cyclin E/Cdk2 kinase, is degraded, preventing the activation of Cdk2. The ATM/Chk2-Cdc25A-Cdk2 axis accounts for the activation of the G1/S checkpoint via a p53-independent mechanism. In the G2/M checkpoint signaling pathway, the key downstream effector is the Cyclin B/Cdk1 protein complex, whose activation is restrained by ATM/Chk2 and ATR/Chk1 after IR-induced DNA damage. Moreover, Cdc25C phosphatase is also inhibited by Chk1/2 to activate the G2/M checkpoint. Key effectors that interact with ER signaling are marked in blue.

Interestingly, a ligand-independent induction in ERα was observed in breast cancer cells after irradiation, which might be a consequence of the cell cycle arrest and related regulatory proteins [42]. Induced cell cycle arrest by low doses of X-ray could be abolished by 17β-estradiol, increasing survival of tumor cells and restraining cellular senescence by the regulation of p21 and Rb-related pathways, but independent of p53 [43]. Molinari and coworkers [44] found that estrogen treatment of breast cancer cell lines modified the intracellular distribution and functional activity of p53, indicating estradiol-induced inactivation of p53 might contribute to carcinogenesis of estrogen-dependent tumors.

ERα has been reported to bind directly to p53 at target gene promoters, such as *CDKN1A* and *PCNA*, resulting in abrogation of p53 function. Moreover, 17β-estradiol promotes the interaction of ERα and p53, consistent with inhibition of p21 transcription [45]. Several studies have also shown that nuclear factor-κB (NFκB), a transcription factor regulating a variety of cellular processes, is linked to ER signaling in breast cancer. These studies suggest a critical role of a functional crosstalk between ER and NFκB in the resistance of cancer cells against endocrine and irradiation therapies [46,47].

Besides transcriptional effects of estrogen, there are also non-genomic signaling pathways to be considered. Estrogen can induce growth factor signal cascades, including insulin-like growth factor I receptor (IGF-IR), mitogen-activated protein kinase (MAPK), phosphatidylinositol-3-kinase (PI3K), and epidermal growth factor receptor (EGFR) signaling, which trigger increased cell proliferation and enhanced radioresistance (Figure 1B) [48–51]. In ER-positive lung cancer, the EGFR directly phosphorylates ERα at specific serine residues [52]. Vice versa, estrogen triggers MAPK and PI3K/AKT signaling pathways to facilitate tumor metastasis through epithelial-to-mesenchymal transition (EMT) [53,54]. EMT is generally considered to be associated with radioresistance in distinct tumors [55–58]. Thus, estrogens exert a radioprotective function via genomic signal pathways and various classical growth factor pathways, indicating a rationale for anti-estrogen treatment to enhance the radiosensitivity of cancer cells (Figure 3).

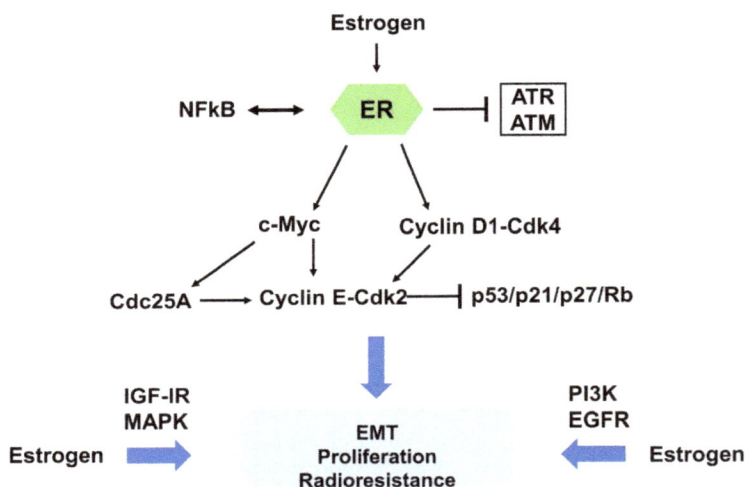

Figure 3. Molecular mechanism of estrogen and ER signaling contributions to radioresistance. The impact of estrogen and ER signaling on cell cycle progression is a critical factor for their contribution to radioresistance. c-Myc and Cyclin D1, two key regulators of cell cycle progression, have significant functions in estrogen and ER signaling mediated radioresistance. In addition, ER can interact with NFκB, a transcription factor, in resistance of cancer cells. Several protein kinase cascades, such as insulin-like growth factor I receptor (IGF-IR), mitogen-activated protein kinase (MAPK), phosphatidylinositol-3-kinase (PI3K), and epidermal growth factor receptor (EGFR) signaling, facilitate EMT, increased cell proliferation and enhanced radioresistance.

3. The Combination of Anti-Estrogen and Irradiation Therapy in Cancer

Anti-estrogen therapy exerts functions by competing with estrogens for binding to ERs, most widely applied for the treatment of women with ER positive breast cancer. In 1971, a new anti-estrogen drug tamoxifen was reported firstly in the management of breast cancer [59]. Until now, tamoxifen reveals a significant clinical benefit and represents the most frequently prescribed anti-estrogen drug [59]. However, only a few studies exist that have addressed the potential value of tamoxifen therapy during or post radiotherapy.

Wazer and colleagues [60] investigated the interaction of tamoxifen and irradiation in the MCF-7 cell line. They observed that growth-inhibitory doses of tamoxifen reduced the radiosensitivity of breast cancer cells, indicating an enhanced repair of irradiation-related DNA damage. In order to unravel the effect of tamoxifen on radiosensitivity, Wazer and coworkers [61] extended the study on the ER-negative cell line MDA-MB-231, in which tamoxifen revealed no alterations in intrinsic radiosensitivity. They hypothesized that the interaction of estrogen, tamoxifen and irradiation in ER-positive breast cancer cells would be achieved by the regulation of the G1/S checkpoint. As introduced above, the G1/S checkpoint can be activated by irradiation to induce cell cycle arrest and to provide the time for DNA repair. This G1 phase block can be enhanced by tamoxifen and attenuated by estrogen. These observations have been confirmed by several studies with similar experimental conditions [62–64]. Interestingly, it has been demonstrated that anti-estrogen therapy can change radiosensitivity independent of the ER status, indicating that hormonal modulators might exert their effect via ERs but also in a non-receptor mode of action. Newton and colleagues [65] observed an enhanced apoptotic cell death in MCF-7 cells which were treated by a combination of irradiation and ZM182780, a pure anti-estrogen, or tamoxifen. In a more recent study, tamoxifen enhanced the radiosensitivity of human glioma cells by inducing cell apoptosis and sustaining G2/M arrest [66]. Moreover, fulvestrant, another pure anti-estrogen drug, revealed a positive effect on radiosensitization of ER-positive breast cancer cells by inducing cell cycle arrest and inhibiting proteins involved in DSB repair [67]. However, in contrast to these findings, Sarkaria and coworkers [68] reported that tamoxifen had no impact on radiosenstitivty of MCF-7 cells. The reason for these conflicting findings might be that MCF-7 cells exhibit an unusual form of apoptosis induction via activation of caspase-7 due to deficient caspase-3 activity [69]. Therefore, the MCF-7 cell line might not be an appropriate in vitro model to investigate irradiation or tamoxifen-induced apoptosis for breast cancer.

In order to further explore the interaction of anti-estrogen and irradiation therapy, in vivo studies have been conducted. Kantorowitz and colleagues [70] observed that a combination of tamoxifen and irradiation leads to a significant reduction of tumor volume and inhibits occurrence of additional tumors in a rat model of breast cancer induced by 1-methy-1-nitrosourea. Additional animal studies suggested that tamoxifen inhibits the initiation and promotion of irradiation-induced mammary tumors [71,72]. In line with these findings, anti-hormonal drugs, such as mifepristone, ICI182780 and Letrozole showed a sensitizing activity on chemo-radiotherapy in vitro and in vivo by increasing G2/M arrest in cervical cancer [73,74]. These data support the assumption that the interaction of anti-estrogen treatment and irradiation might be related to the regulation of cell cycle checkpoints. Although activation of cell cycle arrest by anti-estrogen drugs can facilitate DNA repair and eliminate irradiation-induced genomic lesions, its therapeutic activity might be due to the suppression of tumor cells repopulation after the irradiation interval. It is worth noting that concurrent treatment with anti-estrogen and irradiation therapy revealed an increased risk of lung fibrosis, cardiac damage and pneumonitis, which could be caused by the induced levels of transforming growth factor beta (TGF-β) [75–78].

In contrast to several in vitro studies, in vivo studies indicate a synergistic effect of concurrent tamoxifen and radiotherapy, which might be due to alterations in the tumor microenvironment. Several randomized controlled trials have shown that concurrent radiotherapy with tamoxifen achieved higher local control in breast cancer patients after lumpectomy compared to the treatment without tamoxifen, indicating that the combination of anti-estrogen and irradiation therapy is effective in the control of invasive cancer [79,80]. However, the optimal sequencing of endocrine therapy relative to radiotherapy remains elusive. An increasing number of retrospective clinical studies in breast cancer suggest that concurrent anti-estrogen and irradiation therapy shows no clear improved local control or favorable clinical outcome (Table 1) [81–87]. Moreover, the question of sequencing of hormonal and irradiation therapy for breast cancer has been addressed by several large randomized trials [87,88].

No significant difference between concurrent or sequential anti-estrogen therapy with irradiation was observed concerning clinical outcome of patients with breast cancer. Again, breast, lung or cardiac fibrosis was detected in patients with concurrent hormonal and irradiation therapy [89–93]. Therefore, it is reasonable for patients to receive hormonal and irradiation therapy sequentially to avoid the risk of toxicities. However, in view of the uncertainty and complexity of these trials, this conclusion should be treated with a great deal of caution. Furthermore, the complexity of the human immune response could explain the discrepancy between preclinical and clinical studies. Preclinical and clinical studies taking into account the paracrine interaction of tumor and immune cells, including novel immune-modulating therapies, are urgently needed to further address and to confirm this conclusion.

Although there is some evidence for a benefit of concurrent radiotherapy and anti-hormonal therapy in breast cancer, it might be more appropriate in other cancer entities. However, Dahhan and colleagues [94] reported a single case of low grade endometrial stromal sarcoma (ESS), where radiotherapy and anti-hormonal therapy were administered at the same time. This treatment resulted in tumor progression and was discontinued. Apart from that single case, no study has been published for other cancer entities that used concurrent radiotherapy and anti-hormonal therapy. Recently, a preclinical study conducted in our group provided experimental evidence for a causal link between ERβ expression and radioresistance in head and neck cancer [95].

Table 1. Summary of clinical studies comparing concurrent and sequential anti-estrogen and irradiation therapy in breast cancer.

Type	Treatment Groups (n)	Tamoxifen or Aromatase Inhibitors	Radiotherapy	Chemotherapy (n)	Follow-up	Outcome	Reference
Retrospective 1976–1999	Concurrent (254) vs. Sequential (241)	generally for 5 years	48 Gy in 2 Gy Fractions with boost to primary tumor bed median total dose 64 Gy	CMF based (71) Adriamycin (42) other (16) none (371)	10.4 years	No difference in overall survival (OS), HR, 1.234; 95% CI, 0.42 to 2.05; No difference in local recurrence, HR, 0.932; 95% CI 0.42 to 2.05	[81]
Retrospective 1980–1995	Concurrent (174) vs. Sequential (104)	20 mg OD or 10 mg BID	Tangents only (182) or tangents and nodal (95) median total dose 64 Gy	Methotrexate-based (67) Doxorubicin-based (44) None (167)	8.6 years	No difference in OS, HR 1.56; 95% CI, 0.87 to 2.79; No difference in relapse-free survival, HR 1.23; 95% CI, 0.63 to 2.41; No difference in local recurrence, HR 1.22; 95% CI, 0.33 to 4.49; No difference in cosmesis, or significant complications.	[82]
Retrospective 1989–1993	Concurrent (202) vs. Sequential (107)	20 mg daily for 5 years	45–50 Gy to whole breast	cyclophosphamide, methotrexate, and fluorouracil (CMF) (156) cyclophosphamide, doxorubicin, and fluorouracil (CAF) (153)	10.3 years	No difference in OS, HR 0.84; 95% CI 0.40 to 1.78; No difference in local recurrence, HR 0.73; 95% CI, 0.26 to 2.04; No difference in grade 3 or 4 hematologic toxicity.	[83]
Retrospective 2001–2008	Concurrent (113) vs. Sequential (151)	anastrozole 1 mg or letrozole 2.5 mg daily for 5 years	50 Gy in 2 Gy Fractions with boost to primary tumor bed median total dose 63.2 Gy	CMF (1) Taxane-based (7) Anthracycline-based (31) Combination of anthracycline and taxane (6)	2.9 years	No differences in clinical outcome and treatment-related complications	[84]
Retrospective 2001–2009	Concurrent (158) vs. Sequential (157)	anastrozole 1 mg or letrozole 2.5 mg daily for 5 years	50 Gy in 2 Gy fractions with a boost of up to 63.2 Gy	Yes (57) None (258)	5.6 years	No difference in disease-free survival. No difference in Grade 3 or 5 toxicities	[85]
Retrospective 1998–2008	Concurrent (57) vs. Sequential (126)	Anastrozole or Tamoxifen	45–54 Gy over an average of 49.5 days	anthracycline or taxane (51) none (132)	2.3 years (Con) 2.6 years (Seq)	No difference in detectable breast fibrosis Concurrent (1.8%) vs. Sequential (4%) in Local recurrence	[86]
Randomized 2005–2007	Concurrent (75) vs. Sequential (75)	2.5 mg Letrozole daily for 5 years	A total dose of 50 Gy in 2 Gy fractions	FEC (28) None (122)	2.2 years	No difference in subcutaneous fibrosis, lung fibrosis and quality of life	[87]

4. ER Signaling in Head and Neck Cancer

In contrast to breast cancer, the role of estrogen and ER-related signaling is less well established in head and neck cancer (HNC). HNC is one of the most common human malignancies with around 600,000 new cases per year worldwide [96]. More than 90% of cases are diagnosed as head and neck squamous cell carcinoma (HNSCC), which develop from the mucosal epithelium of the upper aerodigestive tract. This includes oral cavity, nasopharynx, oropharynx, larynx and hypopharynx. Main risk factors are tobacco and alcohol consumption, human papilloma virus (HPV) and to a lesser extent Epstein–Barr virus (EBV) infection [97,98]. Current treatment options mainly consist of surgery, radiotherapy and chemotherapy, mostly platinum based [99].

Healthy tissues of human oral epithelium and salivary glands express mainly ERβ [100]. However, one study focusing on parotid gland pleomorphic adenoma found ERα and ERβ expression in normal tissue of the parotid gland especially in ductal cells. ERβ expression was enhanced in pleomorphic adenoma compared to normal tissue suggesting a possible role in tumor development [101].

There is also a possible role of ER signaling during HNSCC carcinogenesis. In a cell culture model, premalignant cells showed prominent ERβ but not ERα expression. Estradiol (E2) treatment induced Cytochrome P450 1B1 (CYP1B1), an enzyme that causes formation of carcinogenic metabolites from E2. E2 inhibited apoptosis but did not alter proliferation. In human tissue sections, ERβ showed distinct expression in normal tissue, dysplasia and squamous cell carcinoma (SCC) with most prominent staining in SCC [102]. In contrast, ERα expression was almost absent, suggesting a more prominent role of ERβ-related signaling. On the contrary, when using an artificial overexpression system for ERα combining with E2-treatment, Sumida et al reported an increase in proliferation and expression of EMT-markers [103].

First in vivo evidence for a causal role of ER-related signaling in the pathogenesis of HNSCC emerged in the late 1980s. In a mouse xenograft model for laryngeal cancer estradiol treatment enhanced the kinetic of tumor formation and tumor size [104]. Although there are some studies suggesting that HNSCCs express mostly ERα rather than ERβ [105], most research points in the opposite direction, claiming ERβ outweighing ERα. Two studies reported a more favorable outcome of ERβ positive tumors as compared to ERβ negative tumors [95,106]. Tumors positive for ERα, the effects of which are reported to be counteracted by ERβ, have been associated with slightly poorer survival [102,107]. However, it is worth noting that those results were produced investigating HNSCC from different primary sites and were based on different detection methods for ER expression.

It has also been shown that estrogen signaling exerts different biological effects in HNSCC tumor cells. E2-stimulation activates MAPK signaling in an additive fashion with EGF and induce invasion of HNSCC tumor cells [48]. This activity could be mediated by ERs or GPER1 (G protein-coupled estrogen receptor 1), which was shown to trigger proliferation and migration in laryngeal squamous cell carcinoma (LaSCC) via Interleukin-6(IL-6) and signal transducer and activator of transcription 3 (STAT3) [108]. There are also experimental data indicating that ERβ causes an increase of NOTCH1 expression and thereby favors differentiation in SCCs, including HNSCC. In line with this assumption, ERβ overexpression or treatment with specific agonists inhibited proliferation of SCC cell lines, including HNSCC [109].

More recently, our laboratory demonstrated an accumulation of ERβ-positive HNSCC cells after fractionated irradiation in vitro, suggesting a critical role of ERβ-related functions in radioresistance. Indeed, tamoxifen or Fulvestrant treatment revealed a sensitization of these cells to irradiation, which was accompanied by augmented apoptosis. Radioresistant tumor cells were also positive for submaxillary gland androgen-regulated protein 3A (SMR3A), which is a putative ERβ downstream target and was shown to serve as a prognostic biomarker for HNSCC patients. Accordingly, HNSCC with a high ERβ and SMR3A expression pattern were significantly associated with an unfavorable progression-free survival and disease specific survival [95].

Int. J. Mol. Sci. **2018**, *19*, 713

5. Conclusions and Perspectives

The controversial findings of in vitro and in vivo preclinical studies in breast cancer indicate an enormous complexity and context-dependency with a strong impact on the efficacy of anti-estrogen treatment in combination with radiotherapy. Many questions are waiting to be resolved including better understandings of estrogen and estrogen modulators action on normal and cancer cells, precise mechanisms of interaction of estrogen signaling and irradiation, the development of novel estrogen modulators, as well as effective therapeutic strategies of combination of endocrine therapy and radiotherapy. Moreover, the distinct expression of ER subtypes in various cancer tissues, components of the ER-related signaling cascade, and regulation of many transcription factors, all contribute to a complex situation that impedes the therapeutic efficiency of endocrine therapy and radiotherapy. Therefore, not only breakthroughs from basic and preclinical studies but also translational clinical trials are urgently required to further explore and develop the combination therapies in distinct malignancies. Finally, the oncology research community including academic, hospital, industry and government will need to overcome challenges and achieve an encouraging therapeutic outcome for cancer patients.

Acknowledgments: We acknowledge the financial support of the China Scholarship Council for a PhD fellowship to Chao Rong, Heinrich F.C. Behr-fellowship of the German Cancer Research Center (DKFZ) to Étienne Fasolt Richard Corvin Meinert.

Author Contributions: Chao Rong and Jochen Hess conceived and designed the review; Chao Rong drew the figures; Chao Rong and Étienne Fasolt Richard Corvin Meinert wrote the manuscript; Jochen Hess revised and approved the final manuscript.

Conflicts of Interest: The authors declare no conflict of interest.

References

1. Heldring, N.; Pike, A.; Andersson, S.; Matthews, J.; Cheng, G.; Hartman, J.; Tujague, M.; Strom, A.; Treuter, E.; Warner, M.; et al. Estrogen receptors: How do they signal and what are their targets. *Physiol. Rev.* **2007**, *87*, 905–931. [CrossRef] [PubMed]
2. Thomas, C.; Gustafsson, J.A. The different roles of er subtypes in cancer biology and therapy. *Nat. Rev. Cancer* **2011**, *11*, 597–608. [CrossRef] [PubMed]
3. Osborne, C.K.; Zhao, H.; Fuqua, S.A. Selective estrogen receptor modulators: Structure, function, and clinical use. *J. Clin. Oncol.* **2000**, *18*, 3172–3186. [CrossRef] [PubMed]
4. Osborne, C.K.; Schiff, R.; Fuqua, S.A.; Shou, J. Estrogen receptor: Current understanding of its activation and modulation. *Clin. Cancer Res.* **2001**, *7*, 4338s–4342s. [PubMed]
5. Bjornstrom, L.; Sjoberg, M. Mechanisms of estrogen receptor signaling: Convergence of genomic and nongenomic actions on target genes. *Mol. Endocrinol.* **2005**, *19*, 833–842. [CrossRef] [PubMed]
6. Burns, K.A.; Korach, K.S. Estrogen receptors and human disease: An update. *Arch. Toxicol.* **2012**, *86*, 1491–1504. [CrossRef] [PubMed]
7. Ellem, S.J.; Risbridger, G.P. Treating prostate cancer: A rationale for targeting local oestrogens. *Nat. Rev. Cancer* **2007**, *7*, 621–627. [CrossRef] [PubMed]
8. Ali, S.; Coombes, R.C. Endocrine-responsive breast cancer and strategies for combating resistance. *Nat. Rev. Cancer* **2002**, *2*, 101–112. [CrossRef] [PubMed]
9. Whitaker, S.J. DNA damage by drugs and radiation: What is important and how is it measured? *Eur. J. Cancer* **1992**, *28*, 273–276. [CrossRef]
10. Ward, J.F. DNA damage as the cause of ionizing radiation-induced gene activation. *Radiat. Res.* **1994**, *138*, S85–S88. [CrossRef] [PubMed]
11. Santivasi, W.L.; Xia, F. Ionizing radiation-induced DNA damage, response, and repair. *Antioxid. Redox Signal.* **2014**, *21*, 251–259. [CrossRef] [PubMed]
12. Leadon, S.A. Repair of DNA damage produced by ionizing radiation: A minireview. *Semin. Radiat. Oncol.* **1996**, *6*, 295–305. [CrossRef]
13. Borrego-Soto, G.; Ortiz-Lopez, R.; Rojas-Martinez, A. Ionizing radiation-induced DNA injury and damage detection in patients with breast cancer. *Genet. Mol. Biol.* **2015**, *38*, 420–432. [CrossRef] [PubMed]

14. Henderson, B.W.; Miller, A.C. Effects of scavengers of reactive oxygen and radical species on cell survival following photodynamic treatment in vitro: Comparison to ionizing radiation. *Radiat. Res.* **1986**, *108*, 196–205. [CrossRef] [PubMed]

15. Okoh, V.; Deoraj, A.; Roy, D. Estrogen-induced reactive oxygen species-mediated signalings contribute to breast cancer. *Biochim. Biophys. Acta* **2011**, *1815*, 115–133. [CrossRef] [PubMed]

16. Bartkova, J.; Horejsi, Z.; Koed, K.; Kramer, A.; Tort, F.; Zieger, K.; Guldberg, P.; Sehested, M.; Nesland, J.M.; Lukas, C.; et al. DNA damage response as a candidate anti-cancer barrier in early human tumorigenesis. *Nature* **2005**, *434*, 864–870. [CrossRef] [PubMed]

17. Mahaney, B.L.; Meek, K.; Lees-Miller, S.P. Repair of ionizing radiation-induced DNA double-strand breaks by non-homologous end-joining. *Biochem. J.* **2009**, *417*, 639–650. [CrossRef] [PubMed]

18. Kasparek, T.R.; Humphrey, T.C. DNA double-strand break repair pathways, chromosomal rearrangements and cancer. *Semin. Cell Dev. Biol.* **2011**, *22*, 886–897. [CrossRef] [PubMed]

19. Pallis, A.G.; Karamouzis, M.V. DNA repair pathways and their implication in cancer treatment. *Cancer Metastasis Rev.* **2010**, *29*, 677–685. [CrossRef] [PubMed]

20. Helleday, T.; Petermann, E.; Lundin, C.; Hodgson, B.; Sharma, R.A. DNA repair pathways as targets for cancer therapy. *Nat. Rev. Cancer* **2008**, *8*, 193–204. [CrossRef] [PubMed]

21. Malu, S.; Malshetty, V.; Francis, D.; Cortes, P. Role of non-homologous end joining in V(D)J recombination. *Immunol. Res.* **2012**, *54*, 233–246. [CrossRef] [PubMed]

22. Wan, R.; Wu, J.; Baloue, K.K.; Crowe, D.L. Regulation of the nijmegen breakage syndrome 1 gene *NBS1* by *c-Myc*, *p53* and coactivators mediates estrogen protection from DNA damage in breast cancer cells. *Int. J. Oncol.* **2013**, *42*, 712–720. [CrossRef] [PubMed]

23. Rothkamm, K.; Kruger, I.; Thompson, L.H.; Lobrich, M. Pathways of DNA double-strand break repair during the mammalian cell cycle. *Mol. Cell. Biol.* **2003**, *23*, 5706–5715. [CrossRef] [PubMed]

24. Fang, M.; Xia, F.; Mahalingam, M.; Virbasius, C.M.; Wajapeyee, N.; Green, M.R. MEN1 is a melanoma tumor suppressor that preserves genomic integrity by stimulating transcription of genes that promote homologous recombination-directed DNA repair. *Mol. Cell. Biol.* **2013**, *33*, 2635–2647. [CrossRef] [PubMed]

25. Urbanska, K.; Pannizzo, P.; Lassak, A.; Gualco, E.; Surmacz, E.; Croul, S.; Del Valle, L.; Khalili, K.; Reiss, K. Estrogen receptor β-mediated nuclear interaction between IRS-1 and Rad51 inhibits homologous recombination directed DNA repair in medulloblastoma. *J. Cell. Physiol.* **2009**, *219*, 392–401. [CrossRef] [PubMed]

26. Schiewer, M.J.; Knudsen, K.E. Linking DNA damage and hormone signaling pathways in cancer. *Trends Endocrinol. Metab.* **2016**, *27*, 216–225. [CrossRef] [PubMed]

27. Zhou, B.B.; Sausville, E.A. Drug discovery targeting Chk1 and Chk2 kinases. *Prog. Cell Cycle Res.* **2003**, *5*, 413–421. [PubMed]

28. Harper, J.W.; Elledge, S.J. The DNA damage response: Ten years after. *Mol. Cell* **2007**, *28*, 739–745. [CrossRef] [PubMed]

29. Krempler, A.; Deckbar, D.; Jeggo, P.A.; Lobrich, M. An imperfect G_2M checkpoint contributes to chromosome instability following irradiation of s and G_2 phase cells. *Cell Cycle* **2007**, *6*, 1682–1686. [CrossRef] [PubMed]

30. Feigelson, H.S.; Ross, R.K.; Yu, M.C.; Coetzee, G.A.; Reichardt, J.K.; Henderson, B.E. Genetic susceptibility to cancer from exogenous and endogenous exposures. *J. Cell. Biochem. Suppl.* **1996**, *25*, 15–22. [CrossRef]

31. Foster, J.S.; Henley, D.C.; Bukovsky, A.; Seth, P.; Wimalasena, J. Multifaceted regulation of cell cycle progression by estrogen: Regulation of Cdk inhibitors and Cdc25A independent of cyclin D1-Cdk4 function. *Mol. Cell. Biol.* **2001**, *21*, 794–810. [CrossRef] [PubMed]

32. Wang, W.J.; Wu, S.P.; Liu, J.B.; Shi, Y.S.; Huang, X.; Zhang, Q.B.; Yao, K.T. Myc regulation of Chk1 and Chk2 promotes radioresistance in a stem cell-like population of nasopharyngeal carcinoma cells. *Cancer Res.* **2013**, *73*, 1219–1231. [CrossRef] [PubMed]

33. Cowling, V.H.; Cole, M.D. Turning the tables: Myc activates wnt in breast cancer. *Cell Cycle* **2007**, *6*, 2625–2627. [CrossRef] [PubMed]

34. Dubik, D.; Shiu, R.P. Mechanism of estrogen activation of c-Myc oncogene expression. *Oncogene* **1992**, *7*, 1587–1594. [PubMed]

35. Watson, P.H.; Pon, R.T.; Shiu, R.P. Inhibition of c-Myc expression by phosphorothioate antisense oligonucleotide identifies a critical role for c-Myc in the growth of human breast cancer. *Cancer Res.* **1991**, *51*, 3996–4000. [PubMed]

36. Prall, O.W.; Rogan, E.M.; Sutherland, R.L. Estrogen regulation of cell cycle progression in breast cancer cells. *J. Steroid Biochem. Mol. Biol.* **1998**, *65*, 169–174. [CrossRef]

37. Watts, C.K.; Brady, A.; Sarcevic, B.; deFazio, A.; Musgrove, E.A.; Sutherland, R.L. Antiestrogen inhibition of cell cycle progression in breast cancer cells in associated with inhibition of cyclin-dependent kinase activity and decreased retinoblastoma protein phosphorylation. *Mol. Endocrinol.* **1995**, *9*, 1804–1813. [PubMed]

38. Altucci, L.; Addeo, R.; Cicatiello, L.; Dauvois, S.; Parker, M.G.; Truss, M.; Beato, M.; Sica, V.; Bresciani, F.; Weisz, A. 17β-estradiol induces cyclin D1 gene transcription, p36D1-p34cdk4 complex activation and p105Rb phosphorylation during mitogenic stimulation of G_1-arrested human breast cancer cells. *Oncogene* **1996**, *12*, 2315–2324. [PubMed]

39. Sabbah, M.; Courilleau, D.; Mester, J.; Redeuilh, G. Estrogen induction of the cyclin D1 promoter: Involvement of a camp response-like element. *Proc. Natl. Acad. Sci. USA* **1999**, *96*, 11217–11222. [CrossRef] [PubMed]

40. Castro-Rivera, E.; Samudio, I.; Safe, S. Estrogen regulation of cyclin D1 gene expression in ZR-75 breast cancer cells involves multiple enhancer elements. *J. Biol. Chem.* **2001**, *276*, 30853–30861. [CrossRef] [PubMed]

41. Wilcken, N.R.; Prall, O.W.; Musgrove, E.A.; Sutherland, R.L. Inducible overexpression of cyclin D1 in breast cancer cells reverses the growth-inhibitory effects of antiestrogens. *Clin. Cancer Res.* **1997**, *3*, 849–854. [PubMed]

42. Toillon, R.A.; Magne, N.; Laios, I.; Lacroix, M.; Duvillier, H.; Lagneaux, L.; Devriendt, D.; Van Houtte, P.; Leclercq, G. Interaction between estrogen receptor α, ionizing radiation and (anti-) estrogens in breast cancer cells. *Breast Cancer Res. Treat.* **2005**, *93*, 207–215. [CrossRef] [PubMed]

43. Toillon, R.A.; Magne, N.; Laios, I.; Castadot, P.; Kinnaert, E.; Van Houtte, P.; Desmedt, C.; Leclercq, G.; Lacroix, M. Estrogens decrease γ-ray-induced senescence and maintain cell cycle progression in breast cancer cells independently of p53. *Int. J. Radiat. Oncol. Biol. Phys.* **2007**, *67*, 1187–1200. [CrossRef] [PubMed]

44. Molinari, A.M.; Bontempo, P.; Schiavone, E.M.; Tortora, V.; Verdicchio, M.A.; Napolitano, M.; Nola, E.; Moncharmont, B.; Medici, N.; Nigro, V.; et al. Estradiol induces functional inactivation of p53 by intracellular redistribution. *Cancer Res.* **2000**, *60*, 2594–2597. [PubMed]

45. Liu, W.; Konduri, S.D.; Bansal, S.; Nayak, B.K.; Rajasekaran, S.A.; Karuppayil, S.M.; Rajasekaran, A.K.; Das, G.M. Estrogen receptor-alpha binds p53 tumor suppressor protein directly and represses its function. *J. Biol. Chem.* **2006**, *281*, 9837–9840. [CrossRef] [PubMed]

46. Magne, N.; Toillon, R.A.; Bottero, V.; Didelot, C.; Houtte, P.V.; Gerard, J.P.; Peyron, J.F. NFκB modulation and ionizing radiation: Mechanisms and future directions for cancer treatment. *Cancer Lett.* **2006**, *231*, 158–168. [CrossRef] [PubMed]

47. Sas, L.; Lardon, F.; Vermeulen, P.B.; Hauspy, J.; Van Dam, P.; Pauwels, P.; Dirix, L.Y.; Van Laere, S.J. The interaction between er and NFκB in resistance to endocrine therapy. *Breast Cancer Res. BCR* **2012**, *14*, 212. [CrossRef] [PubMed]

48. Egloff, A.M.; Rothstein, M.E.; Seethala, R.; Siegfried, J.M.; Grandis, J.R.; Stabile, L.P. Cross-talk between estrogen receptor and epidermal growth factor receptor in head and neck squamous cell carcinoma. *Clin. Cancer Res.* **2009**, *15*, 6529–6540. [CrossRef] [PubMed]

49. Siegfried, J.M.; Hershberger, P.A.; Stabile, L.P. Estrogen receptor signaling in lung cancer. *Semin. Oncol.* **2009**, *36*, 524–531. [CrossRef] [PubMed]

50. Hsu, L.H.; Chu, N.M.; Kao, S.H. Estrogen, estrogen receptor and lung cancer. *Int. J. Mol. Sci.* **2017**, *18*. [CrossRef] [PubMed]

51. Mawson, A.; Lai, A.; Carroll, J.S.; Sergio, C.M.; Mitchell, C.J.; Sarcevic, B. Estrogen and insulin/IGF-1 cooperatively stimulate cell cycle progression in MCF-7 breast cancer cells through differential regulation of c-Myc and cyclin D1. *Mol. Cell. Endocrinol.* **2005**, *229*, 161–173. [CrossRef] [PubMed]

52. Marquez-Garban, D.C.; Chen, H.W.; Fishbein, M.C.; Goodglick, L.; Pietras, R.J. Estrogen receptor signaling pathways in human non-small cell lung cancer. *Steroids* **2007**, *72*, 135–143. [CrossRef] [PubMed]

53. Zhao, X.Z.; Liu, Y.; Zhou, L.J.; Wang, Z.Q.; Wu, Z.H.; Yang, X.Y. Role of estrogen in lung cancer based on the estrogen receptor-epithelial mesenchymal transduction signaling pathways. *OncoTargets Ther.* **2015**, *8*, 2849–2863. [CrossRef] [PubMed]

54. Zhao, G.; Nie, Y.; Lv, M.; He, L.; Wang, T.; Hou, Y. ERβ-mediated estradiol enhances epithelial mesenchymal transition of lung adenocarcinoma through increasing transcription of midkine. *Mol. Endocrinol.* **2012**, *26*, 1304–1315. [CrossRef] [PubMed]

55. Johansson, A.C.; La Fleur, L.; Melissaridou, S.; Roberg, K. The relationship between EMT, CD44(high) /EGFR(low) phenotype, and treatment response in head and neck cancer cell lines. *J. Oral Pathol. Med.* **2016**, *45*, 640–646. [CrossRef] [PubMed]

56. Theys, J.; Jutten, B.; Habets, R.; Paesmans, K.; Groot, A.J.; Lambin, P.; Wouters, B.G.; Lammering, G.; Vooijs, M. E-cadherin loss associated with emt promotes radioresistance in human tumor cells. *Radiother. Oncol.* **2011**, *99*, 392–397. [CrossRef] [PubMed]

57. Chang, L.; Graham, P.H.; Hao, J.; Ni, J.; Bucci, J.; Cozzi, P.J.; Kearsley, J.H.; Li, Y. Acquisition of epithelial-mesenchymal transition and cancer stem cell phenotypes is associated with activation of the PI3K/Akt/mTOR pathway in prostate cancer radioresistance. *Cell Death Dis.* **2013**, *4*, e875. [CrossRef] [PubMed]

58. Zhang, H.; Luo, H.; Jiang, Z.; Yue, J.; Hou, Q.; Xie, R.; Wu, S. Fractionated irradiation-induced EMT-like phenotype conferred radioresistance in esophageal squamous cell carcinoma. *J. Radiat. Res.* **2016**, *57*, 370–380. [CrossRef] [PubMed]

59. Cole, M.P.; Jones, C.T.; Todd, I.D. A new anti-oestrogenic agent in late breast cancer. An early clinical appraisal of ICI46474. *Br. J. Cancer* **1971**, *25*, 270–275. [CrossRef] [PubMed]

60. Wazer, D.E.; Tercilla, O.F.; Lin, P.S.; Schmidt-Ullrich, R. Modulation in the radiosensitivity of MCF-7 human breast carcinoma cells by 17β-estradiol and tamoxifen. *Br. J. Radiol.* **1989**, *62*, 1079–1083. [CrossRef] [PubMed]

61. Wazer, D.E.; Joyce, M.; Jung, L.; Band, V. Alterations in growth phenotype and radiosensitivity after fractionated irradiation of breast carcinoma cells from a single patient. *Int. J. Radiat. Oncol. Biol. Phys.* **1993**, *26*, 81–88. [CrossRef]

62. Villalobos, M.; Aranda, M.; Nunez, M.I.; Becerra, D.; Olea, N.; Ruiz de Almodovar, M.; Pedraza, V. Interaction between ionizing radiation, estrogens and antiestrogens in the modification of tumor microenvironment in estrogen dependent multicellular spheroids. *Acta Oncol.* **1995**, *34*, 413–417. [CrossRef] [PubMed]

63. Villalobos, M.; Becerra, D.; Nunez, M.I.; Valenzuela, M.T.; Siles, E.; Olea, N.; Pedraza, V.; Ruiz de Almodovar, J.M. Radiosensitivity of human breast cancer cell lines of different hormonal responsiveness. Modulatory effects of oestradiol. *Int. J. Radiat. Biol.* **1996**, *70*, 161–169. [CrossRef] [PubMed]

64. Paulsen, G.H.; Strickert, T.; Marthinsen, A.B.; Lundgren, S. Changes in radiation sensitivity and steroid receptor content induced by hormonal agents and ionizing radiation in breast cancer cells in vitro. *Acta Oncol.* **1996**, *35*, 1011–1019. [CrossRef] [PubMed]

65. Newton, C.J.; Schlatterer, K.; Stalla, G.K.; Von Angerer, E.; Wowra, B. Pharmacological enhancement of radiosurgery response: Studies on an in vitro model system. *J. Radiosurg.* **1998**, *1*, 51–56. [CrossRef]

66. Yang, L.; Yuan, X.; Wang, J.; Gu, C.; Zhang, H.; Yu, J.; Liu, F. Radiosensitization of human glioma cells by tamoxifen is associated with the inhibition of PKC-ı activity in vitro. *Oncol. lett.* **2015**, *10*, 473–478. [CrossRef] [PubMed]

67. Wang, J.; Yang, Q.; Haffty, B.G.; Li, X.; Moran, M.S. Fulvestrant radiosensitizes human estrogen receptor-positive breast cancer cells. *Biochem. Biophys. Res. Commun.* **2013**, *431*, 146–151. [CrossRef] [PubMed]

68. Sarkaria, J.N.; Miller, E.M.; Parker, C.J.; Jordan, V.C.; Mulcahy, R.T. 4-hydroxytamoxifen, an active metabolite of tamoxifen, does not alter the radiation sensitivity of MCF-7 breast carcinoma cells irradiated in vitro. *Breast Cancer Res. Treat.* **1994**, *30*, 159–165. [CrossRef] [PubMed]

69. Mc Gee, M.M.; Hyland, E.; Campiani, G.; Ramunno, A.; Nacci, V.; Zisterer, D.M. Caspase-3 is not essential for DNA fragmentation in MCF-7 cells during apoptosis induced by the pyrrolo-1,5-benzoxazepine, pbox-6. *FEBS Lett.* **2002**, *515*, 66–70. [CrossRef]

70. Kantorowitz, D.A.; Thompson, H.J.; Furmanski, P. Effect of conjoint administration of tamoxifen and high-dose radiation on the development of mammary carcinoma. *Int. J. Radiat. Oncol. Biol. Phys.* **1993**, *26*, 89–94. [CrossRef]

71. Inano, H.; Onoda, M. Prevention of radiation-induced mammary tumors. *Int. J. Radiat. Oncol. Biol. Phys.* **2002**, *52*, 212–223. [CrossRef]

72. Inano, H.; Onoda, M.; Suzuki, K.; Kobayashi, H.; Wakabayashi, K. Prevention of radiation-induced mammary tumours in rats by combined use of WR-2721 and tamoxifen. *Int. J. Radiat. Biol.* **2000**, *76*, 1113–1120. [CrossRef] [PubMed]

73. Segovia-Mendoza, M.; Jurado, R.; Mir, R.; Medina, L.A.; Prado-Garcia, H.; Garcia-Lopez, P. Antihormonal agents as a strategy to improve the effect of chemo-radiation in cervical cancer: In vitro and in vivo study. *BMC Cancer* **2015**, *15*, 21. [CrossRef] [PubMed]

74. Azria, D.; Larbouret, C.; Cunat, S.; Ozsahin, M.; Gourgou, S.; Martineau, P.; Evans, D.B.; Romieu, G.; Pujol, P.; Pelegrin, A. Letrozole sensitizes breast cancer cells to ionizing radiation. *Breast Cancer Res. BCR* **2005**, *7*, R156–R163. [CrossRef] [PubMed]

75. Anscher, M.S.; Kong, F.M.; Andrews, K.; Clough, R.; Marks, L.B.; Bentel, G.; Jirtle, R.L. Plasma transforming growth factor β1 as a predictor of radiation pneumonitis. *Int. J. Radiat. Oncol. Biol. Phys.* **1998**, *41*, 1029–1035. [CrossRef]

76. Anscher, M.S.; Kong, F.M.; Jirtle, R.L. The relevance of transforming growth factor β 1 in pulmonary injury after radiation therapy. *Lung Cancer* **1998**, *19*, 109–120. [CrossRef]

77. Chen, Y.; Williams, J.; Ding, I.; Hernady, E.; Liu, W.; Smudzin, T.; Finkelstein, J.N.; Rubin, P.; Okunieff, P. Radiation pneumonitis and early circulatory cytokine markers. *Semin. Radiat. Oncol.* **2002**, *12*, 26–33. [CrossRef] [PubMed]

78. Butta, A.; MacLennan, K.; Flanders, K.C.; Sacks, N.P.; Smith, I.; McKinna, A.; Dowsett, M.; Wakefield, L.M.; Sporn, M.B.; Baum, M.; et al. Induction of transforming growth factor β 1 in human breast cancer in vivo following tamoxifen treatment. *Cancer Res.* **1992**, *52*, 4261–4264. [PubMed]

79. Dalberg, K.; Johansson, H.; Johansson, U.; Rutqvist, L.E. A randomized trial of long term adjuvant tamoxifen plus postoperative radiation therapy versus radiation therapy alone for patients with early stage breast carcinoma treated with breast-conserving surgery. Stockholm breast cancer study group. *Cancer* **1998**, *82*, 2204–2211. [CrossRef]

80. Fisher, B.; Dignam, J.; Wolmark, N.; Wickerham, D.L.; Fisher, E.R.; Mamounas, E.; Smith, R.; Begovic, M.; Dimitrov, N.V.; Margolese, R.G.; et al. Tamoxifen in treatment of intraductal breast cancer: National surgical adjuvant breast and bowel project B-24 randomised controlled trial. *Lancet* **1999**, *353*, 1993–2000. [CrossRef]

81. Ahn, P.H.; Vu, H.T.; Lannin, D.; Obedian, E.; DiGiovanna, M.P.; Burtness, B.; Haffty, B.G. Sequence of radiotherapy with tamoxifen in conservatively managed breast cancer does not affect local relapse rates. *J. Clin. Oncol.* **2005**, *23*, 17–23. [CrossRef] [PubMed]

82. Harris, E.E.; Christensen, V.J.; Hwang, W.T.; Fox, K.; Solin, L.J. Impact of concurrent versus sequential tamoxifen with radiation therapy in early-stage breast cancer patients undergoing breast conservation treatment. *J. Clin. Oncol.* **2005**, *23*, 11–16. [CrossRef] [PubMed]

83. Pierce, L.J.; Hutchins, L.F.; Green, S.R.; Lew, D.L.; Gralow, J.R.; Livingston, R.B.; Osborne, C.K.; Albain, K.S. Sequencing of tamoxifen and radiotherapy after breast-conserving surgery in early-stage breast cancer. *J. Clin. Oncol.* **2005**, *23*, 24–29. [CrossRef] [PubMed]

84. Ishitobi, M.; Komoike, Y.; Motomura, K.; Koyama, H.; Nishiyama, K.; Inaji, H. Retrospective analysis of concurrent vs. Sequential administration of radiotherapy and hormone therapy using aromatase inhibitor for hormone receptor-positive postmenopausal breast cancer. *Anticancer Res.* **2009**, *29*, 4791–4794. [PubMed]

85. Ishitobi, M.; Shiba, M.; Nakayama, T.; Motomura, K.; Koyama, H.; Nishiyama, K.; Tamaki, Y. Treatment sequence of aromatase inhibitors and radiotherapy and long-term outcomes of breast cancer patients. *Anticancer Res.* **2014**, *34*, 4311–4314. [PubMed]

86. Valakh, V.; Trombetta, M.G.; Werts, E.D.; Labban, G.; Khalid, M.K.; Kaminsky, A.; Parda, D. Influence of concurrent anastrozole on acute and late side effects of whole breast radiotherapy. *Am. J. Clin. Oncol.* **2011**, *34*, 245–248. [CrossRef] [PubMed]

87. Azria, D.; Belkacemi, Y.; Romieu, G.; Gourgou, S.; Gutowski, M.; Zaman, K.; Moscardo, C.L.; Lemanski, C.; Coelho, M.; Rosenstein, B.; et al. Concurrent or sequential adjuvant letrozole and radiotherapy after conservative surgery for early-stage breast cancer (CO-HO-RT): A phase 2 randomised trial. *Lancet Oncol.* **2010**, *11*, 258–265. [CrossRef]

88. Whelan, T.; Levine, M. Radiation therapy and tamoxifen: Concurrent or sequential? That is the question. *J. Clin. Oncol.* **2005**, *23*, 1–4. [CrossRef] [PubMed]

89. Chargari, C.; Toillon, R.A.; Macdermed, D.; Castadot, P.; Magne, N. Concurrent hormone and radiation therapy in patients with breast cancer: What is the rationale? *Lancet Oncol.* **2009**, *10*, 53–60. [CrossRef]

90. Varga, Z.; Cserhati, A.; Kelemen, G.; Boda, K.; Thurzo, L.; Kahan, Z. Role of systemic therapy in the development of lung sequelae after conformal radiotherapy in breast cancer patients. *Int. J. Radiat. Oncol. Biol. Phys.* **2011**, *80*, 1109–1116. [CrossRef] [PubMed]

91. Koc, M.; Polat, P.; Suma, S. Effects of tamoxifen on pulmonary fibrosis after cobalt-60 radiotherapy in breast cancer patients. *Radiother. Oncol.* **2002**, *64*, 171–175. [CrossRef]

92. Bentzen, S.M.; Skoczylas, J.Z.; Overgaard, M.; Overgaard, J. Radiotherapy-related lung fibrosis enhanced by tamoxifen. *J. Natl. Cancer Inst.* **1996**, *88*, 918–922. [CrossRef] [PubMed]

93. Fowble, B.; Fein, D.A.; Hanlon, A.L.; Eisenberg, B.L.; Hoffman, J.P.; Sigurdson, E.R.; Daly, M.B.; Goldstein, L.J. The impact of tamoxifen on breast recurrence, cosmesis, complications, and survival in estrogen receptor-positive early-stage breast cancer. *Int. J. Radiat. Oncol. Biol. Phys.* **1996**, *35*, 669–677. [CrossRef]

94. Dahhan, T.; Fons, G.; Buist, M.R.; Ten Kate, F.J.; van der Velden, J. The efficacy of hormonal treatment for residual or recurrent low-grade endometrial stromal sarcoma. A retrospective study. *Eur. J. Obstet. Gynecol. Reprod. Biol.* **2009**, *144*, 80–84. [CrossRef] [PubMed]

95. Grünow, J.; Rong, C.; Hischmann, J.; Zaoui, K.; Flechtenmacher, C.; Weber, K.J.; Plinkert, P.; Hess, J. Regulation of submaxillary gland androgen-regulated protein 3A via estrogen receptor 2 in radioresistant head and neck squamous cell carcinoma cells. *J. Exp. Clin. Cancer Res.* **2017**, *36*, 25. [CrossRef] [PubMed]

96. Ferlay, J.; Soerjomataram, I.; Ervik, M.; Dikshit, R.; Eser, S.; Mathers, C.; Rebelo, M.; Parkin, D.M.; Forman, D.; Bray, F. Globocan 2012 v1.0, Cancer Incidence and Mortality Worldwide: Iarc Cancerbase No. 11. International Agency for Research on Cancer: Lyon, France. Available online: http://globocan.iarc.fr (accessed on 12 November 2017).

97. Vineis, P.; Alavanja, M.; Buffler, P.; Fontham, E.; Franceschi, S.; Gao, Y.T.; Gupta, P.C.; Hackshaw, A.; Matos, E.; Samet, J.; et al. Tobacco and cancer: Recent epidemiological evidence. *J. Natl. Cancer Inst.* **2004**, *96*, 99–106. [CrossRef] [PubMed]

98. Gillison, M.L.; Chaturvedi, A.K.; Anderson, W.F.; Fakhry, C. Epidemiology of human papillomavirus-positive head and neck squamous cell carcinoma. *J. Clin. Oncol.* **2015**, *33*, 3235–3242. [CrossRef] [PubMed]

99. Argiris, A.; Karamouzis, M.V.; Raben, D.; Ferris, R.L. Head and neck cancer. *Lancet* **2008**, *371*, 1695–1709. [CrossRef]

100. Valimaa, H.; Savolainen, S.; Soukka, T.; Silvoniemi, P.; Makela, S.; Kujari, H.; Gustafsson, J.A.; Laine, M. Estrogen receptor-β is the predominant estrogen receptor subtype in human oral epithelium and salivary glands. *J. Endocrinol.* **2004**, *180*, 55–62. [CrossRef] [PubMed]

101. Wong, M.H.W.; Dobbins, T.A.; Tseung, J.; Tran, N.; Lee, C.S.; O'Brien, C.J.; Clark, J.; Rose, B.R. Oestrogen receptor β expression in pleomorphic adenomas of the parotid gland. *J. Clin. Pathol.* **2009**, *62*, 789–793. [CrossRef] [PubMed]

102. Shatalova, E.G.; Klein-Szanto, A.J.; Devarajan, K.; Cukierman, E.; Clapper, M.L. Estrogen and cytochrome P450 1B1 contribute to both early- and late-stage head and neck carcinogenesis. *Cancer Prev. Res. (Phila.)* **2011**, *4*, 107–115. [CrossRef] [PubMed]

103. Sumida, T.; Ishikawa, A.; Mori, Y. Stimulation of the estrogen axis induces epithelial-mesenchymal transition in human salivary cancer cells. *Cancer Genom. Proteom.* **2016**, *13*, 305–310.

104. Somers, K.D.; Koenig, M.; Schechter, G.L. Growth of head and neck squamous cell carcinoma in nude mice: Potentiation of laryngeal carcinoma by 17β-estradiol. *J. Natl. Cancer Inst.* **1988**, *80*, 688–691. [CrossRef] [PubMed]

105. Lukits, J.; Remenar, E.; Rásó, E.; Ladányi, A.; Kásler, M.; Tímár, J. Molecular identification, expression and prognostic role of estrogen- and progesterone receptors in head and neck cancer. *Int. J. Oncol.* **2007**, *30*, 155–160. [CrossRef] [PubMed]

106. Grsic, K.; Opacic, I.L.; Sitic, S.; Milkovic Perisa, M.; Suton, P.; Sarcevic, B. The prognostic significance of estrogen receptor β in head and neck squamous cell carcinoma. *Oncol. Lett.* **2016**, *12*, 3861–3865. [CrossRef] [PubMed]

107. Fei, M.; Zhang, J.; Zhou, J.; Xu, Y.; Wang, J. Sex-related hormone receptor in laryngeal squamous cell carcinoma: Correlation with androgen estrogen-α and prolactin receptor expression and influence of prognosis. *Acta Otolaryngol.* **2018**, *138*, 66–72. [CrossRef] [PubMed]

108. Li, S.; Wang, B.; Tang, Q.; Liu, J.; Yang, X. Bisphenol a triggers proliferation and migration of laryngeal squamous cell carcinoma via gper mediated upregulation of IL-6. *Cell. Biochem. Funct.* **2017**, *35*, 209–216. [CrossRef] [PubMed]

109. Brooks, Y.S.; Ostano, P.; Jo, S.H.; Dai, J.; Getsios, S.; Dziunycz, P.; Hofbauer, G.F.; Cerveny, K.; Chiorino, G.; Lefort, K.; et al. Multifactorial ERβ and NOTCH1 control of squamous differentiation and cancer. *J. Clin. Investig.* **2014**, *124*, 2260–2276. [CrossRef] [PubMed]

International Journal of
Molecular Sciences

MDPI

Review

Estrogen, Angiogenesis, Immunity and Cell Metabolism: Solving the Puzzle

Annalisa Trenti [1,†], Serena Tedesco [2,†], Carlotta Boscaro [1,†], Lucia Trevisi [1], Chiara Bolego [1] and Andrea Cignarella [3,*]

[1]   Department of Pharmaceutical and Pharmacological Sciences, University of Padua, 35131 Padua, Italy; annalisa.trenti@gmail.com (A.T.); carlotta.boscaro@phd.unipd.it (C.B.); lucia.trevisi@unipd.it (L.T.); chiara.bolego@unipd.it (C.B.)
[2]   Venetian Institute of Molecular Medicine, 35129 Padua, Italy; serena.tedesco1988@gmail.com
[3]   Department of Medicine, University of Padua, 35128 Padua, Italy
*   Correspondence: andrea.cignarella@unipd.it; Tel.: +39-049-8275101
†   These authors contributed equally to this work.

Received: 19 February 2018; Accepted: 13 March 2018; Published: 15 March 2018

Abstract: Estrogen plays an important role in the regulation of cardiovascular physiology and the immune system by inducing direct effects on multiple cell types including immune and vascular cells. Sex steroid hormones are implicated in cardiovascular protection, including endothelial healing in case of arterial injury and collateral vessel formation in ischemic tissue. Estrogen can exert potent modulation effects at all levels of the innate and adaptive immune systems. Their action is mediated by interaction with classical estrogen receptors (ERs), ERα and ERβ, as well as the more recently identified G-protein coupled receptor 30/G-protein estrogen receptor 1 (GPER1), via both genomic and non-genomic mechanisms. Emerging data from the literature suggest that estrogen deficiency in menopause is associated with an increased potential for an unresolved inflammatory status. In this review, we provide an overview through the puzzle pieces of how 17β-estradiol can influence the cardiovascular and immune systems.

Keywords: estrogen; 17β-estradiol; angiogenesis; metabolism; endothelium; macrophages; immune response

---

1. Setting the Stage: Estrogen, the Cardiovascular System and the Immune Response

In addition to its essential role in sexual development and reproduction in females, estrogen is involved in a wide range of physiological processes in different tissues [1], even in male subjects. Evidence accumulated over the years demonstrated that estrogen has protective effects on the cardiovascular system [2–4], mainly related to interaction with multiple cell types including immune cells, such as B lymphocytes and macrophages [5] and vessel wall cells, including smooth muscle [6,7] and endothelial cells [8–11].

In women, estrogen circulating levels fluctuate during the menstrual cycle and its concentration changes in relation to age [12]. The most important estrogen circulating from menarche to menopause is 17β-estradiol (E2). Close to menopause, estrogen plasma levels decrease compared to those present in fertile women [13] and become equivalent to those present in men. However, E2 continues to be synthesized, starting from androgens, in extragonadal sites such as breast, brain, muscle, bone and adipose tissue where it acts locally as a paracrine or autocrine factor [14]. Declining estrogen levels are associated with a variety of metabolic changes and cardiovascular diseases [15]. The metabolic effects mediated by estrogen take place in multiple tissues including skeletal muscle and liver [16].

E2 prevents endothelial dysfunction, vascular inflammation and atherosclerosis [15]. In addition, available evidence points to E2 as a key factor in promoting endothelial healing and angiogenesis [8–10,17] through endothelial progenitor cells, immune inflammatory cells and platelet mobilization, which contribute synergistically to endothelial repair [18–20]. The important role of E2 in the angiogenic process is also noticed in ischemia-reperfusion tissue injury, where E2 induces the formation of collateral vessels [21]. Angiogenesis stimulation by E2 accelerates functional endothelial recovery after arterial injury, which could be beneficial in coronary artery disease, peripheral arterial disease, cerebral ischemia and congestive heart failure [21]. The direct actions of E2 on endothelial cells contribute to accelerate re-endothelialization in vivo [22]. This process following endothelial damage [20,22] is accompanied by reduced neointima formation as a result of inhibition of smooth muscle cell proliferation and migration [23]. Furthermore, E2 promotes the natural resolution of inflammation in wound healing [24].

The immune system demonstrates remarkable sex differences: females tend to have a more responsive immune system compared to their male counterparts. The outcome and survival rates from e.g., infections or sepsis are sometimes better in females than in males [25]. Females, however, respond more aggressively to self-antigens and are more susceptible to autoimmune diseases [26]. The body of human data on gender differences in immune response is rapidly growing. Amadori and colleagues [27] were the first to demonstrate that circulating T lymphocytes in fertile women are more abundant than those in men; this also occurs in other female mammals, suggesting a common trait in different species that endows females with a more rapid and efficient immune response [28,29]. It has long been recognized that steroid hormones play a role in the regulation of the immune response to infection or tissue damage and modulate all levels of the innate (neutrophils, macrophages/monocytes, natural killer cells, dendritic cells) and adaptive immune systems (T and B cells) (reviewed in [30]). Estrogen has been shown to regulate neutrophil number and function, and the production of chemokines such as monocyte chemoattractant protein (MCP)-1 and cytokines including tumor necrosis factor (TNF)-α, interleukin (IL)-6 and IL-1β. On the other hand, since ovarian activity decreases and eventually stops with aging, several disease conditions may show up, which are characterized by a strong inflammatory component associated with the post-menopausal state [26,31].

The aim of this review is to discuss the multifaceted role of estrogen in vascular biology and in the immune response, particularly in the monocyte/macrophage system, and to further integrate available evidence (i.e., solve the puzzle) regarding the estrogenic control of double-edged processes such as angiogenesis and metabolism.

2. Estrogen Receptors

The effects induced by estrogen in different tissues are the result of the activation of transcriptional and non-transcriptional signal pathways. Estrogen exerts both rapid and long-term actions through their binding with ERs [1]. Several ER subtypes have been identified: the nuclear isoforms, ERα and ERβ, and the transmembrane G-protein-coupled receptor 30/G-protein estrogen receptor 1 (GPER1). ERα and ERβ act as transcription factors responsible for many genomic effects, modulating gene expression by direct binding to DNA at specific estrogen response elements (EREs) [32]. In contrast, GPER1 is mainly involved in mediating rapid intracellular responses induced by estrogen [33,34].

The genes encoding ERα, ERβ and GPER1 are *ESR1*, *ESR2* and *GPER1*, respectively. The two intracellular receptors have different molecular weights; in particular, ERα consists of 595 and ERβ of 530 amino acids, respectively [35]. Their structure consists of two main domains: the carboxy-terminal domain for interaction with the ligand (ligand-binding domain, LBD), which contains the activator factor-2 (AF-2), mediating a wide range of functional responses, and the central DNA-binding domain, responsible for binding to EREs [36]. Other regions are involved in transcriptional activation: the transcriptional regulatory domain (constitutively active amino-terminal domain, AF-1) and a hinge domain between the DNA-binding domain and LBD, which gives flexibility to the protein [1,37].

ERα is a ligand-dependent transcription factor that exerts its genomic, also called nuclear actions through binding to chromatin and mobilization of cofactors to influence the transcription of its target genes. A fraction of ERα can elicit membrane signaling (non-genomic effects) by association with the plasma membrane [38,39]. Rapid changes in adenylate cyclase, mitogen-activated protein kinases (MAPK) and phosphatidylinositol 3 kinase (PI3K) activities or in cytoplasmic calcium concentration and endothelial nitric oxide synthase (eNOS) activation constitute established non-genomic effects. By using mice expressing ERα proteins with inactivated genomic or nongenomic signaling, it has been shown that the preserved arterial actions of E2 were membrane-dependent [40], whilst the estrogenic responses of uteri were highly dependent upon the genomic actions of ERα [41]. These studies thus demonstrated for the first time that the respective contributions of nuclear/genomic and membrane effects towards the estrogenic response are tissue-specific. Accordingly, we showed that administration of a selective ERα agonist confers cardiovascular protection dissected from unwanted uterotrophic effects [3], suggesting that ERα-selective agonists represent a potential safer alternative to natural hormones.

The estrogenic membrane receptor GPER1 belongs to the family of G protein-coupled receptors and is characterized by the presence of seven transmembrane helices. The organization of the seven helices involves the amino-terminal portion located outside the cell and the carboxy-terminal portion in the cytoplasm. Cytoplasmic loops are involved in the selective binding and activation of various heterotrimeric proteins [1]. This receptor is expressed at the endoplasmic reticulum and in the plasma membrane [33,42]. E2 binds to GPER1 with nanomolar affinity, in the range of 3–6 nM [33], while its affinity for nuclear receptors is ten times higher, in the range of 0.1–0.4 nM [1]. However, in cells expressing both ERα and GPER1, coordinated signaling is likely to occur, with some evidence supporting this in monocytes [43], ovarian cancer cells [44], uterine stromal cells [45] and coronary vessels [46]. Accordingly, the emerging notion that GPER acts as an autonomous ER in vivo and also interacts with intracellular ERs has been recently reviewed by Romano and Gorelik [47].

3. Estrogen Receptors and Endothelial Function

ERα is expressed in the vascular tissue [48,49]; although ERβ distribution in vascular tissues is less characterized, human endothelial cells do express ERβ [50]. ERα has been long recognized to mediate most beneficial cardiovascular effects of E2 [2–4], but it is also involved in pathologic cell proliferation in the setting of cancer [51].

Several of estrogen cardiovascular actions are actually mediated by direct effects on the vessel wall resulting in the control of endothelial function and plasma lipid profile. In particular, estrogen increases the synthesis and release of nitric oxide (NO) and prostacyclin, well-known endothelial-derived vasodilators and anti-platelet agents, and negatively regulates production of several pro-inflammatory mediators in situations of vascular injury [52]. More specifically, estrogen upregulates the expression of enzymes involved in prostacyclin biosynthesis, i.e., cyclooxygenase (COX)-1 and prostacyclin synthase, thereby increasing systemic prostacyclin levels in rodents [53]. Moreover, estrogen increases both COX2 expression and prostacyclin generation in ovariectomized low-density lipoprotein receptor null (LDLR−/−) mice and substantially reduces atherosclerotic lesion size [54]. Accordingly, the protective effects of estrogen were abrogated by disruption of the prostacyclin receptor (IP) gene in the double LDLR−/−/IP−/− null mouse, suggesting that the protective actions of estrogen within the cardiovascular system are, at least in part, mediated by endothelial prostacyclin and its receptor, the IP. Notably, a physiological concentration of E2 induces transcription but not translation of COX-2 in human endothelial cells exposed to laminar shear stress [4]. E2 increases NO levels in cerebral and peripheral endothelial cells in vitro via eNOS activation and ER-mediated mechanisms [55]. NO is essential for vascular endothelial growth factor (VEGF)-induced angiogenesis in vitro [56] and in vivo [57]. Recent studies have shown that changes in the relative expression of ERβ/ERα may influence some E2 effects, such as the modulation of vascular NO bioavailability in aging rodents [58].

Endothelial cells also express GPER1 [59], which mediates nongenomic rapid effects including calcium influx, cAMP synthesis or kinase (such as PI3K) activation. These events are involved in the regulation of vascular tone [33,34,60]. Interestingly, a novel role for GPER has emerged in regulating the expression of NADPH oxidase 1 (NOX1), which is essential for reactive oxygen species generation in the cardiovascular system [61].

In conclusion, estrogen mediates both rapid and longer-term effects on the vessel wall. Novel vascular target genes regulated by ER subtypes are being identified, thereby providing potential opportunities for pharmacological intervention.

4. Estrogen, Angiogenesis and Metabolism

Additional puzzle pieces that need to fit together include estrogen, angiogenesis and metabolism. Migration and proliferation of endothelial cells are closely involved in re-endothelialization and angiogenesis. Angiogenesis consists of a number of subsequent biological events and is a tightly regulated process. In adult organisms, angiogenesis is virtually absent under normal conditions, except in the female reproductive tract, where it is routinely observed in the uterus in association with E2 fluctuations [62]. E2 stimulates endothelial cell proliferation in vitro [8] and in vivo [8,9,17], and inhibits spontaneous, as well as TNF-α-induced, apoptosis [63,64]. Furthermore, E2 enhances adhesion of HUVECs to various matrix proteins and increases cell migration, thus promoting angiogenesis [8,10]. The mechanisms responsible for the proangiogenic effect of E2 have been widely investigated and appear to be largely mediated by ERα activation [65]; accordingly, angiogenesis is impaired in ERα knockout mice [66]. In HUVECs, E2 has been shown to enhance cyclins A and B1 gene expression through involvement of the classical ER pathway [67]. E2 treatment also promotes proliferation and increases RhoA gene expression and activity in an ERα-dependent manner [10,68]. Through a rapid, non-genomic pathway ligand activated by ERα, E2 promotes rearrangements of actin cytoskeleton that allow the formation of specialized cell membrane structures, such as focal adhesion complexes, pseudopodia and membrane ruffles [50]. Estrogen also stimulates VEGF production in uterine and vascular tissue [69,70]. The rapid re-endothelialization induced by estrogen after vascular injury may be due, in part, to increased local expression of VEGF [9,17]. E2-induced increases in VEGF receptor-2 expression on human myometrial microvascular endothelial cells appears to be mediated primarily by ERα [71]. In addition, E2 promotes increased β1, α5 and α6 integrin expression on endothelial cell surface [72] and induces phosphorylation of focal adhesion kinase (FAK) followed by its translocation toward membrane sites, where focal adhesion complexes are assembled [65].

In pathological circumstances, such as breast cancer, a clear association has been made between estrogen, ER expression by endothelial cells, angiogenic activity and/or tumor invasiveness [73]. In this context, transient E2 induction of VEGF results from E2-induced upregulation of the oncogenic nuclear transcription factor c-Myc via ERα activation, whereas estrogen withdrawal in tumors induces hypoxic conditions responsible for VEGF upregulation [74]. Because the expression of glucose transporter 1 (GLUT1) is regulated by c-myc [75], it is conceivable that estrogen interaction with ERα activates c-myc, which in turn up-regulates GLUT-1 expression, thereby affecting tumor perfusion and glucose transport and metabolism through glycolysis. However, recent in vitro and in vivo observations indicate that membrane ERα signaling effects could mediate, or at least potentiate, the beneficial actions of estrogen on energy balance, insulin sensitivity, and glucose metabolism. Indeed, selective activation of the extranuclear ERα pool appears to induce endothelial actions and limit adipose tissue and fatty liver accumulation [76]. Moreover, preliminary data obtained from a mouse model with membrane-specific loss of function of ERα support a significant role of membrane ERα pool and membrane-derived signaling effects in the metabolic protective effects of estrogen [77].

During angiogenesis, endothelial cells must increase their metabolic activity to generate energy quickly and to facilitate the incorporation of nutrients into biomass. De Bock and colleagues [78] demonstrated that phosphofructokinase-2/fructose-2,6-bisphosphatase-3 (PFKFB3)-driven glycolysis regulates vessel branching. PFKFB3 is a direct target of E2 action; in ER-responsive breast cancer cells (MCF-7), E2 promotes PFKFB3 mRNA transcription and up-regulates PFKFB3 protein expression through ERα via direct binding to PFKFB3 promoter [79]. Recently, we demonstrated that the increased angiogenic response in E2-stimulated HUVEC is mediated by enhanced PFKFB3 expression peaking after 3 h, consistent with the activation of a membrane receptor considering that a nuclear/genomic effect would require a longer time. Treatment with the selective GPER1 agonist G-1 mimics the chemotactic and proangiogenic effect of E2 and also increases PFKFB3 expression, suggesting that E2-induced angiogenesis is mediated, at least in part, by the membrane receptor GPER1 [11]. Hence, even if steroid hormones have been classically described to mediate biological effects via intracellular receptors, non-genomic mechanisms of activation through membrane receptors responsible for endothelial cell motility, proliferation, and angiogenesis have also been demonstrated. Additional mechanisms for GPER1-mediated angiogenic stimulation may include the up-regulation of acid ceramidase expression, the increase of X-linked inhibitor of apoptosis protein (XIAP) and the regulation of Na^+/H^+ exchanger-1 (NHE-1) activity as reviewed recently by De Francesco et al. [80].

Experimental evidence accumulated over the past decade indicates that the direct effect of E2 on endothelial cells explains some cardiovascular benefits of the ovarian sex steroid hormone (Figure 1), but the specific pathways they influence remain to be elucidated. We have unraveled a previously unrecognized mechanism of estrogen-dependent endocrine-metabolic crosstalk in HUVECs which may have implications in angiogenesis occurring in ischemic or hypoxic tissues [11]. However, fitting these puzzle pieces together would require dissecting the molecular mechanisms of estrogen's proangiogenic effect in different disease contexts such as cancer. Thus, tissue-specific pharmacological control of endocrine-metabolic crosstalk appears to be a rewarding therapeutic strategy.

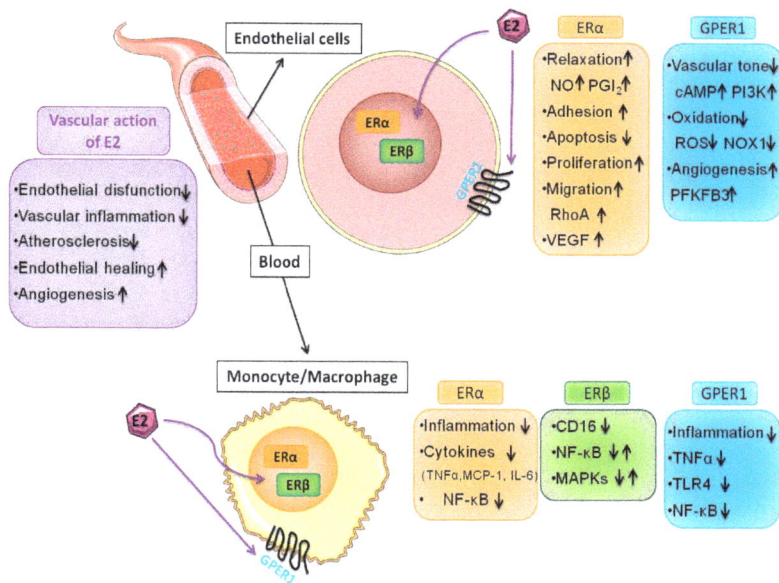

Figure 1. Multiple effects of 17β-estradiol (E2) in endothelial cells and macrophages. E2 induces protective effects on the cardiovascular system by promoting endothelial healing and angiogenesis through various pathways including the acceleration of re-endothelialization in vivo, the induction of proliferation and rearrangements of the actin cytoskeleton. E2 regulates the induction of chemokines and cytokines, and modulates macrophage immune phenotypes. These events are mediated by intracellular and membrane ER subtypes that are operatively linked in several cell types. The interaction between endothelial cells and macrophages is relevant in multiple disease settings such as atherosclerosis and cancer.

5. Estrogen and Macrophage Function

New data are redefining macrophages as diverse, polyfunctional and plastic cells that respond to the needs of the tissue at steady state and during disturbed homeostasis. Inflammation plays a critical role in the onset and progression of degenerative diseases, and is characterized by activation of tissue-resident macrophages as well as monocyte-derived macrophages that originate and renew from adult bone marrow. Under normal conditions, these cells provide immune surveillance and host defense in tissues to maintain homeostasis. However, upon sensing changes in the microenvironment, macrophages become activated, undergoing a morphological and functional switch [81]. Activation of these cells is not an "all-or-none" process, but rather a continuum characterized by a wide spectrum of molecular and functional phenotypes ranging from the "classical" M1 activated phenotype, with a highly pro-inflammatory profile, to the "alternative" M2 phenotype, associated with a beneficial, less inflammatory, protective profile [82,83]. Accordingly, these new models of activation and classification account for the functional diversity of macrophages that is relevant in vivo both in health and disease conditions including obesity, autoimmunity and neurodegeneration [84]. For instance, since a prominent feature of tissue remodeling is neoangiogenesis, macrophage polarization could affect the angiogenic process [85,86], which in turn is a determinant of adipose tissue expansion during obesity [87].

Estrogen has been shown to act as regulator of the immune function of the monocyte-macrophage system, especially regarding the production of cytokines. For instance, estrogen treatment in ovariectomized animals reduces expression of vascular MCP-1 and leukocyte infiltration into injured tissues, such as arteries and lung [88,89]. Estrogen affects the activation of nuclear factor kappa-light-chain-enhancer of activated B cells (NF-κB) in monocytes derived from umbilical cord blood, suggesting that high E2 concentrations during gestation affect the immune response in newborns [90]. Later in life, the production of cytokines by monocyte/macrophages is heavily influenced by the ovarian cycle, oral contraceptive use and estrogen replacement [91,92]. In vitro pre-treatment with E2 of human macrophages inhibits the NF-κB signaling pathway and the production of TNF-α induced by lipopolysaccharide (LPS) [93]. Estrogen has been also shown to enhance production [94] and prevent degradation of the endogenous NF-κB inhibitor IκB-α [95]. Other authors have reported the inhibitory effect of E2 on the production of pro-inflammatory cytokines [96]. By contrast, chronic exposure of murine macrophages to E2 in vivo increases production of pro-inflammatory cytokines (e.g., IL-1β, IL-6, TNF-α) [97,98].

Macrophages have long been recognized as crucial regulators of vascularization and healing [99,100]; in particular, the macrophage switch from the inflammatory to resolving phenotype is an essential step. In fact, in patients with non-healing and diabetic venous ulcers, failure in the M1-to-M2 switch results in local chronic inflammation with impaired healing progression [101]. Interestingly, gene regulation by estrogen is a key mediator of age-related delayed human wound healing [24]. The beneficial effects of estrogen on cutaneous healing are, in part, mediated through macrophage ERα, and estrogen fails to promote alternative macrophage activation in the absence of ERα in vitro [102]. Thus, we propose that estrogen acts as a reprogramming stimulus that accelerates macrophage transition towards a resolving, reparative phenotype [93,103].

Local and systemic metabolism is integrated at the cellular level to regulate immune cell function. By interacting with ER subtypes as discussed in Section 6 below, estrogen also affects metabolic reprogramming in macrophages, which accompanies different activation pathways in response to microenvironmental cues [15,32,81]. It is worth noting that similarities in metabolic reprogramming of macrophages, other immune cells and endothelial cells are emerging [104]. Hence, new insights in immunometabolism can be translated to the clinic to improve current treatments and develop novel therapies for metabolic diseases, inflammation, autoimmunity, and cancer.

These findings point to a complex and partially unresolved role of estrogen in immune and inflammatory responses [98]. Here we suggest that the duality in the action of estrogen on monocyte/macrophages cytokine production depends on many factors including the stimulus triggering the inflammatory response (endogenous or exogenous antigens), the target organ, the different estrogen concentration and ER expression patterns in tissues.

6. Estrogen Receptors in the Monocyte/Macrophage System

Recently, ER expression in human monocytes and macrophages has been investigated, increasing the number of pieces of this already complex puzzle. Both cell types express all ERs (Figure 1). Human primary monocytes express the ERα 36-kDa splice variant and GPER1 in a sex-independent manner [43], and these are physically associated. Macrophages have a higher ERα expression and lower ERβ expression than monocytes, and treatment with E2 in monocytes and in human macrophages in vitro induces an increase in ERα expression in macrophages, but not in monocytes [5]. Deficiency of ERα, but not of ERβ, increased TNF-α production by mouse peritoneal macrophages in response to bacterial stimuli, suggesting a prominent role of ERα in mediating the anti-inflammatory effects of estrogen [32,93,96]. Moreover, treatment with the selective GPER1 agonist G-1 is able to inhibit LPS-induced TNF-α production in human macrophages [105]. GPER1 also affects macrophage function via decreasing the expression of TLR4 [106]. In another recent study, it has been demonstrated that E2 confers protection against LPS/NF-κB–induced inflammation, with a role for ERα and GPER1 in mediating these anti-inflammatory properties [43]. In this study, treatment with both ICI 182,780, an ER antagonist/GPER agonist, and G15, a GPER antagonist, blocked the effects of E2. Studies about ERβ and macrophage function are limited; Kramer and colleagues [107] showed that ERβ suppresses CD16 expression with no effect on the activation of MAPKs and NF-κB, while Xing et al. [94] demonstrated an opposite effect showing the ability of selective ERβ activation to inhibit expression of inflammatory mediators. A recent study in human macrophages demonstrated that LPS is able to increase ERα phosphorylation but has no effect on ERβ activation [108]. This study also showed that macrophages isolated from males are more sensitive to the LPS effects than those from females.

As noted above, E2 is able to modulate the activation of different macrophage immune phenotypes [103,109]. The deletion of ERα in hematopoietic cells in mice causes an inability to induce the alternative phenotype in IL-4-stimulated macrophages, and induces high levels of inflammation and insulin resistance, suggesting that ERα is involved in the control of inflammation [110]. Defects in macrophage function due to myeloid-specific ERα deletion also lead to a variety of metabolic disorders including obesity and increased atherosclerosis [110]. Toniolo and colleagues demonstrated that in vitro isolated macrophages stimulated for 48 h with LPS and interferon (IFN)-γ show decreased ERα expression (with unchanged ERβ and GPER-1), and that pre-treatment with E2 counteract the LPS/IFNγ-mediated down-regulation of M2 markers, suggesting that female hormones modulate macrophage immune phenotypes [93]. The observation of a transient up-regulation of ERα mRNA in human macrophages following treatment with IL-4/IL-13 [93] as well as in mouse macrophages treated with IL-4 [103] suggests that this IL-4 effect is well conserved in mammals and may be functionally relevant to the inhibition of the pro-inflammatory response. By using a transcriptomic approach in peritoneal mouse macrophages, Pepe and colleagues recently reported that E2 promotes an anti-inflammatory and pro-resolving macrophage phenotype, which converges on the induction of genes related to macrophage alternative activation and on IL-10 expression in vivo [109].

The regulation of the immune response to infection or tissue damage is a complex interplay of multiple factors, but it has long been recognized that estrogen steers the innate and adaptive immune systems at various levels. Thus, we believe that pharmacological targeting of macrophage estrogen pathways may restore the impaired resolution of inflammation associated with aging and chronic inflammatory disease.

7. Estrogen in Women's Health

It has been reported that young women generally have much lower rates of cardiometabolic disease than men. However, midlife women lose this apparent protection during the menopausal transition, so that cardiometabolic disease is most common in post-menopause than any other stage of a woman's lifespan. In fact, fundamental aspects of metabolic homeostasis are regulated differently in males and females [16,31,111], and influence both the development of disease and the response to pharmacological intervention. Estrogen effects on the cardiovascular system include the modulation of inflammatory response and immune cell function. Aging is characterized by systemic inflammatory changes and organ dysfunction. In females, loss of estrogen makes these changes more intense [112]. Menopause is associated with an increased risk of cardiovascular and metabolic disease largely due to post-menopausal estrogen reduction. For instance, changes in the metabolism of sex hormones lead to accumulation of excess fat in intra-abdominal adipose tissue [15,113]. Post-menopausal women have an abrupt acceleration of atherosclerosis. Although restoration of estrogen would seem to be protective, double-blind clinical studies on the use of estrogen replacement have not shown a benefit in terms of e.g., reduced mortality (reviewed in [114]).

Sex steroid hormones alter the biology of vessel wall cells and the inflammatory cells that accrue as atherosclerosis progresses differently in the early versus later stages of the disease [52]. Hence, the beneficial effects of menopausal hormone therapy in preventing atherosclerotic cardiovascular disease occur only if therapy is initiated before the development of advanced atherosclerosis. Proof of this concept has come from a randomized trial showing that initiation of menopausal hormone therapy in women early after menopause significantly reduces the risk of the combined endpoint of mortality, myocardial infarction or heart failure without resulting in an increased risk of breast cancer or stroke [115]. This suggests that inflammatory pathways should remain an important therapeutic target of estrogen for treating women close to the onset of menopause.

An age relationship of estrogen–monocyte/macrophage number and function has long been identified, which may have several implications for postmenopausal health [112,116]. Studies in human macrophages derived from men and post-menopausal women treated in vitro with E2 highlight that E2 has no influence on the expression of TNF-α, IL-6 and IL-1β, regardless of gender [117,118]. However, the work of Toniolo and colleagues on macrophages derived from women in fertile or menopausal state showed that the response to M2-associated stimuli (IL-4/IL-13) is markedly impaired in macrophages from post- vs. pre-menopausal women, while the response to M1-associated stimuli (LPS/IFNγ) is similar. This results in an increased M1/M2 response ratio in menopausal state, associated with the loss of circulating estrogen [93].

The role of E2 in regulating macrophage function is still an evolving topic. In particular, there is interest in understanding how E2 levels in vivo influence the activation of macrophage phenotypes in physiological conditions at different stages of the menstrual cycle as well as in pathological conditions associated with changes in circulating estrogen levels. This piece fits into the broader puzzle of how estrogen pathways impact on macrophage function and, consequently, on immune response, angiogenesis, wound healing and metabolism (Figure 1). Further research on gender differences in the immune response and the onset and progression of autoimmune disease will allow the identification of new preventive strategies and personalized therapeutic approaches for treatment of these immuno-mediated disorders.

8. Conclusions

The role of estrogen and its multiple receptors in health and disease is heterogeneous. This makes trying and putting the numerous puzzle pieces together a rather complex task. The protection against cardiovascular disease in women during reproductive age is related, at least in part, to estrogen since endogenous E2 levels and ER expression differ considerably between sexes. Estrogen prevents endothelial dysfunction and atherosclerosis by promoting endothelial healing and increasing angiogenesis. The number of puzzle pieces and with them our knowledge of the mechanisms of estrogen action is growing (Figure 1). Today, it is clear that the combined rapid and genomic effects of estrogen are critical to its overall function; however, these interactions are complex and involve multiple receptor subtypes, both intracellular and membrane-associated. Pharmacological research is poised to design ER ligands that can drive specific transcriptional outcomes, including pathway- and tissue-selective signaling. Targeting specific ERs in the cardiovascular system and fitting together the entire puzzle may result in novel and possibly safer therapeutic options for cardiovascular protection.

Acknowledgments: This study was supported by institutional funding from the University of Padova to Chiara Bolego, Andrea Cignarella and Lucia Trevisi.

Author Contributions: Annalisa Trenti, Serena Tedesco and Carlotta Boscaro performed comprehensive literature search and drafted the paper. Lucia Trevisi and Chiara Bolego critically revised the paper. Andrea Cignarella selected the topic, performed additional literature search and revised the paper.

Conflicts of Interest: The authors declare no conflict of interest.

Abbreviations

AF	activator factor
cAMP	cyclic adenosine monophosphate
E2	17β-estradiol
eNOS	endothelial nitric oxide synthase
ERE	estrogen response element
ER	estrogen receptor
FAK	focal adhesion kinase
G-1	(±)-1-[(3aR*,4S*,9bS*)-4-(6-Bromo-1,3-benzodioxol-5-yl)-3a,4,5,9b-tetrahydro-3H -cyclopenta[c]quinolin-8-yl]-ethanone
GPER1	G-protein-coupled ER
HUVEC	human umbilical vein endothelial cells
IL	interleukin
IFNγ	interferon-γ
LBD	Ligand-Binding Domain
LPS	lipopolysaccharide
MAPK	mitogen-activated protein kinase
MCP-1	monocyte chemoattractant protein-1
NF-κB	nuclear factor kappa-light-chain-enhancer of activated B cells
NO	nitric oxide
PFKFB3	phosphofructokinase-2/fructose-2,6-bisphosphatase-3
PI3K	phosphatidylinositol-3-kinases
TNF-α	tumor necrosis factor α
VEGF	vascular endothelial growth factor

References

1. Prossnitz, E.R.; Arterburn, J.B. International Union of Basic and Clinical Pharmacology. XCVII. G protein-coupled estrogen receptor and its pharmacologic modulators. *Pharmacol. Rev.* **2015**, *67*, 505–540. [CrossRef] [PubMed]
2. Pare, G.; Krust, A.; Karas, R.H.; Dupont, S.; Aronovitz, M.; Chambon, P.; Mendelsohn, M.E. Estrogen receptor-α mediates the protective effects of estrogen against vascular injury. *Circ. Res.* **2002**, *90*, 1087–1092. [CrossRef] [PubMed]
3. Bolego, C.; Rossoni, G.; Fadini, G.P.; Vegeto, E.; Pinna, C.; Albiero, M.; Boscaro, E.; Agostini, C.; Avogaro, A.; Gaion, R.M.; et al. Selective estrogen receptor-α agonist provides widespread heart and vascular protection with enhanced endothelial progenitor cell mobilization in the absence of uterotrophic action. *FASEB J.* **2010**, *24*, 2262–2272. [CrossRef] [PubMed]
4. Marcantoni, E.; Di Francesco, L.; Totani, L.; Piccoli, A.; Evangelista, V.; Tacconelli, S.; Patrignani, P. Effects of estrogen on endothelial prostanoid production and cyclooxygenase-2 and heme oxygenase-1 expression. *Prostaglandins Other Lipid Mediat.* **2012**, *98*, 122–128. [CrossRef] [PubMed]
5. Murphy, A.J.; Guyre, P.M.; Wira, C.R.; Pioli, P.A. Estradiol regulates expression of estrogen receptor ERα46 in human macrophages. *PLoS ONE* **2009**, *4*, e5539. [CrossRef] [PubMed]

6. Geraldes, P.; Sirois, M.G.; Bernatchez, P.N.; Tanguay, J.F. Estrogen regulation of endothelial and smooth muscle cell migration and proliferation: Role of p38 and p42/44 mitogen-activated protein kinase. *Arterioscler. Thromb. Vasc. Biol.* **2002**, *22*, 1585–1590. [CrossRef] [PubMed]

7. Maggi, A.; Cignarella, A.; Brusadelli, A.; Bolego, C.; Pinna, C.; Puglisi, L. Diabetes undermines estrogen control of inducible nitric oxide synthase function in rat aortic smooth muscle cells through overexpression of estrogen receptor-β. *Circulation* **2003**, *108*, 211–217. [CrossRef] [PubMed]

8. Morales, D.E.; McGowan, K.A.; Grant, D.S.; Maheshwari, S.; Bhartiya, D.; Cid, M.C.; Kleinman, H.K.; Schnaper, H.W. Estrogen promotes angiogenic activity in human umbilical vein endothelial cells in vitro and in a murine model. *Circulation* **1995**, *91*, 755–763. [CrossRef] [PubMed]

9. Concina, P.; Sordello, S.; Barbacanne, M.A.; Elhage, R.; Pieraggi, M.T.; Fournial, G.; Plouet, J.; Bayard, F.; Arnal, J.F. The mitogenic effect of 17β-estradiol on in vitro endothelial cell proliferation and on in vivo reendothelialization are both dependent on vascular endothelial growth factor. *J. Vasc. Res.* **2000**, *37*, 202–208. [CrossRef] [PubMed]

10. Simoncini, T.; Scorticati, C.; Mannella, P.; Fadiel, A.; Giretti, M.S.; Fu, X.D.; Baldacci, C.; Garibaldi, S.; Caruso, A.; Fornari, L.; et al. Estrogen receptor α interacts with Gα13 to drive actin remodeling and endothelial cell migration via the RhoA/Rho kinase/moesin pathway. *Mol. Endocrinol.* **2006**, *20*, 1756–1771. [CrossRef] [PubMed]

11. Trenti, A.; Tedesco, S.; Boscaro, C.; Ferri, N.; Cignarella, A.; Trevisi, L.; Bolego, C. The glycolytic enzyme PFKFB3 is involved in estrogen-mediated angiogenesis via GPER1. *J. Pharmacol. Exp. Ther.* **2017**, *361*, 398–407. [CrossRef] [PubMed]

12. Bao, A.M.; Liu, R.Y.; van Someren, E.J.; Hofman, M.A.; Cao, Y.X.; Zhou, J.N. Diurnal rhythm of free estradiol during the menstrual cycle. *Eur. J. Endocrinol.* **2003**, *148*, 227–232. [CrossRef] [PubMed]

13. Wildman, R.P.; Colvin, A.B.; Powell, L.H.; Matthews, K.A.; Everson-Rose, S.A.; Hollenberg, S.; Johnston, J.M.; Sutton-Tyrrell, K. Associations of endogenous sex hormones with the vasculature in menopausal women: The Study of Women's health Across the Nation (SWAN). *Menopause* **2008**, *15*, 414–421. [CrossRef] [PubMed]

14. Mauvais-Jarvis, F.; Clegg, D.J.; Hevener, A.L. The role of estrogens in control of energy balance and glucose homeostasis. *Endocr. Rev.* **2013**, *34*, 309–338. [CrossRef] [PubMed]

15. Cignarella, A.; Kratz, M.; Bolego, C. Emerging role of estrogen in the control of cardiometabolic disease. *Trends Pharmacol. Sci.* **2010**, *31*, 183–189. [CrossRef] [PubMed]

16. Della Torre, S.; Benedusi, V.; Fontana, R.; Maggi, A. Energy metabolism and fertility: A balance preserved for female health. *Nat. Rev. Endocrinol.* **2014**, *10*, 13–23. [CrossRef] [PubMed]

17. Krasinski, K.; Spyridopoulos, I.; Asahara, T.; van der Zee, R.; Isner, J.M.; Losordo, D.W. Estradiol accelerates functional endothelial recovery after arterial injury. *Circulation* **1997**, *95*, 1768–1772. [CrossRef] [PubMed]

18. Iwakura, A.; Luedemann, C.; Shastry, S.; Hanley, A.; Kearney, M.; Aikawa, R.; Isner, J.M.; Asahara, T.; Losordo, D.W. Estrogen-mediated, endothelial nitric oxide synthase-dependent mobilization of bone marrow-derived endothelial progenitor cells contributes to reendothelialization after arterial injury. *Circulation* **2003**, *108*, 3115–3121. [CrossRef] [PubMed]

19. Strehlow, K.; Werner, N.; Berweiler, J.; Link, A.; Dirnagl, U.; Priller, J.; Laufs, K.; Ghaeni, L.; Milosevic, M.; Bohm, M.; et al. Estrogen increases bone marrow-derived endothelial progenitor cell production and diminishes neointima formation. *Circulation* **2003**, *107*, 3059–3065. [CrossRef] [PubMed]

20. Toutain, C.E.; Filipe, C.; Billon, A.; Fontaine, C.; Brouchet, L.; Guery, J.C.; Gourdy, P.; Arnal, J.F.; Lenfant, F. Estrogen receptor α expression in both endothelium and hematopoietic cells is required for the accelerative effect of estradiol on reendothelialization. *Arterioscler. Thromb. Vasc. Biol.* **2009**, *29*, 1543–1550. [CrossRef] [PubMed]

21. Rubanyi, G.M.; Johns, A.; Kauser, K. Effect of estrogen on endothelial function and angiogenesis. *Vasc. Pharmacol.* **2002**, *38*, 89–98. [CrossRef]

22. Filipe, C.; Lam Shang Leen, L.; Brouchet, L.; Billon, A.; Benouaich, V.; Fontaine, V.; Gourdy, P.; Lenfant, F.; Arnal, J.F.; Gadeau, A.P.; et al. Estradiol accelerates endothelial healing through the retrograde commitment of uninjured endothelium. *Am. J. Physiol. Heart Circ. Physiol.* **2008**, *294*, H2822–H2830. [CrossRef] [PubMed]

23. Brouchet, L.; Krust, A.; Dupont, S.; Chambon, P.; Bayard, F.; Arnal, J.F. Estradiol accelerates reendothelialization in mouse carotid artery through estrogen receptor-α but not estrogen receptor-β. *Circulation* **2001**, *103*, 423–428. [CrossRef] [PubMed]

24. Hardman, M.J.; Ashcroft, G.S. Estrogen, not intrinsic aging, is the major regulator of delayed human wound healing in the elderly. *Genome Biol.* **2008**, *9*, R80. [CrossRef] [PubMed]
25. Libert, C.; Dejager, L.; Pinheiro, I. The X chromosome in immune functions: When a chromosome makes the difference. *Nat. Rev. Immunol.* **2010**, *10*, 594–604. [CrossRef] [PubMed]
26. Gubbels Bupp, M.R. Sex, the aging immune system, and chronic disease. *Cell. Immunol.* **2015**, *294*, 102–110. [CrossRef] [PubMed]
27. Amadori, A.; Zamarchi, R.; De Silvestro, G.; Forza, G.; Cavatton, G.; Danieli, G.A.; Clementi, M.; Chieco-Bianchi, L. Genetic control of the CD4/CD8 T-cell ratio in humans. *Nat. Med.* **1995**, *1*, 1279–1283. [CrossRef] [PubMed]
28. Scotland, R.S.; Stables, M.J.; Madalli, S.; Watson, P.; Gilroy, D.W. Sex differences in resident immune cell phenotype underlie more efficient acute inflammatory responses in female mice. *Blood* **2011**, *118*, 5918–5927. [CrossRef] [PubMed]
29. Benedek, G.; Zhang, J.; Nguyen, H.; Kent, G.; Seifert, H.; Vandenbark, A.A.; Offner, H. Novel feedback loop between M2 macrophages/microglia and regulatory B cells in estrogen-protected EAE mice. *J. Neuroimmunol.* **2017**, *305*, 59–67. [CrossRef] [PubMed]
30. Nadkarni, S.; McArthur, S. Oestrogen and immunomodulation: New mechanisms that impact on peripheral and central immunity. *Curr. Opin. Pharmacol.* **2013**, *13*, 576–581. [CrossRef] [PubMed]
31. Della Torre, S.; Maggi, A. Sex differences: A resultant of an evolutionary pressure? *Cell. Metab.* **2017**, *25*, 499–505. [CrossRef] [PubMed]
32. Bolego, C.; Vegeto, E.; Pinna, C.; Maggi, A.; Cignarella, A. Selective agonists of estrogen receptor isoforms: New perspectives for cardiovascular disease. *Arterioscler. Thromb. Vasc. Biol.* **2006**, *26*, 2192–2199. [CrossRef] [PubMed]
33. Revankar, C.M.; Cimino, D.F.; Sklar, L.A.; Arterburn, J.B.; Prossnitz, E.R. A transmembrane intracellular estrogen receptor mediates rapid cell signaling. *Science* **2005**, *307*, 1625–1630. [CrossRef] [PubMed]
34. Chakrabarti, S.; Davidge, S.T. G-protein coupled receptor 30 (GPR30): A novel regulator of endothelial inflammation. *PLoS ONE* **2012**, *7*, e52357. [CrossRef] [PubMed]
35. Kumar, R.; Zakharov, M.N.; Khan, S.H.; Miki, R.; Jang, H.; Toraldo, G.; Singh, R.; Bhasin, S.; Jasuja, R. The dynamic structure of the estrogen receptor. *J. Amino Acids* **2011**, *2011*, 812540. [CrossRef] [PubMed]
36. Heldring, N.; Pike, A.; Andersson, S.; Matthews, J.; Cheng, G.; Hartman, J.; Tujague, M.; Strom, A.; Treuter, E.; Warner, M.; et al. Estrogen receptors: How do they signal and what are their targets. *Physiol. Rev.* **2007**, *87*, 905–931. [CrossRef] [PubMed]
37. Leitman, D.C.; Paruthiyil, S.; Vivar, O.I.; Saunier, E.F.; Herber, C.B.; Cohen, I.; Tagliaferri, M.; Speed, T.P. Regulation of specific target genes and biological responses by estrogen receptor subtype agonists. *Curr. Opin. Pharmacol.* **2010**, *10*, 629–636. [CrossRef] [PubMed]
38. Kim, K.H.; Bender, J.R. Membrane-initiated actions of estrogen on the endothelium. *Mol. Cell. Endocrinol.* **2009**, *308*, 3–8. [CrossRef] [PubMed]
39. Acconcia, F.; Ascenzi, P.; Bocedi, A.; Spisni, E.; Tomasi, V.; Trentalance, A.; Visca, P.; Marino, M. Palmitoylation-dependent estrogen receptor α membrane localization: Regulation by 17β-estradiol. *Mol. Biol. Cell* **2005**, *16*, 231–237. [CrossRef] [PubMed]
40. Billon-Gales, A.; Krust, A.; Fontaine, C.; Abot, A.; Flouriot, G.; Toutain, C.; Berges, H.; Gadeau, A.P.; Lenfant, F.; Gourdy, P.; et al. Activation function 2 (AF2) of estrogen receptor-α is required for the atheroprotective action of estradiol but not to accelerate endothelial healing. *Proc. Natl. Acad. Sci. USA* **2011**, *108*, 13311–13316. [CrossRef] [PubMed]
41. Adlanmerini, M.; Solinhac, R.; Abot, A.; Fabre, A.; Raymond-Letron, I.; Guihot, A.L.; Boudou, F.; Sautier, L.; Vessieres, E.; Kim, S.H.; et al. Mutation of the palmitoylation site of estrogen receptor α in vivo reveals tissue-specific roles for membrane versus nuclear actions. *Proc. Natl. Acad. Sci. USA* **2014**, *111*, E283–E290. [CrossRef] [PubMed]
42. Filardo, E.J.; Thomas, P. Minireview: G protein-coupled estrogen receptor-1, GPER-1: Its mechanism of action and role in female reproductive cancer, renal and vascular physiology. *Endocrinology* **2012**, *153*, 2953–2962. [CrossRef] [PubMed]

43. Pelekanou, V.; Kampa, M.; Kiagiadaki, F.; Deli, A.; Theodoropoulos, P.; Agrogiannis, G.; Patsouris, E.; Tsapis, A.; Castanas, E.; Notas, G. Estrogen anti-inflammatory activity on human monocytes is mediated through cross-talk between estrogen receptor ERα36 and GPR30/GPER1. *J. Leukoc. Biol.* **2016**, *99*, 333–347. [CrossRef] [PubMed]

44. Albanito, L.; Madeo, A.; Lappano, R.; Vivacqua, A.; Rago, V.; Carpino, A.; Oprea, T.I.; Prossnitz, E.R.; Musti, A.M.; Ando, S.; et al. G protein-coupled receptor 30 (GPR30) mediates gene expression changes and growth response to 17β-estradiol and selective GPR30 ligand G-1 in ovarian cancer cells. *Cancer Res.* **2007**, *67*, 1859–1866. [CrossRef] [PubMed]

45. Gao, F.; Ma, X.; Ostmann, A.B.; Das, S.K. GPR30 activation opposes estrogen-dependent uterine growth via inhibition of stromal ERK1/2 and estrogen receptor alpha (ERα) phosphorylation signals. *Endocrinology* **2011**, *152*, 1434–1447. [CrossRef] [PubMed]

46. Traupe, T.; Stettler, C.D.; Li, H.; Haas, E.; Bhattacharya, I.; Minotti, R.; Barton, M. Distinct roles of estrogen receptors α and β mediating acute vasodilation of epicardial coronary arteries. *Hypertension* **2007**, *49*, 1364–1370. [CrossRef] [PubMed]

47. Romano, S.N.; Gorelick, D.A. Crosstalk between nuclear and G protein-coupled estrogen receptors. *Gen. Comp. Endocrinol.* **2017**. [CrossRef] [PubMed]

48. Venkov, C.D.; Rankin, A.B.; Vaughan, D.E. Identification of authentic estrogen receptor in cultured endothelial cells—A potential mechanism for steroid hormone regulation of endothelial function. *Circulation* **1996**, *94*, 727–733. [CrossRef] [PubMed]

49. Karas, R.H.; Patterson, B.L.; Mendelsohn, M.E. Human vascular smooth muscle cells contain functional estrogen receptor. *Circulation* **1994**, *89*, 1943–1950. [CrossRef] [PubMed]

50. Simoncini, T.; Maffei, S.; Basta, G.; Barsacchi, G.; Genazzani, A.R.; Liao, J.K.; De Caterina, R. Estrogens and glucocorticoids inhibit endothelial vascular cell adhesion molecule-1 expression by different transcriptional mechanisms. *Circ. Res.* **2000**, *87*, 19–25. [CrossRef] [PubMed]

51. Harris, H.A.; Katzenellenbogen, J.A.; Katzenellenbogen, B.S. Characterization of the biological roles of the estrogen receptors, ERα and ERβ, in estrogen target tissues in vivo through the use of an ERα-selective ligand. *Endocrinology* **2002**, *143*, 4172–4177. [CrossRef] [PubMed]

52. Mendelsohn, M.E.; Karas, R.H. Molecular and cellular basis of cardiovascular gender differences. *Science* **2005**, *308*, 1583–1587. [CrossRef] [PubMed]

53. Ospina, J.A.; Krause, D.N.; Duckles, S.P. 17β-estradiol increases rat cerebrovascular prostacyclin synthesis by elevating cyclooxygenase-1 and prostacyclin synthase. *Stroke* **2002**, *33*, 600–605. [CrossRef] [PubMed]

54. Egan, K.M.; Lawson, J.A.; Fries, S.; Koller, B.; Rader, D.J.; Smyth, E.M.; Fitzgerald, G.A. COX-2-derived prostacyclin confers atheroprotection on female mice. *Science* **2004**, *306*, 1954–1957. [CrossRef] [PubMed]

55. Nevzati, E.; Shafighi, M.; Bakhtian, K.D.; Treiber, H.; Fandino, J.; Fathi, A.R. Estrogen induces nitric oxide production via nitric oxide synthase activation in endothelial cells. *Acta Neurochir. Suppl.* **2015**, *120*, 141–145. [CrossRef] [PubMed]

56. Papapetropoulos, A.; Garcia-Cardena, G.; Madri, J.A.; Sessa, W.C. Nitric oxide production contributes to the angiogenic properties of vascular endothelial growth factor in human endothelial cells. *J. Clin. Investig.* **1997**, *100*, 3131–3139. [CrossRef] [PubMed]

57. Ziche, M.; Morbidelli, L.; Choudhuri, R.; Zhang, H.T.; Donnini, S.; Granger, H.J.; Bicknell, R. Nitric oxide synthase lies downstream from vascular endothelial growth factor-induced but not basic fibroblast growth factor-induced angiogenesis. *J. Clin. Investig.* **1997**, *99*, 2625–2634. [CrossRef] [PubMed]

58. Novensa, L.; Novella, S.; Medina, P.; Segarra, G.; Castillo, N.; Heras, M.; Hermenegildo, C.; Dantas, A.P. Aging negatively affects estrogens-mediated effects on nitric oxide bioavailability by shifting ERα/ERβ balance in female mice. *PLoS ONE* **2011**, *6*, e25335. [CrossRef] [PubMed]

59. Rowlands, D.J.; Chapple, S.; Siow, R.C.; Mann, G.E. Equol-stimulated mitochondrial reactive oxygen species activate endothelial nitric oxide synthase and redox signaling in endothelial cells: Roles for F-actin and GPR30. *Hypertension* **2011**, *57*, 833–840. [CrossRef] [PubMed]

60. Thomas, P.; Pang, Y.; Filardo, E.J.; Dong, J. Identity of an estrogen membrane receptor coupled to a G protein in human breast cancer cells. *Endocrinology* **2005**, *146*, 624–632. [CrossRef] [PubMed]

61. Meyer, M.R.; Fredette, N.C.; Daniel, C.; Sharma, G.; Amann, K.; Arterburn, J.B.; Barton, M.; Prossnitz, E.R. Obligatory role for GPER in cardiovascular aging and disease. *Sci. Signal.* **2016**, *9*, ra105. [CrossRef] [PubMed]

62. Reynolds, L.P.; Killilea, S.D.; Redmer, D.A. Angiogenesis in the female reproductive system. *FASEB J.* **1992**, *6*, 886–892. [CrossRef] [PubMed]

63. Alvarez, R.J.; Gips, S.J.; Moldovan, N.; Wilhide, C.C.; Milliken, E.E.; Hoang, A.T.; Hruban, R.H.; Silverman, H.S.; Dang, C.V.; Goldschmidt-Clermont, P.J. 17β-estradiol inhibits apoptosis of endothelial cells. *Biochem. Biophys. Res. Commun.* **1997**, *237*, 372–381. [CrossRef] [PubMed]

64. Spyridopoulos, I.; Sullivan, A.B.; Kearney, M.; Isner, J.M.; Losordo, D.W. Estrogen-receptor-mediated inhibition of human endothelial cell apoptosis. Estradiol as a survival factor. *Circulation* **1997**, *95*, 1505–1514. [CrossRef] [PubMed]

65. Sanchez, A.M.; Flamini, M.I.; Zullino, S.; Gopal, S.; Genazzani, A.R.; Simoncini, T. Estrogen receptor-α promotes endothelial cell motility through focal adhesion kinase. *Mol. Hum. Reprod.* **2011**, *17*, 219–226. [CrossRef] [PubMed]

66. Johns, A.; Freay, A.D.; Fraser, W.; Korach, K.S.; Rubanyi, G.M. Disruption of estrogen receptor gene prevents 17 beta estradiol-induced angiogenesis in transgenic mice. *Endocrinology* **1996**, *137*, 4511–4513. [CrossRef] [PubMed]

67. Oviedo, P.J.; Hermenegildo, C.; Tarin, J.J.; Cano, A. Raloxifene increases proliferation of human endothelial cells in association with increased gene expression of cyclins A and B1. *Fertil. Steril.* **2007**, *88*, 326–332. [CrossRef] [PubMed]

68. Oviedo, P.J.; Sobrino, A.; Laguna-Fernandez, A.; Novella, S.; Tarin, J.J.; Garcia-Perez, M.A.; Sanchis, J.; Cano, A.; Hermenegildo, C. Estradiol induces endothelial cell migration and proliferation through estrogen receptor-enhanced RhoA/ROCK pathway. *Mol. Cell. Endocrinol.* **2011**, *335*, 96–103. [CrossRef] [PubMed]

69. Karas, R.H.; Gauer, E.A.; Bieber, H.E.; Baur, W.E.; Mendelsohn, M.E. Growth factor activation of the estrogen receptor in vascular cells occurs via a mitogen-activated protein kinase-independent pathway. *J. Clin. Investig.* **1998**, *101*, 2851–2861. [CrossRef] [PubMed]

70. Shifren, J.L.; Tseng, J.F.; Zaloudek, C.J.; Ryan, I.P.; Meng, Y.G.; Ferrara, N.; Jaffe, R.B.; Taylor, R.N. Ovarian steroid regulation of vascular endothelial growth factor in the human endometrium: Implications for angiogenesis during the menstrual cycle and in the pathogenesis of endometriosis. *J. Clin. Endocrinol. Metab.* **1996**, *81*, 3112–3118. [PubMed]

71. Gargett, C.E.; Zaitseva, M.; Bucak, K.; Chu, S.; Fuller, P.J.; Rogers, P.A. 17β-estradiol up-regulates vascular endothelial growth factor receptor-2 expression in human myometrial microvascular endothelial cells: Role of estrogen receptor-α and -β. *J. Clin. Endocrinol. Metab.* **2002**, *87*, 4341–4349. [CrossRef] [PubMed]

72. Cid, M.C.; Esparza, J.; Schnaper, H.W.; Juan, M.; Yague, J.; Grant, D.S.; Urbano-Marquez, A.; Hoffman, G.S.; Kleinman, H.K. Estradiol enhances endothelial cell interactions with extracellular matrix proteins via an increase in integrin expression and function. *Angiogenesis* **1999**, *3*, 271–280. [CrossRef] [PubMed]

73. Haran, E.F.; Maretzek, A.F.; Goldberg, I.; Horowitz, A.; Degani, H. Tamoxifen enhances cell death in implanted MCF7 breast cancer by inhibiting endothelium growth. *Cancer Res.* **1994**, *54*, 5511–5514. [PubMed]

74. Dadiani, M.; Seger, D.; Kreizman, T.; Badikhi, D.; Margalit, R.; Eilam, R.; Degani, H. Estrogen regulation of vascular endothelial growth factor in breast cancer in vitro and in vivo: The role of estrogen receptor α and c-Myc. *Endocr. Relat. Cancer* **2009**, *16*, 819–834. [CrossRef] [PubMed]

75. Osthus, R.C.; Shim, H.; Kim, S.; Li, Q.; Reddy, R.; Mukherjee, M.; Xu, Y.; Wonsey, D.; Lee, L.A.; Dang, C.V. Deregulation of glucose transporter 1 and glycolytic gene expression by c-Myc. *J. Biol. Chem.* **2000**, *275*, 21797–21800. [CrossRef] [PubMed]

76. Pedram, A.; Razandi, M.; O'Mahony, F.; Harvey, H.; Harvey, B.J.; Levin, E.R. Estrogen reduces lipid content in the liver exclusively from membrane receptor signaling. *Sci. Signal.* **2013**, *6*, ra36. [CrossRef] [PubMed]

77. Guillaume, M.; Montagner, A.; Fontaine, C.; Lenfant, F.; Arnal, J.F.; Gourdy, P. Nuclear and membrane actions of estrogen receptor alpha: Contribution to the regulation of energy and glucose homeostasis. *Adv. Exp. Med. Biol.* **2017**, *1043*, 401–426. [CrossRef] [PubMed]

78. De Bock, K.; Georgiadou, M.; Schoors, S.; Kuchnio, A.; Wong, B.W.; Cantelmo, A.R.; Quaegebeur, A.; Ghesquiere, B.; Cauwenberghs, S.; Eelen, G.; et al. Role of PFKFB3-driven glycolysis in vessel sprouting. *Cell* **2013**, *154*, 651–663. [CrossRef] [PubMed]

79. Imbert-Fernandez, Y.; Clem, B.F.; O'Neal, J.; Kerr, D.A.; Spaulding, R.; Lanceta, L.; Clem, A.L.; Telang, S.; Chesney, J. Estradiol stimulates glucose metabolism via 6-phosphofructo-2-kinase (PFKFB3). *J. Biol. Chem.* **2014**, *289*, 9440–9448. [CrossRef] [PubMed]

80. De Francesco, E.M.; Sotgia, F.; Clarke, R.B.; Lisanti, M.P.; Maggiolini, M. G protein-coupled receptors at the crossroad between physiologic and pathologic angiogenesis: Old paradigms and emerging concepts. *Int J. Mol. Sci.* **2017**, *18*, 2713. [CrossRef] [PubMed]

81. Sica, A.; Mantovani, A. Macrophage plasticity and polarization: In vivo veritas. *J. Clin. Investig.* **2012**, *122*, 787–795. [CrossRef] [PubMed]

82. Mosser, D.M.; Edwards, J.P. Exploring the full spectrum of macrophage activation. *Nat. Rev. Immunol.* **2008**, *8*, 958–969. [CrossRef] [PubMed]

83. Mantovani, A.; Biswas, S.K.; Galdiero, M.R.; Sica, A.; Locati, M. Macrophage plasticity and polarization in tissue repair and remodelling. *J. Pathol.* **2013**, *229*, 176–185. [CrossRef] [PubMed]

84. Xue, J.; Schmidt, S.V.; Sander, J.; Draffehn, A.; Krebs, W.; Quester, I.; De Nardo, D.; Gohel, T.D.; Emde, M.; Schmidleithner, L.; et al. Transcriptome-based network analysis reveals a spectrum model of human macrophage activation. *Immunity* **2014**, *40*, 274–288. [CrossRef] [PubMed]

85. Qian, B.Z.; Pollard, J.W. Macrophage diversity enhances tumor progression and metastasis. *Cell* **2010**, *141*, 39–51. [CrossRef] [PubMed]

86. Brecht, K.; Weigert, A.; Hu, J.; Popp, R.; Fisslthaler, B.; Korff, T.; Fleming, I.; Geisslinger, G.; Brune, B. Macrophages programmed by apoptotic cells promote angiogenesis via prostaglandin E2. *FASEB J.* **2011**, *25*, 2408–2417. [CrossRef] [PubMed]

87. Bruemmer, D. Targeting angiogenesis as treatment for obesity. *Arterioscler. Thromb. Vasc. Biol.* **2012**, *32*, 161–162. [CrossRef] [PubMed]

88. Miller, A.P.; Feng, W.; Xing, D.; Weathington, N.M.; Blalock, J.E.; Chen, Y.F.; Oparil, S. Estrogen modulates inflammatory mediator expression and neutrophil chemotaxis in injured arteries. *Circulation* **2004**, *110*, 1664–1669. [CrossRef] [PubMed]

89. Hsieh, Y.C.; Frink, M.; Hsieh, C.H.; Choudhry, M.A.; Schwacha, M.G.; Bland, K.I.; Chaudry, I.H. Downregulation of migration inhibitory factor is critical for estrogen-mediated attenuation of lung tissue damage following trauma-hemorrhage. *Am. J. Physiol. Lung Cell. Mol. Physiol.* **2007**, *292*, L1227–L1232. [CrossRef] [PubMed]

90. Giannoni, E.; Guignard, L.; Knaup Reymond, M.; Perreau, M.; Roth-Kleiner, M.; Calandra, T.; Roger, T. Estradiol and progesterone strongly inhibit the innate immune response of mononuclear cells in newborns. *Infect. Immun.* **2011**, *79*, 2690–2698. [CrossRef] [PubMed]

91. Campesi, I.; Sanna, M.; Zinellu, A.; Carru, C.; Rubattu, L.; Bulzomi, P.; Seghieri, G.; Tonolo, G.; Palermo, M.; Rosano, G.; et al. Oral contraceptives modify DNA methylation and monocyte-derived macrophage function. *Biol. Sex. Differ.* **2012**, *3*, 4. [CrossRef] [PubMed]

92. Pechenino, A.S.; Lin, L.; Mbai, F.N.; Lee, A.R.; He, X.M.; Stallone, J.N.; Knowlton, A.A. Impact of aging vs. estrogen loss on cardiac gene expression: Estrogen replacement and inflammation. *Physiol. Genom.* **2011**, *43*, 1065–1073. [CrossRef] [PubMed]

93. Toniolo, A.; Fadini, G.P.; Tedesco, S.; Cappellari, R.; Vegeto, E.; Maggi, A.; Avogaro, A.; Bolego, C.; Cignarella, A. Alternative activation of human macrophages is rescued by estrogen treatment in vitro and impaired by menopausal status. *J. Clin. Endocrinol. Metab.* **2015**, *100*, E50–E58. [CrossRef] [PubMed]

94. Xing, D.; Oparil, S.; Yu, H.; Gong, K.; Feng, W.; Black, J.; Chen, Y.F.; Nozell, S. Estrogen modulates NFκB signaling by enhancing IκBα levels and blocking p65 binding at the promoters of inflammatory genes via estrogen receptor-β. *PLoS ONE* **2012**, *7*, e36890. [CrossRef] [PubMed]

95. Murphy, A.J.; Guyre, P.M.; Pioli, P.A. Estradiol suppresses NF-κB activation through coordinated regulation of let-7a and miR-125b in primary human macrophages. *J. Immunol.* **2010**, *184*, 5029–5037. [CrossRef] [PubMed]

96. Lambert, K.C.; Curran, E.M.; Judy, B.M.; Lubahn, D.B.; Estes, D.M. Estrogen receptor-α deficiency promotes increased TNF-α secretion and bacterial killing by murine macrophages in response to microbial stimuli in vitro. *J. Leukoc. Biol.* **2004**, *75*, 1166–1172. [CrossRef] [PubMed]

97. Calippe, B.; Douin-Echinard, V.; Delpy, L.; Laffargue, M.; Lelu, K.; Krust, A.; Pipy, B.; Bayard, F.; Arnal, J.F.; Guery, J.C.; et al. 17β-estradiol promotes TLR4-triggered proinflammatory mediator production through direct estrogen receptor α signaling in macrophages in vivo. *J. Immunol.* **2010**, *185*, 1169–1176. [CrossRef] [PubMed]

98. Straub, R.H. The complex role of estrogens in inflammation. *Endocr. Rev.* **2007**, *28*, 521–574. [CrossRef] [PubMed]

99. Murray, P.J.; Wynn, T.A. Protective and pathogenic functions of macrophage subsets. *Nat. Rev. Immunol.* **2011**, *11*, 723–737. [CrossRef] [PubMed]

100. Spiller, K.L.; Anfang, R.R.; Spiller, K.J.; Ng, J.; Nakazawa, K.R.; Daulton, J.W.; Vunjak-Novakovic, G. The role of macrophage phenotype in vascularization of tissue engineering scaffolds. *Biomaterials* **2014**, *35*, 4477–4488. [CrossRef] [PubMed]

101. Sindrilaru, A.; Peters, T.; Wieschalka, S.; Baican, C.; Baican, A.; Peter, H.; Hainzl, A.; Schatz, S.; Qi, Y.; Schlecht, A.; et al. An unrestrained proinflammatory M1 macrophage population induced by iron impairs wound healing in humans and mice. *J. Clin. Investig.* **2011**, *121*, 985–997. [CrossRef] [PubMed]

102. Campbell, L.; Emmerson, E.; Williams, H.; Saville, C.R.; Krust, A.; Chambon, P.; Mace, K.A.; Hardman, M.J. Estrogen receptor-alpha promotes alternative macrophage activation during cutaneous repair. *J. Investig. Dermatol.* **2014**, *134*, 2447–2457. [CrossRef] [PubMed]

103. Villa, A.; Rizzi, N.; Vegeto, E.; Ciana, P.; Maggi, A. Estrogen accelerates the resolution of inflammation in macrophagic cells. *Sci. Rep.* **2015**, *5*, 15224. [CrossRef] [PubMed]

104. Tang, C.-Y.; Mauro, C. Similarities in the metabolic reprogramming of immune system and endothelium. *Front. Immunol.* **2017**, *8*, 837. [CrossRef] [PubMed]

105. Blasko, E.; Haskell, C.A.; Leung, S.; Gualtieri, G.; Halks-Miller, M.; Mahmoudi, M.; Dennis, M.K.; Prossnitz, E.R.; Karpus, W.J.; Horuk, R. Beneficial role of the GPR30 agonist G-1 in an animal model of multiple sclerosis. *J. Neuroimmunol.* **2009**, *214*, 67–77. [CrossRef] [PubMed]

106. Rettew, J.A.; McCall, S.H.T.; Marriott, I. GPR30/GPER-1 mediates rapid decreases in TLR4 expression on murine macrophages. *Mol. Cell. Endocrinol.* **2010**, *328*, 87–92. [CrossRef] [PubMed]

107. Kramer, P.R.; Winger, V.; Kramer, S.F. 17β-estradiol utilizes the estrogen receptor to regulate CD16 expression in monocytes. *Mol. Cell. Endocrinol.* **2007**, *279*, 16–25. [CrossRef] [PubMed]

108. Campesi, I.; Marino, M.; Montella, A.; Pais, S.; Franconi, F. Sex differences in estrogen receptor α and β levels and activation status in LPS-stimulated human macrophages. *J. Cell. Physiol.* **2017**, *232*, 340–345. [CrossRef] [PubMed]

109. Pepe, G.; Braga, D.; Renzi, T.A.; Villa, A.; Bolego, C.; D'Avila, F.; Barlassina, C.; Maggi, A.; Locati, M.; Vegeto, E. Self-renewal and phenotypic conversion are the main physiological responses of macrophages to the endogenous estrogen surge. *Sci. Rep.* **2017**, *7*, 44270. [CrossRef] [PubMed]

110. Ribas, V.; Drew, B.G.; Le, J.A.; Soleymani, T.; Daraei, P.; Sitz, D.; Mohammad, L.; Henstridge, D.C.; Febbraio, M.A.; Hewitt, S.; et al. Myeloid-specific estrogen receptor α deficiency impairs metabolic homeostasis and accelerates atherosclerotic lesion development. *Proc. Natl. Acad. Sci. USA* **2011**, *108*, 16457–16462. [CrossRef] [PubMed]

111. Della Torre, S.; Mitro, N.; Fontana, R.; Gomaraschi, M.; Favari, E.; Recordati, C.; Lolli, F.; Quagliarini, F.; Meda, C.; Ohlsson, C.; et al. An essential role for liver ERα in coupling hepatic metabolism to the reproductive cycle. *Cell Rep.* **2016**, *15*, 360–371. [CrossRef] [PubMed]

112. Bowling, M.R.; Xing, D.; Kapadia, A.; Chen, Y.F.; Szalai, A.J.; Oparil, S.; Hage, F.G. Estrogen effects on vascular inflammation are age dependent: Role of estrogen receptors. *Arterioscler. Thromb. Vasc. Biol.* **2014**, *34*, 1477–1485. [CrossRef] [PubMed]

113. Palmer, B.F.; Clegg, D.J. The sexual dimorphism of obesity. *Mol. Cell. Endocrinol.* **2015**, *402*, 113–119. [CrossRef] [PubMed]

114. Manson, J.E.; Aragaki, A.K.; Rossouw, J.E.; Anderson, G.L.; Prentice, R.L.; LaCroix, A.Z.; Chlebowski, R.T.; Howard, B.V.; Thomson, C.A.; Margolis, K.L.; et al. Menopausal hormone therapy and long-term all-cause and cause-specific mortality: The Women's Health Initiative randomized trials. *JAMA* **2017**, *318*, 927–938. [CrossRef] [PubMed]

115. Schierbeck, L.L.; Rejnmark, L.; Tofteng, C.L.; Stilgren, L.; Eiken, P.; Mosekilde, L.; Kober, L.; Jensen, J.E. Effect of hormone replacement therapy on cardiovascular events in recently postmenopausal women: Randomised trial. *BMJ* **2012**, *345*, e6409. [CrossRef] [PubMed]

116. Stopinska-Gluszak, U.; Waligora, J.; Grzela, T.; Gluszak, M.; Jozwiak, J.; Radomski, D.; Roszkowski, P.I.; Malejczyk, J. Effect of estrogen/progesterone hormone replacement therapy on natural killer cell cytotoxicity and immunoregulatory cytokine release by peripheral blood mononuclear cells of postmenopausal women. *J. Reprod. Immunol.* **2006**, *69*, 65–75. [CrossRef] [PubMed]

117. Corcoran, M.P.; Meydani, M.; Lichtenstein, A.H.; Schaefer, E.J.; Dillard, A.; Lamon-Fava, S. Sex hormone modulation of proinflammatory cytokine and C-reactive protein expression in macrophages from older men and postmenopausal women. *J. Endocrinol.* **2010**, *206*, 217–224. [CrossRef] [PubMed]

118. Campesi, I.; Carru, C.; Zinellu, A.; Occhioni, S.; Sanna, M.; Palermo, M.; Tonolo, G.; Mercuro, G.; Franconi, F. Regular cigarette smoking influences the transsulfuration pathway, endothelial function, and inflammation biomarkers in a sex-gender specific manner in healthy young humans. *Am. J. Transl. Res.* **2013**, *5*, 497–509. [PubMed]

International Journal of
Molecular Sciences

MDPI

Review

miRNA as a New Regulatory Mechanism of Estrogen Vascular Action

Daniel Pérez-Cremades [1,2], **Ana Mompeón** [1,2], **Xavier Vidal-Gómez** [1,2], **Carlos Hermenegildo** [1,2] **and Susana Novella** [1,2,]*

[1]   Department of Physiology, Faculty of Medicine and Dentistry, University of Valencia, 46010 Valencia, Spain; daniel.perez@uv.es (D.P.-C.); ana.mompeon@uv.es (A.M.); xavier.vidal@uv.es (X.V.-G.); carlos.hermenegildo@uv.es (C.H.)
[2]   INCLIVA Biomedical Research Institute, 46010 Valencia, Spain
*   Correspondence: susana.novella@uv.es; Tel.: +34-963-864-730

Received: 22 December 2017; Accepted: 1 February 2018; Published: 6 February 2018

Abstract: The beneficial effects of estrogen on the cardiovascular system have been reported extensively. In fact, the incidence of cardiovascular diseases in women is lower than in age-matched men during their fertile stage of life, a benefit that disappears after menopause. These sex-related differences point to sexual hormones, mainly estrogen, as possible cardiovascular protective factors. The regulation of vascular function by estrogen is mainly related to the maintenance of normal endothelial function and is mediated by both direct and indirect gene transcription through the activity of specific estrogen receptors. Some of these mechanisms are known, but many remain to be elucidated. In recent years, microRNAs have been established as non-coding RNAs that regulate the expression of a high percentage of protein-coding genes in mammals and are related to the correct function of human physiology. Moreover, within the cardiovascular system, miRNAs have been related to physiological and pathological conditions. In this review, we address what is known about the role of estrogen-regulated miRNAs and their emerging involvement in vascular biology.

Keywords: miRNA; estradiol; estrogen receptors; epigenetic regulation; endothelial cells

1. Introduction

Estrogen is involved in many physiological processes, including sexual development and reproduction, regulation of skeletal homeostasis, lipid and carbohydrate metabolism, electrolyte balance, central nervous system function (including cognition and behavior), and cardiovascular system regulation [1,2]. In addition to its physiological relevance, the effects of estrogen (or its absence) on target tissues are related to the development of numerous diseases, which include various types of well-known hormone-dependent cancers including breast, ovarian, endometrial, and prostate cancer, among others. However, estrogen is also implicated in the progression of osteoporosis, neurodegenerative diseases, metabolic disorders (insulin resistance and obesity), autoimmune diseases (lupus erythematosus, multiple sclerosis, and rheumatoid arthritis), endometriosis, and cardiovascular diseases [3].

Sex differences in cardiovascular diseases have been extensively reported [4], suggesting that sex hormones have an important influence on the cardiovascular system. Indeed, statistical data have shown that women develop cardiovascular disease 7–10 years later than men [5]. In addition, epidemiological studies have provided evidence that cardiovascular diseases are more frequent in men than in premenopausal women of the same age. However, during the fifth decade of a woman's life, the decrease in estrogen levels that occurs in menopause is accompanied by an increase in the incidence of cardiovascular diseases [6,7], suggesting that estrogen plays a beneficial role in cardiovascular system.

Int. J. Mol. Sci. **2018**, 19, 473; doi:10.3390/ijms19020473   127   www.mdpi.com/journal/ijms

Based on the beneficial role of estrogen, hormonal replacement therapies (HRT) have been used in postmenopausal women with controversial findings [8,9]. The current consensus on HRT indicates that the vascular protective effects of estrogen depend on the onset of treatment after menopause, which has been recently reviewed in depth elsewhere [10]. The phenomenon, referred to as the "timing hypothesis", postulates that the beneficial effects of hormonal replacement in the prevention of cardiovascular disease may occur only when hormonal supplementation is initiated before the detrimental effects that aging has on the cardiovascular system have become established [11]. In this regard, it has been reported that age moderates the vasodilatory [12] and anti-inflammatory [13] effects that estrogen have on vascular tissue in postmenopausal women.

Estrogen can modulate the cardiovascular system by acting directly on vascular cells or indirectly by systemic effects. Endothelial cells, vascular smooth muscle cells (VSMCs), and cardiomyocytes are estrogen targets because they express estrogen receptors (ER) [14]. In addition, ER expression described in monocytes, macrophages, and dendritic cells suggests that modulation of inflammatory processes, a key event in the initiation and development of cardiovascular diseases, may also be estrogen-dependent [15,16].

ERs function through two predominant mechanisms. In the "classical" mechanism, estrogen diffuses into the cell and binds the ERs, creating a complex that then binds to specific DNA motifs called estrogen response elements (EREs) in the promoter region of estrogen-responsive genes [17]. Classical mechanisms are mediated by two main ER isoforms, ERα and ERβ, which form homo- or heterodimers before binding to EREs, and which induce changes in gene expression. Several studies have provided evidence that ERα and ERβ have different physiological functions [18]. Indeed, these subtypes can have opposing gene-expression regulatory effects [19,20] and also have redundant mediatory roles [21,22]. In addition, estrogen signaling is selectively regulated by the relative balance between ERα and ERβ expression in target organs [23], although studies using ERα and ERβ knockout mice revealed that the beneficial effects estrogen has on the vascular system are mainly mediated by ERα [24,25].

Besides their classic genomic action, ERs can also trigger faster responses (in minutes) through plasma membrane receptors. Indeed, ERα and ERβ are present in plasma membranes and other cytoplasmic organelles such as mitochondria and endoplasmic reticulum membranes [26]. In addition, the recently described G protein-coupled ER (GPER) is also expressed in vascular tissues [27]. Indeed, many of the beneficial effects of estrogen seen in human and animal models, such as reduced myocardial pro-inflammatory cytokine expression, inhibition of VSMC proliferation, and nitric oxide (NO)-dependent vasodilation [28], have been recently attributed to the presence of GPER in the cardiovascular system.

2. Role of Estrogen in Vascular Physiology

As described above, vascular tissues are targets for sex hormones because specific receptors are expressed in both endothelial cells and VSMCs [14] and clinical and experimental data have demonstrated that estrogen has beneficial effects at the cardiovascular level [29,30]. In general, these protective effects have been attributed to their role in increasing arterial vasodilation and inhibiting inflammatory processes, which, in turn, prevent the development of atherosclerosis [6]. Moreover, estrogen can also indirectly influence plaque progression by modulating systemic lipid metabolism [31] and oxidative status [32].

The regulation of vascular reactivity by estrogen is mainly related to the maintenance of normal endothelial function [33]. Indeed, enhanced acetylcholine-induced vasodilation mediated by NO release in arteries isolated from estrogen-treated ovariectomized rabbits was one of the first evidence indicating the role of estrogen in vascular tone [34]. In endothelial cells, the modulation of NO bioavailability by estrogen has been extensively studied and is attributed to both genomic and non-genomic effects [35–37]. In addition to NO, the action of estrogen has also been implicated in the release of other endothelial-derived molecules such as prostacyclin [38] and angiotensin (Ang)

1–7 [39] and a decrease in endothelin-1 bioavailability [40] and Ang II receptor type 1 expression [41], thus reducing vasoconstriction and promoting vasodilation.

Besides their effect on vasomotor regulation, the anti-inflammatory responses induced by estrogen have been described in in vitro assays as well as in different vascular-injury models [42–45]. In this regard, estrogen reduces cell adhesion molecule expression in endothelial cells exposed to pro-inflammatory stimuli [46,47], and significantly decreases the cytokine-induced adhesion of monocytes to endothelium [48,49]. Moreover, the modulation of neutrophil chemotaxis [44] and leukocyte infiltration [45] by estradiol has been described in rat carotid arteries after acute injury. Estrogen treatment after rat carotid artery damage [50] also attenuates neointima formation by increasing endothelial cell growth and decreasing VSMC proliferation.

Estrogen also participates in the regulation of lipid accumulation in the vascular wall by modulating the plasma lipid profile and inhibiting the direct action of lipids on the vascular system. On the one hand, estrogen reduces the level of circulating cholesterol [51] and the rate of conversion of hepatic low-density lipoprotein (LDL) into bile acids [52] while on the other, it increases high-density lipoprotein (HDL) levels [53]. In addition, estrogen is associated with reduced lipid loading in human monocyte-derived macrophages [54] and VSMCs [55], preventing foam cell formation. Furthermore, estradiol exposure inhibits cellular permeability [56] and apoptosis [57] in LDL-exposed endothelial cells. Finally, estrogen attenuates the oxidative stress-mediated increase in LDL modifications, which accelerates lipid accumulation in arterial walls [58].

Although the antioxidant properties of steroids were first attributed to their phenolic structure [59], estrogen can also modulate antioxidant enzyme expression [60,61]. For instance, estradiol attenuates Ang II-induced superoxide production by increasing superoxide dismutase activity and protein expression in VSMCs [60] and endothelial cells [61]. Estradiol also reduces superoxide production by inhibiting nicotinamide adenine dinucleotide phosphate (NADPH) oxidase expression, thus reducing adhesion molecule and cytokine expression in VSMCs [62] and, in an experimental murine model of menopause, by reverting cyclooxygenase (COX) 2-dependent superoxide production in aortic tissue [63].

3. miRNA as Epigenetic Regulatory Mechanism

As previously described, classical regulation of physiological processes by estradiol includes estrogen signaling induced by direct and indirect target gene transcription. However, epigenetic mechanisms have recently emerged as another important source of gene expression regulation and are being widely studied. At the molecular level, epigenetics is based on three main pathways: (1) DNA methylation; (2) histone density, variants, and post-translational modifications; and (3) RNA-based mechanisms [64]. Together, these pathways are characterized by their ability to influence gene expression without changing the DNA sequence and many have been established as fundamental determinants of cardiovascular health and disease [65,66].

There is some evidence that epigenetic estrogen-regulation mechanisms are implicated in the regulation of cardiovascular function. For example, genes encoding ERs are more methylated (denoting the suppression of estrogenic activity) in atherosclerotic plaques compared to non-plaque regions in vascular tissues [67,68], thus suggesting that epigenetic ER inhibition plays an important role in atherosclerosis formation. On the other hand, histone modifications and chromatin remodeling also likely have estrogen-dependent effects on the vasculature [69,70]. Indeed, divergent estrogen-dependent gene expression in endothelial cells and VSMCs is linked to differential target-gene promoter histone acetylation [69]. Moreover, the vascular dysfunction prevented by estradiol is associated with histone 3 acetylation in a post-menopausal metabolic syndrome experimental model [70]. Finally, RNA-based epigenetic gene-expression regulatory mechanisms mediated by sequence-specific interactions have more recently been described and are our main focus in this review.

Regulatory non-coding RNA can be classified depending on the RNA length. Long non-coding RNA (lncRNA) is a heterogenic class of RNA that includes intergenic lncRNA, antisense transcripts, and enhancer RNA. All of them are described as non-protein-coding transcripts larger than 200 nucleotides (nt) so as to differentiate them from small non-coding RNAs [71]. These include microRNA (miRNA), small interfering RNA (siRNA), and Piwi-interacting RNA (piRNA), and are defined as small (20–30 nt) RNAs, which are associated with Argonaute (AGO) family proteins [72]. Moreover, a new class of non-coding RNAs derived from sequences located adjacent to miRNAs, termed miRNA offset RNA (moR), has been described [73]. Although moRs were first considered a by-product of miRNA biogenesis, recent studies have provided evidence that are biologically active and can alter gene expression to regulate cell proliferation in VSMCs [74].

miRNAs about 20–22 nt long are the dominant class of small non-coding RNA in most tissues and are derived from nuclear transcripts with characteristic stem–loop structures (pri-miRNAs). The first step in miRNA biosynthesis is pri-miRNA cleavage, mediated by a processing complex comprising the RNase III Drosha and DiGeorge syndrome critical region 8 (DGCR8), also known as the microprocessor complex. Nuclear processing involves cropping the stem–loop to release a small hairpin-shaped RNA (pre-miRNA), which is then transported into the cytoplasm through exportin 5 where maturation can be completed. The second processing step is mediated by the RNase III, DICER1, which cleaves the pre-miRNA into 22-nt miRNA duplexes. Usually, one strand from the cleavage products remains as a mature miRNA due to a selective process that depends on thermodynamic stability. Finally, RNA generated is loaded into an AGO protein to form the effector RNA-induced silencing complex (RISC) along with other component such as TAR RNA-binding protein (TRBP) or protein kinase R-activating protein (PACT). miRNAs function as a guide by base pairing with their target messenger RNAs (mRNAs), while AGO proteins recruit factors that induce this translational repression; miRNA-binding sites are usually located at the $3'$-untranslated region (UTR) of the target mRNA [75]. Figure 1 shows a schematic of the miRNA biosynthesis pathway along with most of the relevant implicated molecules.

Although no specific research has so far focused on the influence estrogen might exert on miRNA biosynthesis in vascular tissues, our group's work on human endothelial cells treated with estradiol produced mRNA microarray data revealing the deregulation of key miRNA biosynthesis pathway genes [76]. Our data shows DGCR8 upregulation and DICER1 and AGO-2 downregulation in estradiol-treated cells (Table 1), suggesting that estrogen regulates endothelial miRNA production machinery.

Table 1. Microarray expression data for key miRNA biosynthesis pathway molecules. mRNA expression data were obtained from previously published mRNA microarray data obtained for human umbilical vein endothelial cells (HUVECs) treated with 1 nmol/L estradiol for 24 h. The probe set ID, gene symbol, official full name, *p*-value, and fold change are shown. These mRNA microarray data are deposited in NCBI's Gene Expression Omnibus (http://www.ncbi.nlm.nih.gov/geo), accessible through GEO series accession number GSE16683.

Probe Set ID	Symbol	Official Full Name	Fold Change	*p* Value
218269_at	DROSHA	drosha, ribonuclease type III	−1.117	0.586
64474_g_at	DGCR8	DiGeorge syndrome critical region gene 8	2.376	0.016
223056_s_at	XPO5	exportin 5	1.514	0.259
213229_at	DICER1	dicer 1, ribonuclease type III	−1.979	0.012
225569_at	AGO-2	argonaute-2	−1.290	0.002

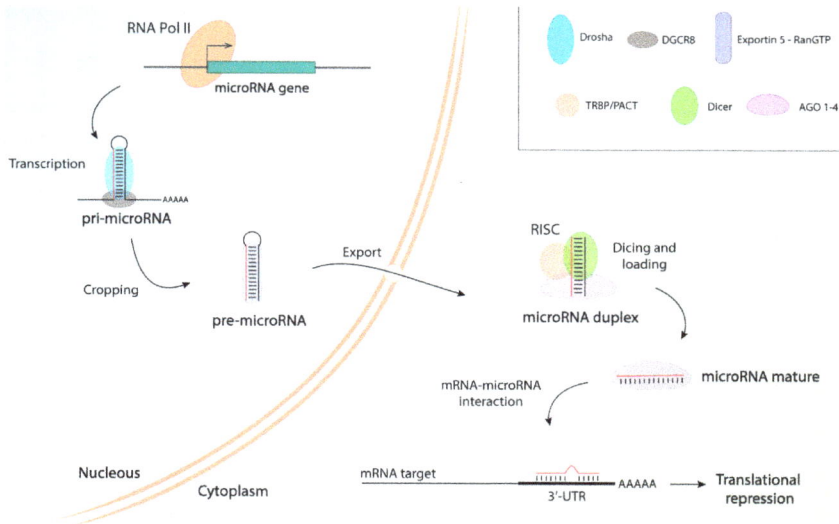

Figure 1. MicroRNA biosynthesis pathway. MicroRNAs (miRNAs) are transcribed by RNA polymerase II (Pol II) activity to generate the primary transcripts (pri-miRNAs). miRNA production is a two-step process involving nuclear cropping and cytosolic dicing processes. First, pri-miRNA cleavage is mediated by a processing complex comprising the RNase III, Drosha, and DiGeorge syndrome critical region 8 (DGCR8), which is also known as the microprocessor complex. This generates a hairpin-shaped pre-miRNA, which is recognized by nuclear exportin 5 and is exported to the cytoplasm where the RNase III, Dicer, cleaves pre-miRNA into 22-nucleotide miRNA duplexes. One strand from the cleavage products remains as a mature miRNA on the Argonaute (AGO) 1–4 proteins, whereas the other strand is degraded. Dicer, TAR RNA-binding protein (TRBP), protein kinase R-activating protein (PACT), and AGO 1–4 proteins mediate the assembly of the RISC (RNA-induced silencing complex). Finally, miRNAs guide translational repression by base-pairing with their target mRNAs, while AGO proteins recruit factors that induce this translational repression.

In addition to data obtained in estradiol-treated endothelial cells, the relationship between estrogen action and miRNA biosynthesis has been extensively described in breast cancer samples, where differences in key miRNA-processing genes have been observed between ER+ and ER− breast cancer cells [77,78]. Specifically, the expression of DICER1, DGCR8, and DROSHA was higher, and that of AGO-2 lower, in ER+ breast tumors. In addition, of the miRNA processing genes this group studied, only DICER1 contains an ERα binding site in its regulatory region [79]. Indeed, miRNAs that are differentially expressed between ERα− and ERα+ breast cancer cells negatively control DICER1 expression [80], suggesting that a regulatory loop exists between ERs and miRNAs. In addition, other studies suggest that ERs interact with DROSHA to modulate its activity in breast cancer cells [81] and that a significant increase in Exportin-5 mRNA is induced in the mouse uterus by the action of estrogen [82].

Specific miRNAs target ERs and could therefore act as important ER-dependent gene expression modulators. Indeed, some estrogen-induced miRNAs such as miR-18a, miR-19b, and miR-20b target and regulate ERα expression, thus forming a negative feedback loop [83]. Other miRNAs, including miR-18a, miR-22, miR-206, and miR-221/222 have also been implicated in ERα targeting [84]. Finally, the only miRNAs identified as targeting ERβ [85] and GPER [86], respectively, are miR-92 and miR-424.

4. Vascular miRNA and Estrogen Action

The importance of miRNAs in vascular biology was first observed in 2005 by Yang et al., who described impaired vascular formation in DICER1 knockout mice [87]. In endothelial cells, DICER1 knockdown resulted in impaired proliferation and vessel formation, as well as altered expression of key proteins implicated in vascular tone regulation and angiogenesis, such as vascular endothelial growth factor receptor 2 (VEGFR), interleukin 8 (IL-8), and endothelial NO synthases (eNOS) [88,89], thus suggesting a role of miRNAs production in endothelial and vascular function.

Sex differences in miRNA expression have also been described in different physiological and pathological conditions [90,91], providing evidence for a role for sex hormones in miRNA regulation. Nevertheless, the relationship between sex-dependent miRNA expression and cardiovascular diseases has so far been little explored [90], although regulation of miRNA expression by estrogen was observed in different cell types and tissues [92]. In addition, the role of estrogen in the circulating miRNA profile has been described in both ovariectomized rats and postmenopausal women receiving hormone replacement treatments [93,94]; based on these results, different authors have proposed using these miRNA profiles as possible biomarkers for pathologies involving estrogen.

4.1. Estrogen-Dependent miRNA and Cardiovascular Function

Different studies have proposed that estrogen exerts its vascular protective effects, at least in part, via miRNA activity. For instance, the role of estrogen-induced miRNAs in heart tissue, VSMCs, and endothelial cells has been described; Table 2 summarizes the main miRNAs involved in the action of estrogen at cardiovascular level. Additionally, sex-dimorphic miRNA expression in heart tissue from males versus females has been noted, including for miR-222. As previously mentioned, this miRNA is involved in ERα regulation [84] and is implicated in modulating eNOS expression in cardiomyocytes by directly inhibiting the transcription factor ets-1 [95]. These results suggest that estrogen plays a role in regulating both the miRNA expression profile in cardiac tissues as well as the key molecules involved in cardiac function. In addition, miR-21, miR-24, miR-27a/b, and miR-106a/b were among the sex-specific miRNAs expressed via ERβ modulation in a murine model of pressure overload-induced cardiac fibrosis [96] and could help explain the differences in adaptation to pressure overload and vascular remodeling observed between women and men [97].

Important roles for miR-23a and miR-22 have also been described in cardiac function involving the action of estrogen. Specifically, miR-23a has regulatory regions containing ERα binding sites and plays a protective role in estrogen deficiency-induced cardiac gap-junction damage in rats [98]. The authors showed that estradiol inhibits miR-23-dependent downregulation of connexin 43 in a menopausal rat model, and provide new mechanisms of post-menopause-related arrhythmia [99]. In addition to its role in cardiac function, miR-23a levels also differ in males and females after cerebral ischemia and are related to accelerating apoptosis by regulating X-linked inhibitor of apoptosis (XIAP) expression and XIAP-caspase complex formation [100]. Thus, this evidence provides new insights into the molecular mechanisms underlying the sex-dependent responses observed following stroke [101]. Moreover, miR-22 provides estrogenic cardioprotection in female rats by controlling myocardial oxidative stress [102]. This same study also described a reciprocal feedback loop between ERα and miR-22, suggesting that estrogen action is closely regulated via post-transcriptional control of ERα expression. Similarly, the sex-specific regulation of miR-22 processing in muscle lipid metabolism has also recently been described and may contribute to understanding the well-described differences in muscle metabolism and body weight between males and females [103].

Considering vascular tissue, some studies show that VSMC proliferation is affected by miRNAs and highlight their potential as therapeutic agents in the treatment of proliferative cardiovascular diseases. In the case of mouse aorta, miR-203 contributes to the inhibition of VSMC proliferation because its upregulation is ER-dependent [104]. Estradiol induces miR-143 and miR-145 expression in pulmonary artery VSMCs via specific ER binding sites located in their promoter regions [105].

Moreover, estradiol-treated VSMCs secrete exosomes enriched with miR-143 and miR-145 which regulate VSMC-endothelium crosstalk in pulmonary arterial hypertension [105].

Focusing on the endothelium, microarrays were recently used to reveal that physiological (1 nmol/L) estradiol concentrations induce changes in the miRNA expression profile of endothelial cells [106]; among these, the miRNAs with the strongest differential expression were miR-30b-5p, miR-487a-5p, miR-4710, miR-501-3p, miR-378h, and miR-1244. Functional analysis using bioinformatic tools revealed that estradiol-modulated miRNAs were associated with key molecular pathways such as extracellular signaling from signal-regulated kinase/mitogen activated protein kinase (ERK/MAPK), integrins, and actin cytoskeleton signaling, which are important pathways in the regulation of vascular physiology in health and disease [106]. Additionally, most validated estradiol-regulated miRNAs were modulated by ERα, and to a lesser extent, by ERβ and GPER [106], thus lending weight to the idea that ERα plays a crucial role in estradiol-dependent effects on vascular tissues. On the other hand, estradiol is also implicated in the increased miR-126-3p expression observed in endothelial cells, resulting in increased cell migration, proliferation, and tube formation and decreased monocyte adhesion [107].

As previously described, estrogen plays a key role in modulating the immune system and this is probably the underlying cause of the sex differences observed in the inflammatory processes of atherosclerosis [108]. For instance, estradiol is involved in nuclear factor-kB (NF-kB) activity inhibition by regulating let-7a and miR-125b expression in stimulated macrophages [109]. Moreover, specific estradiol-regulated miRNAs—miR-146a and miR-223—have been described as key regulators of lipopolysaccharide-induced interferon-gamma (IFNγ) in lymphocytes [110]. Therefore, selective miRNA expression regulated by estrogen in immune cells could also be involved in the sex dimorphism observed in vascular diseases.

Table 2. miRNA-dependent estrogen actions. Focusing on the role of estrogen in cardiovascular system and in HRT, estrogen-dependent effect and its associated estrogen-related miRNA are shown.

Estrogen Action	miRNA	References
Sex differences in heart	miR-1 miR-106b miR-720 miR-29b miR-144 miR-34b-5p miR-205 miR-222	[95]
Sex differences in cardiac fibrosis	miR-21 miR-24 miR-27a/b miR-106a/b	[96]
Cardiac gap junction regulation	miR-23a	[98]
Regulation of oxidative stress in the myocardium	miR-22	[102]
Inhibition of VSMC proliferation	miR-203	[104]
VSMC and endothelial cell communication	miR-143 miR-145	[105]
Endothelial cell proliferation	miR-126-3p	[107]
miRNA expression profile in estradiol-treated endothelial cells	miR-30b-5p miR487a-5p miR-4710 miR-501-3p miR-378h miR-1244	[106]
Regulation of NF-kB pathway in macrophages	let-7a and miR-125b	[109]
Regulation of IFNγ released in lymphocytes	miR-146a miR-223	[110]
Regulation of Insulin/IGF-1 pathway in skeletal muscle	miR-182 and miR-223	[111]
Circulating Inflammation markers	miR-21 miR-146a	[105]
Negative regulation of bone mass.	miR-127 and miR-136	[112]
Serum biomarker in osteoporosis	miR-30b-5p	[93]
Circulating miRNA	miR-106-5p miR-148a-3p miR-27-3p miR-126-5p miR-28-3p miR-30a-5p	[94]

4.2. miRNA and Hormone Replacement Therapy

The use of HRT has recently been associated with the miRNA content of circulating exosomes in women [94]. In addition, the miRNA-mediated effects of this type of estrogenic therapy appear to improve the parameters of some disorders such as osteoporosis and sarcopenia and help to reduce the inflammation markers associated with these phenomena in postmenopausal women using HRT.

Although the relationship between estrogen levels and osteoporosis has been established for decades [113], changes in the miRNA expression profile in bone tissue from ovariectomy-induced osteoporotic mice and in postmenopausal women have recently been described [93,112]. Specifically, from among the miRNAs that are differentially expressed in estrogen-depleted mice, miR-127 and miR-136 negatively regulate bone mass [112], whereas miR-30b-5p may be a suitable serum biomarker for osteoporosis and osteopenia in postmenopausal women [93]. Moreover, suppressing the expression of miR-182 and miR-223, both implicated in regulating the insulin/insulin-like growth factor (IGF-1) pathway, in the skeletal muscle of postmenopausal women using HRT plays a central role in muscle mass regulation [111]. Therefore, the identification of estrogen-regulated miRNAs could be used as possible therapeutic targets to provide new insights into aging-related disorders such as sarcopenia. In addition, a study in monozygotic twin pairs revealed a relationship between changes in serum inflammatory markers and inflammatory-related miRNAs such as miR-21 and miR-146a, in postmenopausal women using HRT [114]. Thus, estrogen-sensitive miRNAs could be used as potential biomarkers for specific physiological deteriorations associated with female aging.

In another study, in premenopausal women and their monozygotic postmenopausal twins using estrogenic HRT, other circulating miRNAs included in exosomes, such as miR-148a-3p, miR-27-3p, miR-28-3p, miR-30a-5p, miR-106b-5p, and miR-126-5p were associated with serum estradiol levels [94]. miR-148a is related to regulation of plasma LDL/HDL ratio by directly regulating hepatic LDL receptor (LDLR) [115]. This effect could be related to the previously demonstrated effects of estrogen on circulating cholesterol levels as estrogen is implicated in the reduction of circulating cholesterol by increasing LDLR expression [116]. Another estrogen-related miRNA, miR-27, is also implicated in LDLR expression without producing changes in plasma cholesterol levels [117]; this miRNA is also related to angiogenic processes [89] and was recently suggested as a biomarker for stenotic progression in asymptomatic carotid stenosis [118]. In this regard, there are sex-related differences in patients with this pathology [119] that may be partly related to the role of estrogen-regulated miRNAs. Therefore, a better understanding of the mechanisms underlying these processes could improve new sex-specific therapeutic approaches.

MiR-106b-5p decreases tumor necrosis factor (TNF) α-induced apoptosis by repressing phosphatase and tensin homolog (PTEN)-caspase activity in vascular endothelial cells [120]. Moreover, these effects correlate with the repressive effects that estrogen have on PTEN and apoptosis [121,122]. miR-126-5p is required to produce correct vascular integrity and is key in angiogenic processes [123,124] and also decreases leukocyte-endothelium interactions by suppressing vascular cell adhesion molecule (VCAM)-1 [125]. In line with the aforementioned studies, miR-126-5p is among the estradiol-regulated miRNAs present in endothelial cells [107]. Therefore, the estradiol-sensitive miRNAs described could provide insight into the mechanisms by which estrogen modulates important endothelial processes such as apoptosis or angiogenesis to provide correct vascular physiology.

5. Conclusions

The differences observed in cardiovascular diseases between the sexes attribute a protective role to estrogen, which is mediated through the regulation of transcription processes and, in turn, cellular physiology. Indeed, sex-biased gene expression in the cardiovascular system and mediated by estrogen has already been reported. It is estimated that miRNAs regulate the expression of approximately 30% of all protein-coding genes in mammals, implying their importance in correctly functioning human physiology, including that of the cardiovascular system. However, although there is increasing evidence to establish epigenetic mechanisms, including miRNAs, as crucial regulators of vascular function, the role of miRNAs in estrogen-mediated vascular functions must still be elucidated. Therefore, future research focused on characterizing the role of specific estradiol-mediated miRNAs involved in vascular function will be required to provide new knowledge about how the levels of sex hormones can contribute to sex-related differences in cardiovascular diseases.

Acknowledgments: This work was supported by Spanish Ministerio de Economía y Competitividad, Instituto de Salud Carlos III—FEDER-ERDF (grants FIS PI13/00617 and PI16/00229). Ana Mompeón is a "Formación de Profesorado Universitario" fellow (grant number FPU13/02235 Spanish Ministerio de Educación, Cultura y Deporte).

Author Contributions: All authors wrote and revised the manuscript, and agreed to be accountable for all aspects of this study.

Conflicts of Interest: The authors declare no conflict of interest.

References

1. Burns, K.A.; Korach, K.S. Estrogen receptors and human disease: An update. *Arch. Toxicol.* **2012**, *86*, 1491–1504. [CrossRef] [PubMed]
2. Vrtacnik, P.; Ostanek, B.; Mencej-Bedrac, S.; Marc, J. The many faces of estrogen signaling. *Biochem. Med.* **2014**, *24*, 329–342. [CrossRef] [PubMed]
3. Deroo, B.J.; Korach, K.S. Estrogen receptors and human disease. *J. Clin. Investig.* **2006**, *116*, 561–570. [CrossRef] [PubMed]
4. Mendelsohn, M.E.; Karas, R.H. Molecular and Cellular Basis of Cardiovascular Gender Differences. *Science* **2005**, *308*, 1583–1587. [CrossRef] [PubMed]
5. Hayward, C.S.; Kelly, R.P.; Collins, P. The roles of gender, the menopause and hormone replacement on cardiovascular function. *Cardiovasc. Res.* **2000**, *46*, 28–49. [CrossRef]
6. Mendelsohn, M.E.; Karas, R.H. The protective effects of estrogen on the cardiovascular system. *N. Engl. J. Med.* **1999**, *340*, 1801–1811. [CrossRef] [PubMed]
7. Mikkola, T.S.; Gissler, M.; Merikukka, M.; Tuomikoski, P.; Ylikorkala, O. Sex differences in age-related cardiovascular mortality. *PLoS ONE* **2013**, *8*, e63347. [CrossRef] [PubMed]
8. Simon, J.A.; Hsia, J.; Cauley, J.A.; Richards, C.; Harris, F.; Fong, J.; Barrett-Connor, E.; Hulley, S.B. Postmenopausal hormone therapy and risk of stroke: The Heart and Estrogen-progestin Replacement Study (HERS). *Circulation* **2001**, *103*, 638–642. [CrossRef] [PubMed]
9. Rossouw, J.E.; Prentice, R.L.; Manson, J.E.; Wu, L.; Barad, D.; Barnabei, V.M.; Ko, M.; LaCroix, A.Z.; Margolis, K.L.; Stefanick, M.L. Postmenopausal hormone therapy and risk of cardiovascular disease by age and years since menopause. *JAMA* **2007**, *297*, 1465–1477. [CrossRef] [PubMed]
10. Lobo, R.A. Hormone-replacement therapy: Current thinking. *Nat. Rev. Endocrinol.* **2017**, *13*, 220–231. [CrossRef] [PubMed]
11. Clarkson, T.B.; Melendez, G.C.; Appt, S.E. Timing hypothesis for postmenopausal hormone therapy: Its origin, current status, and future. *Menopause* **2013**, *20*, 342–353. [CrossRef] [PubMed]
12. Sherwood, A.; Bower, J.K.; McFetridge-Durdle, J.; Blumenthal, J.A.; Newby, L.K.; Hinderliter, A.L. Age moderates the short-term effects of transdermal 17β-estradiol on endothelium-dependent vascular function in postmenopausal women. *Arterioscler. Thromb. Vasc. Biol.* **2007**, *27*, 1782–1787. [CrossRef] [PubMed]
13. Novella, S.; Heras, M.; Hermenegildo, C.; Dantas, A.P. Effects of estrogen on vascular inflammation: A matter of timing. *Arterioscler. Thromb. Vasc. Biol.* **2012**, *32*, 2035–2042. [CrossRef] [PubMed]
14. Khalil, R.A. Estrogen, vascular estrogen receptor and hormone therapy in postmenopausal vascular disease. *Biochem. Pharmacol.* **2013**, *86*, 1627–1642. [CrossRef] [PubMed]
15. Härkönen, P.L.; Väänänen, H.K. Monocyte—Macrophage System as a Target for Estrogen and Selective Estrogen Receptor Modulators. *Ann. N. Y. Acad. Sci.* **2006**, *1089*, 218–227. [CrossRef] [PubMed]
16. Kovats, S. Estrogen receptors regulate innate immune cells and signaling pathways. *Cell. Immunol.* **2015**, *294*, 63–69. [CrossRef] [PubMed]
17. Klinge, C.M. Estrogen receptor interaction with estrogen response elements. *Nucleic Acids Res.* **2001**, *29*, 2905–2919. [CrossRef] [PubMed]
18. O'Lone, R.; Knorr, K.; Jaffe, I.Z.; Schaffer, M.E.; Martini, P.G.; Karas, R.H.; Bienkowska, J.; Mendelsohn, M.E.; Hansen, U. Estrogen receptors α and β mediate distinct pathways of vascular gene expression, including genes involved in mitochondrial electron transport and generation of reactive oxygen species. *Mol. Endocrinol.* **2007**, *21*, 1281–1296. [CrossRef] [PubMed]

19. Lindberg, M.K.; Moverare, S.; Skrtic, S.; Gao, H.; Dahlman-Wright, K.; Gustafsson, J.A.; Ohlsson, C. Estrogen receptor (ER)-β reduces ERα-regulated gene transcription, supporting a "ying yang" relationship between ERα and ERβ in mice. *Mol. Endocrinol.* **2003**, *17*, 203–208. [CrossRef] [PubMed]

20. Tsutsumi, S.; Zhang, X.; Takata, K.; Takahashi, K.; Karas, R.H.; Kurachi, H.; Mendelsohn, M.E. Differential regulation of the inducible nitric oxide synthase gene by estrogen receptors 1 and 2. *J. Endocrinol.* **2008**, *199*, 267–273. [CrossRef] [PubMed]

21. Lahm, T.; Crisostomo, P.R.; Markel, T.A.; Wang, M.; Wang, Y.; Tan, J.; Meldrum, D.R. Selective estrogen receptor-α and estrogen receptor-β agonists rapidly decrease pulmonary artery vasoconstriction by a nitric oxide-dependent mechanism. *Am. J. Physiol. Regul. Integr. Comp. Physiol.* **2008**, *295*, R1486–R1493. [CrossRef] [PubMed]

22. Arias-Loza, P.A.; Hu, K.; Dienesch, C.; Mehlich, A.M.; Konig, S.; Jazbutyte, V.; Neyses, L.; Hegele-Hartung, C.; Heinrich Fritzemeier, K.; Pelzer, T. Both estrogen receptor subtypes, α and β, attenuate cardiovascular remodeling in aldosterone salt-treated rats. *Hypertension* **2007**, *50*, 432–438. [CrossRef] [PubMed]

23. Murphy, E.; Steenbergen, C. Estrogen regulation of protein expression and signaling pathways in the heart. *Biol. Sex Differ.* **2014**, *5*, 6. [CrossRef] [PubMed]

24. Pare, G.; Krust, A.; Karas, R.H.; Dupont, S.; Aronovitz, M.; Chambon, P.; Mendelsohn, M.E. Estrogen receptor-α mediates the protective effects of estrogen against vascular injury. *Circ. Res.* **2002**, *90*, 1087–1092. [CrossRef] [PubMed]

25. Arnal, J.-F.; Lenfant, F.; Metivier, R.; Flouriot, G.; Henrion, D.; Adlanmerini, M.; Fontaine, C.; Gourdy, P.; Chambon, P.; Katzenellenbogen, B.; et al. Membrane and Nuclear Estrogen Receptor α Actions: From Tissue Specificity to Medical Implications. *Physiol. Rev.* **2017**, *97*, 1045. [CrossRef] [PubMed]

26. Levin, E.R. Plasma membrane estrogen receptors. *Trends Endocrinol. Metab.* **2009**, *20*, 477–482. [CrossRef] [PubMed]

27. Revankar, C.M.; Cimino, D.F.; Sklar, L.A.; Arterburn, J.B.; Prossnitz, E.R. A transmembrane intracellular estrogen receptor mediates rapid cell signaling. *Science* **2005**, *307*, 1625–1630. [CrossRef] [PubMed]

28. Prossnitz, E.R.; Barton, M. The G protein-coupled estrogen receptor GPER in health and disease. *Nat. Rev. Endocrinol.* **2011**, *7*, 715–726. [CrossRef] [PubMed]

29. Vitale, C.; Mendelsohn, M.E.; Rosano, G.M. Gender differences in the cardiovascular effect of sex hormones. *Nat. Rev. Cardiol.* **2009**, *6*, 532–542. [CrossRef] [PubMed]

30. Miller, V.M.; Duckles, S.P. Vascular actions of estrogens: Functional implications. *Pharmacol. Rev.* **2008**, *60*, 210–241. [CrossRef] [PubMed]

31. Barton, M. Cholesterol and atherosclerosis: Modulation by oestrogen. *Curr. Opin. Lipidol.* **2013**, *24*, 214–220. [CrossRef] [PubMed]

32. Kondo, T.; Hirose, M.; Kageyama, K. Roles of Oxidative Stress and Redox Regulation in Atherosclerosis. *J. Atheroscler. Thromb.* **2009**, *16*, 532–538. [CrossRef] [PubMed]

33. Miller, V.M.; Mulvagh, S.L. Sex steroids and endothelial function: Translating basic science to clinical practice. *Trends Pharmacol. Sci.* **2007**, *28*, 263–270. [CrossRef] [PubMed]

34. Gisclard, V.; Flavahan, N.A.; Vanhoutte, P.M. α adrenergic responses of blood vessels of rabbits after ovariectomy and administration of 17 β-estradiol. *J. Pharmacol. Exp. Ther.* **1987**, *240*, 466–470. [PubMed]

35. Chambliss, K.L.; Shaul, P.W. Estrogen modulation of endothelial nitric oxide synthase. *Endocr. Rev.* **2002**, *23*, 665–686. [CrossRef] [PubMed]

36. Novella, S.; Laguna-Fernández, A.; Lázaro-Franco, M.; Sobrino, A.; Bueno-Betí, C.; Tarín, J.J.; Monsalve, E.; Sanchís, J.; Hermenegildo, C. Estradiol, acting through estrogen receptor α, restores dimethylarginine dimethylaminohydrolase activity and nitric oxide production in oxLDL-treated human arterial endothelial cells. *Mol. Cell. Endocrinol.* **2013**, *365*, 11–16. [CrossRef] [PubMed]

37. Sobrino, A.; Vallejo, S.; Novella, S.; Lazaro-Franco, M.; Mompeon, A.; Bueno-Beti, C.; Walther, T.; Sanchez-Ferrer, C.; Peiro, C.; Hermenegildo, C. Mas receptor is involved in the estrogen-receptor induced nitric oxide-dependent vasorelaxation. *Biochem. Pharmacol.* **2017**, *129*, 67–72. [CrossRef] [PubMed]

38. Sobrino, A.; Oviedo, P.J.; Novella, S.; Laguna-Fernandez, A.; Bueno, C.; Garcia-Perez, M.A.; Tarin, J.J.; Cano, A.; Hermenegildo, C. Estradiol selectively stimulates endothelial prostacyclin production through estrogen receptor-α. *J. Mol. Endocrinol.* **2010**, *44*, 237–246. [CrossRef] [PubMed]

39. Mompeón, A.; Lázaro-Franco, M.; Bueno-Betí, C.; Pérez-Cremades, D.; Vidal-Gómez, X.; Monsalve, E.; Gironacci, M.M.; Hermenegildo, C.; Novella, S. Estradiol, acting through ERα, induces endothelial non-classic renin-angiotensin system increasing angiotensin 1–7 production. *Mol. Cell. Endocrinol.* **2016**, *422*, 1–8. [CrossRef] [PubMed]

40. Dubey, R.K.; Jackson, E.K.; Keller, P.J.; Imthurn, B.; Rosselli, M. Estradiol metabolites inhibit endothelin synthesis by an estrogen receptor-independent mechanism. *Hypertension* **2001**, *37*, 640–644. [CrossRef] [PubMed]

41. Nickenig, G.; Bäumer, A.T.; Grohè, C.; Kahlert, S.; Strehlow, K.; Rosenkranz, S.; Stäblein, A.; Beckers, F.; Smits, J.F.M.; Daemen, M.J.A.P.; et al. Estrogen modulates AT 1 receptor gene expression in vitro and in vivo. *Circulation* **1998**, *97*, 2197. [CrossRef] [PubMed]

42. Straub, R.H. The complex role of estrogens in inflammation. *Endocr. Rev.* **2007**, *28*, 521–574. [CrossRef] [PubMed]

43. Bakir, S.; Mori, T.; Durand, J.; Chen, Y.F.; Thompson, J.A.; Oparil, S. Estrogen-induced vasoprotection is estrogen receptor dependent: Evidence from the balloon-injured rat carotid artery model. *Circulation* **2000**, *101*, 2342–2344. [CrossRef] [PubMed]

44. Miller, A.P.; Feng, W.; Xing, D.; Weathington, N.M.; Blalock, J.E.; Chen, Y.F.; Oparil, S. Estrogen modulates inflammatory mediator expression and neutrophil chemotaxis in injured arteries. *Circulation* **2004**, *110*, 1664–1669. [CrossRef] [PubMed]

45. Xing, D.; Miller, A.; Novak, L.; Rocha, R.; Chen, Y.F.; Oparil, S. Estradiol and progestins differentially modulate leukocyte infiltration after vascular injury. *Circulation* **2004**, *109*, 234–241. [CrossRef] [PubMed]

46. Caulin-Glaser, T.; Watson, C.A.; Pardi, R.; Bender, J.R. Effects of 17β-estradiol on cytokine-induced endothelial cell adhesion molecule expression. *J. Clin. Investig.* **1996**, *98*, 36–42. [CrossRef] [PubMed]

47. Alvarez, A.; Hermenegildo, C.; Issekutz, A.C.; Esplugues, J.V.; Sanz, M.J. Estrogens inhibit angiotensin II-induced leukocyte-endothelial cell interactions in vivo via rapid endothelial nitric oxide synthase and cyclooxygenase activation. *Circ. Res.* **2002**, *91*, 1142–1150. [CrossRef] [PubMed]

48. Mikkola, T.S.; St Clair, R.W. Estradiol reduces basal and cytokine induced monocyte adhesion to endothelial cells. *Maturitas* **2002**, *41*, 313–319. [CrossRef]

49. Abu-Taha, M.; Rius, C.; Hermenegildo, C.; Noguera, I.; Cerda-Nicolas, J.-M.; Issekutz, A.C.; Jose, P.J.; Cortijo, J.; Morcillo, E.J.; Sanz, M.-J. Menopause and Ovariectomy Cause a Low Grade of Systemic Inflammation that May Be Prevented by Chronic Treatment with Low Doses of Estrogen or Losartan. *J. Immunol.* **2009**, *183*, 1393. [CrossRef] [PubMed]

50. Mori, T.; Durand, J.; Chen, Y.; Thompson, J.A.; Bakir, S.; Oparil, S. Effects of short-term estrogen treatment on the neointimal response to balloon injury of rat carotid artery. *Am. J. Cardiol.* **2000**, *85*, 1276–1279. [CrossRef]

51. Terauchi, M.; Honjo, H.; Mizunuma, H.; Aso, T. Effects of oral estradiol and levonorgestrel on cardiovascular risk markers in postmenopausal women. *Arch. Gynecol. Obstet.* **2012**, *285*, 1647–1656. [CrossRef] [PubMed]

52. Kushwaha, R.S.; Born, K.M. Effect of estrogen and progesterone on the hepatic cholesterol 7-α-hydroxylase activity in ovariectomized baboons. *Biochim. Biophys. Acta* **1991**, *1084*, 300–302. [CrossRef]

53. Walsh, B.W.; Schiff, I.; Rosner, B.; Greenberg, L.; Ravnikar, V.; Sacks, F.M. Effects of postmenopausal estrogen replacement on the concentrations and metabolism of plasma lipoproteins. *N. Engl. J. Med.* **1991**, *325*, 1196–1204. [CrossRef] [PubMed]

54. McCrohon, J.A.; Nakhla, S.; Jessup, W.; Stanley, K.K.; Celermajer, D.S. Estrogen and progesterone reduce lipid accumulation in human monocyte-derived macrophages: A sex-specific effect. *Circulation* **1999**, *100*, 2319–2325. [CrossRef] [PubMed]

55. Wang, H.; Liu, Y.; Zhu, L.; Wang, W.; Wan, Z.; Chen, F.; Wu, Y.; Zhou, J.; Yuan, Z. 17β-estradiol promotes cholesterol efflux from vascular smooth muscle cells through a liver X receptor α-dependent pathway. *Int. J. Mol. Med.* **2014**, *33*, 550–558. [CrossRef] [PubMed]

56. Gardner, G.; Banka, C.L.; Roberts, K.A.; Mullick, A.E.; Rutledge, J.C. Modified LDL-mediated increases in endothelial layer permeability are attenuated with 17 β-estradiol. *Arterioscler. Thromb. Vasc. Biol.* **1999**, *19*, 854–861. [CrossRef] [PubMed]

57. Florian, M.; Magder, S. Estrogen decreases TNF-α and oxidized LDL induced apoptosis in endothelial cells. *Steroids* **2008**, *73*, 47–58. [CrossRef] [PubMed]

58. Walsh, B.A.; Mullick, A.E.; Walzem, R.L.; Rutledge, J.C. 17β-estradiol reduces tumor necrosis factor-α-mediated LDL accumulation in the artery wall. *J. Lipid Res.* **1999**, *40*, 387–396. [PubMed]

59. Mooradian, A.D. Antioxidant properties of steroids. *J. Steroid Biochem. Mol. Biol.* **1993**, *45*, 509–511. [CrossRef]
60. Strehlow, K.; Rotter, S.; Wassmann, S.; Adam, O.; Grohe, C.; Laufs, K.; Bohm, M.; Nickenig, G. Modulation of antioxidant enzyme expression and function by estrogen. *Circ. Res.* **2003**, *93*, 170–177. [CrossRef] [PubMed]
61. Liu, Z.; Gou, Y.; Zhang, H.; Zuo, H.; Zhang, H.; Liu, Z.; Yao, D. Estradiol improves cardiovascular function through up-regulation of SOD2 on vascular wall. *Redox Biol.* **2014**, *3*, 88–99. [CrossRef] [PubMed]
62. Wagner, A.H.; Schroeter, M.R.; Hecker, M. 17β-estradiol inhibition of NADPH oxidase expression in human endothelial cells. *FASEB J.* **2001**, *15*, 2121–2130. [CrossRef] [PubMed]
63. Vidal-Gomez, X.; Novella, S.; Perez-Monzo, I.; Garabito, M.; Dantas, A.P.; Segarra, G.; Hermenegildo, C.; Medina, P. Decreased bioavailability of nitric oxide in aorta from ovariectomized senescent mice. Role of cyclooxygenase. *Exp. Gerontol.* **2016**, *76*, 1–8. [CrossRef] [PubMed]
64. Yan, M.S.; Marsden, P.A. Epigenetics in the Vascular Endothelium: Looking From a Different Perspective in the Epigenomics Era. *Arterioscler. Thromb. Vasc. Biol.* **2015**, *35*, 2297–2306. [CrossRef] [PubMed]
65. Abi Khalil, C. The emerging role of epigenetics in cardiovascular disease. *Ther. Adv. Chronic Dis.* **2014**, *5*, 178–187. [CrossRef] [PubMed]
66. Voelter-Mahlknecht, S. Epigenetic associations in relation to cardiovascular prevention and therapeutics. *Clin. Epigenetics* **2016**, *8*, 4. [CrossRef] [PubMed]
67. Post, W.S.; Goldschmidt-Clermont, P.J.; Wilhide, C.C.; Heldman, A.W.; Sussman, M.S.; Ouyang, P.; Milliken, E.E.; Issa, J.-P.J. Methylation of the estrogen receptor gene is associated with aging and atherosclerosis in the cardiovascular system. *Cardiovasc. Res.* **1999**, *43*, 985–991. [CrossRef]
68. Kim, J.; Kim, J.Y.; Song, K.S.; Lee, Y.H.; Seo, J.S.; Jelinek, J.; Goldschmidt-Clermont, P.J.; Issa, J.-P.J. Epigenetic changes in estrogen receptor β gene in atherosclerotic cardiovascular tissues and in-vitro vascular senescence. *Biochim. Biophys. Acta* **2007**, *1772*, 72–80. [CrossRef] [PubMed]
69. Kawagoe, J.; Ohmichi, M.; Tsutsumi, S.; Ohta, T.; Takahashi, K.; Kurachi, H. Mechanism of the Divergent Effects of Estrogen on the Cell Proliferation of Human Umbilical Endothelial Versus Aortic Smooth Muscle Cells. *Endocrinology* **2007**, *148*, 6092–6099. [CrossRef] [PubMed]
70. Bendale, D.S.; Karpe, P.A.; Chhabra, R.; Shete, S.P.; Shah, H.; Tikoo, K. 17-β Oestradiol prevents cardiovascular dysfunction in post-menopausal metabolic syndrome by affecting SIRT1/AMPK/H3 acetylation. *Br. J. Pharmacol.* **2013**, *170*, 779–795. [CrossRef] [PubMed]
71. Mercer, T.R.; Dinger, M.E.; Mattick, J.S. Long non-coding RNAs: Insights into functions. *Nat. Rev. Genet.* **2009**, *10*, 155–159. [CrossRef] [PubMed]
72. Kaikkonen, M.U.; Lam, M.T.; Glass, C.K. Non-coding RNAs as regulators of gene expression and epigenetics. *Cardiovasc. Res.* **2011**, *90*, 430–440. [CrossRef] [PubMed]
73. Shi, W.; Hendrix, D.; Levine, M.; Haley, B. A distinct class of small RNAs arises from pre-miRNA-proximal regions in a simple chordate. *Nat. Struct. Mol. Biol.* **2009**, *16*, 183–189. [CrossRef] [PubMed]
74. Zhao, J.; Schnitzler, G.R.; Iyer, L.K.; Aronovitz, M.J.; Baur, W.E.; Karas, R.H. MicroRNA-Offset RNA Alters Gene Expression and Cell Proliferation. *PLoS ONE* **2016**, *11*, e0156772. [CrossRef] [PubMed]
75. Kim, V.N.; Han, J.; Siomi, M.C. Biogenesis of small RNAs in animals. *Nat. Rev. Mol. Cell Biol.* **2009**, *10*, 126–139. [CrossRef] [PubMed]
76. Sobrino, A.; Mata, M.; Laguna-Fernandez, A.; Novella, S.; Oviedo, P.J.; Garcia-Perez, M.A.; Tarin, J.J.; Cano, A.; Hermenegildo, C. Estradiol stimulates vasodilatory and metabolic pathways in cultured human endothelial cells. *PLoS ONE* **2009**, *4*, e8242. [CrossRef] [PubMed]
77. Cheng, C.; Fu, X.; Alves, P.; Gerstein, M. mRNA expression profiles show differential regulatory effects of microRNAs between estrogen receptor-positive and estrogen receptor-negative breast cancer. *Genome Biol.* **2009**, *10*, R90. [CrossRef] [PubMed]
78. Cizeron-Clairac, G.; Lallemand, F.; Vacher, S.; Lidereau, R.; Bieche, I.; Callens, C. MiR-190b, the highest up-regulated miRNA in ERα-positive compared to ERα-negative breast tumors, a new biomarker in breast cancers? *BMC Cancer* **2015**, *15*, 499. [CrossRef] [PubMed]
79. Bhat-Nakshatri, P.; Wang, G.; Collins, N.R.; Thomson, M.J.; Geistlinger, T.R.; Carroll, J.S.; Brown, M.; Hammond, S.; Srour, E.F.; Liu, Y.; et al. Estradiol-regulated microRNAs control estradiol response in breast cancer cells. *Nucleic Acids Res.* **2009**, *37*, 4850–4861. [CrossRef] [PubMed]
80. Cochrane, D.R.; Cittelly, D.M.; Howe, E.N.; Spoelstra, N.S.; McKinsey, E.L.; LaPara, K.; Elias, A.; Yee, D.; Richer, J.K. MicroRNAs link estrogen receptor α status and Dicer levels in breast cancer. *Horm. Cancer* **2010**, *1*, 306–319. [CrossRef] [PubMed]

81. Paris, O.; Ferraro, L.; Grober, O.M.; Ravo, M.; De Filippo, M.R.; Giurato, G.; Nassa, G.; Tarallo, R.; Cantarella, C.; Rizzo, F.; et al. Direct regulation of microRNA biogenesis and expression by estrogen receptor beta in hormone-responsive breast cancer. *Oncogene* **2012**, *31*, 4196–4206. [CrossRef] [PubMed]

82. Nothnick, W.B.; Healy, C.; Hong, X. Steroidal regulation of uterine miRNAs is associated with modulation of the miRNA biogenesis components Exportin-5 and Dicer1. *Endocrine* **2010**, *37*, 265–273. [CrossRef] [PubMed]

83. Castellano, L.; Giamas, G.; Jacob, J.; Coombes, R.C.; Lucchesi, W.; Thiruchelvam, P.; Barton, G.; Jiao, L.R.; Wait, R.; Waxman, J.; et al. The estrogen receptor-α-induced microRNA signature regulates itself and its transcriptional response. *Proc. Natl. Acad. Sci. USA* **2009**, *106*, 15732–15737. [CrossRef] [PubMed]

84. Klinge, C.M. miRNAs and estrogen action. *Trends Endocrinol. Metab.* **2012**, *23*, 223–233. [CrossRef] [PubMed]

85. Al-Nakhle, H.; Burns, P.A.; Cummings, M.; Hanby, A.M.; Hughes, T.A.; Satheesha, S.; Shaaban, A.M.; Smith, L.; Speirs, V. Estrogen receptor {β}1 expression is regulated by miR-92 in breast cancer. *Cancer Res.* **2010**, *70*, 4778–4784. [CrossRef] [PubMed]

86. Zhang, H.; Wang, X.; Chen, Z.; Wang, W. MicroRNA-424 suppresses estradiol-induced cell proliferation via targeting GPER in endometrial cancer cells. *Cell. Mol. Biol.* **2015**, *61*, 96–101. [PubMed]

87. Yang, W.J.; Yang, D.D.; Na, S.; Sandusky, G.E.; Zhang, Q.; Zhao, G. Dicer is required for embryonic angiogenesis during mouse development. *J. Biol. Chem.* **2005**, *280*, 9330–9335. [CrossRef] [PubMed]

88. Suarez, Y.; Fernandez-Hernando, C.; Pober, J.S.; Sessa, W.C. Dicer dependent microRNAs regulate gene expression and functions in human endothelial cells. *Circ. Res.* **2007**, *100*, 1164–1173. [CrossRef] [PubMed]

89. Kuehbacher, A.; Urbich, C.; Zeiher, A.M.; Dimmeler, S. Role of Dicer and Drosha for endothelial microRNA expression and angiogenesis. *Circ. Res.* **2007**, *101*, 59–68. [CrossRef] [PubMed]

90. Sharma, S.; Eghbali, M. Influence of sex differences on microRNA gene regulation in disease. *Biol. Sex Differ.* **2014**, *5*, 3. [CrossRef] [PubMed]

91. Guo, L.; Zhang, Q.; Ma, X.; Wang, J.; Liang, T. miRNA and mRNA expression analysis reveals potential sex-biased miRNA expression. *Sci. Rep.* **2017**, *7*, 39812. [CrossRef] [PubMed]

92. Klinge, C.M. Estrogen action: Receptors, transcripts, cell signaling, and non-coding RNAs in normal physiology and disease. *Mol. Cell. Endocrinol.* **2015**, *418*, 191–192. [CrossRef] [PubMed]

93. Chen, J.; Li, K.; Pang, Q.; Yang, C.; Zhang, H.; Wu, F.; Cao, H.; Liu, H.; Wan, Y.; Xia, W.; et al. Identification of suitable reference gene and biomarkers of serum miRNAs for osteoporosis. *Sci. Rep.* **2016**, *6*, 36347. [CrossRef] [PubMed]

94. Kangas, R.; Tormakangas, T.; Fey, V.; Pursiheimo, J.; Miinalainen, I.; Alen, M.; Kaprio, J.; Sipila, S.; Saamanen, A.M.; Kovanen, V.; et al. Aging and serum exomiR content in women-effects of estrogenic hormone replacement therapy. *Sci. Rep.* **2017**, *7*, 42702. [CrossRef] [PubMed]

95. Evangelista, A.M.; Deschamps, A.M.; Liu, D.; Raghavachari, N.; Murphy, E. miR-222 contributes to sex-dimorphic cardiac eNOS expression via ets-1. *Physiol. Genom.* **2013**, *45*, 493–498. [CrossRef] [PubMed]

96. Queirós, A.M.; Eschen, C.; Fliegner, D.; Kararigas, G.; Dworatzek, E.; Westphal, C.; Sanchez Ruderisch, H.; Regitz-Zagrosek, V. Sex- and estrogen-dependent regulation of a miRNA network in the healthy and hypertrophied heart. *Int. J. Cardiol.* **2013**, *169*, 331–338. [CrossRef] [PubMed]

97. Petrov, G.; Regitz-Zagrosek, V.; Lehmkuhl, E.; Krabatsch, T.; Dunkel, A.; Dandel, M.; Dworatzek, E.; Mahmoodzadeh, S.; Schubert, C.; Becher, E.; et al. Regression of Myocardial Hypertrophy After Aortic Valve Replacement. *Circulation* **2010**, *122*, S23. [CrossRef] [PubMed]

98. Wang, N.; Sun, L.-Y.; Zhang, S.-C.; Wei, R.; Xie, F.; Liu, J.; Yan, Y.; Duan, M.-J.; Sun, L.-L.; Sun, Y.-H.; et al. MicroRNA-23a Participates in Estrogen Deficiency Induced Gap Junction Remodeling of Rats by Targeting GJA1. *Int. J. Biol. Sci.* **2015**, *11*, 390–403. [CrossRef] [PubMed]

99. Gowd, B.M.P.; Thompson, P.D. Effect of Female Sex on Cardiac Arrhythmias. *Cardiol. Rev.* **2012**, *20*, 297–303. [CrossRef] [PubMed]

100. Siegel, C.; Li, J.; Liu, F.; Benashski, S.E.; McCullough, L.D. miR-23a regulation of X-linked inhibitor of apoptosis (XIAP) contributes to sex differences in the response to cerebral ischemia. *Proc. Natl. Acad. Sci. USA* **2011**, *108*, 11662–11667. [CrossRef] [PubMed]

101. Chauhan, A.; Moser, H.; McCullough, L.D. Sex differences in ischaemic stroke: Potential cellular mechanisms. *Clin. Sci.* **2017**, *131*, 533. [CrossRef] [PubMed]

102. Wang, L.; Tang, Z.P.; Zhao, W.; Cong, B.H.; Lu, J.Q.; Tang, X.L.; Li, X.H.; Zhu, X.Y.; Ni, X. MiR-22/Sp-1 Links Estrogens With the Up-Regulation of Cystathionine gamma-Lyase in Myocardium, Which Contributes to Estrogenic Cardioprotection Against Oxidative Stress. *Endocrinology* **2015**, *156*, 2124–2137. [CrossRef] [PubMed]

103. Schweisgut, J.; Schutt, C.; Wust, S.; Wietelmann, A.; Ghesquiere, B.; Carmeliet, P.; Drose, S.; Korach, K.S.; Braun, T.; Boettger, T. Sex-specific, reciprocal regulation of ERα and miR-22 controls muscle lipid metabolism in male mice. *EMBO J.* **2017**, *36*, 1199–1214. [CrossRef] [PubMed]

104. Zhao, J.; Imbrie, G.A.; Baur, W.E.; Iyer, L.K.; Aronovitz, M.; Kershaw, T.B.; Haselmann, G.M.; Lu, Q.; Karas, R.H. Estrogen receptor-mediated regulation of microRNA inhibits proliferation of vascular smooth muscle cells. *Arterioscler. Thromb. Vasc. Biol.* **2013**, *33*, 257–265. [CrossRef] [PubMed]

105. Deng, L.; Blanco, F.J.; Stevens, H.; Lu, R.; Caudrillier, A.; McBride, M.; McClure, J.D.; Grant, J.; Thomas, M.; Frid, M.; et al. MicroRNA-143 Activation Regulates Smooth Muscle and Endothelial Cell Crosstalk in Pulmonary Arterial Hypertension. *Circ. Res.* **2015**, *117*, 870–883. [CrossRef] [PubMed]

106. Vidal-Gomez, X.; Pérez-Cremades, D.; Mompeón, A.; Dantas, A.P.; Novella, S.; Hermenegildo, C. microRNA as crucial regulators of gene expression in estradiol-treated human endothelial cells. *Cell. Physiol. Biochem.* **2018**, in press.

107. Li, P.; Wei, J.; Li, X.; Cheng, Y.; Chen, W.; Cui, Y.; Simoncini, T.; Gu, Z.; Yang, J.; Fu, X. 17β-estradiol enhances vascular endothelial Ets-1/miR-126-3p expression: The possible mechanism for attenuation of atherosclerosis. *J. Clin. Endocrinol. Metab.* **2016**, *102*, 594–603. [CrossRef] [PubMed]

108. Fairweather, D. Sex Differences in Inflammation during Atherosclerosis. *Clin. Med. Insights Cardiol.* **2014**, *8* (Suppl. 3), 49–59. [CrossRef] [PubMed]

109. Murphy, A.J.; Guyre, P.M.; Pioli, P.A. Estradiol suppresses NF-κB activation through coordinated regulation of let-7a and miR-125b in primary human macrophages. *J. Immunol.* **2010**, *184*, 5029–5037. [CrossRef] [PubMed]

110. Dai, R.; Phillips, R.A.; Zhang, Y.; Khan, D.; Crasta, O.; Ahmed, S.A. Suppression of LPS-induced Interferon-γ and nitric oxide in splenic lymphocytes by select estrogen-regulated microRNAs: A novel mechanism of immune modulation. *Blood* **2008**, *112*, 4591–4597. [CrossRef] [PubMed]

111. Olivieri, F.; Ahtiainen, M.; Lazzarini, R.; Pöllänen, E.; Capri, M.; Lorenzi, M.; Fulgenzi, G.; Albertini, M.C.; Salvioli, S.; Alen, M.J.; et al. Hormone replacement therapy enhances IGF-1 signaling in skeletal muscle by diminishing miR-182 and miR-223 expressions: A study on postmenopausal monozygotic twin pairs. *Aging Cell* **2014**, *13*, 850–861. [CrossRef] [PubMed]

112. An, J.H.; Ohn, J.H.; Song, J.A.; Yang, J.-Y.; Park, H.; Choi, H.J.; Kim, S.W.; Kim, S.Y.; Park, W.-Y.; Shin, C.S. Changes of MicroRNA Profile and MicroRNA-mRNA Regulatory Network in Bones of Ovariectomized Mice. *J. Bone Miner. Res.* **2014**, *29*, 644–656. [CrossRef] [PubMed]

113. Richelson, L.S.; Wahner, H.W.; Melton, L.J.I.; Riggs, B.L. Relative Contributions of Aging and Estrogen Deficiency to Postmenopausal Bone Loss. *N. Engl. J. Med.* **1984**, *311*, 1273–1275. [CrossRef] [PubMed]

114. Kangas, R.; Pollanen, E.; Rippo, M.R.; Lanzarini, C.; Prattichizzo, F.; Niskala, P.; Jylhava, J.; Sipila, S.; Kaprio, J.; Procopio, A.D.; et al. Circulating miR-21, miR-146a and Fas ligand respond to postmenopausal estrogen-based hormone replacement therapy-a study with monozygotic twin pairs. *Mech. Ageing Dev.* **2014**, *143–144*, 1–8. [CrossRef] [PubMed]

115. Goedeke, L.; Rotllan, N.; Canfrán-Duque, A.; Aranda, J.F.; Ramírez, C.M.; Araldi, E.; Lin, C.-S.; Anderson, N.N.; Wagschal, A.; de Cabo, R.; et al. Identification of miR-148a as a novel regulator of cholesterol metabolism. *Nat. Med.* **2015**, *21*, 1280–1289. [CrossRef] [PubMed]

116. Windler, E.E.; Kovanen, P.T.; Chao, Y.S.; Brown, M.S.; Havel, R.J.; Goldstein, J.L. The estradiol-stimulated lipoprotein receptor of rat liver. A binding site that membrane mediates the uptake of rat lipoproteins containing apoproteins B and E. *J. Biol. Chem.* **1980**, *255*, 10464–10471. [PubMed]

117. Goedeke, L.; Rotllan, N.; Ramírez, C.M.; Aranda, J.F.; Canfrán-Duque, A.; Araldi, E.; Fernández-Hernando, A.; Langhi, C.; de Cabo, R.; Baldán, Á.; et al. miR-27b inhibits LDLR and ABCA1 expression but does not influence plasma and hepatic lipid levels in mice. *Atherosclerosis* **2015**, *243*, 499–509. [CrossRef] [PubMed]

118. Dolz, S.; Górriz, D.; Tembl, J.I.; Sánchez, D.; Fortea, G.; Parkhutik, V.; Lago, A. Circulating MicroRNAs as Novel Biomarkers of Stenosis Progression in Asymptomatic Carotid Stenosis. *Stroke* **2016**, *48*, 10–16. [CrossRef] [PubMed]

119. Buratti, L.; Balestrini, S.; Avitabile, E.; Altamura, C.; Vernieri, F.; Viticchi, G.; Falsetti, L.; Provinciali, L.; Silvestrini, M. Sex-associated differences in the modulation of vascular risk in patients with asymptomatic carotid stenosis. *J. Cereb. Blood Flow Metab.* **2015**, *35*, 684–688. [CrossRef] [PubMed]

120. Zhang, J.; Li, S.-F.; Chen, H.; Song, J.-X. MiR-106b-5p Inhibits Tumor Necrosis Factor-induced Apoptosis by Targeting Phosphatase and Tensin Homolog Deleted on Chromosome 10 in Vascular Endothelial Cells. *Chin. Med. J.* **2016**, *129*, 1406–1412. [PubMed]

121. Noh, E.M.; Lee, Y.R.; Chay, K.O.; Chung, E.Y.; Jung, S.H.; Kim, J.S.; Youn, H.J. Estrogen receptor α induces down-regulation of PTEN through PI3-kinase activation in breast cancer cells. *Mol. Med. Rep.* **2011**, *4*, 215–219. [PubMed]

122. Smith, J.A.; Zhang, R.; Varma, A.K.; Das, A.; Ray, S.K.; Banik, N.L. Estrogen partially down-regulates PTEN to prevent apoptosis in VSC4.1 motoneurons following exposure to IFN-γ. *Brain Res.* **2009**, *1301* (Suppl. C), 163–170. [CrossRef] [PubMed]

123. Fish, J.E.; Santoro, M.M.; Morton, S.U.; Yu, S.; Yeh, R.F.; Wythe, J.D.; Ivey, K.N.; Bruneau, B.G.; Stainier, D.Y.; Srivastava, D. miR-126 regulates angiogenic signaling and vascular integrity. *Dev. Cell* **2008**, *15*, 272–284. [CrossRef] [PubMed]

124. Wang, S.; Aurora, A.B.; Johnson, B.A.; Qi, X.; McAnally, J.; Hill, J.A.; Richardson, J.A.; Bassel-Duby, R.; Olson, E.N. The endothelial-specific microRNA miR-126 governs vascular integrity and angiogenesis. *Dev. Cell* **2008**, *15*, 261–271. [CrossRef] [PubMed]

125. Harris, T.A.; Yamakuchi, M.; Ferlito, M.; Mendell, J.T.; Lowenstein, C.J. MicroRNA-126 regulates endothelial expression of vascular cell adhesion molecule 1. *Proc. Natl. Acad. Sci. USA* **2008**, *105*, 1516–1521. [CrossRef] [PubMed]

International Journal of
Molecular Sciences

MDPI

Review

Steroid and Xenobiotic Receptor Signalling in Apoptosis and Autophagy of the Nervous System

Agnieszka Wnuk and Małgorzata Kajta *

Institute of Pharmacology, Polish Academy of Sciences, Department of Experimental Neuroendocrinology,
Smetna Street 12, 31-343 Krakow, Poland; wnuk@if-pan.krakow.pl
* Correspondence: kajta@if-pan.krakow.pl; Tel.: +48-12-662-3235; Fax: +48-12-637-4500

Received: 12 October 2017; Accepted: 9 November 2017; Published: 11 November 2017

Abstract: Apoptosis and autophagy are involved in neural development and in the response of the nervous system to a variety of insults. Apoptosis is responsible for cell elimination, whereas autophagy can eliminate the cells or keep them alive, even in conditions lacking trophic factors. Therefore, both processes may function synergistically or antagonistically. Steroid and xenobiotic receptors are regulators of apoptosis and autophagy; however, their actions in various pathologies are complex. In general, the estrogen (ER), progesterone (PR), and mineralocorticoid (MR) receptors mediate anti-apoptotic signalling, whereas the androgen (AR) and glucocorticoid (GR) receptors participate in pro-apoptotic pathways. ER-mediated neuroprotection is attributed to estrogen and selective ER modulators in apoptosis- and autophagy-related neurodegenerative diseases, such as Alzheimer's and Parkinson's diseases, stroke, multiple sclerosis, and retinopathies. PR activation appeared particularly effective in treating traumatic brain and spinal cord injuries and ischemic stroke. Except for in the retina, activated GR is engaged in neuronal cell death, whereas MR signalling appeared to be associated with neuroprotection. In addition to steroid receptors, the aryl hydrocarbon receptor (AHR) mediates the induction and propagation of apoptosis, whereas the peroxisome proliferator-activated receptors (PPARs) inhibit this programmed cell death. Most of the retinoid X receptor-related xenobiotic receptors stimulate apoptotic processes that accompany neural pathologies. Among the possible therapeutic strategies based on targeting apoptosis via steroid and xenobiotic receptors, the most promising are the selective modulators of the ER, AR, AHR, PPARγ agonists, flavonoids, and miRNAs. The prospective therapies to overcome neuronal cell death by targeting autophagy via steroid and xenobiotic receptors are much less recognized.

Keywords: apoptosis; autophagy; steroid receptors; xenobiotic receptors; nervous system; estrogen receptors

---

1. Introduction

It is generally accepted that mechanisms of neuronal demise involve apoptosis and autophagy. Apoptosis ("self-killing") and autophagy ("self-eating") are involved in neural development, as well as in the response of the nervous system to a variety of insults. Apoptosis is mainly responsible for cell elimination, whereas autophagy can either eliminate the cells or keep them alive, even in conditions lacking trophic factors. Therefore, both processes may act in the same direction or oppositely [1]. An excess of apoptosis or a defect in autophagy have been implicated in neurodegeneration. Autophagy is a basic cellular process that is crucial for postmitotic neurons, whereas apoptosis occurs at each stage of neural development and affects mitotically active and differentiated cells. Last year's Nobel Laureate in Physiology or Medicine, Yoshinori Ohsumi, discovered and elucidated mechanisms underlying autophagy, a fundamental process for degrading and recycling cellular components, using baker's yeast (The Nobel Prize in Medicine, 2016).

Apoptosis is considered to be the form of programmed cell death that is mediated via specific DNA fragmentation and apoptotic body formation. After initially being cut into pieces of 300–50,000 base pairs, DNA is cleaved by endonucleases (e.g., CAD - caspase-activated DNase) and Nuc-18 - endonuclease II) into pieces of 180–200 base pairs. In addition to apoptosis, this ladder-type DNA fragmentation is also found in some cells dying of necrosis [2]. Recent revelations suggest that apoptosis shares characteristics with necrosis, and these phenomena are interlinked in necroptosis. Apart from specific DNA fragmentation and apoptotic body formation, apoptosis is characterized by cell rounding, membrane blebbing, cytoskeletal collapse, cytoplasmic condensation and fragmentation, nuclear pyknosis, and individual cell death without inflammatory response to damage. Apoptosis has been documented to be involved in etiology of various types of neural degenerations, particularly these related to mitochondrial dysfunctions.

Autophagy is another form of programmed cell death that regulates lysosomal turnover of organelles and proteins via sequential events including double-membrane formation, elongation, vesicle maturation, and delivery of the targeted materials to the lysosome. This process is deleterious in acute neural disorders, such as stroke and hypoxic/ischemic injury [3]. However, autophagy appears protective in chronic neurodegenerative diseases such as Alzheimer's disease (AD), Parkinson's disease (PD), Huntington's disease (HD), amyotrophic lateral sclerosis (ALS), and encephalopathy, where it is responsible for degrading not only damaged organelles, but also misfolded proteins [4]. Neural degeneration has been postulated to be associated with acceleration of apoptosis and impairment of autophagy, except for in cases of acute neural injury where both processes are stimulated.

The development of the nervous system is a highly complex process in which progenitor and stem cells differentiate into neurons, astrocytes and oligodendrocytes. During this process, not only does differentiation occur, but also the decisions of cell survival or cell death. The appropriate interplay between apoptosis and autophagy is believed to be essential for the normal development of the nervous system in mammals; therefore, neural development requires the degradation or subsistence of different organelles and proteins. Both major types of programmed cell death play roles in regulating neural cell numbers, tissue remodelling processes, and homeostasis. Apoptosis that occurs physiologically during the period of the growth spurt eliminates excessive neurons during the developmental process termed pruning [5]. Studies revealed an essential role of autophagy in the development and maturation of axons, dendrites and synapses [6].

Steroid and so-called xenobiotic receptors are involved in neural development; however, their actions as regulators of apoptosis and autophagy are complex. In general, estrogen, progesterone, and mineralocorticoid receptors mediate anti-apoptotic signalling, whereas androgen and glucocorticoid receptors participate in pro-apoptotic pathways. Estrogen receptors (ERs) play crucial roles in neurogenesis, astroglial proliferation and synaptogenesis. Neural progenitor cells (NPCs) express robust levels of aryl hydrocarbon receptor (AHR) that participate in NPC expansion and their differentiation into neurons. Retinoid X receptor (RXR) exerts its action by binding to gene sequences as either a homodimer or heterodimer and by regulating the transcription of specific genes. It forms RXR homodimers, such as RXRα, RXRβ, and RXRγ, and heterodimers, including the pregnane X receptor (PXR), constitutive androstane receptor (CAR), liver X receptor (LXR), and peroxisome proliferator-activated receptors (PPARs). All these receptors act as transcription factors, including the RXR-related xenobiotic receptors, which regulate neuronal differentiation both during development and adult neurogenesis. The RXR/nuclear receptor related 1 protein (NURR1) heterodimer has been postulated to be essential for the differentiation of the midbrain dopamine neurons [7].

Although steroid and xenobiotic receptors are essential for proper brain development, this review focused on the roles of steroid and xenobiotic receptors in apoptosis- and autophagy-related pathologies of the nervous system.

2. Molecular Mechanisms of Apoptosis and Autophagy

2.1. Mechanisms of Apoptosis

To identify potential drug targets, a major focus of neuroscience research is to examine the mechanisms involved in neuronal loss. Neurotrophins, such as nerve growth factors (NGFs), brain-derived neurotrophic factors (BDNFs), and neurotrophin-3 (NT-3), has been found to promote neuronal survival via RAS and PI-3K (3-phosphatydylinosytol kinase) pathways. Deficiency of neurotrophic factors inhibits PI-3K and promotes reactive oxygen species (ROS) production, which activates JNK (c-Jun N-terminal kinase)/SAPK (stress-activated protein kinase)-dependent apoptosis [2,8]. Genetic studies have demonstrated that the removal of specific *JNK* genes can reduce the neuronal death associated with cerebral ischemia [9]. A controversy has emerged regarding the question of whether limited neurotrophic factors are associated with the absence of inhibitors of cell death or if they are active signals of apoptosis. In general, apoptotic processes have been classified as extrinsic or intrinsic apoptotic pathways. The extrinsic pathway is induced by specific cell damage and is mediated through so-called "death receptors", e.g., FAS, TNF-R1 (tumour necrosis factor receptor-1), TRAMP (death receptor 3/APO-3/LARD/wsl-1), TRAILR2 (death receptor 5/DR5), and DR6 (death receptor 6). The intrinsic pathway is initiated by non-specific cell damage that leads to the loss of the mitochondrial membrane potential, cytochrome c release from mitochondria and activation of the evolutionarily conserved cysteine-aspartic acid proteases-caspases. Mitochondrial membrane permeability to cytochrome c is primarily regulated by proteins from the BCL2 family, including anti-apoptotic (BCL2, BCLw, and BCLxL) and pro-apoptotic (BAX, BID, BAK, BAD, BOX, and BCLxS) proteins [10,11].

Apoptosis is usually a caspase-dependent process that depends on either the interaction of a death receptor with its ligand and subsequent activation of procaspase-8 or on the participation of mitochondria and the activation of procaspase-9. The main executioner protease of the apoptotic cascade is caspase-3, which activates CAD after cleavage of ICAD (inhibitor of caspase-activated DNase), thereby inducing apoptotic DNA fragmentation and apoptotic cell death [12,13]. In addition to their roles in apoptosis, executioner caspases (e.g., caspases-3, -6, and -7) have been recognized as important regulators of an array of cellular activities in the nervous system, including axonal pathfinding and branching, axonal degeneration, dendritic pruning, and microglial activation in the absence of death. Caspase activation has been postulated to be coordinated at multiple levels, which might underlie apoptotic and non-apoptotic roles of caspases in the nervous system. It has been shown that apoptosis may also be mediated by other cysteine-dependent proteases such as calpains, which are calcium-activated neutral proteases [14].

The intrinsic and extrinsic apoptotic pathways are regulated by p53, which is a cellular sensor for cell cycle and genomic stability. The most commonly inactivated tumour suppressor gene *p53* causes loss of p53 function, inhibits apoptosis, and promotes tumour progression and chemoresistance. Several proteins have been shown to interact with the p53 to regulate its functions. One of these regulatory proteins is glycogen synthase kinase 3 beta (GSK-3β), which binds to p53 and promotes p53-induced apoptosis [15]. GSK-3β is involved in modulating a variety of functions, including cell signalling, growth metabolism, DNA damage, hypoxia, and endoplasmic reticulum stress [16]. GSK-3β has been recognized as a primary kinase involved in tau hyperphosphorylation, and thus, it is responsible for neurodegenerative tauopathies, such as AD [17]. RNA interference silencing of GSK-3β has been found to inhibit the phosphorylation of tau protein, which may have a therapeutic effect on the pathological progression of AD [18]. Moreover, GSK-3β is involved in the accumulation of α-synuclein aggregates, oxidative stress and mitochondrial dysfunction, which make this kinase an attractive therapeutic target for neurodegenerative disorders, such as AD or PD [19].

Apart from the intrinsic and extrinsic apoptotic pathways, there are also other pathways such as the caspase-12-mediated pathway, which is activated by calcium ions stored in the endoplasmic reticulum. Chronic or unresolved endoplasmic reticulum stress can induce neuronal apoptosis by activating JNK, GSK-3β, and the caspase-12 pathway [20]. The activated caspase cleaves procaspase-3 to induce classical apoptosis. Endoplasmic reticulum stress can be induced by a variety of physiological conditions, including perturbations in calcium homeostasis, glucose/energy deprivation, redox changes, ischemia, hyperhomocysteinemia, viral infections and mutations that impair protein folding. The endoplasmic reticulum stress response, also called the unfolded protein response, activates autophagy to remove aggregates of misfolded proteins that cannot be degraded. A previous study suggested that autophagy can provide neuroprotection by enhancing the clearance of these aggregates [21]. Recently, it has been shown that the tumour suppressor p53 can regulate cell death and autophagic activity, particularly mitophagy [22]. The p53 protein was found to be a cellular sensor of various stresses, including apoptotic stimuli. The induction of p53-dependent apoptosis leads to the activation of the intrinsic and extrinsic pathways to trigger cell death through transcription-dependent and transcription-independent mechanisms, among which is mitochondrial ROS production. In neurons, the extracellular signal-regulated kinases (ERKs) have been observed to mediate apoptosis; however, the majority of studies have demonstrated an anti-apoptotic role of ERK signalling. These processes have been visualized in Figure 1.

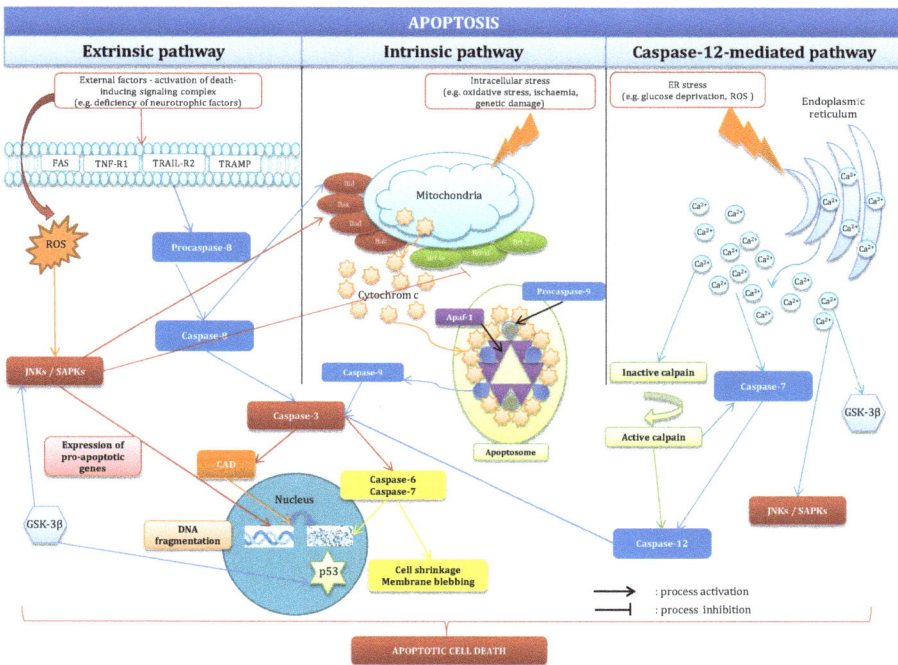

Figure 1. Mechanisms of apoptosis. Apoptosis has been classified as external, internal, and caspase-12-dependent processes. Additional details have been provided in part 2.1. CAD: caspase-activated DNase; GSK-3β: glycogen synthase kinase 3 beta; ROS: reactive oxygen species; JNK: c-Jun N-terminal kinase; SAPK: stress-activated protein kinase.

2.2. Mechanisms of Autophagy

Autophagy is a system of cellular degradation that ensures adequate digestion of cell debris and toxic material. The cytoplasmic debris is first enclosed in an autophagosome, which then fuses with the lysosome and forms an autophagolysosome for full digestion of the sequestered material. The formation of autophagosomes depends on several core Atg proteins, such as the following: ULK1 complex, Beclin1: Vps34/Atg14 L complex, and WIPIs, Atg12 and LC3 conjugation systems, and Atg9. In addition to the degrading function, autophagy promotes the recycling of cellular components, cellular renovation and homeostasis. Due to its important role, it is not surprising that autophagy dysregulation is found in many human diseases, such as cancer and neurodevelopmental and neurodegenerative diseases [23]. Recognition of the molecular mechanisms of autophagy processes was achieved through broad genetic research using mutagenesis in yeast cells, *Saccharomyces cerevisiae*, which allowed for the understanding of this complicated process in mammalian cells [24]. Autophagy can be divided into 3 stages: (1) formation of the autophagosome; (2) formation of the autophagolysosome; and (3) digestion the contents of follicles.

Formation of the autophagosome involves induction, nucleation and elongation. Induction of autophagosome formation has its origins in co-assemblies of ULK1/2, FIP200, and Atg13 and Atg101 creating a so-called ULK1/2 complex. The initiation phase mainly depends on mammalian target of rapamycin (mTOR) kinase, which is a kind of sensor watching over the cell's condition; however, it can also be induced by diverse input signals such as nutrients, growth factors, Ca^{2+}, ATP, cAMP, hormones, and protein accumulation. Under nutrient-rich conditions, phosphoinositide 3-kinase (PIK3C1) activates mTORC1, which then phosphorylates ULK1 and Atg13 and inhibits autophagy. Under autophagy-inducing cases, such as starvation or hypoxia, mTORC1 is inactivated, and this results in activation of ULK1/2, which phosphorylates Atg13 and FIP200 and regulates proper localization of Atg9 and PIK3C3, both of which are essential in further steps of autophagy. For nucleation, at this stage, the most important factor is the complex of PIK3C3: Beclin1: p150-serine kinase. This complex is located at the phagophore and recruits other Atg proteins. PIK3C3 induces phosphatidyl-inositol-3-phosphate (PI3P) and the activity of this kinase is regulated by Beclin1, which interacts with multiple modulators. Elongation is a step of autophagosome formation that is based on two ubiquitin-like conjugation systems: Atg12-Atg5-Atg16 L (consisting of Atg5, Atg7, Atg10, Atg12, Atg16) and LC3-PE (composed of LC3, 3-phosphatidyl ethanolamine, Atg4, Atg7, Atg3). After formation of the autophagosome, it undergoes a process of maturation, i.e., a fusion with lysosome, thus forming the autophagolysosome [25]. It is a complicated and unclear action which requires several proteins such as LAMP-2, RAB proteins, SNAREs, ESCRT, and HOPS [26]. The fusion with lysosomes leads to the start of degradation of vesicle content by hydrolases and lipases. The final stage of autophagy is the efflux of autophagolysosome content into the cytoplasm. Scheme of these processes is presented in Figure 2.

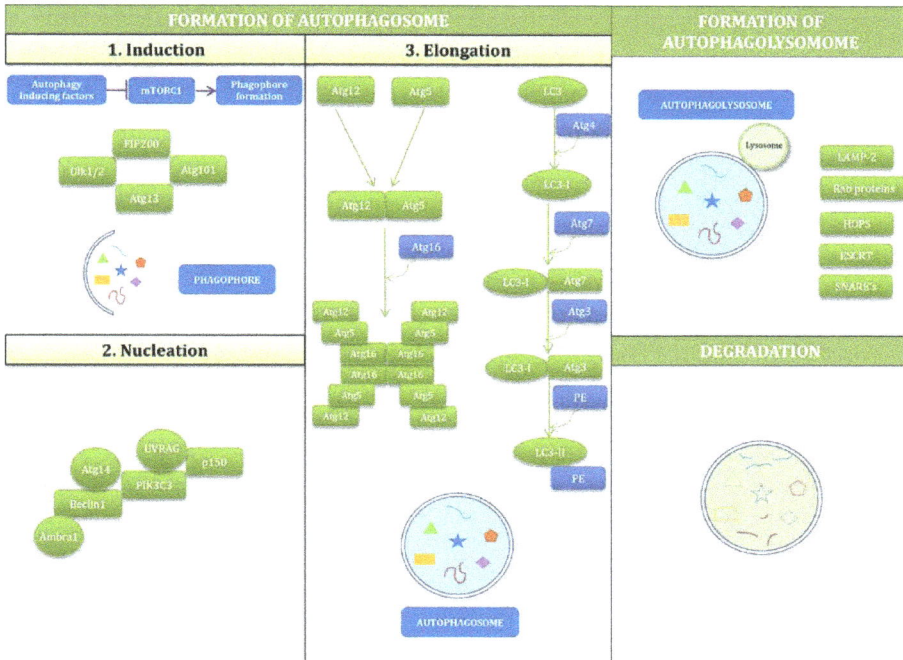

Figure 2. Mechanisms of autophagy. Autophagy can be divided into 3 stages: (1) formation of the autophagosome; (2) formation of the autophagolysosome; and (3) digestion the contents of follicles. Phagophore/autophagosome content: unneeded/misfolded proteins, carbohydrates, lipids, nucleic acids, whole organelles. More information has been provided in part 2.2.

2.3. Crosstalk between Apoptosis and Autophagy

The crosstalk between autophagy and apoptosis has only been partially uncovered. These processes interfere with themselves, mainly with regard to the BCL2 protein family, p53, and Atg5. Autophagy-related Beclin-1 inhibits or stimulates apoptosis, depending on the pro- or anti-apoptotic nature of the members of the BCL2 protein family with which it interacts [27]. Pro-apoptotic p53 initiates or inhibits autophagy depending on the cellular localization of the protein [22]. Atg5 participates in the formation of autophagosomes, translocates to the mitochondria and becomes a pro-apoptotic factor after catalytic incision by calpains [28]. In addition, mTOR, which is an inhibitor of autophagy, also acts on apoptosis-related p53, BAD and BCL2 [29]. Furthermore, caspases mediate cleavage of autophagy-related proteins, including Beclin-1, Atg4D, Atg8, and Atg5 [30]. The latest studies have demonstrated that the anti-apoptotic protein FLIP prevents elongation of autophagosomes by competing with LC3 for binding to Atg3 [31]. However, many questions regarding the interconnecting regulators of apoptosis and autophagy remain unanswered [1]. Crosstalk between apoptosis and autophagy has been outlined in Figure 3.

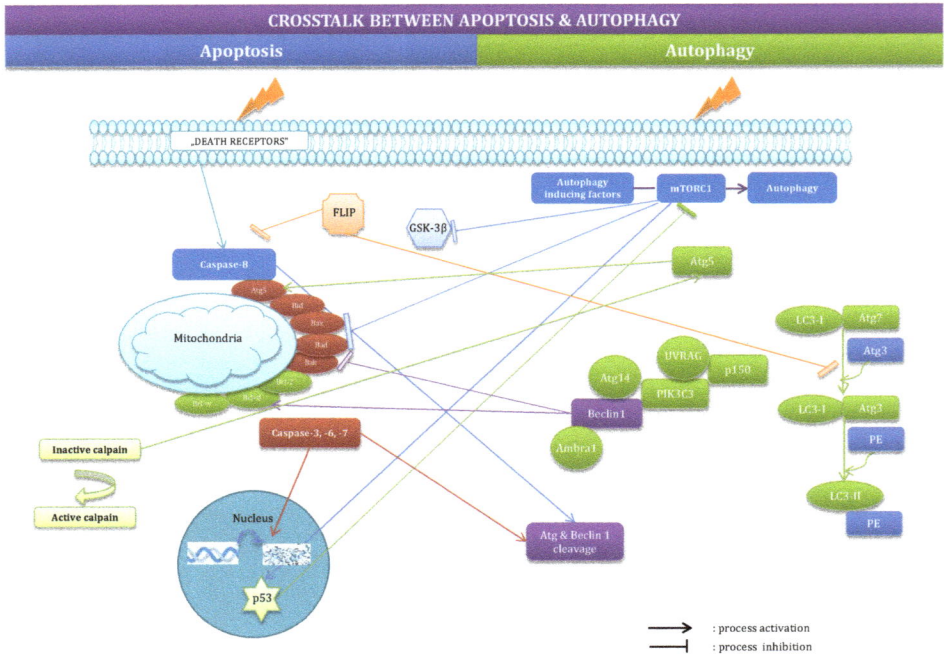

Figure 3. Crosstalk between apoptosis and autophagy. These processes interfere with themselves, mainly with regard to the BCL2 protein family, p53, and Atg5. The crosstalk has been described in detail in part 2.3.

3. Interactions of Apoptosis and Autophagy with Steroid and Xenobiotic Signalling

3.1. Interactions with Estrogen Receptors

ER-dependent pathways stimulate neurotrophin expression, e.g., BDNF and glial cell-derived neurotrophic factor (GDNF), concomitantly with dendritic growth and spinogenesis [10]. In addition to beneficial effects of estrogens on cognitive dysfunction, estrogens were found to protect the functioning of GABAergic neurons [32]. A decline in ERα has been reported in the brains (hippocampus and frontal cortex) of individuals with schizophrenia and AD [33,34]. There was a greater accumulation of amyloid-beta (Aβ) in the brain of ERα knockout mice, and this accumulation greatly worsened memory when compared to control mice [35]. In rapidly autopsied human brain tissue, the frontal cortices of female AD patients exhibited significantly reduced mitochondrial ERβ compared to that in normal controls [36]. In the embryonic brain, ERβ appeared to be necessary for the development of calretinin-immunoreactive GABAergic interneurons and for neuronal migration in the cortex [37]. It became evident that a membrane-bound ER, GPR30, modulates synaptic plasticity in the hippocampus, and its deficiency in male mice results in insulin resistance, dyslipidemia, and a pro-inflammatory state [38].

Most of the biological effects of estrogens are mediated by the classical ERs: ERα and ERβ. The best recognized effects are the interactions between the ERs and the intrinsic mitochondrial apoptotic pathway [39]. There is a line of evidence that suggests that ERs directly interfere with BCL2-dependent apoptotic processes, namely, those observed in AD and PD. Neuroprotection against 1-methyl-4-phenyl-1,2,3,6-tetrahydropyridine- (MPTP-) or Aβ-induced toxicity was found to be mediated via ERα and ERβ, as well as by an increase in BCL2/BAD ratio [40,41]. The involvement of ERs in the inhibition of caspase- and GSK-3β-mediated neuronal cell death has also been shown, including in our studies [42–44]. Furthermore, ERs were found to suppress the extrinsic death receptor-mediated apoptotic pathway by decreasing the cell-surface expression of the Fas/Apo-1 receptor in neuroblasts [45]. Studies have shown that GPR30, also known as G-protein-coupled ER1 (GPER1), mediates non-genomic estradiol signalling in a variety of tissues, including the brain, with particularly high expression in the hypothalamus, hippocampus, cortex, and striatum [46,47]. Recently, GPR30 has been noticed to mediate the neuroprotective effects of estradiol in murine hippocampal and cortical cells [48,49]. It has also been found that GPR30 stabilized blood-brain barrier permeability after ischemic stroke, improved cognitive function that was impaired by traumatic brain injury, and inhibited PD-related neuroinflammation [50–52].

Our data demonstrated a key involvement of GPR30 and/or ERβ in the neuroprotective and anti-apoptotic actions of the phytoestrogens daidzein and genistein [53,54]. Similarly, ERβ signalling has been linked to flavonoid troxerutin-induced mitigation of apoptosis in a 6-hydroxydopamine-induced lesion rat model for PD [55]. In addition, we provided evidence for a crucial role of ERα in the neuroprotective function of raloxifene during hypoxia [56]. According to our in vitro data, the protective action of the selective estrogen receptor modulator (SERM) is mediated by ERα, but not by ERβ or GPR30 and involves the inhibition of apoptosis. This is in line with results from Guo et al., 2016 that showed the enhancement of the ERα-mediated transactivation of the BCL2 gene upon ischemic insult and that provided in vitro evidence that metastasis-associated protein 1 (MTA1) enhances the binding of ERα with the BCL2 promoter via recruitment of HDAC2, together with other unidentified coregulators [57]. In male Wistar rats subjected to transient right middle cerebral artery occlusion (tMCAO), Jover-Mengual et al., 2017 detected the increased expression of ERα or ERα and ERβ in response to estradiol and the SERM bazedoxifene, respectively [58]. In the most recent study, Guo et al., 2017 linked estradiol neuroprotection against ischemic brain injury and apoptosis to the SIRT1-dependent adenosine monophosphate (AMP)-activated kinase (AMPK) pathway [59].

The protective action of notoginsenoside R1 against cerebral hypoxic-ischemic injury was found to be the result of the estrogen receptor-dependent inhibition of endoplasmic reticulum stress pathways, involving a caspase-12-mediated apoptosis, and the activation of Akt/Nrf2 pathways [60,61]. The ginsenoside Rg1 protection against Aβ peptide-induced neuronal apoptosis was shown to involve ERα, as well as the up-regulation of ERK1/2 phosphorylation and the reduction of NF-κB nuclear translocation [62]. Additionally, the pharmacological administration of the isoflavone daidzein was shown to stimulate cell proliferation and inhibit high fat diet-induced apoptosis and gliosis in the rat hippocampus [63]. Furthermore, it became evident that the impairments of ERα and/or GPR30 participate in mechanisms of apoptotic actions of DDT and a chemical UV-filter benzophenone-3 (BP-3) [43,64]. Autosomal dominant optic atrophy, a progressive blinding disease featured by retinal ganglion cell degeneration, has recently been linked to the inhibition of the estrogen receptor expression, which promoted apoptosis in mouse females carrying the human recurrent OPA1 mutation [65].

Estrogen receptors are important regulators of neuronal apoptosis; however, little is known about their impact on autophagy. Knockdown of ERα or antagonizing GPR30 has been found to induce autophagy and apoptosis in cancer cells [66,67]. The phytoestrogen gypenoside XVII (GP-17) attenuated the Aβ25-35-induced parallel autophagic and apoptotic death of NGF-differentiated PC12 cells through the ER-dependent activation of the Nrf2/ARE pathways [68]. In SH-SY5Y cells, ERs mediated Aβ degradation via the up-regulation of neprilysin and promoted autophagy as a protective mechanism against chronic minimal peroxide treatment [69]. Moreover, ER agonists activated PI3K/Akt/mTOR signalling in oligodendrocytes to promote remyelination in a mouse model of multiple sclerosis (MS) [70,71], and the up-regulation of the membrane ERα was associated with functional signals that were compatible with autophagic cytoprotection of neuronal cell line SH-SY5Y.

3.2. Interactions with Androgen Receptors

Dehydroepiandrosterone (DHEA), also known as androstenolone, binds with high affinity to androgen receptors; however, it may also activate the ER, PXR, CAR, and PPARα. There is a body of evidence showing that DHEA activates Akt in neural precursors, in association with inhibition of apoptosis [72]. However, its sulfated derivative, DHEA, has the opposite effect of DHEA during neurogenesis. An anabolic-androgen 17β-trenbolone was found to induce apoptosis in primary hippocampal neurons that was accompanied by the up-regulation of β-amyloid peptide 42 (Aβ42) and the activation of caspase-3 [73]. Androgen-induced neurotoxicity has been detected in the dopaminergic cell line N27 [74]. Following treatment with testosterone or dihydrotestosterone, the cells exhibited mitochondrial dysfunction and died due to apoptosis. Androgens appeared to have both neuroprotective and neurotoxic effects depending on age and sex, as evidenced by examining the effect of androgens on cell survival after an excitatory stimulus in the developing hippocampus in both males and females [75]. Pike et al., 2008 discussed the involvement of androgen cell signalling pathways in neuroprotection, showing, as an example, the age-related testosterone loss in men with increased risk for AD [76].

3.3. Interactions with Progesterone Receptors

Many studies have shown that progesterone promotes the viability of neurons in the brain and spinal cord. Progesterone appeared particularly effective in treating traumatic brain and spinal cord injuries, as well as ischemic stroke [77]. Moreover, progesterone has been shown to attenuate Aβ (25–35)-induced neuronal toxicity via JNK inactivation and the inhibition of mitochondrial apoptotic pathway by progesterone receptor membrane component 1 [78]. Studies using yeast as a model system extended the protective actions of progesterone to the reduction of cytosolic calcium and the reduction in ROS production and ATP levels; however, these effects did not depend on the yeast orthologue of the progesterone receptor, Dap1 [79]. Progesterone receptor membrane component 1 (PGRMC1) is a key regulator of apoptosis, and its up-regulation was found in retinal degeneration 10 (rd10) mice, which are a model system for autosomal recessive retinitis pigmentosa [80].

3.4. Interactions with Glucocorticoid and Mineralocorticoid Receptors

Previous studies have shown that dexamethasone-induced apoptosis of primary hippocampal neurons involved GR and was counteracted by an MR agonist aldosterone [81]. In general, GR activation has been implicated in the induction of an endangered neural phenotype, whereas MR expression appeared to be associated with a neuroprotective phenotype [82]. When used therapeutically to treat respiratory dysfunction associated with premature birth, the endogenous rodent glucocorticoid corticosterone has been shown to activate GR and cause progenitor cell apoptosis, as well as neurodevelopmental deficits, particularly in the cerebellum [83]. In contrast, MR overexpression inhibited apoptosis and promoted survival of embryonic stem cell-derived neurons [84]. Unlike in other brain regions, glucocorticoids play a critical role in retinal photoreceptor survival, whereas mifepristone, which has the capacity to block glucocorticoid receptors, promotes photoreceptor death [85]. Furthermore, an MR agonist aldosterone appeared to be a critical mediator of retinal ganglion cell loss that was independent of elevated intraocular pressure [86].

3.5. Interactions with Aryl Hydrocarbon Receptor

There are data, including ours, that have revealed that the aryl hydrocarbon receptor (AHR) regulates apoptosis in the mammalian brain [42,43,87–90]. Previously, we demonstrated that the selective AHR agonist, β-naphthoflavone, induced caspase-3-dependent apoptosis in primary cultures of mouse neurons. This effect was accompanied by the increased expression of AHR that co-localized with ERβ, thus supporting the direct interaction between AHR-mediated apoptosis and ERs signalling [42]. We also showed the involvement of AHR in apoptotic and neurotoxic actions of the pesticide dichlorodiphenyltrichloroethane (DDT) and the antimicrobial agent triclosan [43,90]. In addition, we detected the enhanced mRNA and protein expression levels of AHR in one-month-old mice that were prenatally exposed to DDT [91]. Since enhanced AHR expression was accompanied by DNA hypomethylation, both global DNA and the DNA of the specific *AHR* gene, we hypothesized that AHR signalling leads to apoptosis that underlies the fetal basis of the adult onset of disease. In our study, we showed that by targeting AHR/AHR nuclear translocator (ARNT) signalling, 3,3′-diindolylmethane (DIM) inhibited caspase-3-dependent apoptosis and rescued neurons from hypoxia [89]. This effect was followed by a decrease in the expression levels of AHR, AHR-regulated cytochrome P450 1A1 (CYP1A1), and ARNT.

3.6. Interactions with RXR-Related Xenobiotic Receptors

Recently, we showed that CAR, PXR, and RXRs were involved in nonylphenol-initiated apoptosis in mouse hippocampal neurons in primary cultures [44,92]. We also found that RXRs participated in the propagation of DDE- and BP-3-induced neuronal apoptosis [93,94], whereas PPARγ was impaired in response to apoptotic actions of BP-3, dibutyl-phthalate, and tetrabromobisphenol A [64,95,96]. Interestingly, the activation of CAR by the CITCO agonist increased ABC-transporter expression (Abcb1 and Abcg2) in blood-brain barrier and inhibited growth and expansion of brain tumour stem cells via inducing cell cycle arrest and apoptosis [97,98]. Similarly, a PPARα-selective activator 4-chloro-6-(2,3-xylidino)2-pyrimidinylthioacetic acid (Wy-14,643) enhanced cell death in cultured cerebellar granule cells [99]. On the other hand, RXR activation was essential for docosahexaenoic acid to protect retina photoreceptors against oxidative stress (paraquat, H_2O_2)-induced apoptosis [100,101], and the addition of 9-*cis*-retinoic acid prevented dimerization of RXR and a nerve growth factor-induced clone B (NGFI-B) that rescued cerebellar granule neurons from calcium-induced apoptosis [102]. An RXR agonist bexarotene was found to up-regulate the lncRNA Neat1 and to inhibit apoptosis in mice after traumatic brain injury [103]. Moreover, activated LXR inhibited a 7-ketocholesterol-induced apoptosis in human neuroblastoma SH-SY5Y cells [104]. Mutation of NR4A2 (also called NURR1), a RXR-partner for heterodimerization, is involved in familiar form of PD. In patients with PD, NR4A2 expression is downregulated, whereas in patients with AD reduced NR4A1 level is observed [105]. Additionally, NR4A has been identified as a key regulator of catecholamine production by macrophages and the mediator of CREB-induced neuronal survival [106,107].

4. The Roles of Apoptosis and Autophagy in Pathologies of the Nervous System

4.1. Apoptosis in Pathologies of the Nervous System

Apoptosis is involved in the pathogenesis of neurodegenerative diseases, such as stroke, HD, AD, and PD. The accumulation of misfolded Aβ and α-synuclein, two major toxic proteins in AD and PD, leads to neuronal apoptosis. There is increasing evidence that caspase-6 is highly involved in axon degeneration in HD and AD. Active caspase-6 has been found in the early stages of AD [108], and cleavage at the caspase-6-cleavage site in mutated huntingtin protein is a prerequisite for the development of the features of HD [109]. The PTEN-induced kinase 1 (PINK1) that is linked to the autosomal recessive familial form of PD has been found to protect cells from mitochondrial dysfunction and is a key player in many signalling pathways in response to oxidative stress, including apoptosis [110]. A previous study supports an important role of ceramides in neuronal apoptosis, particularly their role in the stabilization of β-secretase, amyloidogenic processing of Aβ precursor protein (APP), and generation of Aβ, which is the major component of the senile plaques [111].

Currently, there has been considerable effort directed towards GSK-3β as a potential target for the treatment of many diseases, including Type-II diabetes and neurodegenerative diseases [112]. Activated GSK-3β has been reported to induce apoptosis in neurons [113,114] and to regulate tau phosphorylation and Aβ peptide production. Extracellular deposits of Aβ and intracellular deposits of hyperphosphorylated tau protein are major histopathological hallmarks of AD. Because

GSK-3β phosphorylates tau proteins, it is thought that disruption of GSK-3β signalling may contribute to the onset of the disease. Targeting this enzyme has been found to inhibit the symptoms of PD, such as enhanced expression of α-synuclein and the loss of dopaminergic neurons in the substantia nigra pars compacta [115]. We previously showed that the neuroprotective action of genistein is mediated by the inhibition of GSK-3β signalling [53]. Our recent study demonstrated that the pesticide DDT-induced apoptosis of mouse neurons is a caspase-9-, caspase-3-, and GSK-3β-dependent process [43]. By activating GSK-3β and JNK, the endoplasmic reticulum stress can induce caspase-12-dependent apoptosis, which has been implicated in a broad range of human diseases, including neurodegenerative diseases, cancer, diabetes, and vascular disorders.

Preclinical data showed that, apart from the inhibition of excitatory activity of neurons via modulation of two major neurotransmitter receptor groups, i.e., N-methyl-D-aspartate (NMDA) receptors, and γ-aminobutyric acid (GABA) receptors, anasthetics caused apoptosis and neurodegeneration in the developing brains of neonates [116]. These effects could lead to learning and memory deficits, as well as abnormalities in social memory and social activity. Similarly, glucocorticoid therapy, which was invented to accelerate lung maturation and reduce inflammation in newborns, was also found to induce apoptosis in the cerebellar external granule layer. The therapy caused a disruption of cerebellar development that was followed by neuromotor and cognitive deficits [117].

4.2. Autophagy in Pathologies of the Nervous System

Dysregulation of autophagy is involved in the etiology of neurodevelopmental diseases, such as autism and fragile X syndrome [23]. Global knockout of *Beclin 1*, *Ambra*, *Atg5*, or *Atg7* is lethal and causes death within a few days after birth [118]. The impairment of autophagy has been postulated to be involved in the onset of neural degeneration in PD, AD, and HD, as well as the degeneration accompanying ALS and MS [4]. An age-related decrease in the expression of the indispensable autophagy protein, Beclin1, has been suggested to have consequences for mHtt accumulation in HD [119]. Selective inhibition of autophagy with 3-MA and chloroquine has been shown to cause a delay in tau clearance, and *Atg7* knockout mice were shown to exhibit excessive amount of phosphorylated tau that is typical for AD [120]. Moreover, the failure of autophagy or mutations in *Parkin* and *PINK1* have been found to result in abnormal autophagic degradation of α-synuclein aggregates [121].

The dysfunction of mitochondria that intensively influences the autophagy pathway is one of the important factors in the pathogenesis of MS [122]. Administration of rapamycin, an inhibitor of mTOR, ameliorates relapsing-remitting experimental autoimmune encephalomyelitis (EAE). However, ALS-associated autophagy remains controversial, and it is still not known whether activating autophagy is beneficial or harmful for motor neuron degradation. Studies reported that toxic accumulation of mutant superoxide dismutase (SOD1) protein contributes to autophagy by its retardation [123]. In addition, it has been found that autophagy may be pro- or anti-inflammatory. The interplay between the microglial inflammatory response and microglial autophagy is inherent to acute central nervous system (CNS) injury as well as the recovery stage of chronic CNS injury [124].

Studies have suggested that autophagy can provide neuroprotection by enhancing the clearance of misfolded protein aggregates [21]. There are, however, data that indicated that autophagy acts in parallel with neurodegenerative processes initiated by kainic acid and hypoxia [125,126]. Recent advances in neurodegenerative models associated with the formation of protein aggregates, such as PD, HD, and AD, target autophagy by treatment with the mTOR inhibitor rapamycin or analogues to force the degradation of potentially toxic aggregates. Surprisingly, the inhibition of mTOR has been found to cause neural degeneration, whereas excessive activity of the kinase was found to impair neural development and lead to neuroectodermal dysplasia [127]. Recent studies have demonstrated that cellular autophagy markers are up-regulated in humans upon treatment with antidepressants [128]. One may assume that autophagy might be a double-edged sword in major depressive disorder (MDD), which suggests why some MDD patients remain resistant to certain antidepressant medications.

5. Perspectives Related to Targeting Apoptosis and Autophagy via Steroid and Xenobiotic Receptor Signalling

5.1. Targeting Apoptosis

5.1.1. Via Estrogen Receptors

There is a wealth of information indicating that estrogens exert actions involved in neuroprotection. They protect neurons against different insults, such as anoxia, oxidative stress, glutamic acid, hydrogen peroxide, iron, and the Aβ peptide. However, the application of estrogens as neuroprotectants in humans presents numerous limitations mainly due to the endocrine actions of the molecules on peripheral tissues and the increased frequency of hormone-dependent tumours as well as cerebrovascular disease and stroke risk. In addition, the highly oxidative cellular environment present during neurodegeneration stimulates the hydroxylation of estradiol to the catechol-estrogen metabolites, which can undergo reactive oxygen species-producing redox cycling, setting up a self-generating toxic cascade [129].

The conflict between basic scientific evidence for estrogen neuroprotection and the lack of effectiveness in clinical trials is only now being resolved. From birth to menopause, the ovaries produce high circulating levels of estradiol, which correlates with a low incidence of neurodegenerative disease. Once the menopausal transition occurs, the risk for neurodegenerative diseases, including ischemic stroke and AD, increases. Although the Women's Health Initiative Memory Study (WHIMS) pointed to some cognitive adverse effects of postmenopausal hormone therapy, these results were not relevant to peri- and early menopause, since WHIMS recruited women above the age of 65 years [130]. Emerging evidence from basic science and clinical studies suggests that there is a "critical period" for the beneficial effect of estradiol on the brain. The critical window hypothesis suggests that initiating hormone therapy at a younger age in closer temporal proximity to menopause may reduce the risk of AD [131]. This is in accordance with a transcriptome meta-analysis that revealed a central role for sex steroids in the degeneration of hippocampal neurons in AD [132]. Currently, there is a need for a new, safe, and effective ER-dependent therapy.

The possibility of using SERMs to induce estrogen-like protective actions in the brain has emerged as an alternative to estrogen treatment. SERMs have been found to trigger neuroprotective mechanisms that reduce neural damage in different experimental models of neural trauma, brain inflammation, and neural degeneration. Most studies have focused on tamoxifen and raloxifene, which have been used in human clinics for years. Their neuroprotective actions have been assessed in different experimental models of neural dysfunction. These include models of traumatic injury to the central nervous system and peripheral nerves, stroke, MS, and PD and AD [133]. The best documented neuroprotective actions are antioxidant effects and the prevention of excitotoxicity, which is a common cause of neuronal death in neurodegeneration [134]. Some key molecules have been identified, such as mitogen-activated protein kinases (MAPK), PI3K/Akt, cAMP response element binding protein (CREB), and NF-κB [135–137], but the precise molecular targets and mechanisms involved in the neuroprotective actions of SERMs need to be determined. Our recent study demonstrated protective, including anti-apoptotic, effects of raloxifene on brain neurons subjected to hypoxia [56]. We showed that raloxifene-induced neuroprotection almost exclusively depended on ERα but not ERβ and GPR30.

Currently, a major focus of neuroscience research is to examine the mechanisms involved in neuronal loss that will be necessary to identify potential drug targets. These include microRNA, such as miR-7-1, which enhanced the neuroprotective effects of estrogen receptor agonists, i.e., 1,3,5-*tris*(4-hydroxyphenyl)-4-propyl-1*H*-pyrazole (PPT), Way 200070, and estrogen, in preventing apoptosis in A23187 calcium ionophore-exposed VSC4.1 motoneurons [138]. Targeting ERβ, with a combination of natural estrogen-like compounds that bind the receptor with high selectivity, has been proposed as a therapeutic strategy for Leber's hereditary optic neuropathy [139]. ERβ ligand AC186 is a new candidate for neuroprotection in MS [140]. Following HIV-1 Tat (1–86) exposure, soy isoflavones induced anti-apoptotic actions in neurons by targeting ERs [141]. Purple sweet potato colour significantly suppressed endoplasmic reticulum stress-induced apoptosis and promoted ERα-mediated mitochondrial biogenesis in mice challenged with domoic acid [142]. An inverted correlation between the levels of ERα and parkin in the striatum of adult mice suggests a possible role of the receptor in preventing the parkin-related PD in humans [143]. The discovery of a binding site of ERβ in the *Tnfaip1* (Tumour necrosis factor-induced protein 1) promoter region points to a novel regulatory site that could be targeted by estrogen or other selective ligands to protect the brain against AD and apoptosis [144]. A new strategy for neuroprotection against ischemic insults could involve targeting a novel 36-kDa variant of ERα, i.e., ERα36, which is associated with the phosphorylation of Akt in the cells exposed to glucose deprivation [145]. Targeting ERβ by *Cicer microphyllum* seed extract could rescue neurons from global hypoxia [146].

Up-regulation of neuroglobin upon 17β-estradiol (E2) stimulation has recently been linked to neuroprotection. After relocalization to the mitochondria, neuroglobin associates with cytochrome c, which reduces cytochrome c release into the cytosol, and subsequently inhibits caspase-3 activation and apoptotic cell death. Neuroprotection could also be directed to seladin-1 (selective AD indicator-1), which is down-regulated in the brain regions affected by AD and confers protection against β-amyloid-induced toxicity via the inhibition of caspase-3 activity. A seladin-1 gene has been found to be identical to the gene encoding the enzyme, 3-β-hydroxysterol Δ^{24}-reductase, which is up-regulated by estrogen and involved in the cholesterol biosynthetic pathway [147].

5.1.2. Via Androgen, Progesterone, and Corticoid Receptors

It recently became evident that the selective androgen receptor modulator (SARM), RAD140, has the capacity to protect cultured neurons and brain tissue against kainate-induced apoptosis [148]. This action depended on MAPK signalling, including ERK phosphorylation, and showed the relevance of androgen signalling to neural health and resilience to neurodegeneration. Progesterone was found to play a neuroprotective role in various models of neurodegeneration, including AD, through the inhibition of the mitochondrial apoptotic pathway, and by blocking Aβ-induced JNK activation [78]. A synthetic form of the female hormone progesterone, Norgestrel, which acts via progesterone receptor membrane component 1, has been proposed as therapeutic for the treatment of retinitis pigmentosa [80]. In patients with AD, increased expression of the translocator protein (TSPO; the former peripheral benzodiazepine receptor or PBR) shows another possibility for neuroprotection that could be executed via TSPO-mediated stimulation of steroid synthesis; as such, its neuroprotective and neuroregenerative properties resulted in the inhibition of apoptotic cell death. A novel TSPO (18 kDa) ligand, ZBD-2, which is involved in the synthesis of endogenous neurosteroids (e.g., pregnenolone, DHEA, and progesterone), effectively prevented NMDA-induced excitotoxicity and apoptosis and protected mouse brains against focal cerebral ischemia [149]. Recent studies provided evidence that corticosterone-induced injury in rat adrenal pheochromocytoma PC12 cells can be attenuated by HBOB, an HDAC6 inhibitor, by inhibiting mitochondrial GR translocation and the intrinsic apoptosis pathway [150]. Furthermore, agonizing the MR with fludrocortisone promoted cell survival and proliferation of adult hippocampal progenitors, but inhibited apoptotic signalling, including GSK-3β [151].

5.1.3. Via AHR

Flavonoids have been shown to inhibit the development of AD-like pathology, and red wine consumption appeared to reduce age-related macular degeneration, stroke, and cognitive deficits. In addition to well documented free radical scavenging and anti-inflammatory properties, resveratrol, which is one of the key ingredients responsible for the neuropreventive action of red wine, has been shown to act as AHR antagonist and an inhibitor of apoptosis. The recent study of our group has shown the strong neuroprotective capacity of a selective aryl hydrocarbon receptor modulator (SAHRM) DIM against the hypoxia-induced damage in mouse hippocampal cells in primary cultures [89]. DIM-evoked neuroprotection was mediated by the impairment of AHR/ARNT signalling, as evidenced by the use of specific siRNAs, as well as quantitative (qPCR, ELISA) and qualitative (western blot analysis and confocal microscopy) assessments. Neuroprotective properties of DIM have also been shown in the cellular and animal models of PD [152]. Exciting prospects to overcome the cellular mechanisms that lead to neuronal injury involving apoptosis and autophagy have recently been related to Wnt (wingless-type) signalling, which is known to interact with AHR during neural development [153].

5.1.4. Via Xenobiotic Receptors

In addition to the steroid receptor- and AHR-mediated signalling, xenobiotic receptors have become attractive mediators that can cause neuroprotection through interaction with the apoptotic pathways. PPARγ agonists (derivatives of thiazolidinediones, e.g., troglitazone, rosiglitazone and pioglitazone) have been associated with neuroprotection in different neurological pathologies, including AD and PD, cerebral ischemia, MDD and stroke [154–157]. However, the mechanisms involved in PPARγ effects in the nervous system are still unknown. PPARs and many other nuclear receptors form heterodimers with RXRs, and these heterodimers regulate the transcription of various genes. The activation of RXR/PPARγ by bexarotene was found to have neuroprotective potential in mice subjected to focal cerebral ischemia [158]. UAB30, a novel RXR agonist that induces apoptosis in human neuroblastoma, has also been shown to have a potential therapeutic role [159]. Recently, treatments with RXR agonists (bexarotene and fluorobexarotene) were found to promote Aβ degradation and rapidly reversed Aβ-induced behavioural deficits in AD [160–162]. Bexarotene was also found to protect dopaminergic neurons in animal models of PD [163].

5.2. Targeting Autophagy

In comparison to apoptosis, there is very little known about the interactions between the steroid and xenobiotic receptor signalling pathways and the process of autophagy. Therefore, targeting autophagy via steroid and xenobiotic receptors needs to be further elucidated. Estradiol has been shown to inhibit autophagy in the hippocampus CA1 region and to alleviate neurological deficits following cerebral ischemia [164]. The estrogen receptor and the estrogen-related receptor antagonists, tamoxifen, and 4-hydroxytamoxifen, were shown to induce cytotoxic autophagy in glioblastoma. However, the effect of 4-hydroxytamoxifen in malignant peripheral nerve sheath tumour cells did not depend on ER signalling, but on the degradation of the pro-survival protein Kirsten rat sarcoma viral oncogene homologue [165]. According to Felzen et al.2015, ERα-expressing neuroblastoma cells have a higher autophagic activity than cells expressing ERβ or lacking ER expression [166]. This new non-canonical autophagy is mediated by ERα, but it is estrogen response element (ERE)-independent and involves the function of the co-chaperone BCL2-associated athanogene 3 (BAG3). In studies using the SH-SY5Y cell line, mERα has been postulated to promote the maturation of autophagosomes into functional autolysosomes by regulating ERK [167].

6. Perspectives Related to Targeting Specific miRNAs which Interact with Steroid and Xenobiotic Receptor Signalling

MicroRNAs (miRNAs) are small non-coding RNA molecules that are almost exclusively negative regulators of gene expression in the nervous system. It has been shown that miR-218-affected tau phosphorylation is oppositely-regulated by classical estrogen receptors. ERα was found to increase the expression of miR-218 that was followed by diminished protein expression of tyrosine phosphatase alpha (PTPα), as well as by activation of GSK-3β and inactivation of protein phosphatase 2A, the major tau enzymes involved in AD pathology. In contrast, ERβ reduced miR-218 levels that resulted in inhibition of tau phosphorylation [168]. Recently, Micheli et al., 2016 demonstrated that 17β-estradiol-evoked neuroprotection from the Aβ-induced neurotoxicity was mediated by an increase in miR-125b expression and subsequent decrease in mRNA and protein expression of pro-apoptotic factors BAK1 and p53 [169]. Furthermore, an involvement of estrogen receptors in regulation of specific miRNAs in response to cerebral ischemia has been demonstrated. 17β-estradiol alone or in combination with progesterone was found to upregulate miR-375 expression and its target BCL2 in rat model of cerebral ischemia [170]. MiR-375 was also positively regulated by ERα in response to a phytoestrogen-calycosin that caused protection against cerebral ischemia [171]. Recently, the roles of estrogen and glucocorticoid receptors in regulation of cerebral miRNAs have been supported by contribution of miR-23a and miR-210 in response to cerebral ischemia [172,173]. Glucocorticoid-mediated attenuation of BDNF-dependent neuronal function has been linked to reduced expression of miR-132 [174]. The most relevant reports on AHR- and CAR-regulated miRNA showed the inversely co-related expression of AIP (AHR-interacting protein) and miR-107 in pituitary adenomas, as well as CAR and miR-137 in neuroblastoma cells [175,176].

Acknowledgments: Agnieszka Wnuk received scholarships from the KNOW, which was sponsored by the Ministry of Science and Higher Education in Poland. The manuscript has been edited by the American Journal Experts for English language and grammar E459-9740-DE16-84DC-0D41/CBED-3F5F-92FC-0480-D0B4 that was supported by KNOW funds MNiSW-DS-6002-4693-26/WA/12. The publication charge was supported by grant no. 2015/19/B/NZ7/02449 from the National Science Centre, Poland, and the statutory fund of the Institute of Pharmacology Polish Academy of Sciences, Krakow, Poland.

Conflicts of Interest: These authors contributed equally to this work.

Abbreviations

AD	Alzheimer's disease
AHR	aryl hydrocarbon receptor
AIP	AHR-interacting protein
ALS	amyotrophic lateral sclerosis
AMP	adenosine monophosphate
AMPK	adenosine monophosphate activated kinase
APP	Aβ precursor protein
AR	androgen receptor
ARNT	AHR nuclear translocator
Aβ	amyloid-beta
Aβ42	β-amyloid peptide 42
BAG3	BCL2-associated athanogene 3
BDNF	brain-derived neurotrophic factors
BP-3	benzophenone-3
CAD	caspase-activated DNase
CAR	constitutive androstane receptor
CNS	central nervous system
CREB	cAMP response element binding protein
CYP1A1	cytochrome P450 1A1
DDT	dichlorodiphenyltrichloroethane
DHEA	dehydroepiandrosterone
DIM	3,3'-diindolylmethane
DR6	death receptor 6
E2	17β-estradiol
EAE	experimental autoimmune encephalomyelitis
ER	estrogen receptor
ERE	estrogen response element
ERK	extracellular signal-regulated kinase
GABA	g-Aminobutyric acid
GDNF	glial cell-derived neurotrophic factor
GP-17	gypenoside XVII
GPER1	G-protein-coupled ER1, membrane-bound estrogen receptor
GPR30	membrane-bound estrogen receptor, also known as GPER1
GR	glucocorticoid receptor
GSK-3β	glycogen synthase kinase 3 beta

HD	Huntington's disease
HDAC	histone deacetylases
ICAD	inhibitor of caspase-activated DNase
JNK	c-Jun N-terminal kinase
LXR	liver X receptor
MAPK	mitogen-activated protein kinases
MDD	major depressive disorder
MiRNA	microRNA
MPTP	1-Methyl-4-phenyl-1,2,3,6-tetrahydropyridine
MR	mineralocorticoid receptor
MS	multiple sclerosis
MTA1	metastasis-associated protein 1
mTOR	mammalian target of rapamycin kinase
NGF	nerve growth factors
NGFI-B	nerve growth factor-induced clone B
NMDA	*N*-methyl-D-aspartate
NPC	neural progenitor cells
NT-3	neurotrophin-3
NURR1	nuclear receptor related 1 protein
PD	Parkinson's disease
PGRMC1	progesterone receptor membrane component 1
PI-3K	3-Phosphatydylinosytol kinase
PI3P	phosphatidyl-inositol-3-phosphate
PIK3C1	phosphoinositide 3-kinase
PINK1	PTEN-induced kinase 1
PTPα	tyrosine phosphatase alpha
PPAR	peroxisome proliferator-activated receptor
PR	progesterone receptor
PXR	pregnane X receptor
rd10	retinal degeneration 10
ROS	reactive oxygen species
RXR	retinoid X receptor
SAHRM	selective aryl hydrocarbon receptor modulator
SAPK	stress-activated protein kinase
SARM	selective androgen receptor modulator
seladin-1	selective AD indicator-1
SERM	selective estrogen receptor modulator
SOD1	superoxide dismutase
tMCAO	transient right middle cerebral artery occlusion
Tnfaip1	tumour necrosis factor-induced protein 1
TNF-R1	tumour necrosis factor receptor-1
TRAILR2	death receptor 5/DR5
TRAMP	death receptor 3/APO-3/LARD/wsl-1
TSPO	translocator protein
WHIMS	women's health initiative memory study
Wnt	wingless-type

References

1. Wu, H.J.; Pu, J.L.; Krafft, P.R.; Zhang, J.M.; Chen, S. The molecular mechanisms between autophagy and apoptosis: Potential role in central nervous system disorders. *Cell. Mol. Neurobiol.* **2015**, *35*, 85–99. [CrossRef] [PubMed]

2. Fu, H.J.; Hu, Q.S.; Lin, Z.N.; Ren, T.J.; Song, H.; Cai, C.K.; Dong, S.Z. Aluminium induced apoptosis in cultured cortical neurons and its effect on SAPK/JNK signal transduction pathway. *Brain Res.* **2003**, *980*, 11–23. [CrossRef]

3. Chen, W.; Sun, Y.; Liu, K.; Sun, X. Autophagy: A double-edged sword for neuronal survival after cerebral ischemia. *Neural Regen. Res.* **2014**, *9*, 1210–1216. [CrossRef] [PubMed]

4. Kesidou, E.; Lagoudaki, R.; Touloumi, O.; Poulatsidou, K.N.; Simeonidou, C. Autophagy and neurodegenerative disorders. *Neural Regen. Res.* **2013**, *8*, 2275–2283. [CrossRef] [PubMed]

5. Vanderhaeghen, P.; Cheng, H.J. Guidance Molecules in Axon Pruning and Cell Death. *Cold Spring Harb. Perspect. Biol.* **2010**, *2*, a001859. [CrossRef] [PubMed]

6. Yang, Y.; Coleman, M.; Zhang, L.; Zheng, X.; Yue, Z. Autophagy in axonal and dendritic degeneration. *Trends Neurosci.* **2013**, *36*, 418–428. [CrossRef] [PubMed]

7. Wallen-Mackenzie, A.; Mata de Urquiza, A.; Petersson, S.; Rodriguez, F.J.; Friling, S.; Wagner, J.; Ordentlich, P.; Lengqvist, J.; Heyman, R.A.; Arenas, E.; et al. Nurr1-RXR heterodimers mediate RXR ligand-induced signaling in neuronal cells. *Genes Dev.* **2003**, *17*, 3036–3047. [CrossRef] [PubMed]

8. Yuan, J.; Yankner, B.A. Apoptosis in the nervous sytem. *Nature* **2000**, *407*, 802–809. [CrossRef] [PubMed]

9. Davies, C.; Tournier, C. Exploring the function of the JNK (c-Jun N-terminal kinase) signalling pathway in physiological and pathological processes to design novel therapeutic strategies. *Biochem. Soc. Trans.* **2012**, *40*, 85–89. [CrossRef] [PubMed]

10. Kajta, M.; Beyer, C. Cellular strategies of estrogen-mediated neuroprotection during brain development. *Endocrine* **2003**, *1*, 3–9. [CrossRef]

11. Sastry, P.S.; Rao, K.S. Apoptosis in the nervous system. *J. Neurochem.* **2000**, *74*, 1–20. [CrossRef] [PubMed]

12. Pettmann, B.; Henderson, C.E. Neuronal cell death. *Neuron* **1998**, *20*, 633–647. [CrossRef]

13. Chen, D.; Stetler, R.A.; Cao, G.; Pei, W.; O'Horo, C.; Yin, X.M.; Chen, J. Characterization of the rat DNA fragmentation factor 35/Inhibitor of caspase-activated DNase (Short form). The endogenous inhibitor of caspase-dependent DNA fragmentation in neuronal apoptosis. *J. Biol. Chem.* **2000**, *275*, 38508–38517. [CrossRef] [PubMed]

14. Momeni, H.R. Role of Calpain in Apoptosis. *Cell J. Summer* **2011**, *13*, 65–72.

15. Watcharasit, P.; Bijur, G.N.; Song, L.; Zhu, J.; Chen, X.; Jope, R.S. Glycogen synthase kinase-3β (GSK3β) binds to and promotes the actions of p53. *J. Biol. Chem.* **2003**, *278*, 48872–48879. [CrossRef] [PubMed]

16. Jacobs, K.M.; Bhave, S.R.; Ferraro, D.J.; Jaboin, J.J.; Hallahan, D.E.; Thotala, D. GSK-3β: A Bifunctional Role in Cell Death Pathways. *Int. J. Cell Biol.* **2012**, *930710*. [CrossRef]

17. Hooper, C.; Killick, R.; Lovestone, S. The GSK3 hypothesis of Alzheimer's disease. *J. Neurochem.* **2008**, *104*, 1433–1439. [CrossRef] [PubMed]

18. Bian, H.; Bian, W.; Lin, X.; Ma, Z.; Chen, W.; Pu, Y. RNA Interference Silencing of Glycogen Synthase Kinase 3β Inhibites Tau Phosphorylation in Mice with Alzheimer Disease. *Neurochem. Res.* **2016**, *41*, 2470–2480. [CrossRef] [PubMed]

19. Golpich, M.; Amini, E.; Hemmati, F.; Ibrahim, N.M.; Rahmani, B.; Mohamed, Z.; Raymond, A.A.; Dargahi, L.; Ghasemi, R.; Ahmadiani, A. Glycogen synthase kinase-3β (GSK-3β) signaling: Implications for Parkinson's disease. *Pharmacol. Res.* **2015**, *97*, 16–26. [CrossRef] [PubMed]

20. Liu, D.; Zhang, M.; Yin, H. Signaling pathways involved in endoplasmic reticulum stress-induced neuronal apoptosis. *Int. J. Neurosci.* **2013**, *123*, 155–162. [CrossRef] [PubMed]

21. Brown, M.K.; Naidoo, N. The endoplasmic reticulum stress response in aging and age-related diseases. *Front. Physiol.* **2012**, *3*, 263. [CrossRef] [PubMed]

22. Wang, D.B.; Kinoshita, C.; Kinoshita, Y.; Morrison, R.S. p53 and Mitochondrial Function in Neurons. *Biochim. Biophys. Acta* **2014**, *1842*, 1186–1197. [CrossRef] [PubMed]

23. Lee, K.M.; Hwang, S.K.; Lee, J.A. Neuronal Autophagy and Neurodevelopmental Disorders. *Exp. Neurobiol.* **2013**, *22*, 133–142. [CrossRef] [PubMed]

24. Cebollero, E.; Reggiori, F. Regulation of autophagy in yeast Saccharomyces cerevisiae. *Biochim. Biophys. Acta* **2009**, *1793*, 1413–1421. [CrossRef] [PubMed]

25. Glick, D.; Barth, S.; Macleod, K.F. Autophagy: Cellular and molecular mechanisms. *J. Pathol.* **2010**, *221*, 3–12. [CrossRef] [PubMed]

26. Orhon, I.; Dupont, N.; Pampliega, O.; Cuervo, A.M.; Codogno, P. Autophagy and regulation of cilia function and assembly. *Cell Death Differ.* **2015**, *22*, 389–397. [CrossRef] [PubMed]

27. Marquez, R.T.; Xu, L. Bcl-2: Beclin 1 complex: Multiple, mechanisms regulating autophagy/apoptosis toggle switch. *Am. J. Cancer Res.* **2012**, *2*, 214–221. [PubMed]

28. Fan, Y.J.; Zong, W.X. The cellular decision between apoptosis and autophagy. *Chin. J. Cancer* **2013**, *32*, 121–129. [CrossRef] [PubMed]

29. Mariño, G.; Niso-Santano, M.; Baehrecke, E.H.; Kroemer, G. Self-consumption: The interplay of autophagy and apoptosis. *Nat. Rev. Mol. Cell Biol.* **2014**, *15*, 81–94. [CrossRef] [PubMed]

30. Wu, H.; Che, X.; Zheng, Q.; Wu, A.; Pan, K.; Shao, A.; Wu, Q.; Zhang, J.; Hong, Y. Caspases: A Molecular Switch Node in the Crosstalk between Autophagy and Apoptosis. *Int. J. Biol. Sci.* **2014**, *10*, 1072–1083. [CrossRef] [PubMed]

31. Lee, J.S.; Li, Q.; Lee, J.Y.; Lee, S.H.; Jeong, J.H.; Lee, H.R.; Chang, H.; Zhou, F.C.; Gao, S.J.; Liang, C.; et al. FLIP-mediated autophagy regulation in cell death control. *Nat. Cell Biol.* **2009**, *11*, 1355–1362. [CrossRef] [PubMed]

32. McGregor, C.; Riordan, A.; Thornton, J. Estrogens and the cognitive symptoms of schizophrenia: Possible neuroprotective mechanisms. *Front. Neuroendocrinol.* **2017**, *47*, 19–33. [CrossRef] [PubMed]

33. Perlman, W.R.; Tomaskovic-Crook, E.; Montague, D.M.; Webster, M.J.; Rubinow, D.R.; Kleinman, J.E.; Weickert, C.S. Alteration in estrogen receptor α mRNA levels in frontal cortex and hippocampus of patients with major mental illness. *Biol. Psychiatry* **2005**, *58*, 812–824. [CrossRef] [PubMed]

34. Kelly, J.F.; Bienias, J.L.; Shah, A.; Meeke, K.A.; Schneider, J.A.; Soriano, E.; Bennett, D.A. Levels of estrogen receptors α and β in frontal cortex of patients with Alzheimer's disease: Relationship to Mini-Mental State Examination scores. *Curr. Alzheimer Res.* **2008**, *5*, 45–51. [CrossRef] [PubMed]

35. Hwang, C.J.; Yun, H.M.; Park, K.R.; Song, J.K.; Seo, H.O.; Hyun, B.K.; Choi, D.Y.; Yoo, H.S.; Oh, K.W.; Hwang, D.Y.; et al. Memory Impairment in Estrogen Receptor α Knockout Mice Through Accumulation of Amyloid-β Peptides. *Mol. Neurobiol.* **2015**, *52*, 176–186. [CrossRef] [PubMed]

36. Long, J.; He, P.; Shen, Y.; Li, R. New evidence of mitochondria dysfunction in the female Alzheimer's brain: Deficiency of estrogen receptor-β. *J. Alzheimers Dis.* **2012**, *30*, 545–558. [CrossRef] [PubMed]

37. Fan, X.; Warner, M.; Gustafsson, J.A. Estrogen receptor β expression in the embryonic brain regulates development of calretinin-immunoreactive GABAergic interneurons. *Proc. Natl. Acad. Sci. USA* **2006**, *103*, 19338–19343. [CrossRef] [PubMed]

38. Sharma, G.; Hu, C.; Brigman, J.L.; Zhu, G.; Hathaway, H.J.; Prossnitz, E.R. GPER deficiency in male mice results in insulin resistance, dyslipidemia, and a proinflammatory state. *Endocrinology* **2013**, *154*, 4136–4145. [CrossRef] [PubMed]

39. Kajta, M. Apoptosis in the central nervous system: Mechanisms and protective strategies. *Pol. J. Pharmacol.* **2004**, *56*, 689–700. [PubMed]

40. Bourque, M.; Dluzen, D.E.; di Paolo, T. Neuroprotective actions of sex steroids in Parkinson's disease. *Front. Neuroendocrinol.* **2009**, *30*, 142–157. [CrossRef] [PubMed]

41. Napolitano, M.; Costa, L.; Piacentini, R.; Grassi, C.; Lanzone, A.; Gulino, A. 17β-estradiol protects cerebellar granule cells against β-amyloid-induced toxicity via the apoptotic mitochondrial pathway. *Neurosci. Lett.* **2014**, *561*, 134–139. [CrossRef] [PubMed]

42. Kajta, M.; Wójtowicz, A.K.; Maćkowiak, M.; Lasoń, W. Aryl hydrocarbon receptor-mediated apoptosis of neuronal cells: A possible interaction with estrogen receptor signaling. *Neuroscience* **2009**, *158*, 811–822. [CrossRef] [PubMed]

43. Kajta, M.; Litwa, E.; Rzemieniec, J.; Wnuk, A.; Lason, W.; Zelek-Molik, A.; Nalepa, I.; Grzegorzewska-Hiczwa, M.; Tokarski, K.; Golas, A.; et al. Isomer-nonspecific action of dichlorodiphenyltrichloroethane on aryl hydrocarbon receptor and G-protein-coupled receptor 30 intracellular signaling in apoptotic neuronal cells. *Mol. Cell. Endocrinol.* **2014**, *392*, 90–105. [CrossRef] [PubMed]

44. Litwa, E.; Rzemieniec, J.; Wnuk, A.; Lason, W.; Krzeptowski, W.; Kajta, M. Apoptotic and neurotoxic actions of 4-para-nonylphenol are accompanied by activation of retinoid X receptor and impairment of classical estrogen receptor signaling. *J. Steroid Biochem. Mol. Biol.* **2014**, *144*, 334–347. [CrossRef] [PubMed]

45. Cheema, Z.F.; Santillano, D.R.; Wade, S.B.; Newman, J.M.; Miranda, R.C. The extracellular matrix, p53 and estrogen compete to regulate cell-surface Fas/Apo-1 suicide receptor expression in proliferating embryonic cerebral cortical precursors, and reciprocally, Fas-ligand modifies estrogen control of cell-cycle proteins. *BMC Neurosci.* **2004**, *5*, 11. [CrossRef] [PubMed]

46. Brailoiu, E.; Dun, S.L.; Brailoiu, G.C.; Mizuo, K.; Sklar, L.A.; Oprea, T.I.; Prossnitz, E.R.; Dun, N.J. Distribution and characterization of estrogen receptor G protein-coupled receptor 30 in the rat central nervous system. *J. Endocrinol.* **2007**, *193*, 311–321. [CrossRef] [PubMed]

47. Hazell, G.G.; Yao, S.T.; Roper, J.A.; Prossnitz, E.R.; O'Carroll, A.M.; Lolait, S.J. Localisation of GPR30, a novel G protein-coupled oestrogen receptor, suggests multiple functions in rodent brain and peripheral tissues. *J. Endocrinol.* **2009**, *202*, 223–236. [CrossRef] [PubMed]

48. Gingerich, S.; Kim, G.L.; Chalmers, J.A.; Koletar, M.M.; Wang, X.; Wang, Y.; Belsham, D.D. Estrogen receptor α and G-protein coupled receptor 30 mediate the neuroprotective effects of 17β-estradiol in novel murine hippocampal cell models. *Neurosci. Sep.* **2010**, *170*, 54–66. [CrossRef] [PubMed]

49. Liu, S.B.; Zhang, N.; Guo, Y.Y.; Zhao, R.; Shi, T.Y.; Feng, S.F.; Wang, S.Q.; Yang, Q.; Li, X.Q.; Wu, Y.M.; et al. G-protein-coupled receptor 30 mediates rapid neuroprotective effects of estrogen via depression of NR2B-containing NMDA receptors. *J. Neurosci.* **2012**, *32*, 4887–4900. [CrossRef] [PubMed]

50. Lu, D.; Qu, Y.; Shi, F.; Feng, D.; Tao, K.; Gao, G.; He, S.; Zhao, T. Activation of G protein-coupled estrogen receptor 1 (GPER-1) ameliorates blood-brain barrier permeability after global cerebral ischemia in ovariectomized rats. *Biochem. Biophys. Res. Commun.* **2016**, *477*, 209–214. [CrossRef] [PubMed]

51. Wang, Z.F.; Pan, Z.Y.; Xu, C.S.; Li, Z.Q. Activation of G-protein coupled estrogen receptor 1 improves early-onset cognitive impairment via PI3K/Akt pathway in rats with traumatic brain injury. *Biochem. Biophys. Res. Commun.* **2017**, *482*, 948–953. [CrossRef] [PubMed]

52. Guan, J.; Yang, B.; Fan, Y.; Zhang, J. GPER Agonist G1 Attenuates Neuroinflammation and Dopaminergic Neurodegeneration in Parkinson Disease. *Neuroimmunomodulation* **2017**, *24*, 60–66. [CrossRef] [PubMed]

53. Kajta, M.; Domin, H.; Grynkiewicz, G.; Lason, W. Genistein inhibits glutamate-induced apoptotic processes in primary neuronal cell cultures: An involvement of aryl hydrocarbon receptor and estrogen receptor/glycogen synthase kinase-3β intracellular signaling pathway. *Neuroscience* **2007**, *145*, 592–604. [CrossRef] [PubMed]

54. Kajta, M.; Rzemieniec, J.; Litwa, E.; Lason, W.; Lenartowicz, M.; Krzeptowski, W.; Wojtowicz, A.K. The key involvement of estrogen receptor β and G-protein-coupled receptor 30 in the neuroprotective action of daidzein. *Neuroscience* **2013**, *238*, 345–360. [CrossRef] [PubMed]

55. Baluchnejadmojarad, T.; Jamali-Raeufy, N.; Zabihnejad, S.; Rabiee, N.; Roghani, M. Troxerutin exerts neuroprotection in 6-hydroxydopamine lesion rat model of Parkinson's disease: Possible involvement of PI3K/ERβ signaling. *Eur. J. Pharmacol.* **2017**, *801*, 72–78. [CrossRef] [PubMed]

56. Rzemieniec, J.; Litwa, E.; Wnuk, A.; Lason, W.; Gołas, A.; Krzeptowski, W.; Kajta, M. Neuroprotective action of raloxifene against hypoxia-induced damage in mouse hippocampal cells depends on ERα but not ERβ or GPR30 signalling. *J. Steroid Biochem. Mol. Biol.* **2015**, *146*, 26–37. [CrossRef] [PubMed]

57. Guo, J.; Zhang, T.; Yu, J.; Li, H.Z.; Zhao, C.; Qiu, J.; Zhao, B.; Zhao, J.; Li, W.; Zhao, T.Z. Neuroprotective effects of a chromatin modifier on ischemia/reperfusion neurons: Implication of its regulation of BCL2 transactivation by ERα signaling. *Cell Tissue Res.* **2016**, *364*, 475–488. [CrossRef] [PubMed]

58. Jover-Mengual, T.; Castelló-Ruiz, M.; Burguete, M.C.; Jorques, M.; López-Morales, M.A.; Aliena-Valero, A.; Jurado-Rodríguez, A.; Pérez, S.; Centeno, J.M.; Miranda, F.J.; et al. Molecular mechanisms mediating the neuroprotective role of the selective estrogen receptor modulator, bazedoxifene, in acute ischemic stroke: A comparative study with 17β-estradiol. *J. Steroid Biochem. Mol. Biol.* **2017**, *171*, 296–304. [CrossRef] [PubMed]

59. Guo, J.M.; Shu, H.; Wang, L.; Xu, J.J.; Niu, X.C.; Zhang, L. SIRT1-dependent AMPK pathway in the protection of estrogen against ischemic brain injury. *CNS Neurosci. Ther.* **2017**, *23*, 360–369. [CrossRef] [PubMed]

60. Meng, X.; Wang, M.; Wang, X.; Sun, G.; Ye, J.; Xu, H.; Sun, X. Suppression of NADPH oxidase- and mitochondrion-derived superoxide by Notoginsenoside R1 protects against cerebral ischemia-reperfusion injury through estrogen receptor-dependent activation of Akt/Nrf2 pathways. *Free Radic. Res.* **2014**, *48*, 823–838. [CrossRef] [PubMed]

61. Wang, Y.; Tu, L.; Li, Y.; Chen, D.; Wang, S. Notoginsenoside R1 Protects against Neonatal Cerebral Hypoxic-Ischemic Injury through Estrogen Receptor-Dependent Activation of Endoplasmic Reticulum Stress Pathways. *J. Pharmacol. Exp. Ther.* **2016**, *357*, 591–605. [CrossRef] [PubMed]

62. Wu, J.; Pan, Z.; Wang, Z.; Zhu, W.; Shen, Y.; Cui, R.; Lin, J.; Yu, H.; Wang, Q.; Qian, J.; et al. Ginsenoside Rg1 protection against β-amyloid peptide-induced neuronal apoptosis via estrogen receptor α and glucocorticoid receptor-dependent anti-protein nitration pathway. *Neuropharmacology* **2012**, *63*, 349–361. [CrossRef] [PubMed]

63. Rivera, P.; Pérez-Martín, M.; Pavón, F.J.; Serrano, A.; Crespillo, A.; Cifuentes, M.; López-Ávalos, M.D.; Grondona, J.M.; Vida, M.; Fernández-Llebrez, P.; et al. Pharmacological administration of the isoflavone daidzein enhances cell proliferation and reduces high fat diet-induced apoptosis and gliosis in the rat hippocampus. *PLoS ONE* **2013**, *8*, e64750. [CrossRef] [PubMed]

64. Wnuk, A.; Rzemieniec, J.; Lasoń, W.; Krzeptowski, W.; Kajta, M. Apoptosis Induced by the UV Filter Benzophenone-3 in Mouse Neuronal Cells Is Mediated via Attenuation of Erα/Pparγ and Stimulation of Erβ/Gpr30 Signaling. *Mol. Neurobiol.* **2017**. [CrossRef] [PubMed]

65. Sarzi, E.; Seveno, M.; Angebault, C.; Milea, D.; Rönnbäck, C.; Quilès, M.; Adrian, M.; Grenier, J.; Caignard, A.; Lacroux, A.; et al. Increased steroidogenesis promotes early-onset and severe vision loss in females with OPA1 dominant optic atrophy. *Hum. Mol. Genet.* **2016**, *25*, 2539–2551. [CrossRef] [PubMed]

66. Bai, L.Y.; Weng, J.R.; Hu, J.L.; Wang, D.; Sargeant, A.M.; Chiu, C.F. G15, a GPR30 antagonist, induces apoptosis and autophagy in human oral squamous carcinoma cells. *Chem. Biol. Interact.* **2013**, *206*, 375–384. [CrossRef] [PubMed]

67. Cook, K.L.; Clarke, P.A.; Parmar, J.; Hu, R.; Schwartz-Roberts, J.L.; Abu-Asab, M.; Wärri, A.; Baumann, W.T.; Clarke, R. Knockdown of estrogen receptor-α induces autophagy and inhibits antiestrogen-mediated unfolded protein response activation, promoting ROS-induced breast cancer cell death. *FASEB J.* **2014**, *28*, 3891–3905. [CrossRef] [PubMed]

68. Meng, X.; Wang, M.; Sun, G.; Ye, J.; Zhou, Y.; Dong, X.; Wang, T.; Lu, S.; Sun, X. Attenuation of Aβ25-35-induced parallel autophagic and apoptotic cell death by gypenoside XVII through the estrogen receptor-dependent activation of Nrf2/ARE pathways. *Toxicol. Appl. Pharmacol.* **2014**, *279*, 63–75. [CrossRef] [PubMed]

69. Liang, K.; Yang, L.; Yin, C.; Xiao, Z.; Zhang, J.; Liu, Y.; Huang, J. Estrogen stimulates degradation of β-amyloid peptide by up-regulating neprilysin. *J. Biol. Chem.* **2010**, *285*, 935–942. [CrossRef] [PubMed]

70. Barbati, C.; Pierdominici, M.; Gambardella, L.; Malchiodi Albedi, F.; Karas, R.H.; Rosano, G.; Malorni, W.; Ortona, E. Cell surface estrogen receptor α is upregulated during subchronic metabolic stress and inhibits neuronal cell degeneration. *PLoS ONE* **2012**, *7*, e42339. [CrossRef] [PubMed]

71. Kumar, S.; Patel, R.; Moore, S.; Crawford, D.K.; Suwanna, N.; Mangiardi, M.; Tiwari-Woodruff, S.K. Estrogen receptor β ligand therapy activates PI3K/Akt/mTOR signaling in oligodendrocytes and promotes remyelination in a mouse model of multiple sclerosis. *Neurobiol. Dis.* **2013**, *56*, 131–144. [CrossRef] [PubMed]

72. Zhang, L.; Li, B.; Ma, W.; Barker, J.L.; Chang, Y.H.; Zhao, W.; Rubinow, D.R. Dehydroepiandrosterone (DHEA) and its sulfated derivative (DHEAS) regulate apoptosis during neurogenesis by triggering the Akt signaling pathway in opposing ways. *Brain Res. Mol. Brain Res.* **2002**, *98*, 58–66. [CrossRef]

73. Ma, F.; Liu, D. 17β-trenbolone, an anabolic-androgenic steroid as well as an environmental hormone, contributes to neurodegeneration. *Toxicol. Appl. Pharmacol.* **2015**, *282*, 68–76. [CrossRef] [PubMed]

74. Cunningham, R.L.; Giuffrida, A.; Roberts, J.L. Androgens induce dopaminergic neurotoxicity via caspase-3-dependent activation of protein kinase Cdelta. *Endocrinology* **2009**, *150*, 5539–5548. [CrossRef] [PubMed]

75. Zup, S.L.; Edwards, N.S.; McCarthy, M.M. Sex- and age-dependent effects of androgens on glutamate-induced cell death and intracellular calcium regulation in the developing hippocampus. *Neuroscience* **2014**, *281*, 77–87. [CrossRef] [PubMed]

76. Pike, C.J.; Nguyen, T.V.; Ramsden, M.; Yao, M.; Murphy, M.P.; Rosario, E.R. Androgen cell signaling pathways involved in neuroprotective actions. *Horm. Behav.* **2008**, *53*, 693–705. [CrossRef] [PubMed]

77. Schumacher, M.; Guennoun, R.; Stein, D.G.; De Nicola, A.F. Progesterone: Therapeutic opportunities for neuroprotection and myelin repair. *Pharmacol. Ther.* **2007**, *116*, 77–106. [CrossRef] [PubMed]

78. Qin, Y.; Chen, Z.; Han, X.; Wu, H.; Yu, Y.; Wu, J.; Liu, S.; Hou, Y. Progesterone attenuates Aβ(25-35)-induced neuronal toxicity via JNK inactivation and progesterone receptor membrane component 1-dependent inhibition of mitochondrial apoptotic pathway. *J. Steroid Biochem. Mol. Biol.* **2015**, *154*, 302–311. [CrossRef] [PubMed]

79. Stekovic, S.; Ruckenstuhl, C.; Royer, P.; Winkler-Hermaden, C.; Carmona-Gutierrez, D.; Fröhlich, K.U.; Kroemer, G.; Madeo, F. The neuroprotective steroid progesterone promotes mitochondrial uncoupling, reduces cytosolic calcium and augments stress resistance in yeast cells. *Microb. Cell* **2017**, *4*, 191–199. [CrossRef] [PubMed]

80. Jackson, A.C.; Roche, S.L.; Byrne, A.M.; Ruiz-Lopez, A.M.; Cotter, T.G. Progesterone receptor signalling in retinal photoreceptor neuroprotection. *J. Neurochem.* **2016**, *136*, 63–77. [CrossRef] [PubMed]

81. Crochemore, C.; Lu, J.; Wu, Y.; Liposits, Z.; Sousa, N.; Holsboer, F.; Almeida, O.F. Direct targeting of hippocampal neurons for apoptosis by glucocorticoids is reversible by mineralocorticoid receptor activation. *Mol. Psychiatry* **2005**, *10*, 790–798. [CrossRef] [PubMed]

82. Rogalska, J. Mineralocorticoid and glucocorticoid receptors in hippocampus: Their impact on neurons survival and behavioral impairment after neonatal brain injury. *Vitam. Horm.* **2010**, *82*, 391–419. [CrossRef] [PubMed]

83. Noguchi, K.K.; Lau, K.; Smith, D.J.; Swiney, B.S.; Farber, N.B. Glucocorticoid receptor stimulation and the regulation of neonatal cerebellar neural progenitor cell apoptosis. *Neurobiol. Dis.* **2011**, *43*, 356–363. [CrossRef] [PubMed]

84. Munier, M.; Law, F.; Meduri, G.; Le Menuet, D.; Lombès, M. Mineralocorticoid receptor overexpression facilitates differentiation and promotes survival of embryonic stem cell-derived neurons. *Endocrinology* **2012**, *153*, 1330–1340. [CrossRef] [PubMed]

85. Cubilla, M.A.; Bermúdez, V.; Marquioni Ramella, M.D.; Bachor, T.P.; Suburo, A.M. Mifepristone, a blocker of glucocorticoid receptors, promotes photoreceptor death. *Investig. Ophthalmol. Vis. Sci.* **2013**, *54*, 313–322. [CrossRef] [PubMed]

86. Nitta, E.; Hirooka, K.; Tenkumo, K.; Fujita, T.; Nishiyama, A.; Nakamura, T.; Itano, T.; Shiraga, F. Aldosterone: A mediator of retinal ganglion cell death and the potential role in the pathogenesis in normal-tension glaucoma. *Cell Death Dis.* **2013**, *4*, e711. [CrossRef] [PubMed]

87. Collins, L.L.; Williamson, M.A.; Thompson, B.D.; Dever, D.P.; Gasiewicz, T.A.; Opanashuk, L.A. 2,3,7,8-Tetracholorodibenzo-p-dioxin exposure disrupts granule neuron precursor maturation in the developing mouse cerebellum. *Toxicol. Sci.* **2008**, *103*, 125–136. [CrossRef] [PubMed]

88. Sánchez-Martín, F.J.; Fernández-Salguero, P.M.; Merino, J.M. Aryl hydrocarbon receptor-dependent induction of apoptosis by 2,3,7,8-tetrachlorodibenzo-p-dioxin in cerebellar granule cells from mouse. *J. Neurochem.* **2011**, *118*, 153–162. [CrossRef] [PubMed]

89. Rzemieniec, J.; Litwa, E.; Wnuk, A.; Lason, W.; Krzeptowski, W.; Kajta, M. Selective Aryl Hydrocarbon Receptor Modulator 3,3′-Diindolylmethane Impairs AhR and ARNT Signaling and Protects Mouse Neuronal Cells Against Hypoxia. *Mol. Neurobiol.* **2016**, *53*, 5591–5606. [CrossRef] [PubMed]

90. Szychowski, K.A.; Wnuk, A.; Kajta, M.; Wójtowicz, A.K. Triclosan activates aryl hydrocarbon receptor (AhR)-dependent apoptosis and affects Cyp1a1 and Cyp1b1 expression in mouse neocortical neurons. *Environ. Res.* **2016**, *151*, 106–114. [CrossRef] [PubMed]

91. Kajta, M.; Wnuk, A.; Rzemieniec, J.; Litwa, E.; Lason, W.; Zelek-Molik, A.; Nalepa, I.; Rogóż, Z.; Grochowalski, A.; Wojtowicz, A.K. Depressive-like effect of prenatal exposure to DDT involves global DNA hypomethylation and impairment of GPER1/ESR1 protein levels but not ESR2 and AHR/ARNT signaling. *J. Steroid Biochem. Mol. Biol.* **2017**, *171*, 94–109. [CrossRef] [PubMed]

92. Litwa, E.; Rzemieniec, J.; Wnuk, A.; Lason, W.; Krzeptowski, W.; Kajta, M. RXRα, PXR and CAR xenobiotic receptors mediate the apoptotic and neurotoxic actions of nonylphenol in mouse hippocampal cells. *J. Steroid Biochem. Mol. Biol.* **2016**, *156*, 43–52. [CrossRef] [PubMed]

93. Wnuk, A.; Rzemieniec, J.; Litwa, E.; Lasoń, W.; Krzeptowski, W.; Wójtowicz, A.K.; Kajta, M. The Crucial Involvement of Retinoid X Receptors in DDE Neurotoxicity. *Neurotox. Res.* **2016**, *29*, 155–172. [CrossRef] [PubMed]

94. Wnuk, A.; Rzemieniec, J.; Lasoń, W.; Krzeptowski, W.; Kajta, M. Benzophenone-3 Impairs Autophagy, Alters Epigenetic Status, and Disrupts Retinoid X Receptor Signaling in Apoptotic Neuronal Cells. *Mol. Neurobiol.* **2017**. [CrossRef] [PubMed]

95. Wojtowicz, A.K.; Szychowski, K.A.; Kajta, M. PPAR-γ agonist GW1929 but not antagonist GW9662 reduces TBBPA-induced neurotoxicity in primary neocortical cells. *Neurotox. Res.* **2014**, *25*, 311–322. [CrossRef] [PubMed]

96. Wojtowicz, A.K.; Szychowski, K.A.; Wnuk, A.; Kajta, M. Dibutyl Phthalate (DBP)-Induced Apoptosis and Neurotoxicity are Mediated via the Aryl Hydrocarbon Receptor (AhR) but not by Estrogen Receptor A (ERα), Estrogen Receptor B (ERβ), or Peroxisome Proliferator-Activated Receptor Gamma (PPARγ) in Mouse Cortical Neurons. *Neurotox. Res.* **2017**, *31*, 77–89. [CrossRef] [PubMed]

97. Lemmen, J.; Tozakidis, I.E.; Bele, P.; Galla, H.J. Constitutive androstane receptor upregulates Abcb1 and Abcg2 at the blood-brain barrier after CITCO activation. *Brain Res.* **2013**, *1501*, 68–80. [CrossRef] [PubMed]

98. Chakraborty, S.; Kanakasabai, S.; Bright, J.J. Constitutive androstane receptor agonist CITCO inhibits growth and expansion of brain tumour stem cells. *Br. J. Cancer* **2011**, *104*, 448–459. [CrossRef] [PubMed]

99. Smith, S.A.; May, F.J.; Monteith, G.R.; Roberts-Thomson, S.J. Activation of the peroxisome proliferator-activated receptor-α enhances cell death in cultured cerebellar granule cells. *J. Neurosci. Res.* **2001**, *66*, 236–341. [CrossRef] [PubMed]

100. German, O.L.; Monaco, S.; Agnolazza, D.L.; Rotstein, N.P.; Politi, L.E. Retinoid X receptor activation is essential for docosahexaenoic acid protection of retina photoreceptors. *J. Lipid Res.* **2013**, *54*, 2236–2246. [CrossRef] [PubMed]

101. Ayala-Peña, V.B.; Pilotti, F.; Volonté, Y.; Rotstein, N.P.; Politi, L.E.; German, O.L. Protective effects of retinoid x receptors on retina pigment epithelium cells. *Biochim. Biophys. Acta* **2016**, *1863 Pt A*, 1134–1145. [CrossRef] [PubMed]

102. Austdal, L.P.; Mathisen, G.H.; Løberg, E.M.; Paulsen, R.E. Calcium-induced apoptosis of developing cerebellar granule neurons depends causally on NGFI-B. *Int. J. Dev. Neurosci.* **2016**, *55*, 82–90. [CrossRef] [PubMed]

103. Zhong, J.; Jiang, L.; Huang, Z.; Zhang, H.; Cheng, C.; Liu, H.; He, J.; Wu, J.; Darwazeh, R.; Wu, Y.; et al. The long non-coding RNA Neat1 is an important mediator of the therapeutic effect of bexarotene on traumatic brain injury in mice. *Brain Behav. Immun.* **2017**, *65*, 183–194. [CrossRef] [PubMed]

104. Okabe, A.; Urano, Y.; Itoh, S.; Suda, N.; Kotani, R.; Nishimura, Y.; Saito, Y.; Noguchi, N. Adaptive responses induced by 24S-hydroxycholesterol through liver X receptor pathway reduce 7-ketocholesterol-caused neuronal cell death. *Redox Biol.* **2013**, *2*, 28–35. [CrossRef] [PubMed]

105. Montarolo, F.; Perga, S.; Martire, S.; Navone, D.N.; Marchet, A.; Leotta, D.; Bertolotto, A. Altered NR4A Subfamily Gene Expression Level in Peripheral Blood of Parkinson's and Alzheimer's Disease Patients. *Neurotox. Res.* **2016**, *30*, 338–344. [CrossRef] [PubMed]

106. Volakakis, N.; Kadkhodaei, B.; Joodmardi, E.; Wallis, K.; Panman, L.; Silvaggi, J.; Spiegelman, B.M.; Perlmann, T. NR4A orphan nuclear receptors as mediators of CREB-dependent neuroprotection. *Proc. Natl. Acad. Sci. USA* **2010**, *107*, 12317–12322. [CrossRef] [PubMed]

107. Shaked, I.; Hanna, R.N.; Shaked, H.; Chodaczek, G.; Nowyhed, H.N.; Tweet, G.; Tacke, R.; Basat, A.B.; Mikulski, Z.; Togher, S.; et al. Transcription factor Nr4a1 couples sympathetic and inflammatory cues in CNS-recruited macrophages to limit neuroinflammation. *Nat. Immunol.* **2015**, *16*, 1228–1234. [CrossRef] [PubMed]

108. Wang, X.J.; Cao, Q.; Zhang, Y.; Su, X.D. Activation and regulation of caspase-6 and its role in neurodegenerative diseases. *Annu. Rev. Pharmacol. Toxicol.* **2015**, *55*, 553–572. [CrossRef] [PubMed]

109. Sari, Y. Huntington's Disease: From Mutant Huntingtin Protein to Neurotrophic Factor Therapy. *Int. J. Biomed. Sci.* **2011**, *7*, 89–100. [PubMed]

110. Gaki, G.S.; Papavassiliou, A.G. Oxidative Stress-Induced Signaling Pathways Implicated in the Pathogenesis of Parkinson's Disease. *Neuromol. Med.* **2014**, *16*, 217–230. [CrossRef] [PubMed]

111. Jazvinšćak Jembrek, M.; Hof, P.R.; Šimić, G. Ceramides in Alzheimer's Disease: Key Mediators of Neuronal Apoptosis Induced by Oxidative Stress and Aβ Accumulation. *Oxid. Med. Cell. Longev.* **2015**, *346783*. [CrossRef] [PubMed]

112. Martinez, A.; Gil, C.; Perez, D.I. Glycogen Synthase Kinase 3 Inhibitors in the Next Horizon for Alzheimer's Disease Treatment. *Int. J. Alzheimer's Dis.* **2011**, *280502*. [CrossRef] [PubMed]

113. Kim, J.; Yang, M.; Kim, S.H.; Kim, J.C.; Wang, H.; Shin, T.; Moon, C. Possible role of the glycogen synthase kinase-3 signaling pathway in trimethyltin-induced hippocampal neurodegeneration in mice. *PLoS ONE* **2013**, *8*, e70356. [CrossRef] [PubMed]

114. Iqbal, K.; Gong, C.X.; Liu, F. Microtubule-associated protein tau as a therapeutic target in Alzheimer's disease. *Expert Opin. Ther. Targets* **2014**, *18*, 307–318. [CrossRef] [PubMed]

115. Morales-García, J.A.; Susís, C.; Alonso-Gil, S.; Pérez, D.I.; Palomo, V.; Pérez, C.; Conde, S.; Santos, A.; Gil, C.; Martínez, A.; et al. Glycogen Synthase Kinase-3 Inhibitors as Potent Therapeutic Agents for the Treatment of Parkinson Disease. *ACS Chem. Neurosci.* **2013**, *4*, 350–360. [CrossRef] [PubMed]

116. Fredriksson, A.; Pontén, E.; Gordh, T.; Eriksson, P. Neonatal Exposure to a Combination of N-Methyl-D-aspartate and γ-Aminobutyric Acid Type A Receptor Anesthetic Agents Potentiates Apoptotic Neurodegeneration and Persistent Behavioral Deficits. *Anesthesiology* **2007**, *107*, 427–436. [CrossRef] [PubMed]

117. Noguchi, K.N. Glucocorticoid Induced Cerebellar Toxicity in the Developing Neonate: Implications for Glucocorticoid Therapy during Bronchopulmonary Dysplasia. *Cells* **2014**, *3*, 36–52. [CrossRef] [PubMed]

118. Mizushima, N.; Levine, B. Autophagy in mammalian development and differentiation. *Nat. Cell Biol.* **2010**, *12*, 823–830. [CrossRef] [PubMed]

119. Ciechanover, A.; Kwon, Y.T. Degradation of misfolded proteins in neurodegenerative diseases: Therapeutic targets and strategies. *Exp. Mol. Med.* **2015**, *47*, e147. [CrossRef] [PubMed]

120. Inoue, K.; Rispoli, J.; Kaphzan, H.; Klann, E.; Chen, E.I.; Kim, J.; Komatsu, M.; Abeliovich, A. Macroautophagy deficiency mediates age-dependent neurodegeneration through a phospho-tau pathway. *Mol. Neurodegener.* **2012**, *7*, 48. [CrossRef] [PubMed]

121. Cook, C.; Stetler, C.; Petrucelli, L. Disruption of Protein Quality Control in Parkinson's Disease. *Cold Spring Harb. Perspect. Med.* **2012**, *2*, a009423. [CrossRef] [PubMed]

122. Liang, P.; Le, W. Role of autophagy in the pathogenesis of multiple sclerosis. *Neurosci. Bull.* **2015**, *31*, 435–444. [CrossRef] [PubMed]

123. Meissner, F.; Molawi, K.; Zychlinsky, A. Mutant superoxide dismutase 1-induced IL-1β accelerates ALS pathogenesis. *Proc. Natl. Acad. Sci. USA* **2010**, *107*, 13046–13050. [CrossRef] [PubMed]

124. Su, P.; Zhang, J.; Wang, D.; Zhao, F.; Cao, Z.; Aschner, M.; Luo, W. The role of autophagy in modulation of neuroinflammation in microglia. *Neuroscience* **2016**, *319*, 155–167. [CrossRef] [PubMed]

125. Chang, C.F.; Huang, H.J.; Lee, H.C.; Hung, K.C.; Wu, R.T.; Lin, A.M. Melatonin attenuates kainic acid-induced neurotoxicity in mouse hippocampus via inhibition of autophagy and α-synuclein aggregation. *J. Pineal Res.* **2012**, *52*, 312–321. [CrossRef] [PubMed]

126. Ginet, V.; Spiehlmann, A.; Rummel, C.; Rudinskiy, N.; Grishchuk, Y.; Luthi-Carter, R.; Clarke, P.G.H.; Truttmann, A.C.; Puyal, J. Involvement of autophagy in hypoxic-excitotoxic neuronal death. *Autophagy* **2014**, *10*, 846–860. [CrossRef] [PubMed]

127. Takei, N.; Nawa, H. mTOR signaling and its roles in normal and abnormal brain development. *Front. Mol. Neurosci.* **2014**, *7*, 28. [CrossRef] [PubMed]

128. Jia, J.; Le, W. Molecular network of neuronal autophagy in the pathophysiology and treatment of depression. *Neurosci. Bull.* **2015**, *31*, 427–434. [CrossRef] [PubMed]

129. Nilsen, J. Estradiol and neurodegenerative oxidative stress. *Front. Neuroendocrinol.* **2008**, *29*, 463–475. [CrossRef] [PubMed]

130. Pines, A. Surgical menopause and cognitive decline. *Climacteric* **2014**, *17*, 580–582. [CrossRef] [PubMed]

131. Scott, J.S.; Goldberg, F.W.; Turnbull, A.V. Medicinal chemistry of inhibitors of 11β-hydroxysteroid dehydrogenase type 1 (11β-HSD1). *J. Med. Chem.* **2014**, *57*, 4466–4486. [CrossRef] [PubMed]

132. Winkler, J.M.; Fox, H.S. Transcriptome meta-analysis reveals a central role for sex steroids in the degeneration of hippocampal neurons in Alzheimer's disease. *BMC Syst. Biol.* **2013**, *7*, 51. [CrossRef] [PubMed]

133. Morissette, M.; Al Sweidi, S.; Callier, S.; di Paolo, T. Estrogen and SERM neuroprotection in animal models of Parkinson's disease. *Mol. Cell Endocrinol.* **2008**, *290*, 60–69. [CrossRef] [PubMed]

134. Biewenga, E.; Cabell, L.; Audesirk, T. Estradiol and raloxifene protect cultured SN4741 neurons against oxidative stress. *Neurosci. Lett.* **2005**, *373*, 179–183. [CrossRef] [PubMed]

135. Du, B.; Ohmichi, M.; Takahashi, K.; Kawagoe, J.; Ohshima, C.; Igarashi, H.; Mori-Abe, A.; Saitoh, M.; Ohta, T.; Ohishi, A.; et al. Both estrogen and raloxifene protect against β-amyloid-induced neurotoxicity in estrogen receptor α-transfected PC12 cells by activation of telomerase activity via Akt cascade. *J. Endocrinol.* **2004**, *183*, 605–615. [CrossRef] [PubMed]

136. Martin, D.; Song, J.; Mark, C.; Eyster, K. Understanding the cardiovascular actions of soy isoflavones: Potential novel targets for antihypertensive drug development. *Cardiovasc. Hematol. Disord. Drug Targets* **2008**, *8*, 297–312. [CrossRef] [PubMed]

137. Bourque, M.; Morissette, M.; di Paolo, T. Raloxifene activates G protein-coupled estrogen receptor 1/Akt signaling to protect dopamine neurons in 1-methyl-4-phenyl-1,2,3,6-tetrahydropyridine mice. *Neurobiol. Aging* **2014**, *35*, 2347–2356. [CrossRef] [PubMed]

138. Chakrabarti, M.; Ray, S.K. Experimental Procedures for Demonstration of MicroRNA Mediated Enhancement of Functional Neuroprotective Effects of Estrogen Receptor Agonists. *Methods Mol. Biol.* **2016**, *1366*, 359–372. [CrossRef] [PubMed]

139. Pisano, A.; Preziuso, C.; Iommarini, L.; Perli, E.; Grazioli, P.; Campese, A.F.; Maresca, A.; Montopoli, M.; Masuelli, L.; Sadun, A.A.; et al. Targeting estrogen receptor β as preventive therapeutic strategy for Leber's hereditary optic neuropathy. *Hum. Mol. Genet.* **2015**, *24*, 6921–6931. [CrossRef] [PubMed]

140. Itoh, N.; Kim, R.; Peng, M.; DiFilippo, E.; Johnsonbaugh, H.; MacKenzie-Graham, A.; Voskuhl, R.R. Bedside to bench to bedside research: Estrogen receptor β ligand as a candidate neuroprotective treatment for multiple sclerosis. *J. Neuroimmunol.* **2017**, *304*, 63–71. [CrossRef] [PubMed]

141. Adams, S.M.; Aksenova, M.V.; Aksenov, M.Y.; Mactutus, C.F.; Booze, R.M. Soy isoflavones genistein and daidzein exert anti-apoptotic actions via a selective ER-mediated mechanism in neurons following HIV-1 Tat(1-86) exposure. *PLoS ONE* **2012**, *7*, e37540. [CrossRef] [PubMed]

142. Lu, J.; Wu, D.M.; Zheng, Y.L.; Hu, B.; Cheng, W.; Zhang, Z.F. Purple sweet potato color attenuates domoic acid-induced cognitive deficits by promoting estrogen receptor-α-mediated mitochondrial biogenesis signaling in mice. *Free Radic. Biol. Med.* **2012**, *52*, 646–659. [CrossRef] [PubMed]

143. Rodríguez-Navarro, J.A.; Solano, R.M.; Casarejos, M.J.; Gomez, A.; Perucho, J.; de Yébenes, J.G.; Mena, M.A. Gender differences and estrogen effects in parkin null mice. *J. Neurochem.* **2008**, *106*, 2143–2157. [CrossRef] [PubMed]

144. Liu, H.; Yang, L.; Zhao, Y.; Zeng, G.; Wu, Y.; Chen, Y.; Zhang, J.; Zeng, Q. Estrogen is a novel regulator of Tnfaip1 in mouse hippocampus. *Int. J. Mol. Med.* **2014**, *34*, 219–227. [CrossRef] [PubMed]

145. Liang, X.F.; Fang, C.; Ma, Y.N.; Guan, X.; Liu, Y.; Han, C.; Liu, J.; Zou, W. Relationship between ER-α36 and Akt in PC12 cells exposed to glucose deprivation. *Sheng Li Xue Bao* **2013**, *65*, 381–388. [PubMed]

146. Sharma, D.; Biswal, S.N.; Kumar, K.; Bhardwaj, P.; Barhwal, K.K.; Kumar, A.; Hota, S.K.; Chaurasia, O.P. Estrogen Receptor β Mediated Neuroprotective Efficacy of Cicer microphyllum Seed Extract in Global Hypoxia. *Neurochem. Res.* **2017**. [CrossRef] [PubMed]

147. Peri, A.; Serio, M. Estrogen receptor-mediated neuroprotection: The role of the Alzheimer's disease-related gene seladin-1. *Neuropsychiatr. Dis. Treat.* **2008**, *4*, 817–824. [CrossRef] [PubMed]

148. Jayaraman, A.; Christensen, A.; Moser, V.A.; Vest, R.S.; Miller, C.P.; Hattersley, G.; Pike, C.J. Selective androgen receptor modulator RAD140 is neuroprotective in cultured neurons and kainate-lesioned male rats. *Endocrinology* **2014**, *155*, 1398–1406. [CrossRef] [PubMed]

149. Li, X.B.; Guo, H.L.; Shi, T.Y.; Yang, L.; Wang, M.; Zhang, K.; Guo, Y.Y.; Wu, Y.M.; Liu, S.B.; Zhao, M.G. Neuroprotective effects of a novel translocator protein (18 kDa) ligand, ZBD-2, against focal cerebral ischemia and NMDA-induced neurotoxicity. *Clin. Exp. Pharmacol. Physiol.* **2015**, *42*, 1068–1074. [CrossRef] [PubMed]

150. Li, Z.Y.; Li, Q.Z.; Chen, L.; Chen, B.D.; Zhang, C.; Wang, X.; Li, W.P. HPOB, an HDAC6 inhibitor, attenuates corticosterone-induced injury in rat adrenal pheochromocytoma PC12 cells by inhibiting mitochondrial GR translocation and the intrinsic apoptosis pathway. *Neurochem. Int.* **2016**, *99*, 239–251. [CrossRef] [PubMed]

151. Gesmundo, I.; Villanova, T.; Gargantini, E.; Arvat, E.; Ghigo, E.; Granata, R. The Mineralocorticoid Agonist Fludrocortisone Promotes Survival and Proliferation of Adult Hippocampal Progenitors. *Front. Endocrinol.* **2016**, *7*, 66. [CrossRef] [PubMed]

152. De Miranda, B.R.; Popichak, K.A.; Hammond, S.L.; Miller, J.A.; Safe, S.; Tjalkens, R.B. Novel para-phenyl substituted diindolylmethanes protect against MPTP neurotoxicity and suppress glial activation in a mouse model of Parkinson's disease. *Toxicol. Sci.* **2015**, *143*, 360–373. [CrossRef] [PubMed]

153. Schneider, A.J.; Branam, A.M.; Peterson, R.E. Intersection of AHR and Wnt Signaling in Development, Health, and Disease. *Int. J. Mol. Sci.* **2014**, *15*, 17852–17885. [CrossRef] [PubMed]

154. Fong, W.H.; Tsai, H.D.; Chen, Y.C.; Wu, J.S.; Lin, T.N. Anti-apoptotic actions of PPAR-γ against ischemic stroke. *Mol. Neurobiol.* **2010**, *41*, 180–186. [CrossRef] [PubMed]

155. Falcone, R.; Florio, T.M.; di Giacomo, E.; Benedetti, E.; Cristiano, L.; Antonosante, A.; Fidoamore, A.; Massimi, M.; Alecci, M.; Ippoliti, R.; et al. PPARβ/δ and γ in a rat model of Parkinson's disease: Possible involvement in PD symptoms. *J. Cell Biochem.* **2015**, *116*, 844–855. [CrossRef] [PubMed]

156. Chiang, M.C.; Nicol, C.J.; Cheng, Y.C.; Lin, K.H.; Yen, C.H.; Lin, C.H. Rosiglitazone activation of PPARγ-dependent pathways is neuroprotective in human neural stem cells against amyloid-β-induced mitochondrial dysfunction and oxidative stress. *Neurobiol. Aging* **2016**, *40*, 181–190. [CrossRef] [PubMed]

157. Colle, R.; de Larminat, D.; Rotenberg, S.; Hozer, F.; Hardy, P.; Verstuyft, C.; Fève, B.; Corruble, E. PPAR-γ Agonists for the Treatment of Major Depression: A Review. *Pharmacopsychiatry* **2017**, *50*, 49–55. [CrossRef] [PubMed]

158. Certo, M.; Endo, Y.; Ohta, K.; Sakurada, S.; Bagetta, G.; Amantea, D. Activation of RXR/PPARγ underlies neuroprotection by bexarotene in ischemic stroke. *Pharmacol. Res.* **2015**, *102*, 298–307. [CrossRef] [PubMed]

159. Waters, A.M.; Stewart, J.E.; Atigadda, V.R.; Mroczek-Musulman, E.; Muccio, D.D.; Grubbs, C.J.; Beierle, E.A. Preclinical Evaluation of a Novel RXR Agonist for the Treatment of Neuroblastoma. *Mol. Cancer Ther.* **2015**, *14*, 1559–1569. [CrossRef] [PubMed]

160. Bachmeier, C.; Beaulieu-Abdelahad, D.; Crawford, F.; Mullan, M.; Paris, D. Stimulation of the retinoid X receptor facilitates β-amyloid clearance across the blood-brain barrier. *J. Mol. Neurosci.* **2013**, *49*, 270–276. [CrossRef] [PubMed]

161. Koster, K.P.; Smith, C.; Valencia-Olvera, A.C.; Thatcher, G.R.; Tai, L.M.; LaDu, M.J. Rexinoids as Therapeutics for Alzheimer's Disease: Role of APOE. *Curr. Top. Med. Chem.* **2017**, *17*, 708–720. [CrossRef] [PubMed]

162. Mariani, M.M.; Malm, T.; Lamb, R.; Jay, T.R.; Neilson, L.; Casali, B.; Medarametla, L.; Landreth, G.E. Neuronally-directed effects of RXR activation in a mouse model of Alzheimer's disease. *Sci. Rep.* **2017**, *7*, 42270. [CrossRef] [PubMed]

163. McFarland, K.; Spalding, T.A.; Hubbard, D.; Ma, J.N.; Olsson, R.; Burstein, E.S. Low dose bexarotene treatment rescues dopamine neurons and restores behavioral function in models of Parkinson's disease. *ACS Chem. Neurosci.* **2013**, *4*, 1430–1438. [CrossRef] [PubMed]

164. Li, L.; Chen, J.; Sun, S.; Zhao, J.; Dong, X.; Wang, J. Effects of Estradiol on Autophagy and Nrf-2/ARE Signals after Cerebral Ischemia. *Cell. Physiol. Biochem.* **2017**, *41*, 2027–2036. [CrossRef] [PubMed]

165. Graham, C.D.; Kaza, N.; Klocke, B.J.; Gillespie, G.Y.; Shevde, L.A.; Carroll, S.L.; Roth, K.A. Tamoxifen Induces Cytotoxic Autophagy in Glioblastoma. *J. Neuropathol. Exp. Neurol.* **2016**, *75*, 946–954. [CrossRef] [PubMed]

166. Felzen, V.; Hiebel, C.; Koziollek-Drechsler, I.; Reißig, S.; Wolfrum, U.; Kögel, D.; Brandts, C.; Behl, C.; Morawe, T. Estrogen receptor α regulates non-canonical autophagy that provides stress resistance to neuroblastoma and breast cancer cells and involves BAG3 function. *Cell Death Dis.* **2015**, *6*, e1812. [CrossRef] [PubMed]

167. Li, X.Z.; Sui, C.Y.; Chen, Q.; Chen, X.P.; Zhang, H.; Zhou, X.P. Upregulation of cell surface estrogen receptor α is associated with the mitogen-activated protein kinase/extracellular signal-regulated kinase activity and promotes autophagy maturation. *Int. J. Clin. Exp. Pathol.* **2015**, *8*, 8832–8841. [PubMed]

168. Xiong, Y.S.; Liu, F.F.; Liu, D.; Huang, H.; Wei, N.; Tan, L.; Chen, J.G.; Man, H.Y.; Gong, C.X.; Lu, Y.; et al. Opposite effects of two estrogen receptors on tau phosphorylation through disparate effects on the miR-218/PTPA pathway. *Aging Cell* **2015**, *14*, 867–877. [CrossRef] [PubMed]

169. Micheli, F.; Palermo, R.; Talora, C.; Ferretti, E.; Vacca, A.; Napolitano, M. Regulation of proapoptotic proteins Bak1 and p53 by miR-125b in an experimental model of Alzheimer's disease: Protective role of 17β-estradiol. *Neurosci. Lett.* **2016**, *629*, 234–240. [CrossRef] [PubMed]

170. Herzog, R.; Zendedel, A.; Lammerding, L.; Beyer, C.; Slowik, A. Impact of 17β-estradiol and progesterone on inflammatory and apoptotic microRNA expression after ischemia in a rat model. *J. Steroid Biochem. Mol. Biol.* **2017**, *167*, 126–134. [CrossRef] [PubMed]

171. Wang, Y.; Dong, X.; Li, Z.; Wang, W.; Tian, J.; Chen, J. Downregulated RASD1 and upregulated miR-375 are involved in protective effects of calycosin on cerebral ischemia/reperfusion rats. *J. Neurol. Sci.* **2014**, *339*, 144–148. [CrossRef] [PubMed]

172. Siegel, C.; Li, J.; Liu, F.; Benashski, S.E.; McCullough, L.D. miR-23a regulation of X-linked inhibitor of apoptosis (XIAP) contributes to sex differences in the response to cerebral ischemia. *Proc. Natl. Acad. Sci. USA* **2011**, *108*, 11662–11667. [CrossRef] [PubMed]

173. Ma, Q.; Dasgupta, C.; Li, Y.; Bajwa, N.M.; Xiong, F.; Harding, B.; Hartman, R.; Zhang, L. Inhibition of microRNA-210 provides neuroprotection in hypoxic-ischemic brain injury in neonatal rats. *Neurobiol. Dis.* **2016**, *89*, 202–212. [CrossRef] [PubMed]

174. Kawashima, H.; Numakawa, T.; Kumamaru, E.; Adachi, N.; Mizuno, H.; Ninomiya, M.; Kunugi, H.; Hashido, K. Glucocorticoid attenuates brain-derived neurotrophic factor-dependent upregulation of glutamate receptors via the suppression of microRNA-132 expression. *Neuroscience* **2010**, *165*, 1301–1311. [CrossRef] [PubMed]

175. Trivellin, G.; Butz, H.; Delhove, J.; Igreja, S.; Chahal, H.S.; Zivkovic, V.; McKay, T.; Patócs, A.; Grossman, A.B.; Korbonits, M. MicroRNA miR-107 is overexpressed in pituitary adenomas and inhibits the expression of aryl hydrocarbon receptor-interacting protein in vitro. *Am. J. Physiol. Endocrinol. Metab.* **2012**, *303*, E708–E719. [CrossRef] [PubMed]

176. Takwi, A.A.; Wang, Y.M.; Wu, J.; Michaelis, M.; Cinatl, J.; Chen, T. miR-137 regulates the constitutive androstane receptor and modulates doxorubicin sensitivity in parental and doxorubicin-resistant neuroblastoma cells. *Oncogene* **2014**, *33*, 3717–3729. [CrossRef] [PubMed]

International Journal of
Molecular Sciences

MDPI

Review

Phytochemicals Targeting Estrogen Receptors: Beneficial Rather Than Adverse Effects?

Sylvain Lecomte, Florence Demay, François Ferrière and Farzad Pakdel *

Institut de Recherche en Santé-Environnement-Travail (IRSET), UMR 1085 Inserm, TREC Team,
University of Rennes 1, 35000 Rennes, France; sylvain.lecomte@univ-rennes1.fr (S.L.);
florence.demay@univ-rennes1.fr (F.D.); francois.ferriere@univ-rennes1.fr (F.F.)
* Correspondence: farzad.pakdel@univ-rennes1.fr; Tel.: +33-223-235-132; Fax: +33-223-235-055

Received: 4 May 2017; Accepted: 24 June 2017; Published: 28 June 2017

Abstract: In mammals, the effects of estrogen are mainly mediated by two different estrogen receptors, ERα and ERβ. These proteins are members of the nuclear receptor family, characterized by distinct structural and functional domains, and participate in the regulation of different biological processes, including cell growth, survival and differentiation. The two estrogen receptor (ER) subtypes are generated from two distinct genes and have partially distinct expression patterns. Their activities are modulated differently by a range of natural and synthetic ligands. Some of these ligands show agonistic or antagonistic effects depending on ER subtype and are described as selective ER modulators (SERMs). Accordingly, a few phytochemicals, called phytoestrogens, which are synthesized from plants and vegetables, show low estrogenic activity or anti-estrogenic activity with potentially anti-proliferative effects that offer nutraceutical or pharmacological advantages. These compounds may be used as hormonal substitutes or as complements in breast cancer treatments. In this review, we discuss and summarize the in vitro and in vivo effects of certain phytoestrogens and their potential roles in the interaction with estrogen receptors.

Keywords: estrogen receptor; ligand; xenoestrogens; selective estrogen receptor modulators; transcription; epigenetic regulation; cell signaling; cancer

1. Introduction

Estrogens, such as 17 β-estradiol (E2), are steroid hormones derived from cholesterol by the successive action of steroidogenic enzymes. They are involved in multiple physiological processes by acting on various tissues. In particular, they participate in the establishment and regulation of the reproductive organs in both males and females, including the gonads or the mammary gland [1]. Furthermore, estrogens participate in many physiological processes in non-reproductive tissues, such as growth and remodeling of bone, differentiation and protection of the central nervous system, vasodilation of cardiovascular systems and lipid metabolism in the liver [1,2]. At the cellular level, E2 has multiple effects, including proliferation, differentiation and survival. E2 is a small, liposoluble molecule that passively enters the cell through the plasma membrane. E2 actions are mainly mediated by their binding to two estrogen receptors, ERα and ERβ, which are localized in the cytoplasm and in the nucleus (Figure 1B). These receptors are members of the nuclear receptor superfamily, which also includes receptors for androgens, progesterone, glucocorticoids, thyroids, retinoid acids, and vitamin D, as well as more than twenty orphan receptors (Figure 1A) [3]. Many nuclear receptors are activated by specific ligands and generally act as transcription factors by binding to specific DNA sequences in the genome. Similar to the other nuclear receptors, ERs are modular proteins that consist of distinct structural and functional domains. The N-terminal domain contains the ligand-independent transactivation function (AF1). The central domain contains the conserved zinc finger DNA-binding-domain, and the C-terminal domain contains the ligand-dependent transactivation function (AF2), as well as the ligand binding and dimerization sequences (Figure 1).

ER-mediated E2 actions at the transcriptional level of the estrogen-sensitive genes are called "genomic" E2 actions. Direct binding of ERs to the chromatin occurs at the estrogen-responsive-element (ERE) at target gene promoters. This induces the mobilization of the transcription coregulators needed to modify chromatin structure and thereby transcriptional regulation of specific gene (Figure 1B). This represents the classical pathway, but many E2-target genes do not contain the ERE. In this case, ERs modulate transcription through DNA-binding sites for Sp1 (stimulating protein 1) or AP1 (activator protein 1) transcription factors [4]. Furthermore, genome-wide studies performed by ChIP (chromatin immunoprecipitation) experiments in breast cancer cell lines specified that ER preferentially regulates its target genes by binding distal regulatory elements [5]. These distal regulatory sites can interact with the promoters of E2 target genes due to chromatin looping. This mechanism of transcriptional regulation represents more than 90% of E2-target genes [5]. Interestingly, these regulatory elements are capable of interacting with several promoters and other enhancers at the same time and are mainly contained in genomic areas called TADs (topologically associating domains) [6,7].

In contrast to the genomic action, the non-genomic actions of estrogens involve cytoplasmic signaling pathways (Figure 1B) and occur rapidly, on the order of seconds or minutes. This leads to the activation of several intracellular signaling pathways such as MAPK (mitogen activated protein kinase) or PI3K (phosphatidylinositide 3-kinase) [8]. Recent studies reported convergence or cross-talk between the genomic and non-genomic actions of ER, enabling a fine regulation of target genes and increasing the complexity of the estrogenic pathways [9–12].

ERs are generated from two different genes that are localized on chromosome 6, for ERα, and chromosome 14, for ERβ, in humans. The utilization of different promoters and splicing processes results in multiple ER variants that can interfere with the transcriptional activity of wild type ERs in various cell types [13–16].

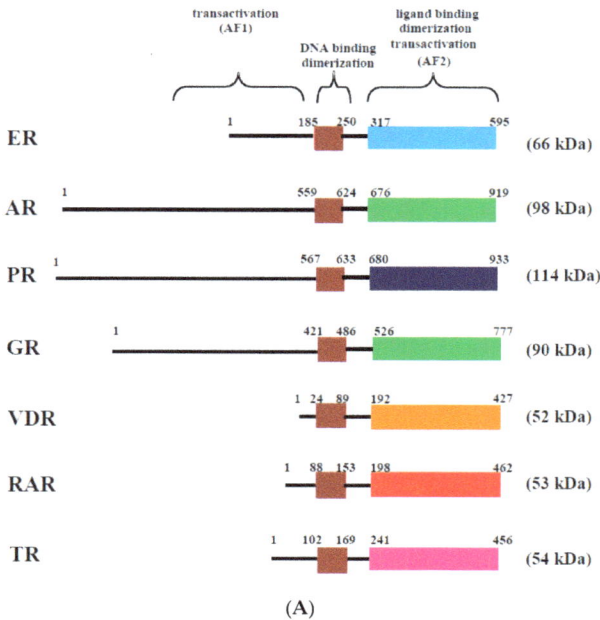

(A)

Figure 1. *Cont.*

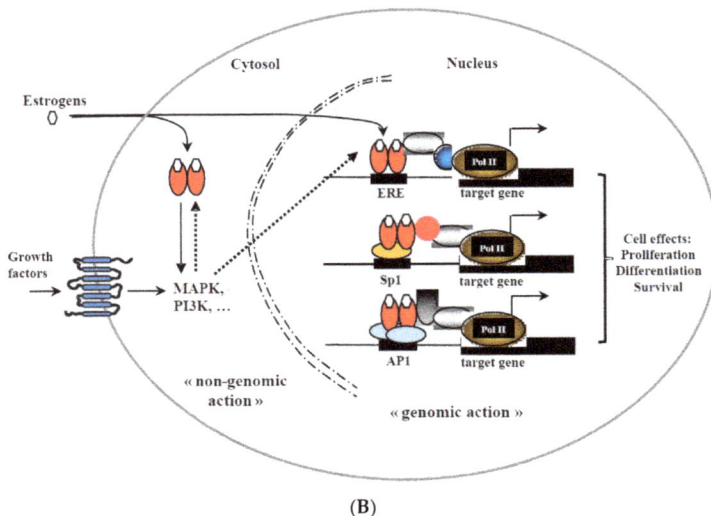

(B)

Figure 1. Structure and mechanisms of action of the estrogen receptor (ER). (**A**) The evolutionarily conserved domains of several nuclear receptors, including ER, AR (androgen receptor), PR (progesterone receptor), GR (glucocorticoid receptor), VDR (vitamin D receptor), RAR (retinoid acid receptor) and TR (thyroid receptor). Domains involved in DNA and ligand binding, as well as in dimerization, ligand-independent transactivation function (AF1) and ligand-dependent transactivation function (AF2) are shown. The number of amino acids for each domain is presented. The approximate molecular weight of each nuclear receptor is also indicated on the right side; (**B**) estradiol (E2) mediates multiple phenotypic changes in cells by binding to its receptor. E2 enter the cell through the lipid membranes and binds ER in the cytoplasm or the nucleus. ER mediates E2 effects through diverse transcriptional mechanisms. In the nucleus, the activated ER forms a dimer to tightly fix DNA directly at the ERE sites or indirectly at Sp1 or Ap1 sites. The activated ER is then able to recruit cofactors and RNA polymerase II (pol. II), which allows the transcription of target genes (ER genomic action). Furthermore, ERs can use rapid non-genomic action through the activation of intracellular kinases related or not to the growth factor signaling.

Many tissues express both ER subtypes, but with variable expression profiles. For instance, ERα is highly expressed in female reproductive tissues (ovary, womb, mammary gland). ERβ is greatly expressed in ovaries but poorly expressed in the mammary gland. In men, ERα is strongly expressed in the testicle (Leydig cells and gubernaculum), whereas ERβ is found in the prostate, germinal cells and epididymis. On the other hand, both receptors with variable expression levels are found in male and female, lung, hepatic, fat, osseous, and nervous tissues and endothelial cells [1,17–19]. Knockout in mice demonstrated crucial roles for both ERα and ERβ during the development of reproductive tissues, gametogenesis, and neuronal growth and differentiation [1,20,21]. The appearance of ER seems to be under spatio-temporal control during development [1,19,20]. For instance, ERα expression has been found in the developing uterus as soon as the 15th day of fetal in mesenchymal cells, while it appears later in the epithelial cells and it rises during the neonatal period. In the cerebral cortex of rodents, ERα expression is greater in postnatal life and decreases substantially during puberty [20]. In the testis under development, ERα is expressed in the gubernaculum, a ligament which differentiates into the cremaster muscle involved in the final positioning of the testis within the scrotum. Its expression is strong between 17 to 20 dpc and barely detectable between 4 and 12 dpp, indicating a role of estrogens and ERα in the right positioning of the testis [19]. However, during mouse brain development, ERβ distribution varies in different areas. ERβ is found mostly in the midbrain and hypothalamus at E12.5, and its expression increases at E15.5 and E16.5. Interestingly, ERβ expression appears intensely and extensively throughout the brain, including in the cerebellum and striatum, at E18.5, whereas very few positive cells may be distinguished in the ventricular region [21].

Many natural and synthetic chemicals in the environment and in food have been reported with hormonal activity, particularly showing estrogenic potency [22]. These compounds are called endocrine disrupting chemicals (EDCs). A lot of EDCs are generated from human activities. For example, polycyclic aromatic hydrocarbons (PAH), such as polychlorinated dibenzo-*p*-dioxins and dibenzofurans, or polychlorinated biphenyls (PCBs), which are the most persistent and widespread in the environment. Bisphenol A, nonylphenol and ethinyl estradiol were also reported to be among the major environmental estrogens. A series of experimental and epidemiological studies over the past decades have suggested that these environmental contaminants can interfere with normal hormonal processes and induce deterioration of the reproduction function in males and females [22–27].

Furthermore, numerous natural molecules present in vegetables and plants possess estrogen- and antiestrogen-mimetic activities. These natural molecules are mainly phytoestrogen isoflavones, the most widely consumed. The most abundant isoflavones are genistein and daidzein, which are present in soybean, in particular, and also found in certain fruits, legumes and nuts [28]. Flavones, such as coumestans and lignans are other classes of phytoestrogens, which are also found in certain fruits and legumes [28]. Moreover, some mushrooms, mosses and fungi produce compounds with estrogenic activity. These compounds are called mycoestrogens, such as zearalenone [28]. In this review, we focus on these phytochemicals interacting with ERs and discuss their molecular actions and their potential effects on human health.

2. Structure and Sources of the Major Dietary Phytoestrogens

Against environmental stresses and aggressions, plants produce secondary metabolites belonging to the large family of polyphenols, which have many biological activities, such as antioxidant, antifungal and antibiotic properties. All of these compounds contain one or several aromatic rings with at least one hydroxyl group. Hydroxyl groups can be free, but most of the time they are engaged in another function with an ester, ether or a glycoside. Among these compounds, phytoestrogens have a structural similarity with 17β-estradiol and could bind both ERs. Phytoestrogens are classified into six groups based on their chemical structures (Figure 2). In this review, we have chosen to present only the aglycone structure of a few phytoestrogens.

2.1. Flavonoids

Flavonoids, from *flavus* (yellow in Latin), are pigments of flowers and fruit, and represent the major group. They are formed by 2 aromatic rings bearing at least one hydroxyl group. The aromatic rings, called A and B, are connected by a carbon bridge consisting of three carbons combined with an oxygen to carbons of the A ring. Together, they formed a new 6-ring structure, called C [29] (Figure 2). Flavonoids could be divided into sub-classes depending on the position of the B ring at position 2 for flavones and derivatives and at position 3 for isoflavones and derivatives. Moreover, depending on hydroxylation degree and/or the position of the hydroxyl group, one can distinguish the flavan-3-ols, the flavanones and the flavonols [29].

Here, we have focused on flavones and isoflavones. Flavones are represented by compounds, such as apigenin, found in parsley or chamomile. Apigenin has a beneficial effect on human health [30]. The daily intake of flavones is very low and estimated between 0.3 and 1.6 mg/day [31]. Isoflavones such as genistein or daidzein are found in large quantities in soybean. The daily intake of isoflavones is low in Western countries (0.1–1.2 mg/day) and higher in Asian countries, where they consume more soy product (up to 47 mg/day) [29,32]. Approximately 30% of the population in Western countries and 60% of the population in Asian countries possess gut microbiota able to metabolize daidzein into the isoflavan equol, which shows a greater affinity for ERs than daidzein. Equol exists through two enantiomers, the R-(+) equol and the S-(−) equol. This latter enantiomer is the natural compound produced by microbiota in human and rat [33].

Class	Subclass	Compounds
Hormone	Natural ligand of estrogen receptor: 17β-estradiol	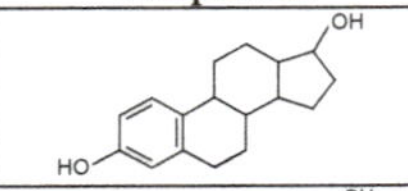
Flavonoids	Flavone Ex: apigenin	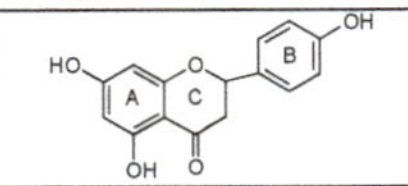
	Flavanone Ex: liquiritigenin	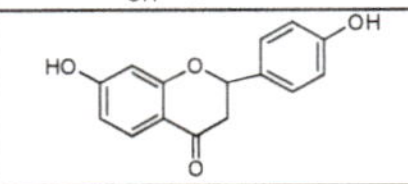
	Isoflavones Ex: daidzein Ex: genistein	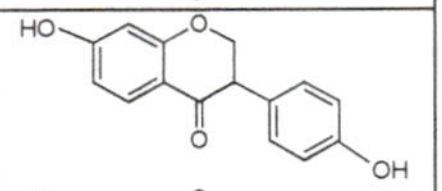 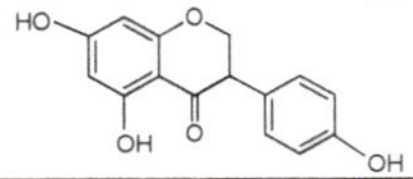
	Isoflavanes Ex: equol	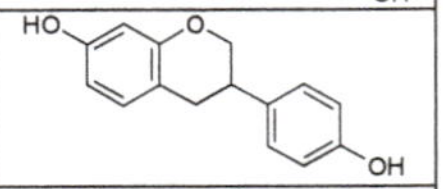
Pterocarpan	Ex: glyceollin I	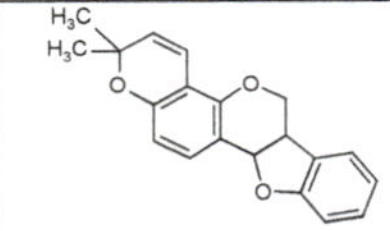
Coumestan	Ex: coumestrol	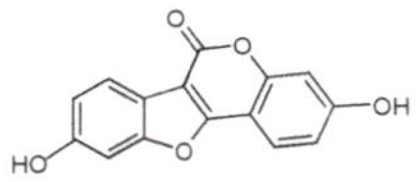
Stilbene	Ex: resveratrol	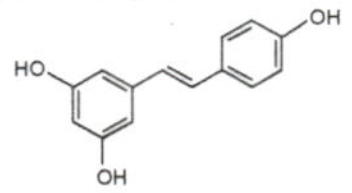
Enterolignan	Ex: enterolactone	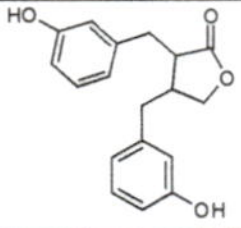
Mycotoxin	Ex: zearalenone	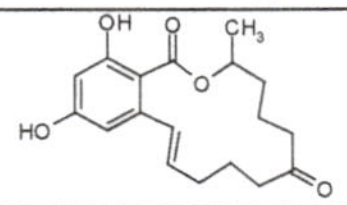

Figure 2. Illustration of the chemical structures of different groups of phytoestrogens. Ex: Example.

2.2. Pterocarpans

Pterocarpans derive from isoflavones. Their structure is described as a benzo-pyrano-furano-benzene, where the B-rings are coupled to position 4-one [34]. Glyceollins, which correspond to prenylated 6a-hydroxy pterocarpans, are the main delegates of this family [35]. They belong to phytoalexin and are produced from daidzein via an enzymatic pathway, mainly in soybean, by the action of a diversity of elicitors such as UV stress, bacterial or fungi infection [36]. These compounds have been known since the 1970s for their involvement in plant defense [37], but the ability of these compounds to act as phytoestrogens was only established in 2000 [38].

2.3. Coumestans

Coumestans are produced by oxidation of pterocarpan [39]. The structure of coumestans consists of a benzoxazole fused to a chromen-2-one [40]. The first, discovered in 1957 by Bickoff et al., and the best documented is coumestrol, which is abundant in alfalfa, soybean and clover [41]. This compound was shown to have a high affinity for both ERs and to induce a response of the same magnitude as that observed with E2 [42].

2.4. Stilbenes

Like glyceollins, stilbene belongs to the phytoalexins and participates in plant defense against injury, stress or infection [29]. Resveratrol, the main representative of the stilbene family, is abundant in grape and red wine, with a concentration up to 12 mg/L [43]. Although resveratrol was reported to interact with ERs, its agonist or antagonistic effects remain controversial [44,45].

2.5. Lignans

The two best-documented lignans are secoisolariciresinol and matairesinol [46]. Lignans are particularly abundant in flaxseed and sesame seed, and at minor concentrations in cereals, vegetables and fruits. The two major metabolites of lignans in human are also produced by gut microbiota. They present a weak estrogenic action and are called enterolactone and enterodiol [47].

2.6. Mycoestrogens

Another family of dietary estrogen, called mycoestrogens, is produced by fungi. In this family, the most documented is zearalenone and its derivatives. Zearalenone is produced by *Fusarium* and is found in poorly stored cereals. Zearalenone structure consists of resorcinol moiety fused with a 14-member macrocyclic lactone [48]. According to the European Safety Authority (EFSA), zearalenone is found in 15% of cereals consumed in Europe [49]. Zearalenone has adverse effects on human health, including reprotoxicity [50,51], genotoxicity, and oxidative stress [49]. This chemical and its metabolites, particularly α-zearalenol, which is used as growth promoter in cattle, are able to bind ERs with high affinity and act as strong ERα agonists [51].

3. In Vitro Effects of Phytoestrogens

The proliferation of ERα-positive breast cancer cells is enhanced by estrogens, which induce multiple growth factors, cyclins and cytokines involved in cell survival and cell cycle progression. Although ERα has a proliferative effect, ERβ acts as a negative regulator of ERα in breast cancer cells, counteracting the mitogenic effect of estrogens [15,52–54]. Interestingly, in many reported ER-selective bioassays, such as the proliferation of breast cancer cell lines, gene reporter assays in mammalian or non-mammalian cells, and ER binding assays, it was found that most phytoestrogens preferentially interact with ERβ and display high specificity toward ERβ transactivation [55–57]. Recently, using a fluorescence resonance energy transfer (FRET) assay, Jiang et al. [57] showed that some phytoestrogens, such as genistein, daidzein, equol and liquiritigenin, recruit the coactivator SRC3 much more efficiently to ERβ than to ERα. These data strengthen the ERβ-selectivity of many phytoestrogens. Hence, a relationship between the ERα/ERβ ratio and phytoestrogen effects exists [58,59]. It is suggested that the presence of ERβ is associated with the "good" effect of phytoestrogen whereas a high concentration of phytoestrogen in cells expressing ERα was associated to the "bad" effect of phytoestrogen [60].

Several in vitro studies showed that genistein, the most abundant isoflavone present in soybean, has antiproliferative effects on various cancer cells, including prostate, ovarian, and breast cancer [61–63]. While genistein effects can be mediated at least in part by ERβ, other molecular mechanisms, for exemple caspase-3 activation, have been reported to explain growth inhibition or proapoptotic effects of genistein. Additionally, by direct inhibition of tyrosine kinase activities, genistein is also able to prevent cancer cell growth. For example, genistein pretreatment could significantly reduce the activation of Akt kinase by epidermal growth factor (EGF). The inhibition of nuclear factor κB (NF-κB) activity by genistein was also reported in prostate, breast, lung, and pancreatic cancer cells [64–68]. An explanation of this effect is that genistein significantly inhibits Akt kinase activity by decreasing its phosphorylation at serine 473, which can inhibit NF-κB activity [65,69]. Another study reported that the inhibition of prostate cancer cell growth exerted by genistein is linked to a reduction of telomerase activity that is pivotal for cellular proliferation capacity and immortality. [70]. Together, these actions of the isoflavone genistein could contribute to its apoptotic effects in different human cancer cells. It is also interesting to note that in ER-positive MCF-7 cells, the biphasic actions of genistein can be observed with growth stimulation at low concentrations and inhibition at high concentrations. These observations indicate the complexity of the actions of genistein and phytoestrogens globally for their anti-cancer properties.

One of the key mechanisms underlying the maintenance of genome stability and gene expression is DNA methylation. This process occurs on the cytosine of cytosine-guanine dinucleotides (CpG regions). In the human genome, the majority of CpG regions are methylated, except for those located within CpG-rich regions, called CpG islands, which are usually found within gene promoters. Methylation of CpG islands could lead to the inactivation of gene expression by inhibiting the recruitment of transcription factors necessary to induce transcription. Indeed, DNA methylation/demethylation is a dynamic process that allows certain genes to switch ON and OFF at different periods of time. This process appears to be particularly crucial during embryonic development, tumorigenesis, cell division and cell differentiation. For instance, the OCT4 gene, which is essential to maintain pluripotency in embryonic stem cells, becomes methylated in differentiated tissues to avoid unsuitable pluripotency [71]. On the other hand, the loss of expression of tumor suppressor genes by DNA methylation is often observed in cancerous cells. Re-expression of these genes by the inhibition of DNA methyltransferases has provided many successes in the treatment of cancers.

Interestingly, a recent study showed that genistein can reduce DNA methylation in the promoter regions of the Wingless-int (Wnt) genes, which induces the expression of Wnt proteins in colon cancer cells [72]. The Wnt signaling pathway includes a large number of proteins involved in organogenesis and cell-cell adhesion, cell proliferation and differentiation. In addition to its importance in normal cellular physiology, Wnt signaling is also closely involved with carcinogenesis. Notably, the loss of the expression of Wnt proteins by promoter hypermethylation or abnormal activation of Wnt signaling have been detected in the majority of colon tumors and colon cell lines [73–75]. It is, therefore, possible that genistein, acting as an inhibitor of the DNA methyltransferase, could be able to induce significant Wnt signaling pathways to protect the development of colon cancer. Another study conducted on the human colon cancer cell lines SW480 and HCT15 showed that genistein blocks cell proliferation in the G2 phase of the cell cycle [76]. The authors showed that the action of genistein-inhibition on cell growth is mediated by overexpression of Dickkopf 1 (DKK1) in SW480 and HCT15 cells treated with genistein. DKK1 is a key regulator of the Wnt signaling pathway that promotes cell differentiation and apoptosis. The repression of the tumor suppressor DKK1 by the hypermethylation of its promoter is reported in various diseases, including colorectal cancer [77–80]. However, DNA methylation of the DKK1 promoter is not affected by genistein treatment in either cell line. However, genistein induces acetylation of histone H3 within the promoter region of the DKK1 gene in SW480 and HCT15 cells. This indicates that genistein induction of the expression of the DKK1 gene is linked to the increase in histone acetylation [76]. Another recent study showed that genistein is able to epigenetically reactivate ERα in ERα-negative breast cancer models, both in vitro and in vivo [66]. Similarly, in the prostate cancer cell lines LNCaP and LAPC-4, genistein was able to increase the expression of ERβ through decreasing the methylation of the ERβ promoter at physiological ranges (0.5–10 μmol/L) [81]. Hence, genistein could increase the sensitivity of these cancers to endocrine therapies, such as the antiestrogen tamoxifen. In this study, the authors showed that genistein significantly increased histone acetylation patterns in the ERα promoter by inhibiting the enzymatic activities of histone deacetylase (HDAC). It is of interest to note that this effect was enhanced in a synergistic manner when ERα-negative MDA-MB-231 breast cancer cells were co-treated with genistein and TSA, an inhibitor of HDACs. Importantly, the anticancer properties of tamoxifen to inhibit cell growth become much more efficient both in vitro and in vivo, in xenograft nude mice as well as in spontaneous breast tumor mouse models, in the presence of genistein [66]. Together, these studies suggest that in addition to DNA methylation, genistein may also modify histone marks of critical genes to prevent cancer development and progression. These epigenetic actions of genistein mediate the activation of tumor suppressor genes in cancer cell lines but also in animal models. The consequences of endocrine disruptors to different cell types have been widely studied. However, in industrial countries, detectable levels of EDCs were found in human, indicating that people are constantly exposed. Hence, studies on acute effects do not reflect the consequences of constant exposure. Chronic exposure of MCF7 cells with genistein induces a down-regulation of the PI3K/Akt signaling pathway, inhibits the growth-promoting activity of E2 or EGF, and reduces histone H3 acetylation without affecting ER expression. This indicates that chronic treatment leads to epigenetic changes in the cells [82,83].

4. In Vivo Effects of Phytoestrogens

The concentration of phytoestrogens in the plasma is considerably different in the human population. For example, in Finnish men the average plasma concentration of genistein is about 0.5 nM whereas it is about 276 nM in Japanese males. However, after absorption of dietary phytoestrogen, a plasmatic peak was detected between 0.2 and 6.5 μM with bioavailability between 5% and 66% [84]. Moreover a pharmacokinetic study performed on postmenopausal women found that the concentration of free genistein could reach 40 nM [85]. Epidemiological studies reveal that a lower risk of breast and prostate cancers is observed in Asians, who consume 20–50 times more soy products than Americans [86]. In vivo experimental studies have also reported that some dietary components, including isoflavones and enterolignans, could inhibit the development of cancers [69,87,88]. This suggests that active molecules in the soybean, such as genistein, daidzein, equol and glycitein, may act as natural chemopreventive agents and could be used against tumor progression in humans. Moreover, clinical studies carried out to assess the effectiveness of isoflavones in patients with prostate cancer found that isoflavone supplementation significantly reduced the expression of the poor prognostic tumoral marker, prostate specific antigen (PSA), and the expression of androgen receptor (AR), but without effecting the expression of ERβ or circulating hormones. These studies have suggested that isoflavones, including genistein and daidzein, may be beneficial in the prevention of prostate cancer by inhibiting the expression of AR and PSA [89,90]. Furthermore, in vivo xenografts in mice model and in vitro studies conducted on androgen-dependent (LNCaP) and androgen-independent (DU145 or PC3) prostate cancer cell lines showed that some phytoestrogens, such as coumestrol, are able to elicit caspase-dependent apoptosis, supporting the hypothesis that phytoestrogens may have anticancer effects in prostate cancer. Conversely, some clinical trials seeking to establish that consumption of phytoestrogens is beneficial in prostate cancer have been inconclusive [91]. For instance, a double-blind trial conducted by Adams and collaborators [92] showed no significant difference in PSA among men who did or did not consume a diet rich in isoflavones for 12 months. Although this study is limited by the relatively small number of patients, it could indicate that the period and duration of treatment may be essential for the anticancer effects of isoflavones. While the exact mechanisms by which isoflavones can prevent the development or progression of prostate cancer remain unclear, many mechanisms have been proposed, including the regulation of genes involved in the cell cycle, such as an upregulation of p21 resulting in cell cycle arrest at the G1/S phase, apoptosis, antioxidant effects, DNA repair, inhibition of angiogenesis and metastasis, and also the antagonism of estrogen and androgen signaling pathways (for review see [93]). It should be noted, therefore, that changes in some signaling pathways or in the expression of key enzymes involved in steroid metabolism during different stages of prostate cancer could play an essential role in the effects of phytoestrogens.

In vivo studies carried out in a true physiological context in humans and animals have indicated that the food content of isoflavones poses no safety issue, as generally consumed in diets based on soy products [94]. Moreover, the concentration of isoflavones varies considerably depending on the place of soybean cultivation (from 85 mg/100 g in Taiwan to 178 mg/100 g in Korea) or according to the culinary process (6 mg/100 g 195 mg/100 g in Foojook soup) [95]. A similar observation was made in lyophilized cabbage compared to fermented cabbage [96]. As the structure of the major isoflavone compounds is close to that of E2 and because these compounds are known to have weak estrogenic activities, the possible effects of some isoflavones on estrogen-target tissues and on reproductive function have been extensively explored [97–100]. However, there are conflicting results regarding the effects of isoflavones on reproductive function because the long-term studies on the impact of these compounds on the development and function of reproductive tissues are not sufficient. In addition, comparisons between different studies are complicated because there are differences in the experimental design, such as the physiological state of the animal, the presence of circulating hormone, and the duration, doses and methods of exposure (injection or gavage). More importantly, differences in the metabolism of isoflavones between animal models and humans can also give inconclusive results. Thus, all these parameters must be considered when assessing the impact of isoflavones on reproductive function. All major soybean isoflavones, genistein, daidzein, equol and glycitein, were reported to be estrogenic in the mouse or rat uterine growth assay [97,98]. For instance, 100 mg/kg body weight of genistein or equol, administered by gavage for 4 successive days (post-natal at day 17–20), was found to significantly increase uterine weights and the expression of ERα in the uterus [97]. Another study compared the estrogenic potential of several phytoestrogens, including genistein, daidzein and coumestrol in immature mice using different morphological and biochemical tests on the uterus. Interestingly, while certain compounds, such as genistein and coumestrol, showed estrogenic activity in all tests, others showed estrogenicity in only a single test or did not show estrogenicity in any test [101].

The estrogenic potency of isoflavones was also assessed, in vivo, in several non-mammalian model organisms. For example, goldfish or medaka fed for several weeks with a diet containing coumestrol or genistein showed an increased production of the vitellogenin, an egg yolk protein precursor, which is normally produced in the liver under estrogenic stimulation. However, no adverse effects on reproduction function (fecundity and fertility) have been reported [102,103], indicating that the production of vitellogenin may serve as an indicator of estrogen exposure but not as an indicator of reproductive dysfunction by estrogen exposure. In contrast, a recent study from Bennetau-Pellisero and collaborators reported [104] that goldfish fed soybean meal for 20 weeks after hatching show a reduction in fertility success and larvae production. Particularly, both male and female fish groups displayed changes in the plasma testosterone and E2 levels, as well as in their spermatogenesis process and oocyte maturation.

Concerning male reproductive function, a study has been conducted on marmosets fed with soy-based milk during the first six weeks of life and compared to animals fed with a standard cow's milk-based diet [105]. This study reported that soy-fed marmosets had body weights, organ weights (prostate, seminal vesicles, pituitary, thymus and spleen) and penis lengths comparable to the other animals. Although lower blood testosterone and higher Sertoli and Leydig cell numbers per testis were observed in soy-fed marmosets, no adverse reproductive consequences were detected in adulthood, including the timing of puberty and overall fertility [105]. On the other hand, Adachi et al. have carried out a toxicogenomic analysis in mice to investigate long-term effects of neonatal exposure to genistein on testicular gene expression. In addition, the authors used diethylstilbestrol (DES), known as a potent estrogen, as a positive control because exposure to DES has been reported to induce morphological changes and alteration of gene expression in reproductive organs. Male mice fed with genistein (1000 µg/mouse/day) from days 1–6 after birth did not show any morphological changes in testes at 12 weeks of age, despite decreased ER and AR gene expression. As expected, DES (50 µg/mouse/day) did show gene expression and morphological changes in testes at 12 weeks of age [106]. This suggests that neonatal exposure to genistein has no long-term effects, according to this analysis.

Following menopause in women, there is more brain-related pathology, incidence of stroke and loss of bone mass observed than in men [107]. Because estrogens are neuroprotective agents that are involved in bone remolding, one possible explanation may be the decline in estrogen levels. Isoflavones are generally considered to have beneficial effects on bone and brain, although controversial results have been published [108]. In ovariectomized rats, genistein showed a weak osteoprotective effect by promoting bone mineral density [109,110]. Similarly, coumestrol showed a neuroprotection effect in ovariectomized rats subjected to global ischemia [111]. Interestingly, using an ER antagonist, the authors showed that the neuroprotective actions of coumestrol are only partially abolished, suggesting that in addition to classical ER signaling, coumestrol may act via other cellular pathways. Thus, the beneficial effects of isoflavones may depend on the quantity and ratio of the expression of ER subtypes, the endogenous steroid hormones and period of life [111]. Moreover, it would be interesting to find out the cellular targets of coumestrol mediating its neuroprotective action. These cellular pathways could be used in the therapeutic potential of coumestrol in the treatment of pathologies related to the central nervous system.

There is growing evidence suggesting that during critical windows of prenatal and postnatal development, environmental chemicals can induce epigenetic modifications, affecting gene expression and consequently impacting developmental pathways. Importantly, it has been suggested that the effects of some environmental chemicals could act across generations, leading to phenotypic and physiological variation in the development and behavior of offspring. The transmission can be a consequence of changes in the transcriptome and epigenome programming within germ cells. While these effects have been recently reported for a number of environmental compounds, such as vinclozolin, atrazine, bisphenol A (BPA), DES, and dioxine [112–118], studies on the potential transgenerational effects of phytoestrogens are very rare, if any, and need additional work [69,115].

Although the situation is different for phytoestrogens produced by plants, it is worth noting that some fungi also produce compounds, called mycotoxins, with estrogenic properties. For example, zearalenone and fusarin C act as estrogen agonists and are classified as mycoestrogens. These compounds, which could contaminate improperly stored grains, have been linked to increased cancer rates. In vitro, fusarin C, as well as zearalenone and its metabolites, can stimulate the growth and proliferation of human breast tumor cells [119–121]. Moreover, in vivo exposure of rats to environmental doses of zearalenone in the last two to three weeks of fetal development and the first days after birth resulted in long-term changes in the development of the mammary gland associated with increased risk for the development of mammary tumors [122].

5. Conclusions

Although more research is needed, it is clear that some natural compounds from plants, such as phytoestrogens, could have beneficial effects on certain diseases, such as cancer or neurodegenerative diseases. However, in vitro or in vivo studies to analyze the final effects of phytoestrogens may be quite different at low (<1 µM) or high concentrations (>10 µM). For instance, at low doses (from 10 nM to 1 µM), genistein showed mitogenic effects on breast cancer cell growth, whereas at higher concentrations (>10 µM), it showed antiproliferative effects [123,124]. Some of these effects are explained by their interactions with ER subtypes. The ratios and the expressions of ERα and ERβ are different in various tissues depending on the period of life. ERα is mostly expressed in tissues such as the mammary gland, uterus, liver and pituitary, while ERβ is expressed in tissues such as the brain, bone and bladder. Moreover, the abilities of ER subtypes to recruit cofactors, regulate gene expression and stimulate or inhibit cell growth are slightly different. Therefore, in vivo, phytoestrogens may have a complex role, acting as weak estrogens and antiestrogens depending on the tissue. Furthermore, it is believed that the signaling pathways induced by phytoestrogens are not completely identical to those induced by estrogens. As illustrated in Figure 3, phytoestrogens may have different mechanisms of action; therefore, some of these compounds could be considered therapeutic agents and used alone or in combination with usual hormone therapies. For example, the protective effect of isoflavones on prostate cancer may be related to their effects on metabolic pathways involved in androgen and estrogen synthesis [125] or to their epigenetic modifications of DNA, such as the demethylation of CpG islands within the promoters of tumor suppressor genes [81,126,127]. On the other hand, the phytoestrogen coumestrol, which exhibits an important cancer-preventive effect in estrogen-responsive carcinomas, was recently reported to inhibit epithelial ovarian cancer proliferation and invasion by modifying AKT, p70S6K and ERK1/2 phosphorylation [128]. Moreover, previous studies showed an antagonistic effect of genistein and apigenin against the association of ERα with the ubiquitous calcium-dependent protein, calmodulin (CaM). By interacting with ERα, CaM plays a key role in the stabilization and transcriptional activity of ERα dimers at the ERE. The agonistic effect of genistein and apigenin in this interaction may also account for the anti-tumor origin of these compounds against ER-positive breast cancers [129,130]. It is, therefore, essential to continue advances in the understanding diverse signaling pathways activated by phytoestrogens, to fully exploit their anticancer properties and/or their potential roles in estrogen-related diseases. Accordingly, it should also be remembered that changes in the expression or activity of nuclear and membrane receptors for steroids and growth factors, as well as key steroid synthesis enzymes, during cancer progression could play crucial roles in the effects of phytoestrogens (Figure 3). Indeed, flavonoids, especially flavones (ex: luteolin) and flavanones (ex: naringenin), are described as potent inhibitors of aromatase activity [131]. Aromatase is the main enzyme that participates in the transformation of testosterone into estradiol and is hence involved in breast cancer pathology. Moreover, luteolin was also shown to downregulate aromatase gene expression [131]. Phytoestrogens are also able to inhibit proteasome [132], which appears to be essential for breast cancer cell survival [133]. For example, apigenin is capable of inhibiting the catalytic activity of proteasomes, leading to stabilization of ERβ and apoptosis of prostate cancer cells [134].

An important application of phytoestrogens is that they could be used as an alternative to the synthetic selective estrogen receptor modulators (SERMs), which exhibit estrogen agonist or antagonist activity in a tissue-specific manner. Indeed, SERMs are used in the treatment of some estrogen-associated pathologies, such as breast cancer, brain diseases, osteoporosis and menopausal symptoms. In other words, the challenge is to minimize the adverse effects of ER (mitogenic effect) without reducing the beneficial effects (protective effects), such as the control of cell differentiation, neuroprotection, anti-osteoporosis effects, and anti-oxidant activity. Our recent study screening the SERM activity of these compounds revealed a beneficial effect of apigenin and resveratrol, whereas zearalenone has been characterized as having a strong ER-agonist property in breast cancer cell lines and having adverse effects in neuritogenesis [135].

Although recent studies have reported that certain environmental agents caused epigenetic effects that could act across generations, leading to physiological changes of the offspring, there are no examples of perinatal exposure to phytoestrogens at environmentally relevant doses. There is still a need to understand the molecular mechanisms and to investigate how these compounds can influence epigenetic patterns during development.

In this review, we have discussed the effects of phytoestrogens used alone. However, populations are exposed to several compounds at the same time. Thus, it might be important to perform studies of the effect of mixtures of botanical estrogen on human health to improve recommendations for public health.

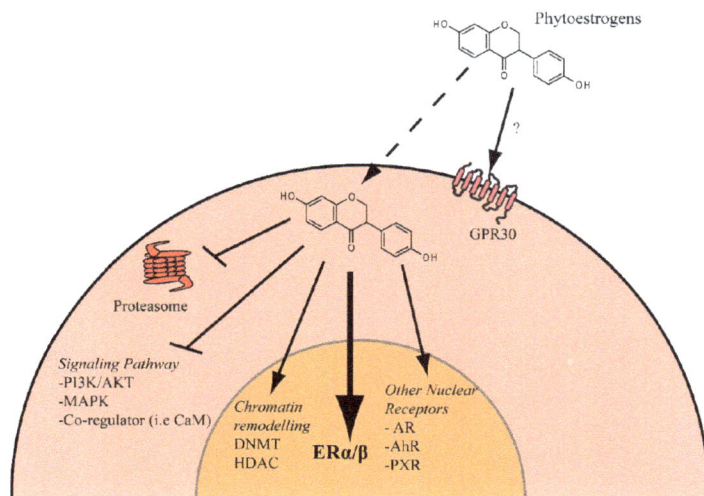

Figure 3. Different targets of phytoestrogens in cells. Cell signaling pathways for estrogens through the nuclear receptors ERα, ERβ and the transmembrane receptor G-protein-coupled ER (GPER; formerly known as GPR30) [136] are shown. Phytoestrogens are able to inhibit mitogenic pathways via ERα or PI3K/MAPK, which in turn inhibit cancer cell proliferation and invasion by modifying AKT, p70S6K and ERK1/2 phosphorylation as well as interaction between ERα with various coregulatory proteins such as calmodulin (CaM). Activation of ERβ inhibits dedifferentiation pathways and induces apoptosis and cell cycle arrest. GPER activation is anti-tumorigenic, as it upregulates p21 and induces cell cycle arrest in prostate cancer [137]. Epigenetic modifications by phytoestrogens, such as demethylation of CpG islands within the promoters of tumor suppressor genes, could contribute to cell growth arrest. Inhibition of proteasomes by phytoestrogens also appears to be another mechanism of phytoestrogen activity in decreasing cancer cell survival.

Acknowledgments: This work was supported by Fond Unique Interministeriel (FUI, project mVolio), the French Ministry of Ecology, Energy and Sustainable Development (PNRPE); La Ligue Contre le Cancer, the INSERM and CNRS.

Author Contributions: Sylvain Lecomte, François Ferrière and Florence Demay were the major contributors in writing the manuscript. Sylvain Lecomte, Florence Demay, François Ferrière and Farzad Pakdel read, corrected and approved the final manuscript

Conflicts of Interest: The authors declare no conflict of interest.

References

1. Couse, J.F.; Korach, K.S. Estrogen receptor null mice: What have we learned and where will they lead us? *Endocr. Rev.* **1999**, *20*, 358–417. [CrossRef] [PubMed]
2. Gustafsson, J.-A. What pharmacologists can learn from recent advances in estrogen signaling. *Trends Pharmacol. Sci.* **2003**, *24*, 479–485. [CrossRef]
3. Germain, P.; Staels, B.; Dacquet, C.; Spedding, M.; Laudet, V. Overview of nomenclature of nuclear receptors. *Pharmacol. Rev.* **2006**, *58*, 685–704. [CrossRef] [PubMed]
4. Safe, S.; Kim, K.; Kim, K. Non-classical genomic estrogen receptor (ER)/specificity protein and ER/activating protein-1 signaling pathways. *J. Mol. Endocrinol.* **2008**, *41*, 263–275. [CrossRef] [PubMed]
5. Carroll, J.S.; Meyer, C.A.; Song, J.; Li, W.; Geistlinger, T.R.; Eeckhoute, J.; Brodsky, A.S.; Keeton, E.K.; Fertuck, K.C.; Hall, G.F.; et al. Genome-wide analysis of estrogen receptor binding sites. *Nat. Genet.* **2006**, *38*, 1289–1297. [CrossRef] [PubMed]

6. Le Dily, F.; Beato, M. TADs as modular and dynamic units for gene regulation by hormones. *FEBS Lett.* **2015**, *589*, 2885–2892. [CrossRef] [PubMed]

7. Le Dily, F.; Baù, D.; Pohl, A.; Vicent, G.P.; Serra, F.; Soronellas, D.; Castellano, G.; Wright, R.H.G.; Ballare, C.; Filion, G.; et al. Distinct structural transitions of chromatin topological domains correlate with coordinated hormone-induced gene regulation. *Genes Dev.* **2014**, *28*, 2151–2162. [CrossRef] [PubMed]

8. Nilsson, S.; Mäkelä, S.; Treuter, E.; Tujague, M.; Thomsen, J.; Andersson, G.; Enmark, E.; Pettersson, K.; Warner, M.; Gustafsson, J.A. Mechanisms of estrogen action. *Physiol. Rev.* **2001**, *81*, 1535–1565. [PubMed]

9. La Rosa, P.; Pesiri, V.; Leclercq, G.; Marino, M.; Acconcia, F. Palmitoylation regulates 17β-estradiol-induced estrogen receptor-α degradation and transcriptional activity. *Mol. Endocrinol.* **2012**, *26*, 762–774. [CrossRef] [PubMed]

10. Pedram, A.; Razandi, M.; Aitkenhead, M.; Hughes, C.C.W.; Levin, E.R. Integration of the non-genomic and genomic actions of estrogen. Membrane-initiated signaling by steroid to transcription and cell biology. *J. Biol. Chem.* **2002**, *277*, 50768–50775. [CrossRef] [PubMed]

11. Vicent, G.P.; Nacht, A.S.; Zaurín, R.; Ballaré, C.; Clausell, J.; Beato, M. Minireview: Role of kinases and chromatin remodeling in progesterone signaling to chromatin. *Mol. Endocrinol.* **2010**, *24*, 2088–2098. [CrossRef] [PubMed]

12. Levin, E.R. Extranuclear estrogen receptor's roles in physiology: Lessons from mouse models. *Am. J. Physiol. Endocrinol. Metab.* **2014**, *307*, E133–E140. [CrossRef] [PubMed]

13. Gosden, J.R.; Middleton, P.G.; Rout, D. Localization of the human oestrogen receptor gene to chromosome 6q24—q27 by in situ hybridization. *Cytogenet. Cell Genet.* **1986**, *43*, 218–220. [CrossRef] [PubMed]

14. Enmark, E.; Pelto-Huikko, M.; Grandien, K.; Lagercrantz, S.; Lagercrantz, J.; Fried, G.; Nordenskjöld, M.; Gustafsson, J.A. Human estrogen receptor β-gene structure, chromosomal localization, and expression pattern. *J. Clin. Endocrinol. Metab.* **1997**, *82*, 4258–4265. [CrossRef] [PubMed]

15. Kerdivel, G.; Flouriot, G.; Pakdel, F. Modulation of estrogen receptor α activity and expression during breast cancer progression. *Vitam. Horm.* **2013**, *93*, 135–160. [PubMed]

16. Wang, Z.; Zhang, X.; Shen, P.; Loggie, B.W.; Chang, Y.; Deuel, T.F. Identification, cloning, and expression of human estrogen receptor-α36, a novel variant of human estrogen receptor-α66. *Biochem. Biophys. Res. Commun.* **2005**, *336*, 1023–1027. [CrossRef] [PubMed]

17. Saunders, P.T. Oestrogen receptor β (ER β). *Rev. Reprod.* **1998**, *3*, 164–171. [CrossRef] [PubMed]

18. Delbès, G.; Levacher, C.; Duquenne, C.; Racine, C.; Pakarinen, P.; Habert, R. Endogenous estrogens inhibit mouse fetal Leydig cell development via estrogen receptor α. *Endocrinology* **2005**, *146*, 2454–2461. [CrossRef] [PubMed]

19. Staub, C.; Rauch, M.; Ferrière, F.; Trépos, M.; Dorval-Coiffec, I.; Saunders, P.T.; Cobellis, G.; Flouriot, G.; Saligaut, C.; Jégou, B. Expression of Estrogen Receptor ESR1 and Its 46-kDa Variant in the Gubernaculum Testis. *Biol. Reprod.* **2005**, *73*, 703–712. [CrossRef] [PubMed]

20. Wilson, M.E.; Westberry, J.M.; Trout, A.L. Estrogen receptor-α gene expression in the cortex: Sex differences during development and in adulthood. *Horm. Behav.* **2011**, *59*, 353–357. [CrossRef] [PubMed]

21. Fan, X.; Kim, H.-J.; Warner, M.; Gustafsson, J.-A. Estrogen receptor β is essential for sprouting of nociceptive primary afferents and for morphogenesis and maintenance of the dorsal horn interneurons. *Proc. Natl. Acad. Sci. USA* **2007**, *104*, 13696–13701. [CrossRef] [PubMed]

22. Kerdivel, G.; Habauzit, D.; Pakdel, F. Assessment and molecular actions of endocrine-disrupting chemicals that interfere with estrogen receptor pathways. *Int. J. Endocrinol.* **2013**, *2013*, 501851. [CrossRef] [PubMed]

23. Sohoni, P.; Sumpter, J.P. Several environmental oestrogens are also anti-androgens. *J. Endocrinol.* **1998**, *158*, 327–339. [CrossRef] [PubMed]

24. Sonnenschein, C.; Soto, A.M. An updated review of environmental estrogen and androgen mimics and antagonists. *J. Steroid Biochem. Mol. Biol.* **1998**, *65*, 143–150. [CrossRef]

25. Rasier, G.; Toppari, J.; Parent, A.-S.; Bourguignon, J.-P. Female sexual maturation and reproduction after prepubertal exposure to estrogens and endocrine disrupting chemicals: A review of rodent and human data. *Mol. Cell. Endocrinol.* **2006**, *254–255*, 187–201. [CrossRef] [PubMed]

26. Toppari, J.; Virtanen, H.; Skakkebaek, N.E.; Main, K.M. Environmental effects on hormonal regulation of testicular descent. *J. Steroid Biochem. Mol. Biol.* **2006**, *102*, 184–186. [CrossRef] [PubMed]

27. Barouki, R.; Coumoul, X.; Fernandez-Salguero, P.M. The aryl hydrocarbon receptor, more than a xenobiotic-interacting protein. *FEBS Lett.* **2007**, *581*, 3608–3615. [CrossRef] [PubMed]

28. Liu, Z.; Kanjo, Y.; Mizutani, S. A review of phytoestrogens: Their occurrence and fate in the environment. *Water Res.* **2010**, *44*, 567–577. [CrossRef] [PubMed]

29. Fantini, M.; Benvenuto, M.; Masuelli, L.; Frajese, G.; Tresoldi, I.; Modesti, A.; Bei, R. In vitro and in vivo antitumoral effects of combinations of polyphenols, or polyphenols and anticancer drugs: Perspectives on cancer treatment. *Int. J. Mol. Sci.* **2015**, *16*, 9236–9282. [CrossRef] [PubMed]

30. Shukla, S.; Gupta, S. Apigenin: A promising molecule for cancer prevention. *Pharm. Res.* **2010**, *27*, 962–978. [CrossRef] [PubMed]

31. Marzocchella, L.; Fantini, M.; Benvenuto, M.; Masuelli, L.; Tresoldi, I.; Modesti, A.; Bei, R. Dietary flavonoids: Molecular mechanisms of action as anti-inflammatory agents. *Recent Pat. Inflamm. Allergy Drug Discov.* **2011**, *5*, 200–220. [CrossRef] [PubMed]

32. Beecher, G.R. Overview of dietary flavonoids: Nomenclature, occurrence and intake. *J. Nutr.* **2003**, *133*, 3248S–3254S. [PubMed]

33. Setchell, K.D.R.; Clerici, C. Equol: History, chemistry, and formation. *J. Nutr.* **2010**, *140*, 1355S–1362S. [CrossRef] [PubMed]

34. National Library of Medecine-MeSH National Library of Medecine-MeSH, 2016. Available online: https://www.nlm.nih.gov/cgi/mesh/2016/MB_cgi (accessed on 12 November 2016).

35. Van de Schans, M.G. M.; Vincken, J.-P.; Bovee, T.F. H.; Cervantes, A.D.; Logtenberg, M.J.; Gruppen, H. Structural changes of 6a-hydroxy-pterocarpans upon heating modulate their estrogenicity. *J. Agric. Food Chem.* **2014**, *62*, 10475–10484. [CrossRef] [PubMed]

36. Zimmermann, M.C.; Tilghman, S.L.; Boué, S.M.; Salvo, V.A.; Elliott, S.; Williams, K.Y.; Skripnikova, E.V.; Ashe, H.; Payton-Stewart, F.; Vanhoy-Rhodes, L.; et al. Glyceollin I, a novel antiestrogenic phytoalexin isolated from activated soy. *J. Pharmacol. Exp. Ther.* **2010**, *332*, 35–45. [CrossRef] [PubMed]

37. Ayers, A.R.; Ebel, J.; Finelli, F.; Berger, N.; Albersheim, P. Host-pathogen interactions: IX. Quantitative assays of elicitor activity and characterization of the elicitor present in the extracellular medium of cultures of *Phytophthora megasperma* var. sojae. *Plant Physiol.* **1976**, *57*, 751–759. [CrossRef] [PubMed]

38. Nikov, G.N.; Hopkins, N.E.; Boue, S.; Alworth, W.L. Interactions of dietary estrogens with human estrogen receptors and the effect on estrogen receptor-estrogen response element complex formation. *Environ. Health Perspect.* **2000**, *108*, 867–872. [CrossRef] [PubMed]

39. Tuskaev, V.A. Synthesis and biological activity of coumestan derivatives. *Pharm. Chem. J.* **2013**, *47*, 1–11. [CrossRef]

40. Nehybová, T.; Šmarda, J.; Beneš, P. Plant coumestans: Recent advances and future perspectives in cancer therapy. *Anticancer Agents Med. Chem.* **2014**, *14*, 1351–1362. [CrossRef] [PubMed]

41. Bickoff, E.M.; Booth, A.N.; Lyman, R.L.; Livingston, A.L.; Thompson, C.R.; Deeds, F. Coumestrol, a new estrogen isolated from forage crops. *Science* **1957**, *126*, 969–970. [CrossRef] [PubMed]

42. Kuiper, G.G.; Lemmen, J.G.; Carlsson, B.; Corton, J.C.; Safe, S.H.; van der Saag, P.T.; van der Burg, B.; Gustafsson, J.A. Interaction of estrogenic chemicals and phytoestrogens with estrogen receptor β. *Endocrinology* **1998**, *139*, 4252–4263. [CrossRef] [PubMed]

43. Stervbo, U.; Vang, O.; Bonnesen, C. A review of the content of the putative chemopreventive phytoalexin resveratrol in red wine. *Food Chem.* **2007**, *101*, 449–457. [CrossRef]

44. Gehm, B.D.; McAndrews, J.M.; Chien, P.Y.; Jameson, J.L. Resveratrol, a polyphenolic compound found in grapes and wine, is an agonist for the estrogen receptor. *Proc. Natl. Acad. Sci. USA* **1997**, *94*, 14138–14143. [CrossRef] [PubMed]

45. Le Corre, L.; Chalabi, N.; Delort, L.; Bignon, Y.-J.; Bernard-Gallon, D.J. Resveratrol and breast cancer chemoprevention: Molecular mechanisms. *Mol. Nutr. Food Res.* **2005**, *49*, 462–471. [CrossRef] [PubMed]

46. Rosmalena, A.; Prasasty, V.D.; Hanafi, M.; Budianto, E.; Elya, B. Lignan derivatives potential as Plasmodium falciparum lactate dehydrogenase inhibitors: Molecular docking approach of antiplasmodial drug design. *Int. J. Pharm. Pharm. Sci.* **2015**, *7*, 394–398.

47. Mueller, S.O.; Simon, S.; Chae, K.; Metzler, M.; Korach, K.S. Phytoestrogens and their human metabolites show distinct agonistic and antagonistic properties on estrogen receptor α (ERα) and ERβ in human cells. *Toxicol. Sci.* **2004**, *80*, 14–25. [CrossRef] [PubMed]

48. Shier, W.T.; Shier, A.C.; Xie, W.; Mirocha, C.J. Structure-activity relationships for human estrogenic activity in zearalenone mycotoxins. *Toxicon* **2001**, *39*, 1435–1438. [CrossRef]

49. EFSA Panel on Contaminants in the Food Chain Scientific. Opinion on the risks for public health related to the presence of zearalenone in food. *EFSA J.* **2011**, *9*, 2197.

50. Lin, P.; Chen, F.; Sun, J.; Zhou, J.; Wang, X.; Wang, N.; Li, X.; Zhang, Z.; Wang, A.; Jin, Y. Mycotoxin zearalenone induces apoptosis in mouse Leydig cells via an endoplasmic reticulum stress-dependent signalling pathway. *Reprod. Toxicol.* **2015**, *52*, 71–77. [CrossRef] [PubMed]

51. Le Guevel, R.; Pakdel, F. Assessment of oestrogenic potency of chemicals used as growth promoter by in vitro methods. *Hum. Reprod.* **2001**, *16*, 1030–1036. [CrossRef] [PubMed]

52. Frasor, J.; Danes, J.M.; Komm, B.; Chang, K.C.N.; Lyttle, C.R.; Katzenellenbogen, B.S. Profiling of estrogen up- and down-regulated gene expression in human breast cancer cells: Insights into gene networks and pathways underlying estrogenic control of proliferation and cell phenotype. *Endocrinology* **2003**, *144*, 4562–4574. [CrossRef] [PubMed]

53. Chang, E.C.; Frasor, J.; Komm, B.; Katzenellenbogen, B.S. Impact of estrogen receptor β on gene networks regulated by estrogen receptor α in breast cancer cells. *Endocrinology* **2006**, *147*, 4831–4842. [CrossRef] [PubMed]

54. Chang, E.C.; Charn, T.H.; Park, S.-H.; Helferich, W.G.; Komm, B.; Katzenellenbogen, J.A.; Katzenellenbogen, B.S. Estrogen Receptors α and β as Determinants of Gene Expression: Influence of Ligand, Dose, and Chromatin Binding. *Mol. Endocrinol.* **2008**, *22*, 1032–1043. [CrossRef] [PubMed]

55. Zhao, L.; Mao, Z.; Brinton, R.D. A select combination of clinically relevant phytoestrogens enhances estrogen receptor β-binding selectivity and neuroprotective activities in vitro and in vivo. *Endocrinology* **2009**, *150*, 770–783. [CrossRef] [PubMed]

56. Shanle, E.K.; Hawse, J.R.; Xu, W. Generation of stable reporter breast cancer cell lines for the identification of ER subtype selective ligands. *Biochem. Pharmacol.* **2011**, *82*, 1940–1949. [CrossRef] [PubMed]

57. Jiang, Y.; Gong, P.; Madak-Erdogan, Z.; Martin, T.; Jeyakumar, M.; Carlson, K.; Khan, I.; Smillie, T.J.; Chittiboyina, A.G.; Rotte, S.C.K.; et al. Mechanisms enforcing the estrogen receptor β selectivity of botanical estrogens. *FASEB J.* **2013**, *27*, 4406–4418. [CrossRef] [PubMed]

58. Pons, D.G.; Nadal-Serrano, M.; Blanquer-Rossello, M.M.; Sastre-Serra, J.; Oliver, J.; Roca, P. Genistein Modulates Proliferation and Mitochondrial Functionality in Breast Cancer Cells Depending on ERα/ERβ Ratio. *J. Cell. Biochem.* **2014**, *115*, 949–958. [CrossRef] [PubMed]

59. Sotoca, A.M.; Ratman, D.; van der Saag, P.; Ström, A.; Gustafsson, J.A.; Vervoort, J.; Rietjens, I.M.C.M.; Murk, A.J. Phytoestrogen-mediated inhibition of proliferation of the human T47D breast cancer cells depends on the ERα/ERβ ratio. *J. Steroid Biochem. Mol. Biol.* **2008**, *112*, 171–178. [CrossRef] [PubMed]

60. Russo, M.; Russo, G.L.; Daglia, M.; Kasi, P.D.; Ravi, S.; Nabavi, S.F.; Nabavi, S.M. Understanding genistein in cancer: The good and the bad effects: A review. *Food Chem.* **2016**, *196*, 589–600. [CrossRef] [PubMed]

61. Kuo, S.M. Antiproliferative potency of structurally distinct dietary flavonoids on human colon cancer cells. *Cancer Lett.* **1996**, *110*, 41–48. [CrossRef]

62. Hwang, K.-A.; Park, M.-A.; Kang, N.-H.; Yi, B.-R.; Hyun, S.-H.; Jeung, E.-B.; Choi, K.-C. Anticancer effect of genistein on BG-1 ovarian cancer growth induced by 17 β-estradiol or bisphenol A via the suppression of the crosstalk between estrogen receptor α and insulin-like growth factor-1 receptor signaling pathways. *Toxicol. Appl. Pharmacol.* **2013**, *272*, 637–646. [CrossRef] [PubMed]

63. Prietsch, R.F.; Monte, L.G.; da Silva, F.A.; Beira, F.T.; Del Pino, F.A.B.; Campos, V.F.; Collares, T.; Pinto, L.S.; Spanevello, R.M.; Gamaro, G.D.; et al. Genistein induces apoptosis and autophagy in human breast MCF-7 cells by modulating the expression of proapoptotic factors and oxidative stress enzymes. *Mol. Cell. Biochem.* **2014**, *390*, 235–242. [CrossRef] [PubMed]

64. Li, Y.; Sarkar, F.H. Inhibition of nuclear factor κB activation in PC3 cells by genistein is mediated via Akt signaling pathway. *Clin. Cancer Res.* **2002**, *8*, 2369–2377. [PubMed]

65. Gong, L.; Li, Y.; Nedeljkovic-Kurepa, A.; Sarkar, F.H. Inactivation of NF-κB by genistein is mediated via Akt signaling pathway in breast cancer cells. *Oncogene* **2003**, *22*, 4702–4709. [CrossRef] [PubMed]

66. Li, Y.; Meeran, S.M.; Patel, S.N.; Chen, H.; Hardy, T.M.; Tollefsbol, T.O. Epigenetic reactivation of estrogen receptor-α (ERα) by genistein enhances hormonal therapy sensitivity in ERα-negative breast cancer. *Mol. Cancer* **2013**, *12*, 9. [CrossRef] [PubMed]

67. Suzuki, R.; Kang, Y.; Li, X.; Roife, D.; Zhang, R.; Fleming, J.B. Genistein potentiates the antitumor effect of 5-Fluorouracil by inducing apoptosis and autophagy in human pancreatic cancer cells. *Anticancer Res.* **2014**, *34*, 4685–4692. [PubMed]

68. Li, C.; Teng, R.-H.; Tsai, Y.-C.; Ke, H.-S.; Huang, J.-Y.; Chen, C.-C.; Kao, Y.-L.; Kuo, C.-C.; Bell, W.R.; Shieh, B. H-Ras oncogene counteracts the growth-inhibitory effect of genistein in T24 bladder carcinoma cells. *Br. J. Cancer* **2005**, *92*, 80–88. [CrossRef] [PubMed]

69. Sarkar, F.H.; Li, Y.; Wang, Z.; Padhye, S. Lesson learned from nature for the development of novel anti-cancer agents: Implication of isoflavone, curcumin, and their synthetic analogs. *Curr. Pharm. Des.* **2010**, *16*, 1801–1812. [CrossRef] [PubMed]

70. Jagadeesh, S.; Kyo, S.; Banerjee, P.P. Genistein represses telomerase activity via both transcriptional and posttranslational mechanisms in human prostate cancer cells. *Cancer Res.* **2006**, *66*, 2107–2115. [CrossRef] [PubMed]

71. Wong, C.J.; Casper, R.F.; Rogers, I.M. Epigenetic changes to human umbilical cord blood cells cultured with three proteins indicate partial reprogramming to a pluripotent state. *Exp. Cell Res.* **2010**, *316*, 927–939. [CrossRef] [PubMed]

72. Zhang, Y.; Chen, H. Genistein attenuates WNT signaling by up-regulating sFRP2 in a human colon cancer cell line. *Exp. Biol. Med.* **2011**, *236*, 714–722. [CrossRef] [PubMed]

73. Gregorieff, A.; Clevers, H. Wnt signaling in the intestinal epithelium: From endoderm to cancer. *Genes Dev.* **2005**, *19*, 877–890. [CrossRef] [PubMed]

74. MacDonald, B.T.; Tamai, K.; He, X. Wnt/β-catenin signaling: Components, mechanisms, and diseases. *Dev. Cell* **2009**, *17*, 9–26. [CrossRef] [PubMed]

75. Qi, J.; Zhu, Y.-Q.; Luo, J.; Tao, W.-H. Hypermethylation and expression regulation of secreted frizzled-related protein genes in colorectal tumor. *World J. Gastroenterol.* **2006**, *12*, 7113–7117. [CrossRef] [PubMed]

76. Wang, H.; Li, Q.; Chen, H. Genistein affects histone modifications on Dickkopf-related protein 1 (DKK1) gene in SW480 human colon cancer cell line. *PLoS ONE* **2012**, *7*, e40955. [CrossRef] [PubMed]

77. Rawson, J.B.; Manno, M.; Mrkonjic, M.; Daftary, D.; Dicks, E.; Buchanan, D.D.; Younghusband, H.B.; Parfrey, P.S.; Young, J.P.; Pollett, A.; et al. Promoter methylation of Wnt antagonists DKK1 and SFRP1 is associated with opposing tumor subtypes in two large populations of colorectal cancer patients. *Carcinogenesis* **2011**, *32*, 741–747. [CrossRef] [PubMed]

78. Aguilera, O.; Fraga, M.F.; Ballestar, E.; Paz, M.F.; Herranz, M.; Espada, J.; García, J.M.; Muñoz, A.; Esteller, M.; González-Sancho, J.M. Epigenetic inactivation of the Wnt antagonist *DICKKOPF-1* (*DKK-1*) gene in human colorectal cancer. *Oncogene* **2006**, *25*, 4116–4121. [CrossRef] [PubMed]

79. Hirata, H.; Hinoda, Y.; Nakajima, K.; Kawamoto, K.; Kikuno, N.; Ueno, K.; Yamamura, S.; Zaman, M.S.; Khatri, G.; Chen, Y.; et al. Wnt antagonist *DKK1* acts as a tumor suppressor gene that induces apoptosis and inhibits proliferation in human renal cell carcinoma. *Int. J. Cancer* **2011**, *128*, 1793–1803. [CrossRef] [PubMed]

80. Ravindranath, M.H.; Muthugounder, S.; Presser, N.; Viswanathan, S. Anticancer therapeutic potential of soy isoflavone, genistein. *Adv. Exp. Med. Biol.* **2004**, *546*, 121–165. [PubMed]

81. Mahmoud, A.M.; Al-Alem, U.; Ali, M.M.; Bosland, M.C. Genistein increases estrogen receptor β expression in prostate cancer via reducing its promoter methylation. *J. Steroid Biochem. Mol. Biol.* **2015**, *152*, 62–75. [CrossRef] [PubMed]

82. Anastasius, N.; Boston, S.; Lacey, M.; Storing, N.; Whitehead, S.A. Evidence that low-dose, long-term genistein treatment inhibits oestradiol-stimulated growth in MCF-7 cells by down-regulation of the PI3-kinase/Akt signalling pathway. *J. Steroid Biochem. Mol. Biol.* **2009**, *116*, 50–55. [CrossRef] [PubMed]

83. Jawaid, K.; Crane, S.R.; Nowers, J.L.; Lacey, M.; Whitehead, S.A. Long-term genistein treatment of MCF-7 cells decreases acetylated histone 3 expression and alters growth responses to mitogens and histone deacetylase inhibitors. *J. Steroid Biochem. Mol. Biol.* **2010**, *120*, 164–171. [CrossRef] [PubMed]

84. Whitten, P.L.; Patisaul, H.B. Cross-species and interassay comparisons of phytoestrogen action. *Environ. Health Perspect.* **2001**. [CrossRef]

85. Soukup, S.T.; Al-Maharik, N.; Botting, N.; Kulling, S.E. Quantification of soy isoflavones and their conjugative metabolites in plasma and urine: An automated and validated UHPLC-MS/MS method for use in large-scale studies. *Anal. Bioanal. Chem.* **2014**, *406*, 6007–6020. [CrossRef] [PubMed]

86. Lee, S.-A.; Shu, X.-O.; Li, H.; Yang, G.; Cai, H.; Wen, W.; Ji, B.-T.; Gao, J.; Gao, Y.-T.; Zheng, W. Adolescent and adult soy food intake and breast cancer risk: Results from the shanghai women's health study. *Am. J. Clin. Nutr.* **2009**, *89*, 1920–1926. [CrossRef] [PubMed]

87. Barnes, S. The chemopreventive properties of soy isoflavonoids in animal models of breast cancer. *Breast Cancer Res. Treat.* **1997**, *46*, 169–179. [CrossRef] [PubMed]

88. Adlercreutz, H. Lignans and human health. *Crit. Rev. Clin. Lab. Sci.* **2007**, *44*, 483–525. [CrossRef] [PubMed]

89. Hamilton-Reeves, J.M.; Rebello, S.A.; Thomas, W.; Slaton, J.W.; Kurzer, M.S. Isoflavone-rich soy protein isolate suppresses androgen receptor expression without altering estrogen receptor-β expression or serum hormonal profiles in men at high risk of prostate cancer. *J. Nutr.* **2007**, *137*, 1769–1775. [PubMed]

90. Pendleton, J.M.; Tan, W.W.; Anai, S.; Chang, M.; Hou, W.; Shiverick, K.T.; Rosser, C.J. Phase II trial of isoflavone in prostate-specific antigen recurrent prostate cancer after previous local therapy. *BMC Cancer* **2008**, *8*, 132. [CrossRef] [PubMed]

91. Goetzl, M.A.; Van Veldhuizen, P.J.; Thrasher, J.B. Effects of soy phytoestrogens on the prostate. *Prostate Cancer Prostatic Dis.* **2007**, *10*, 216–223. [CrossRef] [PubMed]

92. Adams, K.F.; Chen, C.; Newton, K.M.; Potter, J.D.; Lampe, J.W. Soy isoflavones do not modulate prostate-specific antigen concentrations in older men in a randomized controlled trial. *Cancer Epidemiol. Biomark. Prev.* **2004**, *13*, 644–648.

93. Mahmoud, A.M.; Yang, W.; Bosland, M.C. Soy isoflavones and prostate cancer: A review of molecular mechanisms. *J. Steroid Biochem. Mol. Biol.* **2014**, *140*, 116–132. [CrossRef] [PubMed]

94. Munro, I.C.; Harwood, M.; Hlywka, J.J.; Stephen, A.M.; Doull, J.; Flamm, W.G.; Adlercreutz, H. Soy isoflavones: A safety review. *Nutr. Rev.* **2003**, *61*, 1–33. [CrossRef] [PubMed]

95. He, F.-J.; Chen, J.-Q. Consumption of soybean, soy foods, soy isoflavones and breast cancer incidence: Differences between Chinese women and women in Western countries and possible mechanisms. *Food Sci. Hum. Wellness* **2013**, *2*, 146–161. [CrossRef]

96. Ju, Y.H.; Carlson, K.E.; Sun, J.; Pathak, D.; Katzenellenbogen, B.S.; Katzenellenbogen, J.A.; Helferich, W.G. Estrogenic effects of extracts from cabbage, fermented cabbage, and acidified brussels sprouts on growth and gene expression of estrogen-dependent human breast cancer (MCF-7) cells. *J. Agric. Food Chem.* **2000**, *48*, 4628–4634. [CrossRef] [PubMed]

97. Breinholt, V.; Hossaini, A.; Svendsen, G.W.; Brouwer, C.; Nielsen, E. Estrogenic activity of flavonoids in mice. The importance of estrogen receptor distribution, metabolism and bioavailability. *Food Chem. Toxicol.* **2000**, *38*, 555–564. [CrossRef]

98. Owens, W.; Ashby, J.; Odum, J.; Onyon, L. The OECD program to validate the rat uterotrophic bioassay. Phase 2: Dietary phytoestrogen analyses. *Environ. Health Perspect.* **2003**, *111*, 1559–1567. [CrossRef] [PubMed]

99. Phrakonkham, P.; Chevalier, J.; Desmetz, C.; Pinnert, M.-F.; Bergès, R.; Jover, E.; Davicco, M.-J.; Bennetau-Pelissero, C.; Coxam, V.; Artur, Y.; et al. Isoflavonoid-based bone-sparing treatments exert a low activity on reproductive organs and on hepatic metabolism of estradiol in ovariectomized rats. *Toxicol. Appl. Pharmacol.* **2007**, *224*, 105–115. [CrossRef] [PubMed]

100. Cederroth, C.R.; Zimmermann, C.; Nef, S. Soy, phytoestrogens and their impact on reproductive health. *Mol. Cell. Endocrinol.* **2012**, *355*, 192–200. [CrossRef] [PubMed]

101. Jefferson, W.N.; Padilla-Banks, E.; Clark, G.; Newbold, R.R. Assessing estrogenic activity of phytochemicals using transcriptional activation and immature mouse uterotrophic responses. *J. Chromatogr. B Anal. Technol. Biomed. Life Sci.* **2002**, *777*, 179–189. [CrossRef]

102. Inudo, M.; Ishibashi, H.; Matsumura, N.; Matsuoka, M.; Mori, T.; Taniyama, S.; Kadokami, K.; Koga, M.; Shinohara, R.; Hutchinson, T.H.; et al. Effect of estrogenic activity, and phytoestrogen and organochlorine pesticide contents in an experimental fish diet on reproduction and hepatic vitellogenin production in medaka (*Oryzias latipes*). *Comp. Med.* **2004**, *54*, 673–680. [PubMed]

103. Kobayashi, M.; Ishibashi, H.; Moriwaki, T.; Koshiishi, T.; Ogawa, S.; Matsumoto, T.; Arizono, K.; Watabe, S. Production of low-estrogen goldfish diet for in vivo endocrine disrupter test. *Environ. Sci.* **2006**, *13*, 125–136. [PubMed]

104. Bagheri, T.; Imanpoor, M.R.; Jafari, V.; Bennetau-Pelissero, C. Reproductive impairment and endocrine disruption in goldfish by feeding diets containing soybean meal. *Anim. Reprod. Sci.* **2013**, *139*, 136–144. [CrossRef] [PubMed]

105. Tan, K.A. L.; Walker, M.; Morris, K.; Greig, I.; Mason, J.I.; Sharpe, R.M. Infant feeding with soy formula milk: Effects on puberty progression, reproductive function and testicular cell numbers in marmoset monkeys in adulthood. *Hum. Reprod.* **2006**, *21*, 896–904. [CrossRef] [PubMed]

106. Adachi, T.; Ono, Y.; Koh, K.B.; Takashima, K.; Tainaka, H.; Matsuno, Y.; Nakagawa, S.; Todaka, E.; Sakurai, K.; Fukata, H.; et al. Long-term alteration of gene expression without morphological change in testis after neonatal exposure to genistein in mice: Toxicogenomic analysis using cDNA microarray. *Food Chem. Toxicol.* **2004**, *42*, 445–452. [CrossRef] [PubMed]

107. Persky, R.W.; Turtzo, L.C.; McCullough, L.D. Stroke in women: Disparities and outcomes. *Curr. Cardiol. Rep.* **2010**, *12*, 6–13. [CrossRef] [PubMed]

108. Wuttke, W.; Jarry, H.; Westphalen, S.; Christoffel, V.; Seidlová-Wuttke, D. Phytoestrogens for hormone replacement therapy? *J. Steroid Biochem. Mol. Biol.* **2002**, *83*, 133–147. [CrossRef]

109. Sehmisch, S.; Uffenorde, J.; Maehlmeyer, S.; Tezval, M.; Jarry, H.; Stuermer, K.M.; Stuermer, E.K. Evaluation of bone quality and quantity in osteoporotic mice–the effects of genistein and equol. *Phytomedicine* **2010**, *17*, 424–430. [CrossRef] [PubMed]

110. Raghu Nadhanan, R.; Skinner, J.; Chung, R.; Su, Y.-W.; Howe, P.R.; Xian, C.J. Supplementation with fish oil and genistein, individually or in combination, protects bone against the adverse effects of methotrexate chemotherapy in rats. *PLoS ONE* **2013**, *8*, e71592. [CrossRef] [PubMed]

111. Canal Castro, C.; Pagnussat, A.S.; Orlandi, L.; Worm, P.; Moura, N.; Etgen, A.M.; Alexandre Netto, C. Coumestrol has neuroprotective effects before and after global cerebral ischemia in female rats. *Brain Res.* **2012**, *1474*, 82–90. [CrossRef] [PubMed]

112. Anway, M.D.; Leathers, C.; Skinner, M.K. Endocrine disruptor vinclozolin induced epigenetic transgenerational adult-onset disease. *Endocrinology* **2006**, *147*, 5515–5523. [CrossRef] [PubMed]

113. Jirtle, R.L.; Skinner, M.K. Environmental epigenomics and disease susceptibility. *Nat. Rev. Genet.* **2007**, *8*, 253–262. [CrossRef] [PubMed]

114. Salian, S.; Doshi, T.; Vanage, G. Impairment in protein expression profile of testicular steroid receptor coregulators in male rat offspring perinatally exposed to Bisphenol A. *Life Sci.* **2009**, *85*, 11–18. [CrossRef] [PubMed]

115. Guerrero-Bosagna, C.M.; Skinner, M.K. Environmental epigenetics and phytoestrogen/phytochemical exposures. *J. Steroid Biochem. Mol. Biol.* **2014**, *139*, 270–276. [CrossRef] [PubMed]

116. Bruner-Tran, K.L.; Osteen, K.G. Developmental exposure to TCDD reduces fertility and negatively affects pregnancy outcomes across multiple generations. *Reprod. Toxicol.* **2011**, *31*, 344–350. [CrossRef] [PubMed]

117. Manikkam, M.; Tracey, R.; Guerrero-Bosagna, C.; Skinner, M.K. Dioxin (TCDD) induces epigenetic transgenerational inheritance of adult onset disease and sperm epimutations. *PLoS ONE* **2012**, *7*, e46249. [CrossRef] [PubMed]

118. Hao, C.; Gely-Pernot, A.; Kervarrec, C.; Boudjema, M.; Becker, E.; Khil, P.; Tevosian, S.; Jégou, B.; Smagulova, F. Exposure to the widely used herbicide atrazine results in deregulation of global tissue-specific RNA transcription in the third generation and is associated with a global decrease of histone trimethylation in mice. *Nucleic Acids Res.* **2016**. [CrossRef] [PubMed]

119. Sondergaard, T.E.; Hansen, F.T.; Purup, S.; Nielsen, A.K.; Bonefeld-Jørgensen, E.C.; Giese, H.; Sørensen, J.L. Fusarin C acts like an estrogenic agonist and stimulates breast cancer cells in vitro. *Toxicol. Lett.* **2011**, *205*, 116–121. [CrossRef] [PubMed]

120. Khosrokhavar, R.; Rahimifard, N.; Shoeibi, S.; Hamedani, M.P.; Hosseini, M.-J. Effects of zearalenone and α-Zearalenol in comparison with Raloxifene on T47D cells. *Toxicol. Mech. Methods* **2009**, *19*, 246–250. [CrossRef] [PubMed]

121. Parveen, M.; Zhu, Y.; Kiyama, R. Expression profiling of the genes responding to zearalenone and its analogues using estrogen-responsive genes. *FEBS Lett.* **2009**, *583*, 2377–2384. [CrossRef] [PubMed]

122. Belli, P.; Bellaton, C.; Durand, J.; Balleydier, S.; Milhau, N.; Mure, M.; Mornex, J.-F.; Benahmed, M.; Le Jan, C. Fetal and neonatal exposure to the mycotoxin zearalenone induces phenotypic alterations in adult rat mammary gland. *Food Chem. Toxicol.* **2010**, *48*, 2818–2826. [CrossRef] [PubMed]

123. Hsieh, C.-Y.; Santell, R.C.; Haslam, S.Z.; Helferich, W.G. Estrogenic effects of genistein on the growth of estrogen receptor-positive human breast cancer (MCF-7) cells in vitro and in vivo. *Cancer Res.* **1998**, *58*, 3833–3838. [PubMed]

124. Wang, T.T. Y.; Sathyamoorthy, N.; Phang, J.M. Molecular effects of genistein on estrogen receptor mediated pathways. *Carcinogenesis* **1996**, *17*, 271–275. [CrossRef] [PubMed]

125. Rahman, H.P.; Hofland, J.; Foster, P.A. In touch with your feminine side: How oestrogen metabolism impacts prostate cancer. *Endocr. Relat. Cancer* **2016**, *23*, R249–R266. [CrossRef] [PubMed]

126. Majid, S.; Dar, A.A.; Ahmad, A.E.; Hirata, H.; Kawakami, K.; Shahryari, V.; Saini, S.; Tanaka, Y.; Dahiya, A.V.; Khatri, G.; et al. *BTG3* tumor suppressor gene promoter demethylation, histone modification and cell cycle arrest by genistein in renal cancer. *Carcinogenesis* **2009**, *30*, 662–670. [CrossRef] [PubMed]

127. Vardi, A.; Bosviel, R.; Rabiau, N.; Adjakly, M.; Satih, S.; Dechelotte, P.; Boiteux, J.-P.; Fontana, L.; Bignon, Y.-J.; Guy, L.; et al. Soy phytoestrogens modify DNA methylation of *GSTP1, RASSF1A, EPH2* and *BRCA1* promoter in prostate cancer cells. *In Vivo* **2010**, *24*, 393–400. [PubMed]

128. Lim, W.; Jeong, W.; Song, G. Coumestrol suppresses proliferation of ES2 human epithelial ovarian cancer cells. *J. Endocrinol.* **2016**, *228*, 149–160. [CrossRef] [PubMed]

129. Leclercq, G.; Jacquot, Y. Interactions of isoflavones and other plant derived estrogens with estrogen receptors for prevention and treatment of breast cancer-considerations concerning related efficacy and safety. *J. Steroid Biochem. Mol. Biol.* **2014**, *139*, 237–244. [CrossRef] [PubMed]

130. Li, Z.; Zhang, Y.; Hedman, A.C.; Ames, J.B.; Sacks, D.B. Calmodulin lobes facilitate dimerization and activation of estrogen receptor-α. *J. Biol. Chem.* **2017**, *292*, 4614–4622. [CrossRef] [PubMed]

131. Li, F.; Ye, L.; Lin, S.; Leung, L.K. Dietary flavones and flavonones display differential effects on aromatase (CYP19) transcription in the breast cancer cells MCF-7. *Mol. Cell. Endocrinol.* **2011**, *344*, 51–58. [CrossRef] [PubMed]

132. Nakamura, K.; Yang, J.-H.; Sato, E.; Miura, N.; Wu, Y.-X. Effects of hydroxy groups in the A-ring on the anti-proteasome activity of flavone. *Biol. Pharm. Bull.* **2015**, *38*, 935–940. [CrossRef] [PubMed]

133. Marcotte, R.; Sayad, A.; Brown, K.R.; Sanchez-Garcia, F.; Reimand, J.; Haider, M.; Virtanen, C.; Bradner, J.E.; Bader, G.D.; Mills, G.B.; et al. Functional genomic landscape of human breast cancer drivers, vulnerabilities, and resistance. *Cell* **2016**, *164*, 293–309. [CrossRef] [PubMed]

134. Singh, V.; Sharma, V.; Verma, V.; Pandey, D.; Yadav, S.K.; Maikhuri, J.P.; Gupta, G. Apigenin manipulates the ubiquitin–proteasome system to rescue estrogen receptor-β from degradation and induce apoptosis in prostate cancer cells. *Eur. J. Nutr.* **2015**, *54*, 1255–1267. [CrossRef] [PubMed]

135. Lecomte, S.; Lelong, M.; Bourgine, G.; Efstathiou, T.; Saligaut, C.; Pakdel, F. Assessment of the potential activity of major dietary compounds as selective estrogen receptor modulators in two distinct cell models for proliferation and differentiation. *Toxicol. Appl. Pharmacol.* **2017**, *325*, 61–70. [CrossRef] [PubMed]

136. Prossnitz, E.R.; Barton, M. The G-protein-coupled estrogen receptor GPER in health and disease. *Nat. Rev. Endocrinol.* **2011**, *7*, 715–726. [CrossRef] [PubMed]

137. Chan, Q.K.Y.; Lam, H.-M.; Ng, C.-F.; Lee, A.Y.Y.; Chan, E.S.Y.; Ng, H.-K.; Ho, S.-M.; Lau, K.-M. Activation of GPR30 inhibits the growth of prostate cancer cells through sustained activation of Erk1/2, c-jun/c-fos-dependent upregulation of p21, and induction of G(2) cell-cycle arrest. *Cell Death Differ.* **2010**, *17*, 1511–1523. [CrossRef] [PubMed]

International Journal of
Molecular Sciences

MDPI

Article

Variability in DNA Repair Capacity Levels among Molecular Breast Cancer Subtypes: Triple Negative Breast Cancer Shows Lowest Repair

Jaime Matta [1,*], Carmen Ortiz [1], Jarline Encarnación [1], Julie Dutil [1] and Erick Suárez [2]

[1] Department of Basic Sciences, Divisions of Pharmacology, Toxicology, Biochemistry and Cancer Biology, Ponce Health Sciences University—School of Medicine, Ponce Research Institute, Ponce 00716-2348, Puerto Rico; carmenortiz@psm.edu (C.O.); jencarnacion@psm.edu (J.E.); jdutil@psm.edu (J.D.)

[2] Department of Biostatistics and Epidemiology, Graduate School of Public Health, University of Puerto Rico, Medical Sciences Campus, San Juan 00936-5067, Puerto Rico; erick.suarez@upr.edu

* Correspondence: jmatta@psm.edu; Tel.: +1-787-259-7025; Fax: +1-787-259-7018

Received: 21 June 2017; Accepted: 7 July 2017; Published: 12 July 2017

Abstract: Breast cancer (BC) is a heterogeneous disease which many studies have classified in at least four molecular subtypes: Luminal A, Luminal B, HER2-Enriched, and Basal-like (including triple-negative breast cancer, TNBC). These subtypes provide information to stratify patients for better prognostic predictions and treatment selection. Individuals vary in their sensitivities to carcinogens due to differences in their DNA repair capacity (DRC) levels. Although our previous case-control study established low DRC (in terms of NER pathway) as a BC risk factor, we aim to study this effect among the molecular subtypes. Therefore, the objectives of this study include investigating whether DRC varies among molecular subtypes and testing any association regarding DRC. This study comprised 267 recently diagnosed women with BC (cases) and 682 without BC (controls). Our results show a substantial variability in DRC among the molecular subtypes, with TNBC cases ($n = 47$) having the lowest DRC (p-value < 0.05). Almost 80 percent of BC cases had a DRC below the median (4.3%). Low DRC was strongly associated with the TNBC subtype (OR 7.2; 95% CI 3.3, 15.7). In conclusion, our study provides the first report on the variability among the molecular subtypes and provides a hypothesis based on DRC levels for the poor prognosis of TNBC.

Keywords: breast cancer; molecular subtypes; phenotypic variability; DNA repair capacity; multinomial regression analysis; precision medicine

---

1. Introduction

Worldwide, breast cancer (BC) is the most common cancer affecting women [1]. In the U.S. and Puerto Rico, BC now accounts for 30% of all new cancers in women [2,3]. Molecular studies of BC have revealed distinct disease subtypes, each associated with different risk factors, etiology, incidence, prognosis, survival rates, treatment approaches, and responses [4–7]. While high-throughput gene expression analysis continues to reveal more distinctions between BC subtypes, their classification is still in flux. For example, what collectively has been known as triple-negative breast cancer (TNBC) is now yielding distinct genetic profiles for basal-like and claudin-low BC (the latter seeming to be an intermediate between basal-like and luminal BC) [8,9]. Despite gene expression profiling's ability to supply vital genetic information for stratifying patients [10], it is costly, not uniformly available, and the subset of genes analyzed vary by manufacturer. Therefore, immunohistochemistry (IHC) for receptor status and gene expression is a frequently used surrogate tool to classify BC subtypes. Under IHC auspices, Luminal A is ER+, PR+/−, HER2−; Luminal B is also positive for ER and PR+/− but is HER2+. HER2-enriched BC is ER−, PR−, HER2+. Triple-negative BC is negative for all three receptors and has the worst prognosis and treatment response [11,12].

For over a decade, our laboratory has focused, using lymphocytes as surrogate markers, on the role of DNA repair capacity (DRC) as risk factor for BC in women. We have previously shown in our large BC cohort ($n = 824$) that women with BC ($n = 285$) had a mean DRC of 2.40% (range: 0.14–15.00%); whereas the women without BC had a mean DRC level of 6.13% (range: 0.14–19.00%) [13]. In addition, we showed that the likelihood of developing BC increases by 64% for every 1% decrease in DRC measurement [13]. Therefore, our findings support what have been previously published, that low DRC is a marker that correlates with higher cancer risk [14–16]. In BC, the nucleotide excision repair (NER) pathway is particularly affected. NER deficiency is now well established as a DNA repair phenotype in BC, which suggests that it contributes to the etiology of both familial and sporadic BCs [13,16,17]. We also recently showed that ER status is associated with that defective repair phenotype [18]. This current study investigates whether DRC levels vary among BC subtypes and whether that variability is associated with an accompanying variance in risk of developing sporadic BCs.

2. Results

In the analysis of results, first an initial description of the study group was performed. Afterwards, a bivariate analysis was performed for assessing the association between different BC risk factors and molecular subtype of BC, in order to identify the potential confounders for this association. Finally, the magnitude of the association between BC molecular subtype and DRC was estimated, controlling for different confounding variables using a multinomial logistic regression model.

2.1. Description of Study Group

This study comprised 682 controls (women without BC) and 267 treatment-naïve women with recently diagnosed BC: 157 had Luminal A, 41 Luminal B, 22 HER2-positive, and 47 had triple-negative BC (TNBC). At the time of initial diagnosis, the age (years) distribution varied significantly (p-value < 0.05) by subtype; TNBC were the oldest (median 60 years), followed by Luminal A (median 58 years) and Luminal B (median 55 years) (Table 1). This distribution of BC molecular subtypes is consistent with national U.S. statistics [6,7,19–21].

Table 1. Socio-demographic and risk factor distribution by molecular breast cancer subtype.

Socio-demographic Characteristics	Controls n (%)	Luminal A n (%)	Luminal B n (%)	HER2+ n (%)	TN n (%)	p-Value [a]
n = 949	682 (71.9)	157 (16.5)	41 (4.3)	22 (2.3)	47 (4.9)	
DRC						<0.001
<4.3	271 (39.7)	122 (77.7)	26 (63.4)	17 (77.3)	39 (83.0)	
≥4.3	411 (60.3)	35 (22.3)	15 (36.6)	5 (22.7)	8 (17.0)	
Family History of BC						0.082
No	457 (67.0)	118 (75.2)	33 (80.5)	18 (81.8)	32 (68.1)	
Yes	225 (33.0)	39 (24.8)	8 (19.5)	4 (18.2)	15 (31.9)	
Age at Diagnosis						0.040
<45	193 (28.3)	26 (16.6)	6 (14.6)	3 (13.6)	9 (19.1)	
45–55	228 (33.4)	46 (29.3)	15 (36.6)	6 (27.3)	16 (34.0)	
56–65	142 (20.8)	43 (27.4)	15 (36.6)	7 (31.8)	10 (21.3)	
66–75	95 (13.9)	33 (21.0)	4 (9.8)	5 (22.7)	9 (19.1)	
>75	24 (3.5)	9 (5.7)	1 (2.4)	1 (4.6)	3 (6.4)	
Median age [b]	51 (43, 61)	58 (47, 66)	55 (50, 62)	54 (46, 66)	60 (52, 69)	

Table 1. *Cont.*

Socio-demographic Characteristics	Controls n (%)	Luminal A n (%)	Luminal B n (%)	HER2+ n (%)	TN n (%)	*p*-Value [a]
Menopausal Status						0.004
No	258 (37.8)	39 (25.0)	8 (19.5)	6 (27.3)	13 (27.7)	
Yes	424 (62.2)	117 (75.0)	33 (80.5)	16 (72.7)	34 (72.3)	
Not specified	0	1	0	0	0	
Age at Menarche						>0.1
<13	292 (42.9)	74 (48.1)	19 (46.3)	13 (59.1)	22 (46.8)	
≥13	389 (57.1)	80 (51.9)	22 (53.7)	9 (40.9)	25 (53.2)	
Not specified	1	3	0	0	0	
BMI						>0.1
<25	236 (34.6)	43 (27.4)	14 (34.2)	10 (45.5)	13 (27.7)	
25–30	258 (37.8)	52 (33.1)	18 (43.9)	5 (22.7)	20 (42.6)	
>30	188 (27.6)	62 (39.5)	9 (21.9)	7 (31.8)	14 (29.8)	
Alcohol Intake						0.033
Never	555 (82.6)	138 (89.0)	39 (95.1)	17 (77.3)	43 (91.5)	
Ever	117 (17.4)	17 (11.0)	2 (4.9)	5 (22.7)	4 (8.5)	
Not specified	10	2	0	0	0	
Smoking Habit						>0.1
Never	614 (91.1)	137 (87.8)	39 (95.1)	20 (90.9)	43 (91.5)	
Ever	60 (8.9)	19 (12.2)	2 (4.9)	2 (9.1)	4 (8.5)	
Not specified	8	1	0	0	0	
Oral Contraceptives (Pre-Menopausal Women Only)						0.054
Never	303 (45.2)	89 (58.2)	22 (53.7)	12 (54.5)	23 (48.9)	
Ever	367 (54.8)	64 (41.8)	19 (46.3)	10 (45.5)	24 (51.1)	
Not specified	12	4	0	0	0	
Vitamin Intake (Multivitamin or Calcium)						<0.001
No	333 (48.8)	96 (61.1)	28 (68.3)	14 (63.6)	37 (78.7)	
Yes	349 (51.2)	61 (38.9)	13 (31.7)	8 (36.4)	10 (21.3)	
HRT (Post-Menopausal Women Only)						0.009
Never	205 (48.3)	68 (58.1)	22 (66.7)	12 (75.0)	23 (67.6)	
Ever	219 (51.7)	49 (41.9)	11 (33.3)	4 (25.0)	11 (32.4)	

TN: triple negative; DRC: DNA repair capacity; HRT: hormone replacement therapy; SEM: standard error of the mean; [a] *p*-values were computed without missing values ("not specified" information); [b] between parentheses percentile 25, and percentile 75.

2.2. Differential Distribution of DRC

The DNA repair capacity (DRC in %) distribution by molecular subtype of BC showed a positive skew (were highly concentrated at low values) for Luminal A (median 2.0%; p25 = 1.3, p75 = 7.3) and TNBC (median 1.6%; p25 = 1.0, p75 = 2.9). After sorting the DRC (%), the value of the DRC (%) that reached 25% of the women (percentile 25) is different for each subtype: 3.4% for controls, 1.3% for Luminal A, 1.3% for Luminal B, 1.9% for HER2+ and 1.0% for TNBC. These differences were statistically significant (*p*-value < 0.05) (Table 1). In contrast, the DRC % distribution among controls was quite symmetrical (median 5.1%; p25 = 3.4, p75 = 7.3); however, this group showed the largest interquartile range (3.9) (Figure 1).

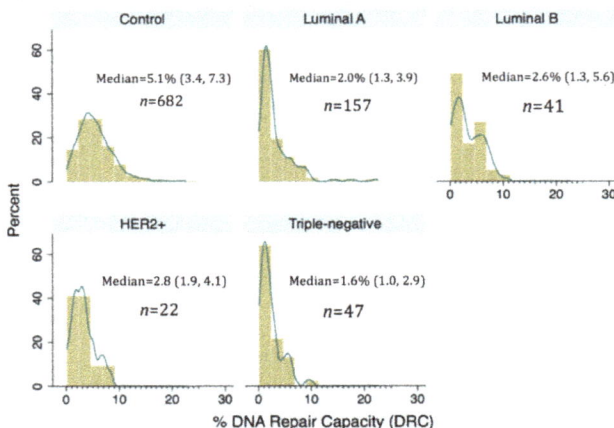

Figure 1. Distribution of DNA repair capacity (DRC) in controls and breast cancer cases stratified using the four principal molecular subtypes of breast cancer. The numbers included on each panel for each group include the median.

2.3. Molecular Subtypes of Breast Cancer by Different Characteristics

The bivariate analysis showed that the following factors had a significant association (p-value < 0.05) with molecular subtype: DRC, age, menopausal status, alcohol intake, vitamin intake, and HRT. Approximately 60% of the controls had DRC values above 4.3%, while only 17% of the women with TNBC women had DRC values above this median value. Among women with Luminal B tumors, only 19.5% reported having a family history of BC, while this distribution in controls was 33%. Around 72% of the women with tumors classified as TNBC were menopausal. Alcohol intake was more frequent in HER2+ (22.7%) followed by women in the control group (17.4%). Approximately 51% of women in the control group had multivitamin consumption (including calcium) and only 21.3% of the TNBC women reported this consumption. Among the post-menopausal women, 51.7% of the controls reported having taken HRT, while only 32.4% of the women with TNBC reported having undergone HRT therapy (Table 1).

2.4. Pathological Characteristics of Breast Tumors

Among women with BC, the results did not show significant differences (p-value > 0.05) regarding the type of BC (in situ, invasive, and mixed invasive) and the molecular subtypes. On the contrary, the grade of the tumor showed different patterns (p-value < 0.001) according to the molecular subtype of BC. Grade II cancer was the most prevalent among Luminal A and B subtypes; however, Grade III was the most prevalent in HER2+ and TN BCs (p-value < 0.001). Consistent with other findings [22,23], the highest prevalence of Grade III (most aggressive) BC occurred in the TNBC patients (30 of 44; 68%) (Table 2).

Table 2. Pathological characteristics of tumors by breast cancer molecular subtype.

Pathological Characteristics	Luminal A *n* (%)	Luminal B *n* (%)	HER2+ *n* (%)	TN *n* (%)	*p*-Value [a]
Type of Breast Cancer					>0.1
Carcinoma in situ	7 (4.6)	3 (7.3)	4 (18.2)	1 (2.2)	
Invasive	130 (85.0)	34 (82.9)	17 (77.3)	42 (93.3)	
Mixed invasive	16 (10.4)	4 (9.8)	1 (4.5)	2 (4.4)	
Not specified	2	0	0	2	
Grade					<0.001
I	28 (20.3)	2 (5.0)	0 (0.0)	0 (0.0)	
II	81 (58.7)	22 (55.0)	9 (42.9)	14 (31.8)	
III	29 (21.0)	16 (40.0)	12 (57.1)	30 (68.2)	
Not specified	19	1	1	3	

[a] *p*-values were computed without missing values. Ductal and lobular tumors were included into each category of carcinoma in situ and invasive BC. Mixed invasive refers to ductal and lobular components within the same tumor.

2.5. Magnitude of the Association

When ranked proportionally, more women with TNBC had a low DRC than any other BC subtype. The proportions of cases with low DRC were: 83% TNBC (39/47), 78% Luminal A (122/157), 77% HER2+ (17/22), 63% Luminal B (26/41). Low DRC was most strongly associated with women who had TNBC (adjusted OR: 7.2; 95% CI 3.3, 15.7). Women with HER2+ and Luminal A BC had virtually the same high association with low DRC (point estimates of the adjusted OR were 5.2 and 5.4, respectively). Women with Luminal B breast cancer had the weakest association between BC subtype and DRC levels (adjusted OR: 2.5; 95% CI 1.3, 4.9) and the least number of cases with a low DRC (Table 3).

Table 3. Multinomial odds ratio regression analysis for association between DNA repair capacity and breast cancer subtype.

Outcome	DRC<4.3%	DRC ≥ 4.3% (Reference)	Crude OR (95% CI)	Adjusted OR (95% CI) [a]
Controls (reference)	271	411	1.0	1.0
Luminal A	122	35	5.3 (3.5, 7.9)	5.4 (3.5, 2.8)
Luminal B	26	15	2.6 (1.4, 5.1)	2.5 (1.3, 4.9)
HER2+	17	5	5.2 (1.9, 14.4)	5.2 (1.9, 14.3)
Triple-negative	39	8	7.4 (3.4, 16.1)	7.2 (3.3, 15.7)
TOTALS	475	474		

[a] Fully adjusted OR: model adjusted for age, menopause status and multivitamin and/or calcium consumption. No significant interaction terms were found in this model using the likelihood ratio test (*p* > 0.05). (*n* = 949).

3. Discussion

This study represents the first report that a low DRC (in relation to controls) is present in the four principal molecular BC subtypes and that is more pronounced in TNBC. Since our study is based on a large sample size, it suggests that significant phenotypic variability in terms of DRC exists amongst the four molecular subtypes studied. Our previous studies [13,16,18] and the study of Latimer et al. (2010) [17] had clearly established the critical importance of low DRC (measured in terms of NER pathway) as a risk factor for BC. Although the focus of our previous work was to study the relationship between DRC and BC, considering the disease as a single entity, we now aim to study this effect in terms of molecular subtypes.

The lowest DRC was associated with TNBC, the molecular subtype associated with the worst prognosis [22,24]. Among the molecular subtypes included, the highest adjusted OR (7.2) was found for

TNBC. Our findings show that recently diagnosed, untreated women with TNBC, had a significantly lower DRC when compared to controls and women with Luminal A, Luminal B, and HER2+ BC. This may provide a hypothesis to at least partially explain why TNBC have a poorer prognosis when compared with other molecular subtypes. Our phenotypic measurement of DRC levels obtained from untreated women with BC confirms what is known about the prognosis of TNBC.

Our results obtained using a phenotypic assay to assess DRC are consonant with the findings reported by Ribeiro et al. (2013) at the gene expression level. This group found significant downregulation of 13 DNA repair genes, including five genes from the NER pathway (*ERCC1, XPA, XPD, XPG, XPF*) in TNBC [25]. Gene expression patterns were obtained following an RNA extraction from formalin-fixed paraffin embedded samples from 70 Luminal A, BC tumors and 80 TNBC tumors obtained from 150 women with BC in Italy. In addition to this group, Alexander et al. (2010) were able to establish prognostic markers for time to recurrence in TNBC through immunohistochemical assessment of key proteins in multiple DNA repair pathways. Among the four markers, a low expression of the NER protein XPF was associated with shorter time to recurrence. Moreover, this group developed a four-antibody model that was able to successfully identify high- and low-risk groups in terms of time to recurrence in TNBC [26]. These two studies, along with our findings, highlight the important role of NER in the biology of TNBC.

Given the research impetus in recent years for a more advanced precision medicine approach, the inclusion of phenotypic variability in DRC levels provides a tool for implementing that goal in BC diagnostic and treatment. Although the main therapeutic focus of DNA repair appears to be in the area of PARP inhibitors, DRC levels might allow us to distinguish subtle differences in BC molecular subtypes and prognosis. Since TNBC patients usually have the worst prognosis and acquire drug resistance more frequently than any other molecular subtype [27], our approach of using lymphocytes as surrogate markers of overall DRC could aid in the prediction of overall therapy response of women with TNBC. A new focus area in our studies is applying phenotypic measurement of DRC levels to study recurrence and metastasis in the large cohort of women with BC that we have studied for the last decade. Our findings may also prove useful in predicting (in terms of DRC levels) which molecular subtypes have a higher risk of recurrence and/or metastasis of BC, an important area in precision medicine.

4. Materials and Methods

4.1. Patient Recruitment

Patients were selected from our larger BC study (1183 patients and controls recruited 2006–2013). This study's cohort comprised 949 Puerto Rican women age 21 or older: 267 with newly diagnosed BC (cases) and 682 without BC (controls). From a previous study [13] in which we recruited 824 Puerto Rican women, age 21 or older (285 newly diagnosed cases and 539 controls) power and sample size calculations were made. Sample size calculations performed initially revealed that a sample size of 824 participants (312 women with BC, 515 women without BC) would allow us to have a statistically significant odds ratio as low as 1.7 when the percent exposed to a low DRC among controls is 15% or higher (e.g., 15% controls are 21 to 30 years of age) with 5% significant level and 80% of statistical power. Selection bias was minimized by recruiting women who were getting routine gynecological screenings in the same clinics and hospitals where they would be treated if they were to develop BC. Those facilities represented 83% of the municipalities (65/78 counties) on the island.

To reduce the likelihood of including undiagnosed BC cases in our controls, only women who had normal results from a clinical breast exam and mammogram within the past six months were included. BC cases were limited to only recently diagnosed, histopathologically confirmed, treatment-naïve BC patients with primary tumors and pathology reports that included hormone receptor information. Because blood transfusions, radio- and chemotherapy can significantly affect DRC [28–30], patients who had received any of those treatments in the past five years were excluded from the study. Also excluded from this analysis were those with metastatic BC, secondary BC, breast metastases from another type of cancer, or any acquired or genetic immunodeficiency.

4.2. Use of Human Subjects

The Ponce Health Sciences University Institutional Review Board approved this study (IRB #130207-JM; Date: 13 February 2013). Each participant signed an informed consent form, giving us permission to draw their blood and review their pathology reports. All participants also completed an epidemiological questionnaire.

4.3. Blood Collection and Isolation of Lymphocytes

With the participants' permission, we drew approximately 30 mL of peripheral blood into heparinized tubes and isolated the lymphocytes using the Ficoll gradient technique. Blood collection was performed during morning hours. Lymphocytes were suspended in 2 mL of freezing media (10% dimethyl sulfoxide, 39% RPMI 1640 medium, 50% fetal bovine serum, 1% antibiotic/antimycotic). Aliquots were stored in a −80 °C freezer for 1–3 weeks until thawed in batches for host-cell reactivation (HCR) assays.

4.4. DNA Repair Capacity Measurements

The isolated lymphocytes were used as surrogate markers of the patients' overall DRC [31,32]. The cells were purified and grown, then the HCR assay was performed on them to measure in vivo DRC, as described in previous studies [13,16,33–35]. Briefly, the lymphocytes were transfected with a plasmid containing the luciferase reporter gene. Plasmids had been damaged with UVC prior to transfection. The cells' ability to repair the foreign DNA was measured via HCR [35] within a specific time frame (40 h) that mirrored the true cellular process [32]. Results reflected the cells' inherent DRC, measured primarily in terms of their NER pathway activity. Details about HCR's sensitivity, specificity, and plasmid transfection efficiency have been published previously [13].

To calculate DRC, the luciferase activity after repair of the UVC-damaged plasmid DNA was compared with the undamaged plasmid DNA. The amount of residual luciferase remaining after the allotted repair time (activity in luminescence units) was a percentage that represented the amount of the individuals' DRC. Because DRC is traditionally low in BC cases, results were analyzed in tertiles, as described in our previous study [13]. However, to perform the proposed statistical analyses, the obtained experimental values of DRC of were dichotomized using the median DRC levels. With a median DRC of 4.3%, this study categorized DRC a dichotomous variable: "low" was <4.3%; "high" was ≥4.3%.

4.5. Hormone Receptor Status

Pathology reports from all cases were reviewed to confirm the diagnosis, tumor grade and size, presence/absence of axillary lymph node metastasis, and ER, PR, and HER2 status, as previously described [13,18]. Receptor status results were provided by 10 private Puerto Rican laboratories, following ASCO (American Society of Clinical Oncology, Alexandria, VA, USA) and CAP (College of American Pathologists, Northfield, IL, USA) guidelines for those immunohistochemistry (IHC) methods. [36,37] ER and PR results included the percentage of positive-staining cells, the intensity of staining (weak, moderate, or strong), and an interpretation. "Receptor positive" meant ≥1% of tumor cells stained positive for ER/PR [37]. HER2 results were reported as 0, 1+, 2+, or 3+. For this study, any 1+ or 2+ result was considered equivocal and was followed up with FISH (Fluorescence In Situ Hybridization) to so we could categorize HER2 as a dichotomous variable (HER2+ = all 3+ results; HER2− = 2+ to 0 results).

4.6. Classification of Tumors Based on IHC Receptor Status Information

Breast cancer subtypes have been defined by others [9,38,39]. Briefly, we used the data collected from the pathology reports on three IHC markers (ER, PR, and HER2) to classify tumors into four groups. Luminal A tumors were ER+, PR+/−, and HER2−. Luminal B tumors differed only by being HER2+. HER2-positive tumors were ER−, PR−, HER2+. Triple-negative tumors were ER−, PR−, HER2.

4.7. Statistical Analysis

Descriptive statistics were used to evaluate categorical data as percentages and continuous variables in terms of mean/standard deviation. Chi-square probability distribution was used to assess the statistical relationship between BC subtypes and the following characteristics: DRC, age, onset of menarche, parity and menopause status, BMI, lactation history, alcohol intake, smoking habits, contraceptive use, vitamin intake (multivitamin/calcium), and hormone replacement therapy (HRT) use. To assess the magnitude of the association between DRC and BC subtype, controlling for potential confounders, the following multinomial logistic regression model was used

$$log\frac{P_k}{P_0} = \beta_{0k} + \beta_{DRC}^k + \sum \beta_j X_j$$

where P_k indicates the prevalence of the kth-category of BC subtype, P_0 indicates the prevalence of the control group under the DRC comparison (<4.3 vs. ≥4.3), β_{DRC}^k is the coefficient associated to DRC, X_j indicates the potential confounders, and β_j is the coefficient associated with X_j. Crude and adjusted odds ratios (OR) were estimated with 95% confidence levels from this model. Statistical analyses were performed using Stata v14 (Stata Corp, College Station, TX, USA).

5. Conclusions

In general, our findings support what is known about the biology of molecular subtypes of BC. However, the ORs in terms of DRC values do not always match with the basic biology of BC. For example, the lowest OR (2.5) corresponded to Luminal B which is a more aggressive BC than Luminal A (OR = 5.4). This suggests that in terms of DNA repair, it might be important (future direction) to look at other pathways in addition to NER. It is possible that double-strand DNA breaks repaired by homologous and non-homologous end joining might also us to distinguish differences in repair between Luminal A and Luminal B.

This study has some limitations. Our study was based only on the NER pathway and we were not able to measure other pathways. We are now standardizing technology in order to be able to obtain a more comprehensive view of the dysregulation of DNA repair in BC. The HCR assay used is very costly and not easily amenable to large scale population studies such as this one. This assay depends on having viable living cells (lymphocytes) from participants (requires a blood sample) versus genetic tests that can be done with DNA isolated from paraffin embedded tumor samples (no need for live cells or draw blood). However, our study provides the first report evidence of the significant phenotypic variability among the four principal molecular subtypes of BC and provides a hypothesis based on DRC levels for the poor prognosis of TNBC. It suggests that significant phenotypic variability in terms of DRC exists amongst the four molecular subtypes studied.

Acknowledgments: Special thanks go to Dr. Luisa Morales for measurements of DNA repair capacity, to Wanda Vargas for patient recruitment and to Dr. Patricia Casbas with help in the preparation of tables. Lana Christian is acknowledged for her technical editing of the manuscript. The study was supported by the National Institutes of Health-Minority Biomedical Research Support (NIH-MBRS) Program grant # S06 GM008239-20 and the MBRS SCORE Program grant 9SC1CA182846-04, awarded to Ponce School of Medicine through Dr. Jaime Matta. Support was also provided by the National Cancer Institute (NCI) Center to Reduce Health Disparities, by the PHSU–U54 CA163071 and MCC–U54 CA163068 grants and by the Molecular and Genomics Core at PHSU through the RCMI Program grant G12-MD007579.

Author Contributions: Erick Suárez and Jaime Matta conceived and designed the study; Jaime Matta and Erick Suárez developed the methodology; Erick Suárez, Jarline Encarnación, Carmen Ortiz, and Julie Dutil performed the data analysis and interpretation; Jaime Matta, Erick Suárez, and all the co-authors wrote and reviewed the manuscript; Carmen Ortiz and Jarline Encarnación provided administrative, technical, and material support; Jaime Matta supervised the study.

Conflicts of Interest: The authors declare no conflict of interest.

Abbreviations

BC	Breast cancer
DRC	DNA repair capacity
NER	Nucleotide excision repair
IHC	Immunohistochemistry
TNBC	Triple-negative breast cancer
BMI	Body mass index
HRT	Hormone replacement therapy
TN	Triple-negative
SEM	Standard error of the mean
OR	Odds ratio

References

1. Ferlay, J.; Soerjomataram, I.; Ervik, M.; Dikshit, R.; Eser, S.; Mathers, C.; Rebelo, M.; Parkin, D.M.; Forman, D.; Bray, F. *GLOBOCAN 2012 v1.0, Cancer Incidence and Mortality Worldwide: IARC CancerBase No. 11*; International Agency for Research on Cancer: Lyon, France, 2013; Available online: http://globocan.iarc.fr (accessed on 18 June 2017).
2. American Cancer Society (ACS). *Cancer Facts & Figures 2017*; American Cancer Society: Atlanta, GA, USA, 2017; Available online: https://www.cancer.org/research/cancer-facts-statistics/all-cancer-facts-figures/cancer-facts-figures-2017.html (accessed on 18 June 2017).
3. PRCCC: Puerto Rico Comprehensive Cancer Control Plan: 2015–2020. In Puerto Rico Cancer Control Coalition and Puerto Rico Comprehensive Control Program. 2014. Available online: http://mindoven.rocks/ccpdf/ (accessed on 15 June 2017).
4. Gaudet, M.M.; Press, M.F.; Haile, R.W.; Lynch, C.F.; Glaser, S.L.; Schildkraut, J.; Gammon, M.D.; Thompson, W.D.; Bernstein, J.L. Risk factors by molecular subtypes of breast cancer across a population-based study of women 56 years or younger. *Breast Cancer Res. Treat.* **2011**, *130*, 587–597. [CrossRef] [PubMed]
5. Anderson, K.N.; Schwab, R.B.; Martinez, M.E. Reproductive risk factors and breast cancer subtypes: A review of the literature. *Breast Cancer Res. Treat.* **2014**, *144*, 1–10. [CrossRef] [PubMed]
6. Banegas, M.P.; Tao, L.; Altekruse, S.; Anderson, W.F.; John, E.M.; Clarke, C.A.; Gomez, S.L. Heterogeneity of breast cancer subtypes and survival among Hispanic women with invasive breast cancer in California. *Breast Cancer Res. Treat.* **2014**, *144*, 625–634. [CrossRef] [PubMed]
7. Singh, M.; Ding, Y.; Zhang, L.Y.; Song, D.; Gong, Y.; Adams, S.; Ross, D.S.; Wang, J.H.; Grover, S.; Doval, D.C.; et al. Distinct breast cancer subtypes in women with early-onset disease across races. *Am. J. Cancer Res.* **2014**, *4*, 337–352. [PubMed]
8. Perou, C.M. Molecular stratification of triple-negative breast cancers. *Oncologist* **2011**, *16*, 61–70. [CrossRef] [PubMed]
9. Prat, A.; Parker, J.; Karginova, O.; Fan, C.; Livasy, C.; Herschkowitz, J.; He, X.; Perou, C. Phenotypic and molecular characterization of the claudin-low intrinsic subtype of breast cancer. *Breast Cancer Res.* **2010**, *12*, R68. [CrossRef] [PubMed]
10. Parker, J.S.; Mullins, M.; Cheang, M.C.U.; Leung, S.; Voduc, D.; Vickery, T.; Davies, S.; Fauron, C.; He, X.; Hu, Z.; et al. Supervised risk predictor of breast cancer based on intrinsic subtypes. *J. Clin. Oncol.* **2009**, *27*, 1160–1167. [CrossRef] [PubMed]
11. Rosa, F.E.; Caldeira, J.R.; Felipes, J.; Bertonha, F.B.; Quevedo, F.C.; Domingues, M.A.; Moraes Neto, F.A.; Rogatto, S.R. Evaluation of estrogen receptor α and β and progesterone receptor expression and correlation with clinicopathologic factors and proliferative marker Ki-67 in breast cancers. *Hum. Pathol.* **2008**, *39*, 720–730. [CrossRef] [PubMed]

12. Partridge, A.H.; Rumble, R.B.; Carey, L.A.; Come, S.E.; Davidson, N.E.; di Leo, A.; Gralow, J.; Hortobagyi, G.N.; Moy, B.; Yee, D.; et al. Chemotherapy and targeted therapy for women with human epidermal growth factor receptor 2-negative (or unknown) advanced breast cancer: American Society of Clinical Oncology Clinical Practice Guideline. *J. Clin. Oncol.* **2014**, *32*, 3307–3329. [CrossRef] [PubMed]

13. Matta, J.; Echenique, M.; Negron, E.; Morales, L.; Vargas, W.; Gaetan, F.S.; Lizardi, E.R.; Torres, A.; Rosado, J.O.; Bolanos, G.; et al. The association of DNA Repair with breast cancer risk in women. A comparative observational study. *BMC Cancer* **2012**, *12*, 490. [CrossRef] [PubMed]

14. Matta, J.L.; Villa, J.L.; Ramos, J.M.; Sanchez, J.; Chompre, G.; Ruiz, A.; Grossman, L. DNA repair and nonmelanoma skin cancer in Puerto Rican populations. *J. Am. Acad. Dermatol.* **2003**, *49*, 433–439. [CrossRef]

15. Wei, Q.; Matanoski, G.M.; Farmer, E.R.; Hedayati, M.A.; Grossman, L. DNA repair and aging in basal cell carcinoma: A molecular epidemiology study. *Proc. Natl. Acad. Sci. USA* **1993**, *90*, 1614–1618. [CrossRef] [PubMed]

16. Ramos, J.M.; Ruiz, A.; Colen, R.; Lopez, I.D.; Grossman, L.; Matta, J.L. DNA repair and breast carcinoma susceptibility in women. *Cancer* **2004**, *100*, 1352–1357. [CrossRef] [PubMed]

17. Latimer, J.J.; Johnson, J.M.; Kelly, C.M.; Miles, T.D.; Beaudry-Rodgers, K.A.; Lalanne, N.A.; Vogel, V.G.; Kanbour-Shakir, A.; Kelley, J.L.; Johnson, R.R.; et al. Nucleotide excision repair deficiency is intrinsic in sporadic stage I breast cancer. *Proc. Natl. Acad. Sci. USA* **2010**, *107*, 21725–21730. [CrossRef] [PubMed]

18. Matta, J.; Morales, L.; Ortiz, C.; Adams, D.; Vargas, W.; Casbas, P.; Dutil, J.; Echenique, M.; Suarez, E. Estrogen receptor expression is associated with DNA repair capacity in breast cancer. *PLoS ONE* **2016**, *11*, e0152422. [CrossRef] [PubMed]

19. Voduc, K.D.; Cheang, M.C.U.; Tyldesley, S.; Gelmon, K.; Nielsen, T.O.; Kennecke, H. Breast cancer subtypes and the risk of local and regional relapse. *J. Clin. Oncol.* **2010**, *28*, 1684–1691. [CrossRef] [PubMed]

20. Keegan, T.H.; DeRouen, M.C.; Press, D.J.; Kurian, A.W.; Clarke, C.A. Occurrence of breast cancer subtypes in adolescent and young adult women. *Breast Cancer Res.* **2012**, *14*, R55. [CrossRef] [PubMed]

21. Ortiz, A.P.; Frias, O.; Perez, J.; Cabanillas, F.; Martinez, L.; Sanchez, C.; Capo-Ramos, D.E.; Gonzalez-Keelan, C.; Mora, E.; Suarez, E. Breast cancer molecular subtypes and survival in a hospital-based sample in Puerto Rico. *Cancer Med.* **2013**, *2*, 343–350. [CrossRef] [PubMed]

22. Ovcaricek, T.; Frkovic, S.G.; Matos, E.; Mozina, B.; Borstnar, S. Triple negative breast cancer—Prognostic factors and survival. *Radiol. Oncol.* **2011**, *45*, 46–52. [CrossRef] [PubMed]

23. Dent, R.; Trudeau, M.; Pritchard, K.I.; Hanna, W.M.; Kahn, H.K.; Sawka, C.A.; Lickley, L.A.; Rawlinson, E.; Sun, P.; Narod, S.A. Triple-negative breast cancer: Clinical features and patterns of recurrence. *Clin. Cancer Res.* **2007**, *13*, 4429–4434. [CrossRef] [PubMed]

24. Pal, S.; Lüchtenborg, M.; Davies, E.A.; Jack, R.H. The treatment and survival of patients with triple negative breast cancer in a London population. *SpringerPlus* **2014**, *3*, 553. [CrossRef] [PubMed]

25. Ribeiro, E.; Ganzinelli, M.; Andreis, D.; Bertoni, R.; Giardini, R.; Fox, S.B.; Broggini, M.; Bottini, A.; Zanoni, V.; Bazzola, L.; et al. Triple negative breast cancers have a reduced expression of DNA repair genes. *PLoS ONE* **2013**, *8*, e66243. [CrossRef] [PubMed]

26. Alexander, B.M.; Sprott, K.; Farrow, D.A.; Wang, X.; D'Andrea, A.D.; Schnitt, S.J.; Collins, L.C.; Weaver, D.T.; Garber, J.E. DNA repair protein biomarkers associated with time to recurrence in triple-negative breast cancer. *Clin. Cancer Res.* **2010**, *16*, 5796–5804. [CrossRef] [PubMed]

27. Bekele, R.T.; Venkatraman, G.; Liu, R.Z.; Tang, X.; Mi, S.; Benesch, M.G.; Mackey, J.R.; Godbout, R.; Curtis, J.M.; McMullen, T.P.; et al. Oxidative stress contributes to the tamoxifen-induced killing of breast cancer cells: Implications for tamoxifen therapy and resistance. *Sci. Rep.* **2016**, *6*, 21164. [CrossRef] [PubMed]

28. Leprat, F.; Alapetite, C.; Rosselli, F.; Ridet, A.; Schlumberger, M.; Sarasin, A.; Suarez, H.G.; Moustacchi, E. Impaired DNA repair as assessed by the "comet" assay in patients with thyroid tumors after a history of radiation therapy: A preliminary study. *Int. J. Radiat Oncol. Biol. Phys.* **1998**, *40*, 1019–1026. [CrossRef]

29. Bosken, C.H.; Wei, Q.; Amos, C.I.; Spitz, M.R. An analysis of DNA repair as a determinant of survival in patients with non-small-cell lung cancer. *J. Natl. Cancer Inst.* **2002**, *94*, 1091–1099. [CrossRef] [PubMed]

30. Van Loon, A.A.; Timmerman, A.J.; van der Schans, G.P.; Lohman, P.H.; Baan, R.A. Different repair kinetics of radiation-induced DNA lesions in human and murine white blood cells. *Carcinogenesis* **1992**, *13*, 457–462. [CrossRef] [PubMed]

31. Mendez, P.; Taron, M.; Moran, T.; Fernandez, M.A.; Requena, G.; Rosell, R. A modified host-cell reactivation assay to quantify DNA repair capacity in cryopreserved peripheral lymphocytes. *DNA Repair.* **2011**, *10*, 603–610. [CrossRef] [PubMed]

32. Athas, W.F.; Hedayati, M.A.; Matanoski, G.M.; Farmer, E.R.; Grossman, L. Development and field-test validation of an assay for DNA repair in circulating human lymphocytes. *Cancer Res.* **1991**, *51*, 5786–5793. [PubMed]

33. Morales, L.; Alvarez-Garriga, C.; Matta, J.; Ortiz, C.; Vergne, Y.; Vargas, W.; Acosta, H.; Ramirez, J.; Perez-Mayoral, J.; Bayona, M. Factors associated with breast cancer in Puerto Rican women. *J. Epidemiol. Glob. Health* **2013**, *3*, 205–215. [CrossRef] [PubMed]

34. Vergne, Y.; Matta, J.; Morales, L.; Vargas, W.; Alvarez-Garriga, C.; Bayona, M. Breast cancer and DNA repair capacity: Association with use of multivitamin and calcium supplements. *Integr. Med.* **2013**, *12*, 38–46.

35. Wang, L.; Wei, Q.; Shi, Q.; Guo, Z.; Qiao, Y.; Spitz, M.R. A modified host-cell reactivation assay to measure repair of alkylating DNA damage for assessing risk of lung adenocarcinoma. *Carcinogenesis* **2007**, *28*, 1430–1436. [CrossRef] [PubMed]

36. Wolff, A.C.; Hammond, M.E.; Hicks, D.G.; Dowsett, M.; McShane, L.M.; Allison, K.H.; Allred, D.C.; Bartlett, J.M.; Bilous, M.; Fitzgibbons, P.; et al. Recommendations for human epidermal growth factor receptor 2 testing in breast cancer: American Society of Clinical Oncology/College of American Pathologists clinical practice guideline update. *Arch. Pathol. Lab. Med.* **2014**, *138*, 241–256. [CrossRef] [PubMed]

37. Hammond, M.E.; Hayes, D.F.; Wolff, A.C.; Mangu, P.B.; Temin, S. American society of clinical oncology/college of american pathologists guideline recommendations for immunohistochemical testing of estrogen and progesterone receptors in breast cancer. *J. Oncol. Pract.* **2010**, *6*, 195–197. [CrossRef] [PubMed]

38. Perou, C.M.; Sorlie, T.; Eisen, M.B.; van de Rijn, M.; Jeffrey, S.S.; Rees, C.A.; Pollack, J.R.; Ross, D.T.; Johnsen, H.; Akslen, L.A.; et al. Molecular portraits of human breast tumours. *Nature* **2000**, *406*, 747–752. [CrossRef] [PubMed]

39. Prat, A.; Perou, C.M. Deconstructing the molecular portraits of breast cancer. *Mol. Oncol.* **2011**, *5*, 5–23. [CrossRef] [PubMed]

International Journal of
Molecular Sciences

MDPI

Article

Estrogen Metabolism-Associated CYP2D6 and IL6-174G/C Polymorphisms in *Schistosoma haematobium* Infection

Rita Cardoso [1], Pedro C. Lacerda [2], Paulo P. Costa [2,3], Ana Machado [3,4], André Carvalho [5], Adriano Bordalo [3,4], Ruben Fernandes [6,7], Raquel Soares [7,8], Joachim Richter [9], Helena Alves [1,10] and Monica C. Botelho [1,7,*]

[1] Department of Health Promotion and Chronic Diseases, National Institute of Health Dr. Ricardo Jorge (INSA), Rua Alexandre Herculano 321, 4000-055 Porto, Portugal; rita.cardoso@insa.min-saude.pt (R.C.); m.helena.alves@insa.min-saude.pt (H.A.)
[2] Department of Human Genetics, National Institute of Health Dr. Ricardo Jorge (INSA), Rua Alexandre Herculano 321, 4000-055 Porto, Portugal; pedro.lacerda@insa.min-saude.pt (P.C.L.); Paulo.costa@insa.min-saude.pt (P.P.C.)
[3] Instituto de Ciências Biomédicas Abel Salazar (ICBAS/UP), Universidade do Porto, Rua Jorge Viterbo Ferreira 228, P 4050-313 Porto, Portugal; ammachado@icbas.up.pt (A.M.); bordalo@icbas.up.pt (A.B.)
[4] Centro Interdisciplinar de Investigação Marinha e Ambiental (CIIMAR/CIMAR), Universidade do Porto, Av. General Norton de Matos s/n, 4450-208 Matosinhos, Portugal
[5] Division of Endocrinology, Diabetes and Metabolism, Santo Antonio Hospital—Centro Hospitalar do Porto (CHP), Largo do Prof. Abel Salazar, 4099-001 Porto, Portugal; andre.carvalho@chporto.min-saude.pt
[6] Escola Superior de Saúde, Instituto Politécnico do Porto, Rua Dr. António Bernardino de Almeida, 400, 4200-079 Porto, Portugal; ruben@ess.ipp.pt
[7] Unit of Metabolism, Nutrition and Endocrinology, Instituto de Investigação e Inovação da Universidade do Porto (i3S), Rua Alfredo Allen, 4200-135, Porto, Portugal; raqsoa@med.up.pt
[8] Departamento de Biomedicina, Unidade de Bioquímica, Faculdade de Medicina, Universidade do Porto, Al. Prof. Hernâni Monteiro, 4200-319 Porto, Portugal
[9] Institute of Tropical Medicine and International Health, Charité—Universitätsmedizin Berlin, Augustenburger Platz 1, 13353 Berlin, Germany; joachim.richter@charite.de
[10] Fundação Professor Ernesto Morais, Rua de Monsanto 512, 4250-288 Porto, Portugal
*   Correspondence: monicabotelho@hotmail.com or monica.botelho@insa.min-saude.pt; Tel.: +351-223-401-114; Fax: +351-223-401-109

Received: 2 October 2017; Accepted: 23 November 2017; Published: 28 November 2017

Abstract: *Schistosoma haematobium* is a human blood fluke causing a chronic infection called urogenital schistosomiasis. Squamous cell carcinoma of the urinary bladder (SCC) constitutes chronic sequelae of this infection, and *S. haematobium* infection is accounted as a risk factor for this type of cancer. This infection is considered a neglected tropical disease and is endemic in numerous countries in Africa and the Middle East. Schistosome eggs produce catechol-estrogens. These estrogenic molecules are metabolized to active quinones that induce modifications in DNA. The cytochrome P450 (CYP) enzymes are a superfamily of mono-oxygenases involved in estrogen biosynthesis and metabolism, the generation of DNA damaging procarcinogens, and the response to anti-estrogen therapies. IL6 Interleukin-6 (IL-6) is a pleiotropic cytokine expressed in various tissues. This cytokine is largely expressed in the female urogenital tract as well as reproductive organs. Very high or very low levels of IL-6 are associated with estrogen metabolism imbalance. In the present study, we investigated the polymorphic variants in the *CYP2D6* gene and the C-174G promoter polymorphism of the *IL-6* gene on *S. haematobium*-infected children patients from Guine Bissau. *CYP2D6* inactivated alleles (28.5%) and *IL6G*-174C (13.3%) variants were frequent in *S. haematobium*-infected patients when compared to previously studied healthy populations (4.5% and 0.05%, respectively). Here we discuss our recent findings on these polymorphisms and whether they can be predictive markers of schistosome infection and/or represent potential biomarkers for urogenital schistosomiasis associated bladder cancer and infertility.

Keywords: estrogen biosynthesis; estrogen metabolism; BMI; *S. haematobium*-associated bladder cancer

1. Introduction

Schistosoma haematobium is a human blood fluke causing a chronic infection called urogenital schistosomiasis. Squamous cell carcinoma of the urinary bladder (SCC) constitutes chronic sequelae of this infection, and *S. haematobium* infection is accounted as a risk factor for this type of cancer. This infection is considered a neglected tropical disease and is endemic in many countries of Africa and the Middle East [1].

S. haematobium is endemic in 53 countries in the Middle East and in most of the African continent, including the islands of Madagascar and Mauritius. Following successful eradication programs, the infection is no longer of public health importance in Egypt, Lebanon, Oman, Syria, Tunisia and Turkey since transmission is low or nonexistent. A borderline and indefinable focus is still in existence in India and requires additional evidence [2].

Infection with *Schistosoma* spp. affects more than 258 million people worldwide. Praziquantel is the main antihelminthic drug currently used to treat this infection. This drug is effective in eliminating adult worms, but is unsuccessful in the prevention of re-infection and does not treat severe liver damage nor bladder cancer [3].

Schistosome eggs produce catechol-estrogens. These estrogenic molecules are metabolized by cytochrome P450 oxygenases to active quinones that cause alterations in DNA, known to promote breast or thyroid cancer [4–6]. Our group has shown that schistosome egg-associated catechol estrogens induce tumor-like phenotypes in urothelial cells, possibly due to the formation of parasite estrogen-host cell chromosomal DNA adducts [5]. These estrogen metabolites also contribute to schistosomiasis-associated infertility [6].

The cytochrome P450 (*CYP*) supergene family encompasses a cluster of oxygenases that play a key role in the metabolism of a miscellaneous group of endogenous substrates such as fatty acids, steroids, and vitamin D as well as exogenous compounds including phytochemicals, environmental pollutants, and pharmaceuticals [7]. Given the vital function of *CYP* genes in the biosynthesis of steroids, especially estrogen, altered expression of *CYP*s might contribute to the development and proliferation of tumor cells and increase tumor growth through the activation of procarcinogens. Specifically *CYP2D6* encodes a critical enzyme on estrogen biosynthesis and metabolism, and the outcome of this *CYP* gene variant can have a downstream cost on patient response [8].

Interleukin-6 (IL-6) is a pleiotropic proinflammatory cytokine, vastly expressed in the female urogenital tract and reproductive organs. It has been given a role in estrogen metabolism imbalance. The promoter region of the *IL-6* gene is dynamically regulated at multiple sites, as well as the 23 base-pair "multiple response element" site, which is activated by interleukin-1, tumor necrosis factor alpha, and other factors [9]. The C-174G promoter polymorphism of the *IL-6* gene has been established to control transcriptional regulation [10] and has been associated with plasma IL-6 levels in patients with systemic-onset juvenile chronic arthritis and in patients with primary Sjögren's syndrome [11].

To the best of our knowledge, despite the established role of *CYP2D6* and *IL6* in estrogen metabolism, there are no studies addressing these gene variants in *S. haematobium*-infected patients. In the present study, we investigated polymorphic variants in *CYP2D6* and the -174 G/C (rs1800795) promoter polymorphism of the *IL-6* gene on a cohort of *S. haematobium*-infected children from Guinea-Bissau.

2. Results

2.1. CYP 2D6 Alleles *3, *4 and *5/*5 in S. haematobium-Infected Patients

From the 18 patients studied, we obtained frequencies of *CYP2D6* for 14 patients. Mutant samples were analyzed in Channel 705 of Light Cycler 2.0 Instrument showing melting peaks for *hCR5* (amplification control) at 47.5 °C. The *CYP 2D6* Alleles *3 and *4 were not found in any of the samples studied. In contrast, we found that 4 of 14 (28.5%) schistosomiasis-haematobia-infected patients are carriers of the inactivated allele *CYP2D6*5*, which is characterized by a deletion of the entire *CYP2D6* gene (Table 1 and Figure 1). Microhaematuria was found in all of the *CYP2D6* inactivated allele carriers and only in 80% of non-carriers. Age and body mass index (BMI) were not significantly different between the two groups.

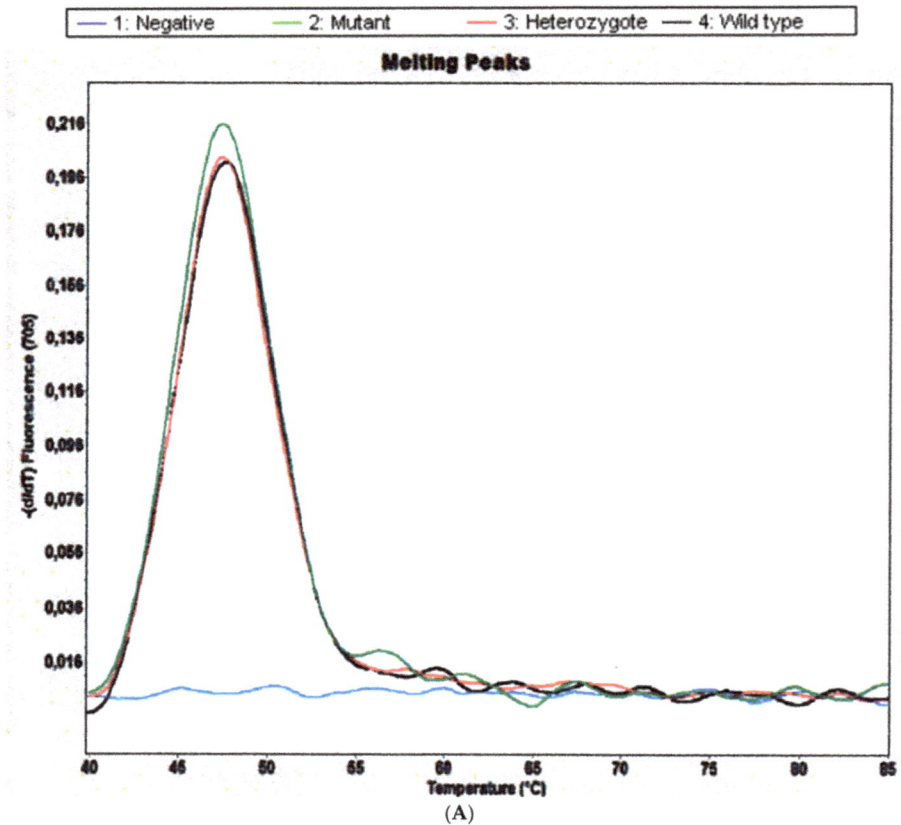

Figure 1. *Cont.*

Melting Peaks

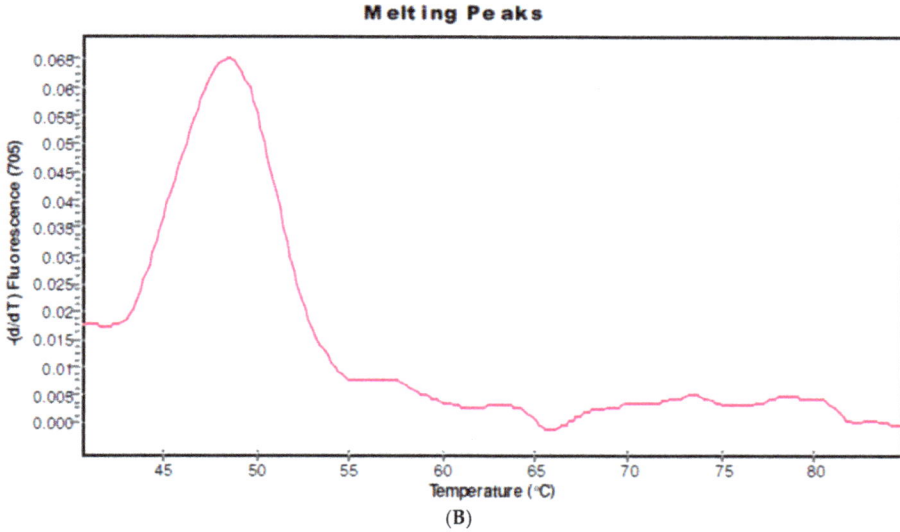

(B)

Figure 1. Genotyping of *CYP2D6 *3*4*5/5*. (**A**) Derivative melting curve plots—dF/dT vs. temperature. **Red** = Wildtype. **Black** = Heterozygote. **Green** = Mutant. (**B**) Genotyping in a mutant patient. In the case where no melting signals are visible, but the control gene has been amplified as shown by the melting point at 48.0 °C, the *CYP 2D6* gene is deleted (*5/*5).

Table 1. Population characteristics of *CYP2D6*5/*5* carriers vs. wild type in children infected with *S. haematobium*.

Population Characteristics	*CYP; n = 4* (28.5%)	WT; *n = 10* (71.4%)	*p* Value	OR	95% CI
Age (years, median ± SD)	10.75	10.8	n.s.		
Female	1	4	n.s.	0.5242	0.01537, 7.015
Male	3	6	n.s.	1.908	0.1426, 65.05
Microhaematuria (%)	4	8	n.s.	1.599	0.07674, 72.45
BMI (median ± SD)	15.2	15.5	n.s.		

CYP—genotype CYP2D6*5/*5; CI—confidence interval; SD—standard deviation; BMI—body mass index; OR—odds ratio; n.s.—not significant; WT—wild type.

2.2. IL6-174C Variant in S. haematobium Infected Patients

Fifteen out of the 18 patients studied presented the *IL6-174C* variant. Mutant samples were analyzed in Channel 640 of the LightCycler 2.0 Instrument showing melting peaks at 57 °C. The *IL6-174C* variant was found in 2 of 15 (13.3%) schistosomiasis-haematobia-infected patients (Table 2 and Figure 2). The two patients carrying the mutant genotype were younger than the ones with the wild type (WT) (6.5 ± 0.7 vs. 10.1 ± 3.1; $p = 0.005$) and presented a lower BMI (10.6 ± 5.9 vs. 14.8 ± 1.8; $p = 0.04$). Microhaematuria was present in one of the mutant carriers (50%) and in 9 of the 13 (69%) WT carrier patients.

Table 2. Population characteristics of *IL6-174C/C* carriers vs. wild type in children infected with *S. haematobium*.

Population Characteristics	IL6; *n* = 4 (28.5%)	WT; *n* = 10 (71.4%)	*p* Value	OR	95% CI
Age (years, median ± SD)	6.5 ± 0.7	10.1 ± 3.1	0.005		
Female	1	5	n.s.	1.549	0.03441, 69.74
Male	1	8	n.s	0.6455	0.01434, 29.06
Microhaematuria (%)	1	9	n.s	0.4714	0.01033, 21.51
BMI (median ± SD)	10.6 ± 5.9	14.8 ± 1.8	0.04		

IL6—genotype *IL6-174C/C*; CI—confidence interval; SD—standard deviation; BMI—body mass index; OR—odds ratio; n.s.—not significant; WT—wild type.

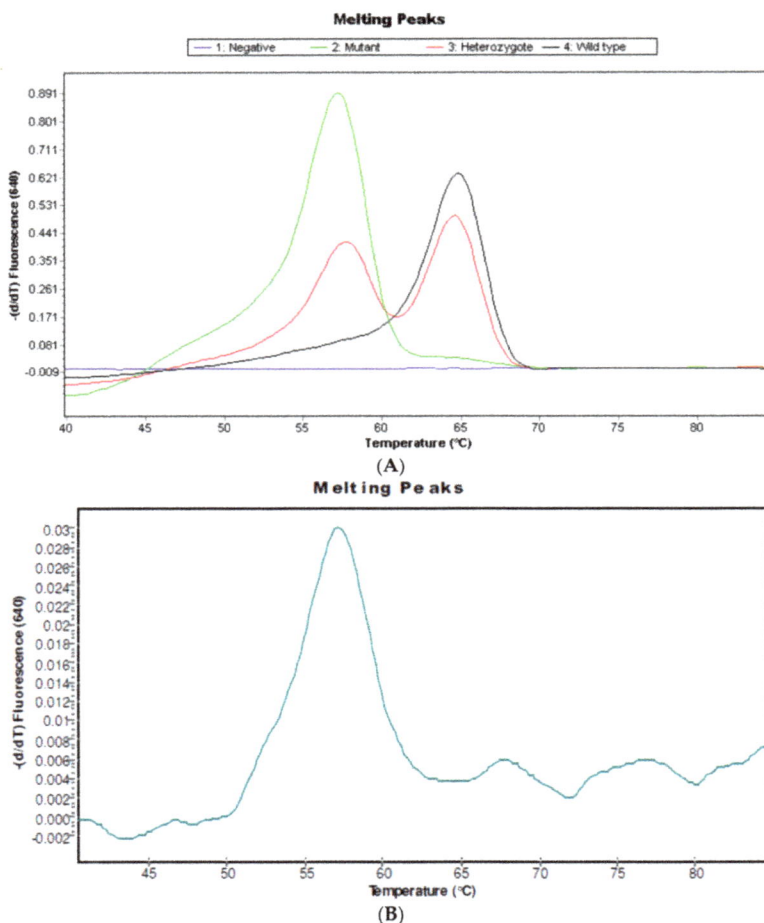

Figure 2. Genotyping of the promoter *IL6 G-174C* region. (**A**) Derivative melting curve plots—dF/dT vs. temperature. **Red** = Wildtype. **Black** = Heterozygote. **Green** = Mutant. (**B**) Genotyping in a mutant patient. The presence of a melting peak at 57 °C indicate a mutant patient corresponding with *IL6-174C7C* genotype.

3. Discussion

This preliminary study with a limited number of patients suggested that *CYP2D6 *5/*5* and *IL6-174C* polymorphisms has an effect on severity and morbidity of schistosomiasis. In the present study, we found that 28.5% of schistosomiasis-infected patients were carriers of *CYP2D6 *5/*5*. To our knowledge, this is the first study of this polymorphism conducted on an African population. According to [12], the frequency of this allele in a healthy population is 4.5%. The cytochrome P450 enzyme debrisoquine 4-hydroxylase (CYP2D6) metabolizes countless diverse classes of universally used drugs and toxins. Due to autosomal recessive inheritance of two mutant *CYP2D6* null alleles, individuals are classified as carriers of the inactivated allele *CYP2D6*5*, which is characterized by a deletion of the entire *CYP2D6* gene and confers the phenotype of poor metabolizers (PM) [13]. Poor metabolizer subjects might acquire toxic plasma concentrations and adverse drug reactions. Additionally, mutant *CYP2D6* alleles have been implicated as a predictor of susceptibility for diseases such as cancer and for neurological disorders [13]. In our study, 28.5% are classified as poor metabolizers. Given the fact that in the present study 4 of 5 (80%) of the poor metabolizers are underweight (BMI < 15), a feature associated in our previous study with *S. haematobium* infection [14], it is likely that this genotype might increase the susceptibility to infection and morbidity of this parasite [13].

Our group has been involved in the identification of parasite-derived substances that might be implicated in the host–parasite interactions of schistosomes [15]. The bulk of these substances are catechol estrogens. The genotoxic effects of these estrogen metabolites are ascribed to oxidation of catechol estrogens to quinones, followed by redox cycling and the formation of reactive oxygen species that sequentially react with DNA [16]. These electrophilic compounds are able to react with DNA to form depurinating adducts [17]. It is conceivable that apurinic sites in chromosomal DNA that result from this reaction generate mutations that might underlie the carcinogenic effect of schistosomes [1,18]. Given the context of the unarguable link between imbalance in the metabolism of estrogens and the production of depurinating estrogen–DNA adducts, the presence of schistosomiasis-derived estrogen metabolites may have practical consequences in the growth development of infected children carriers of the *CYP2D6 *5/*5* allele.

We observed that 174 G/C (rs1800795) promoter polymorphism of the *IL-6* gene was found in 2 of 15 (13.3%) of *S. haematobium*-infected individuals. The frequency of this variant in a healthy African population is 0.05% [10]. These authors also studied Caucasians and Gujarati Indians and found a frequency of 0.4% and 0.1%, respectively, for the mutant *IL6-174 C/C* genotype [10]. The presence of the mutant genotype results in a lower IL-6 expression after a given inflammatory stimulus compared with the wild-type genotype [10]. Therefore, the presence of the mutant genotype in *S. haematobium*-infected patients therefore suggests that this genotype confers a susceptibility influence for the development of the disease. We also found it to be significantly associated with lower BMI (10.6 ± 5.9 vs. 14.8 ± 1.8; $p = 0.04$), indicating that infected carriers of this variant might have an increased risk of developing schistosomiasis-associated chronic sequelae at a much younger age. Concerning the role of IL6 in estrogen metabolism, there is emerging evidence linking IL6 deficiency with reproductive impairment, leading to in how this cytokine contributes to infertility [19]. This is in accordance with our recent new findings that schistosomiasis is associated with infertility and suboptimal fecundity [6].

Altogether, the current survey provides primary data on the frequency of inactivating alleles of *CYP2D6* and *IL6 G-174C* polymorphisms in *S. haematobium*-infected patients. Despite a limited number of patients, we found an appalling increase in the frequencies of *CYP2D6 *5/*5* and *IL6-174C/C* genotypes in comparison to previously studied healthy populations, including populations of healthy African subjects previously studied. The presence of these genotypes could explain schistosomiasis-associated cancer and infertility and may represent potential predictors for growth development and metabolism disorders in these patients. On the other hand, they may have prognostic significance, namely, regarding the development of cancer and infertility, something that will need to be addressed in further studies.

4. Material and Methods

4.1. Study Area, Population and Design

This research (PTDC/AAC-CLI/103539/2008) was carried out in compliance with the Helsinki Declaration and with the approval of the Executive Board of the Institute of Biomedical Sciences Abel Salazar of Porto University.

The study was conducted in a children population from Guinea-Bissau (West Africa) in early September 2011 during the peak of the wet season.

Eighteen schoolchildren aged 6–13 infected with *S. haematobium* were targeted in this study. The purpose of the study was explained to all childrens' parents, and individual informed consent was obtained.

4.2. Urine Collection

Following the anthropometric measurements, each child was asked to urinate in a plastic cup. Urine (50–200 mL) was immediately transferred to 15 mL sterile non-heparinized vacuum tubes, and kept refrigerated in cool boxes.

4.3. Urine Analysis

Upon collection, urine was checked for microhaematuria by means of appropriate reagent strips (Combur®. Roche Diagnostics Division, Basel, Switzerland). In the laboratory, in Portugal, the presence of eggs of *S. haematobium* was detected and quantified by microfiltration of 10 mL of urine through nucleopore filters [20].

4.4. Anthropometric Measurements

Body weight and height were measured using a standardized method of anthropometric techniques (WHO, 1995). Height was measured to the nearest 0.1 cm, and weight was measured to the nearest 0.1 kg using portable digital scales. Body mass index (BMI) of each child was calculated. BMI < 15 kg/m^2 was considered underweight [21].

4.5. DNA Collection and Extraction

Genomic DNA was extracted from urine sediments using High Pure PCR Template Preparation kits (Roche Diagnostics, GmbH, Mannheim, Germany) [22].

4.6. Genetic Analysis

4.6.1. Detection of CYP 2d6 Alleles *3, *4 and *5/*5

We used Lightmix Kit for the detection of CYP 2D6 Alleles *3, *4, and *5/*5. This kit provides a fast, easy, and accurate system to identify CYP 2D6 Alleles *3 and *4 as well as a homozygous deletion of the gene (*5/*5) in a nucleic acid extract according to the manufacturer (TIBMolBiol GmbH, Berlin, Germany) [23]. After amplification with specific primers, the genotypes were identified through specific melting points (Tm) recorded during the melting curve analysis. For identification of Allele *3, a 317 bp fragment from Exon 5 was amplified and analyzed with a SimpleProbe oligomer (Channel 530) depicting a Tm of 60.2 °C for the wild-type allele and 55.0 °C for the deletion 2637delA allele. For analysis of Allele *4, a 336 bp fragment spanning the Intron 3–Exon 4 junction was generated and analyzed with LightCycler Red 640 labeled hybridization probes, exhibiting a Tm of 56.3 °C for the wild-type allele and 64.5 °C for the variant 1934A. The deletion of the entire *CYP2D6* gene (CYP 2D6*5/*5) did not produce any signal in Channels 530 or 640. In this case, to demonstrate the presence of amplifiable DNA in these biological samples, a 234 bp fragment of the human chemokine receptor type 5 (hCR5) was co-amplified with specific primers. The hCR5 amplification was detected using hybridization probes labeled with LightCycler Red 690 (Channel 705), exhibiting a specific melting peak at a Tm of 48 °C.

4.6.2. Detection of IL6 G-174C

We used Lightmix Kit for the detection of IL6 G-174C. This kit provides a fast, easy, and accurate system for identifying the genotype of IL6 G-174C in a nucleic acid extract according to the manufacturer (TIBMolBiol, GmbH, Berlin, Germany) [24]. A 175 bp fragment of the human *IL6* gene spanning the promoter IL6 G-174C region was amplified with specific primers. The resulting PCR fragments were analyzed with hybridization probes labeled with LightCycler Red 640. The genotype was identified by running a melting curve with specific melting points (Tm). The wild-type allele IL6 G-174C exhibited a Tm of 64.0 °C in Channel 640. The allele variant IL6-174C exhibited a Tm of 57.0 °C in Channel 640.

4.6.3. PCR Experiment Protocol

A total of 20 µL of PCR mixture containing 2–5 µL of sample DNA according to Roche's datasheet of LightCycler FastStart DNA Master Plus HybProbe (Roche Diagnostics, Mannheim, Germany) was used [25]. The LC PCR assay was performed on the LightCycler 2.0 Instrument (Roche Diagnostics, Mannheim, Germany) with an initial denaturation at 95 °C for 10 min, followed by 45 cycles with denaturation at 95 °C for 5 s, 60 °C for 10 s, and 72 °C for 15 s. After amplification cycles, the reaction mixture was denatured at 95 °C for 20 s, held at 40 °C for 20 s followed by one step at 40 °C for 30 s, and gradually heated to 85 °C at a rate of 0.2 °C/s. The melting curves were converted to melting peaks by plotting the negative derivative of the fluorescent signal with respect to temperature [d(F2)/dT]. In this way, the presence of a mutant heteroduplex (containing the wild-type sequences and the mutant allele) is easily detectable because of its low melting temperatures.

4.7. Statistical Analysis

For the group comparison, chi-square tests with Yate's correction were used or with Fisher's exact test (two-sided) when expected values were below 5. For independent samples, a Student's *t*-test was used for the comparison of means (OpenEpi software, version 3.03, Atlanta, GA, USA).

5. Conclusions

Altogether, the current survey provides primary data on the frequency of inactivating alleles of *CYP2D6* and *IL6* G-174C polymorphisms in *S. haematobium*-infected patients. Despite a limited number of patients, we found an appalling increase in the frequencies of *CYP2D6* *5/*5 and *IL6*-174C/C genotypes in comparison to previously studied healthy populations, including those of African subjects. The presence of these genotypes could explain schistosomiasis-associated cancer and infertility and may represent potential predictors for growth development and metabolism disorders in these patients. On the other hand, they may have prognostic significance, namely regarding the development of cancer and infertility, something that will need to be addressed in further studies.

Acknowledgments: We would like to thank Jorge Machado (Department of Infectious Diseases, National Institute of Health Ricardo Jorge, Lisbon, Portugal) for the use of LightCycler 2.0. This work was financed by FEDER—Fundo Europeu de Desenvolvimento Regional funds through the COMPETE 2020—Operacional Programme for Competitiveness and Internationalisation (POCI), Portugal 2020, and by Portuguese funds through FCT—Fundação para a Ciência e a Tecnologia/ Ministério da Ciência, Tecnologia e Inovação in the framework of the project "Institute for Research and Innovation in Health Sciences" (POCI-01-0145-FEDER-007274), and by UID/BIM/04293/2013.

Author Contributions: Monica C. Botelho conceived and designed the experiments; Rita Cardoso, Ana Machado, André Carvalho, and Adriano Bordalo performed the experiments; Ruben Fernandes, Helena Alves, and Monica C. Botelho analyzed the data; Pedro C. Lacerda, Paulo P. Costa, Ruben Fernandes, Raquel Soares, and Joachim Richter contributed reagents/materials/analysis tools; Monica C. Botelho wrote the paper.

Conflicts of Interest: The authors declare no conflict of interest.

Abbreviations

SCC   Squamous Cell Carcinoma
CYP   Cytochrome P450
IL-6   Interleukin 6
BMI   Body Mass Index
WT   Wild-Type
PM   Poor Metabolizer

References

1. Botelho, M.C.; Alves, H.; Barros, A.; Rinaldi, G.; Brindley, P.J.; Sousa, M. The role of estrogens and estrogen receptor signaling pathways in cancer and infertility: The case of schistosomes. *Trends Parasitol.* **2015**, *31*, 246–250. [CrossRef] [PubMed]
2. Botelho, M.C.; Machado, J.C.; Brindley, P.J.; Correia da Costa, J.M. Targeting molecular signaling pathways of *Schistosoma haemotobium* infection in bladder cancer. *Virulence* **2011**, *2*, 267–279. [CrossRef] [PubMed]
3. Koslowski, N.; Sombetzki, M.; Loebermann, M.; Engelmann, R.; Grabow, N.; Österreicher, C.H.; Trauner, M.; Mueller-Hilke, B.; Reisinger, E.C. Single-sex infection with female *Schistosoma mansoni* cercariae mitigates hepatic fibrosis after secondary infection. *PLoS Negl. Trop. Dis.* **2017**, *11*, e0005595. [CrossRef] [PubMed]
4. Botelho, M.C.; Soares, R.; Vale, N.; Ribeiro, R.; Camilo, V.; Almeida, R.; Medeiros, R.; Gomes, P.; Machado, J.C.; Correia da Costa, J.M. *Schistosoma haematobium*: Identification of new estrogenic molecules with estradiol antagonistic activity and ability to inactivate estrogen receptor in mammalian cells. *Exp. Parasitol.* **2010**, *126*, 526–535. [CrossRef] [PubMed]
5. Botelho, M.C.; Vale, N.; Gouveia, M.J.; Rinaldi, G.; Santos, J.; Santos, L.L.; Gomes, P.; Brindley, P.J.; Correia da Costa, J.M. Tumour-like phenotypes in urothelial cells after exposure to antigens from eggs of *Schistosoma haematobium*: An oestrogen-DNA adducts mediated pathway? *Int. J. Parasitol.* **2013**, *43*, 17–26. [CrossRef] [PubMed]
6. Santos, J.; Gouveia, M.J.; Vale, N.; Delgado Mde, L.; Gonçalves, A.; da Silva, J.M.; Oliveira, C.; Xavier, P.; Gomes, P.; Santos, L.L.; et al. Urinary estrogen metabolites and self-reported infertility in women infected with *Schistosoma haematobium*. *PLoS ONE* **2014**, *9*, e96774. [CrossRef] [PubMed]
7. Nebert, D.W.; Russell, D.W. Clinical importance of the cytochromes P450. *Lancet* **2002**, *360*, 1155–1162. [CrossRef]
8. Blackburn, H.L.; Ellsworth, D.L.; Shriver, C.D.; Ellsworth, R.E. Role of cytochrome *P450* genes in breast cancer etiology and treatment: Effects on estrogen biosynthesis, metabolism, and response to endocrine therapy. *Cancer Causes Control* **2015**, *26*, 319–332. [CrossRef] [PubMed]
9. Terry, C.F.; Loukaci, V.; Green, F.R. Cooperative influence of genetic polymorphisms on interleukin 6 transcriptional regulation. *J. Biol. Chem.* **2000**, *275*, 18138–18144. [CrossRef] [PubMed]
10. Fishman, D.; Faulds, G.; Jeffery, R.; Mohamed-Ali, V.; Yudkin, J.S.; Humphries, S.; Woo, P. The effect of novel polymorphisms in the interleukin-6 (*IL-6*) gene on IL-6 transcription and plasma IL-6 levels, and an association with systemic-onset juvenile chronic arthritis. *J. Clin. Investig.* **1998**, *102*, 1369–1376. [CrossRef] [PubMed]
11. Hulkkonen, J.; Pertovaara, M.; Antonen, J.; Pasternack, A.; Hurme, M. Elevated interleukin-6 plasma levels are regulated by the promoter region polymorphism of the *IL6* gene in primary Sjogren's syndrome and correlate with the clinical manifestations of the disease. *Rheumatology* **2001**, *40*, 656–661. [CrossRef] [PubMed]
12. Gaedigk, A.; Blum, M.; Gaedigk, R.; Eichelbaum, M.; Meyer, U.A. Deletion of the entire cytochrome P450 *CYP2D6* gene as a cause of impaired drug metabolism in poor metabolizers of the debrisoquine/sparteine polymorphism. *Am. J. Hum. Genet.* **1991**, *48*, 943–950. [PubMed]
13. Steen, V.M.; Molven, A.; Aarskog, N.K.; Gulbrandsen, A.K. Homologous unequal cross-over involving a 2.8 kb direct repeat as a mechanism for the generation of allelic variants of human cytochrome P450 *CYP2D6* gene. *Hum. Mol. Genet.* **1995**, *4*, 2251–2257. [CrossRef] [PubMed]
14. Botelho, M.C.; Machado, A.; Carvalho, A.; Vilaça, M.; Conceição, O.; Rosa, F.; Alves, H.; Richter, J.; Bordalo, A.A. *Schistosoma haematobium* in Guinea-Bissau: Unacknowledged morbidity due to a particularly neglected parasite in a particularly neglected country. *Parasitol. Res.* **2016**, *115*, 1567–1572. [CrossRef] [PubMed]

15. Botelho, M.C.; Ribeiro, R.; Vale, N.; Oliveira, P.; Medeiros, R.; Lopes, C.; Machado, J.C.; Correia da Costa, J.M. Inactivation of estrogen receptor by *Schistosoma haematobium* total antigen in bladder urothelial cells. *Oncol. Rep.* **2012**, *27*, 356–362. [CrossRef] [PubMed]

16. Fussell, K.C.; Udasin, R.G.; Smith, P.J.; Gallo, M.A.; Laskin, J.D. Catechol metabolites of endogenous estrogens induce redox cycling and generate reactive oxygen species in breast epithelial cells. *Carcinogenesis* **2011**, *32*, 1285–1293. [CrossRef] [PubMed]

17. Cavalieri, E.L.; Rogan, E.G. Depurinating estrogen-DNA adducts, generators of cancer initiation: Their minimization leads to cancer prevention. *Clin. Transl. Med.* **2016**, *5*, 12. [CrossRef] [PubMed]

18. Botelho, M.C.; Alves, H.; Richter, J. Estrogen catechols detection as biomarkers in schistosomiasis induced cancer and infertility. *Lett. Drug Des. Discov.* **2017**, *14*, 135–138. [CrossRef] [PubMed]

19. Prins, J.R.; Gomez-Lopez, N.; Robertson, S.A. Interleukin-6 in pregnancy and gestational disorders. *J. Reprod. Immunol.* **2012**, *95*, 1–14. [CrossRef] [PubMed]

20. Botelho, M.C.; Sousa, M. New biomarkers to fight urogenital schistosomiasis: A major neglected tropical disease. *Biomark. Med.* **2014**, *8*, 1061–1063. [CrossRef] [PubMed]

21. World Health Organization. *Physical Status: The Use and Interpretation of Anthropometry*; Technical Report Series 854; WHO Expert Committee: Geneva, Switzerland, 1995; 452p.

22. Vogelstein, B.; Gillespie, D. Preparative and analytical purification of DNA from agarose. *Proc. Natl. Acad. Sci. USA* **1979**, *76*, 615–619. [CrossRef] [PubMed]

23. Sistonen, J.; Sajantila, A.; Lao, O.; Corander, J.; Barbujani, G.; Fuselli, S. CYP2D6 worldwide genetic variation shows high frequency of altered activity variants and no continental structure. *Pharmacogenet. Genom.* **2007**, *17*, 93–101.

24. Sawczenko, A.; Azooz, O.; Paraszczuk, J.; Idestrom, M.; Croft, N.M.; Savage, M.O.; Ballinger, A.B.; Sanderson, I.R. Intestinal inflammation-induced growth retardation acts through IL-6 in rats and depends on the -174 IL-6 G/C polymorphism in children. *Proc. Natl. Acad. Sci. USA* **2005**, *102*, 13260–13265. [CrossRef] [PubMed]

25. Weise, A.; Prause, S.; Eidens, M.; Weber, M.M.; Kann, P.H.; Forst, T.; Pfützner, A. Prevalence of *CYP450* gene variations in patients with type 2 diabetes. *Clin. Lab.* **2010**, *56*, 311–318. [PubMed]

International Journal of
Molecular Sciences

MDPI

Article

Expression Profile of Genes Regulating Steroid Biosynthesis and Metabolism in Human Ovarian Granulosa Cells—A Primary Culture Approach

Wiesława Kranc [1], Maciej Brązert [2], Katarzyna Ożegowska [2], Mariusz J. Nawrocki [1], Joanna Budna [3], Piotr Celichowski [3], Marta Dyszkiewicz-Konwińska [1,4], Maurycy Jankowski [1], Michal Jeseta [5], Leszek Pawelczyk [2], Małgorzata Bruska [1], Michał Nowicki [3], Maciej Zabel [3,6] and Bartosz Kempisty [1,3,5,*]

[1] Department of Anatomy, Poznan University of Medical Sciences, 60-781 Poznan, Poland; wkranc@ump.edu.pl (W.K.); mjnawrocki@ump.edu.pl (M.J.N.); mdyszkiewicz@ump.edu.pl (M.D.-K.); m.jankowski.14@aberdeen.ac.uk (M.J.); mbruska@ump.edu.pl (M.B.)
[2] Division of Infertility and Reproductive Endocrinology, Department of Gynecology, Obstetrics and Gynecological Oncology, Poznan University of Medical Sciences, 60-101 Poznan, Poland; maciejbrazert@ump.edu.pl (M.B.); k.ozegowska@gmail.com (K.O.); pawelczyk.leszek@ump.edu.pl (L.P.)
[3] Department of Histology and Embryology, Poznan University of Medical Sciences, 60-781 Poznan, Poland; joanna.budna@wp.pl (J.B.); pcelichowski@ump.edu.pl (P.C.); mnowicki@ump.edu.pl (M.N.); mazab@ump.edu.pl (M.Z.)
[4] Department of Biomaterials and Experimental Dentistry, Poznan University of Medical Sciences, 60-812 Poznan, Poland
[5] Department of Obstetrics and Gynecology, University Hospital and Masaryk University, 625 00 Brno, Czech Republic; jeseta@gmail.com
[6] Department of Histology and Embryology, Wroclaw Medical University, 50-368 Wroclaw, Poland
[*] Correspondence: bkempisty@ump.edu.pl; Tel.: +48-61-8546-418; Fax: +48-61-8546-440

Received: 8 November 2017; Accepted: 8 December 2017; Published: 9 December 2017

Abstract: Because of the deep involvement of granulosa cells in the processes surrounding the cycles of menstruation and reproduction, there is a great need for a deeper understanding of the ways in which they function during the various stages of those cycles. One of the main ways in which the granulosa cells influence the numerous sex associated processes is hormonal interaction. Expression of steroid sex hormones influences a range of both primary and secondary sexual characteristics, as well as regulate the processes of oogenesis, folliculogenesis, ovulation, and pregnancy. Understanding of the exact molecular mechanisms underlying those processes could not only provide us with deep insight into the regulation of the reproductive cycle, but also create new clinical advantages in detection and treatment of various diseases associated with sex hormone abnormalities. We have used the microarray approach validated by RT-qPCR, to analyze the patterns of gene expression in primary cultures of human granulosa cells at days 1, 7, 15, and 30 of said cultures. We have especially focused on genes belonging to ontology groups associated with steroid biosynthesis and metabolism, namely "Regulation of steroid biosynthesis process" and "Regulation of steroid metabolic process". Eleven genes have been chosen, as they exhibited major change under a culture condition. Out of those, ten genes, namely *STAR, SCAP, POR, SREBF1, GFI1, SEC14L2, STARD4, INSIG1, DHCR7*, and *IL1B*, belong to both groups. Patterns of expression of those genes were analyzed, along with brief description of their functions. That analysis helped us achieve a better understanding of the exact molecular processes underlying steroid biosynthesis and metabolism in human granulosa cells.

Keywords: human; granulosa cells; in vitro culture (IVC); steroid biosynthesis

1. Introduction

The increase of knowledge about the processes underlining the development of human gametes have brought into light the unique interactions between the tissues involved in that process. Granulosa cells, being a part of almost every stage of folliculo and oogenesis, are deeply involved in storage and maturation of the oocytes, expressing a range of reciprocal interactions with the female gamete [1–3]. Additionally, they play a major role in synthesis, expression, and metabolism of a range of hormones, that not only function in maintenance of gamete maturation, regulation of ovulation, and sustenance of pregnancy, but also perform a wide range of secondary functions defining almost every aspect of physiology associated with reproduction [4]. Because of the fact that the mammalian female reproduction cycle is a highly dynamic process, all the processes involved in its regulation and maintenance are extremely complex. Understanding of the exact molecular mechanisms underlining the processes that granulosa cells are a part of directly or indirectly, through endocrinal regulation, not only broadens the knowledge about one of the most essential processes in the mammalian life cycle, but also opens the way for a search for potential clinical possibilities [2,5]. From advances in assisted reproduction, to uncovering the exact mechanism that underline various diseases associated with both the reproduction and hormonal abnormalities, the basis of knowledge that would describe the exact cellular processes that underline those phenomena lies in the patterns of genetic expression associated with all the tissues involved [6–8]. In addition, while the functional analyses of those processes have been long performed and provided us with wide understanding of the ways in which granulosa cells function both by themselves and in relation to the developing oocytes, the knowledge about their exact molecular basis is relatively small [9].

By employing the in vitro cultures of human granulosa cells, we aim to analyze and describe changes in the transcriptome during the long term culture. As regulation of steroid sex hormone synthesis, expression, and metabolism is one of the major functions of the granulosa, we have used microarray assays, together with RT-qPCR validation, to analyze the expression patterns of genes involved in the "Regulation of steroid biosynthesis process" and "Regulation of steroid metabolic process". The genes we aim to identify might later serve for markers for the processes, both normal and pathophysiological, occurring through the whole reproductive cycle, while the expression patterns that they exhibit, together with their mutual relations, might serve as factors to better understand the ways in which granulosa cells behave in vitro and possibly in vivo.

2. Results

Whole transcriptome profiling by Affymetrix microarray allowed us to analyze the expression profile of genes regulating steroid biosynthesis and metabolism in human ovarian granulosa cells in primary culture. By Affymetrix® Human HgU 219 Array (Affymetrix, Santa Clara, CA, USA), we examined the expression of 22,480 transcripts. Genes with fold change higher than |2| and with a corrected p value lower than 0.05 were considered as differentially expressed. This set of genes consisted of 2278 different transcripts. The first detailed analysis based on the gene ontology biological process (GO BP) allowed for the identification of differentially expressed genes belonging to the significantly enriched GO BP terms.

DAVID (Database for Annotation, Visualization and Integrated Discovery) software (Leidos Biomedical Research, Inc., National Cancer Institute, Frederick, MD, USA) was used for extraction of the genes belonging to the analyzed GO BP terms. Up and down regulated gene sets were subjected separately to a DAVID search, and only gene sets where adjusted p values were lower than 0.05 were selected. The DAVID software analysis showed that differently expressed genes belong to 582 gene ontology groups and 45 KEGG pathways. In this paper, we focused on the "regulation of steroid biosynthetic process" (GO:0050810) and "regulation of steroid metabolic process" (GO:0019218) GO BP terms. These sets of genes were subjected to a hierarchical clusterization procedure and presented as heat maps (Figure 1). The gene symbols, fold changes in expression, Entrez gene identifications (IDs), and corrected p values of these genes are shown in Table 1.

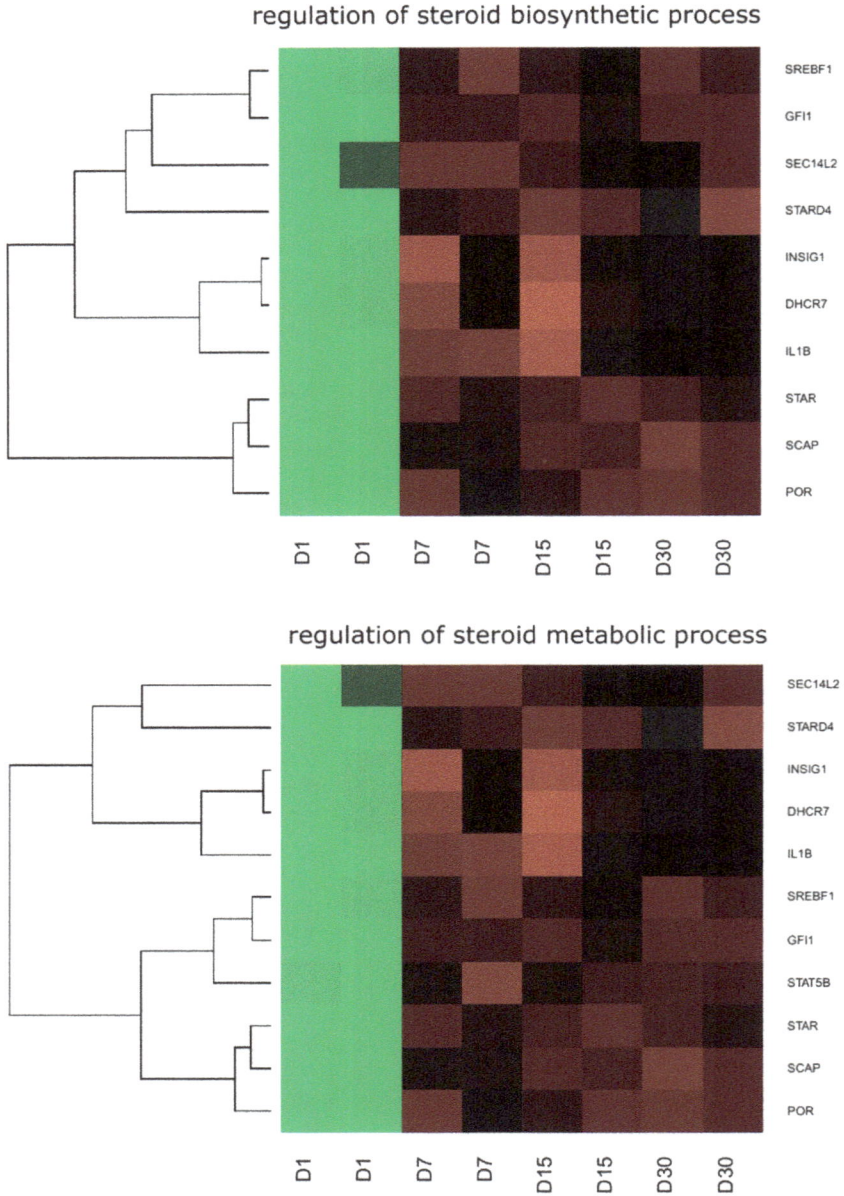

Figure 1. Heat map representation of differentially expressed genes belonging to the "regulation of steroid biosynthetic process" and "regulation of steroid metabolic process" gene ontology biological process (GO BP) terms. Arbitrary signal intensity acquired from microarray analysis is represented by colors (green, higher; red, lower expression). Log2 signal intensity values for any single gene were resized to Row Z-Score scale (from −2, the lowest expression to +2, the highest expression, for a single gene). D: Day of Culture.

Table 1. Gene symbols, fold changes in expression, Entrez gene identifications (IDs), and corrected *p* values of studied genes. Adjusted *p* value = adj.P.Val.

Official Gene Symbol	Fold Change D1/D7	Fold Change D1/D15	Fold Change D1/D30	adj.P.Val. D1/D7	adj.P.Val. D1/D15	adj.P.Val. D1/D30	Entrez Gene ID
DHCR7	0.174701154	0.135435372	0.287884941	0.039383847	0.023551624	0.084599974	1717
GFI1	0.310857826	0.318291331	0.286701236	0.001572194	0.001472925	0.000931728	2672
IL1B	0.068991209	0.095213232	0.183379317	0.027149434	0.035799475	0.083315453	3553
INSIG1	0.112401766	0.111379044	0.231549573	0.040421878	0.035709593	0.102319923	3638
POR	0.297205384	0.246609697	0.226675727	0.02007617	0.011465695	0.008491922	5447
SCAP	0.428465981	0.37245418	0.350801297	0.00234869	0.001405483	0.000934051	22937
SEC14L2	0.393655866	0.47190842	0.474492018	0.049169113	0.080230709	0.075670468	23541
SREBF1	0.342373023	0.40104925	0.351769946	0.004735431	0.007059315	0.004002643	6720
STAR	0.020791115	0.015709262	0.021002203	0.000945846	0.000687578	0.000705428	6770
STARD4	0.192889636	0.155104665	0.215393387	0.037708661	0.023368014	0.037965777	134429
STAT5B	0.343238914	0.381840346	0.358794257	0.008451029	0.010289799	0.007445427	6777

Additionally, using an RT-qPCR assay, we analyzed the relative abundance of *IL1B*, *STAR*, and *POR* in order to quantitatively validate the results obtained through the qualitative microarray analysis. The results were presented in Figure 2. *STAR* and *IL1B* were selected for validation as they presented the most extreme fold changes. *POR* was selected as a mean of control as it presented fold changes on the most intermediate levels. Additionally, *STAR* is a gene that is very important for the described steroid associated processes, while the presence of *IL1B* is especially interesting as it is usually associated with the inflammatory response and not steroid synthesis and metabolism. We have identified the presence of transcripts of the aforementioned genes in human granulosa cells.

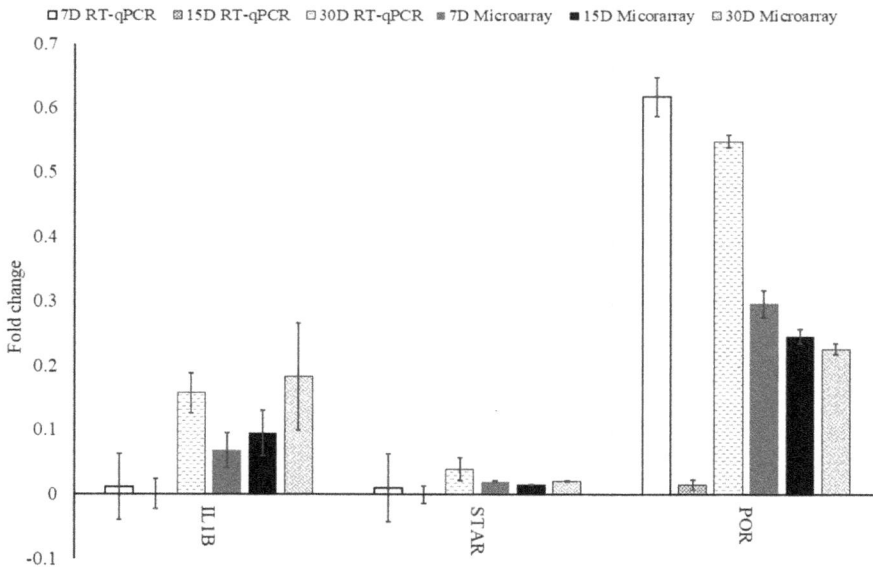

Figure 2. Results from RT-qPCR validation, presented in the form of a bar chart with comparisons to the results obtained with microarray. All the values presented are the relative changes of gene expression, as compared to Day 1 of primary culture. D: Day of Culture.

For both *IL1B* and *STAR*, we found that the levels of transcripts at day 15 were nearly identical as at day 1. In the case of *IL1B*, this does not correspond with the microarray finding. For day 7 there was a slight change in mRNA levels observed, as compared to the entry readings. The most substantial change was observed at day 30 for both of those genes. Moreover, for both *STAR* and *IL1B*, the fold change of expression was relatively the same as in the results shown by microarray analysis, which can be considered as validation of those results. For the *POR* gene, the smallest fold change was observed at day 15, with major changes at both day 7 and 30, as compared to those at the entry point. Again, on day 15, the results of microarray and RT-qPCR analysis varied significantly in scale. Overall, the direction of changes in expression is confirmed by RT-qPCR analysis, while the scale of change often varies, which can be explained by the much higher accuracy of RT-qPCR. It does not interfere with the results of research focused on expression patterns, but indicates the need for extensive validation of microarray results during the studies involving the analysis of the extent of expression changes.

Moreover, in the gene ontology database, genes that formed one particular GO group can also belong to other different GO term categories. For this reason, we explored the gene intersections between selected GO BP terms. The relation between those GO BP terms are presented in a chart (Figure 3).

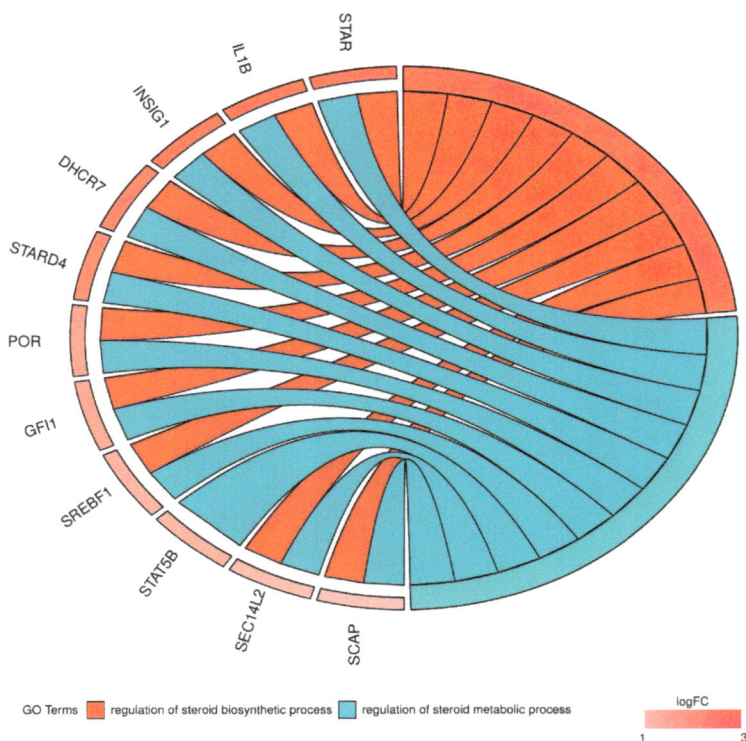

Figure 3. The representation of the mutual relationship between the "regulation of steroid biosynthetic process" and "regulation of steroid metabolic process" GO BP terms. The ribbons indicate which gene belongs to which categories. The genes were sorted by logFC from most to least changed gene, with the most changed gene marked with the most intense color on the side and presented topmost, and the least changed gene marked with the least intense color on the side and presented on the bottom.

We used a STRING (Search Tool for the Retrieval of Interacting Genes/Proteins)-generated network to evaluate the interactions between differentially expressed genes belonging to each of the selected GO BP terms. Using this prediction method provided us with the molecular interaction network which is formed between the protein products of the studied genes (Figure 4).

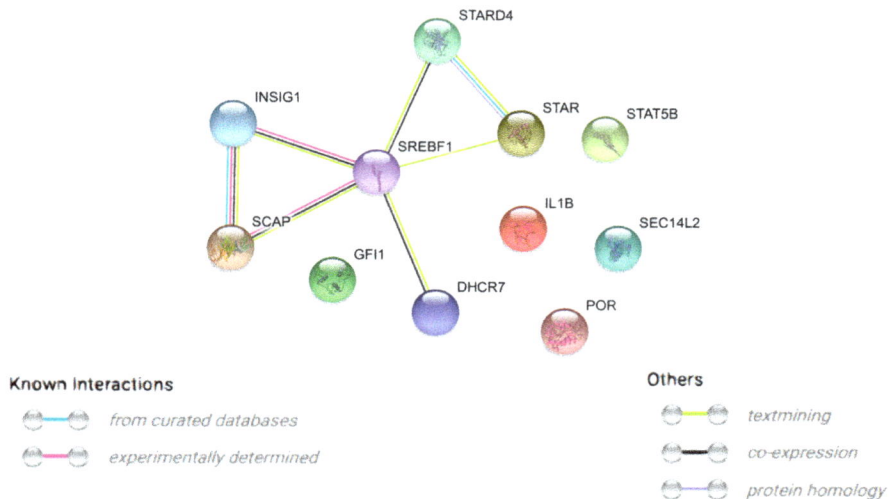

Figure 4. STRING (Search Tool for the Retrieval of Interacting Genes/Proteins)-generated interaction network among differentially expressed genes belonging to the "regulation of steroid biosynthetic process" and "regulation of steroid metabolic process" GO BP terms. The intensity of the edges reflects the strength of the interaction score.

3. Discussion

Granulosa cells are the primary cell type found in the mammalian ovary. They perform multiple functions associated with oocyte storage, maturation, as well as maintaining pregnancy and embryonic development. Because of that, it can be said that they are a somatic cell type that carries the strongest and closest association with the processes underlying development of female gametes and their progression through the process of reproduction. They surround the oocyte all the way from the primary follicle to the moment of ovulation, performing a major role in synthesis and expression of sex hormones [1,10]. The interactions between oocyte and granulosa cells are reciprocal, with both structures expressing signals that influence the development and functioning of the other. After the follicular rupture, granulosa cells that remain in the newly formed corpus luteum undergo transition into granulosa lutein cells, which produce progesterone, taking further part in maintenance of pregnancy. While the function of granulosa cells is fairly well understood, the exact mechanisms and genes expressed through the significant amount of change undergone by these cell types is still full of unknowns [1]. Because of their major significance for the process of follicle development and oogenesis, the complete understanding of gene expression underlining the functioning of granulosa cells and their interactions with oocytes is essential for developing possible clinical applications associated with gene therapy, tissue and cell engineering, as well as prevention, early detection, and treatment of cancers originating in the ovarian region that associated with granulosa cells [11].

By using the microarray approach, we have identified and measured the changes in expression of genes that are part of ontological groups of interest. The RNA analyzed was isolated from primary cell cultures of granulosa cells after 1, 7, 15, and 30 days of maturation in vitro. Two gene ontology groups were inspected: (1) "Regulation of steroid biosynthesis process" and (2) "Regulation of steroid metabolic process". Eleven genes were identified, out of which ten, namely *STAR, SCAP, POR, SREBF1, GFI1, SEC14L2, STARD4, INSIG1, DHCR7* and *IL1B*, belonged to both groups. One of the genes, *STAT5B*, only belonged to the "Regulation of Steroid Metabolic Process" ontological group. Most of the genes presented a uniform pattern of expression. Large upregulation of expression was observed at day 1 of cell culture, with significant downregulation at days 7, 15, and 30. Three genes showed a different pattern at day 30: *INSIG1* and *DHCR7* had very little downregulation and *IL1B* had no change in expression. Despite that, we can assume that the pattern of expression was fairly unified for all the genes, with upregulation at day 1 and general downregulation through the other days. There are several readings that could indicate a slight upregulation in expression in several genes in different days of cell cultures apart from day 1, however, all of the readings do not appear consistently in both reads performed and are contradicted by strong downregulation recorded in the other read, which indicates that they are probably no more than an error in measurement.

Two of the genes identified are very closely associated with processes underlying steroid intracellular transport. They are both hosts to the *STAR*-related transfer domain (START), with the main function of binding lipids, including sterols [12]. STAR steroid acute regulatory protein, otherwise called *STARD1*, encodes a protein that is a key factor in regulation of steroid hormone synthesis. By mediating the transport of cholesterol between mitochondrial membranes, the protein allows cholesterol to be converted into pregnenolone, completing the first major step in the process of steroidogenesis [13,14]. Disruption of the *STAR* gene has been proven to cause congenital lipoid adrenal hyperplasia (CLAH) [15], an endocrine disorder, causing mineralocorticoid deficiency which impairs the synthesis of all categories of adrenal steroids [16]. The other gene that hosts the START domain is the *STAR*-related lipid transfer protein 4 (*STARD4*). It is a gene encoding a soluble protein involved in cholesterol transport. It is a cholesterol regulated gene with two known homologues belonging in its family: *STARD5* and *STARD6*. While being expressed in most tissues, the highest levels of *STARD4* are observed in the liver and kidneys [17]. There are strong suggestions of *STARD4*'s particular involvement in movement of cholesterol to the endoplasmic reticulum [18].

Three of the genes identified work closely for maintenance of intracellular cholesterol homeostasis. The sterol regulatory element binding transcription factor 1 (*SREBF1*) gene encodes a basic helix-loop-helix leucine zipper transcription factor (*SREBP1*) binding the sterol regulatory element (SRE1) [19]. *SREBP1* is stored as a membrane bound precursor and is released by proteolytic cleavage in sterol depleted cells. The loose fragment generated through the cleavage translocates to the nucleus, activating transcription [20]. It has been proven to play a major role in the induction of lipogenesis, as well as an auxiliary factor in synthesis of fatty-acids [21]. *SCAP* is closely associated with *SREBP1*, the protein encoded by *SREBF1*. It is an escort protein that regulates the effects of *SREBP1* expression, detecting the changes of cholesterol concentration. It is the effector of *SREBP1* cleavage, allowing that transcription factor to abandon its membrane-bound state, transduce to the nucleus, and allow transcription of the genes activated by the SREBP1 [22]. The cleavage occurs in the Golgi apparatus, which requires *SCAP* to escort *SREBP1* to the Golgi apparatus from the endoplasmic reticulum, where *SREBP1* is synthesized [23]. *SREBP1* transcriptional regulation effects also cause negative feedback by causing degradation of *SCAP* to prevent sterol overexpression [24]. Large concentrations of sterols prevent the exit of the *SREBP1/SCAP* complex from the endoplasmic reticulum in order to prevent accumulation of cholesterol inside of the cell [22]. This regulative step is achieved with involvement of another gene, insulin induced gene 1 (*INSIG1*). This gene encodes a membrane protein found in the endoplasmic reticulum. In the presence of high sterol concentrations, *INSIG1* binds the sterol-sensing domain of *SCAP*. This causes retention of the *SREBP1/SCAP* complex in the endoplasmic reticulum and prevents its translocation to the Golgi apparatus, therefore, prevents cleavage of *SREBP1* and cholesterol synthesis upregulation [25].

The genes described above are closely associated in the process involved in lipid synthesis and metabolism. However, there are several more genes that were identified in our study that possess individual involvement with lipid metabolic and synthetic processes. *POR* is a gene encoding cytochrome P450 oxidoreductase. This enzyme plays a critical role in the synthesis of cholesterol and steroid hormones as an obligatory intermediate which transfers electrons from NADPH (nicotinamide adenine dinucleotide phosphate carrying electrons and bonded with a hydrogen) to all cytochrome P450 enzymes [26]. It is also involved in the metabolism of ingested substances in the liver, and it has been suggested that *POR* gene polymorphisms are involved in differences in drug metabolism across the population [27]. Defects in *POR* genes have been proven to cause a variety of symptoms, from embryonic lethality to disordered steroidogenesis, congenital adrenal hyperplasia, ambiguous genitalia, and Antley-Bixler syndrome [26,28,29]. The next gene, *DHCR7*, encodes an enzyme, 7-dehydrocholesterol reductase, that catalyzes the conversion of 7-dehydrocholesterol to cholesterol. This fact, makes the enzyme essential in mammalian sterol biosynthesis, as it is necessary in the final step of the synthesis pathway [30]. Defects in *DHCR7* gene were proven to be the cause of Smith-Lemli-Opitz syndrome, which may cause mental retardation, dysmorphism of the face, syndactyly, and holoprosencephaly as a result of insufficient sterol synthesis and 7-dehydrocholesterol accumulation [30,31]. The *SEC14L2* gene encodes SEC14 like protein 2. This lipid-binding carrier protein facilitates transport of hydrophobic molecules between different cellular sites [32]. *SEC14L2* exhibits high affinity to alpha-tocopherol and weaker affinity to other tocopherols and tocotrienols. It has been proven to stimulate squalene monooxygenase, a downstream enzyme in cholesterol biosynthesis [33]. *IL-1B* encodes an interleukin 1 cytokine family member protein, which functions as an important mediator of the inflammatory response [34]. It is usually expressed by activated macrophages in its inactive form, which is later activated by proteolytical processing [35]. The cytokine encoded by *IL-1B* is also involved in other cellular processes including proliferation, differentiation, and apoptosis [36–38]. Proto-oncogene *GFI1* (growth factor independent 1 transcriptional repressor) encodes a nuclear zinc finger protein, functioning as a transcriptional repressor. The repression is achieved by *GFI1* working with other cofactors as a part of a complex that controls histone modifications and silences target gene promoters [39]. It has been proven to play a role in processes such as hematopoiesis and oncogenesis [40]. Mutations in that gene can cause neutropenia in both the congenital and nonimmune chronic idiopathic forms, both being autosomal dominant disorders causing predispositions to leukemia and infections [41]. The final gene described in this study is *STAT5B*, signal transducer and activator of transcription 5B. This gene is only a member of the "Regulation of steroid metabolic process" gene ontology group. It encodes a STAT family transcription factor, expression of which is activated by cytokine activity. Transcription factors belonging to that family function when phosphorylated by the receptor associated kinases. After phosphorylation, hetero or homodimerization occurs, with the resulting complex functioning as a transcription activator after translocating to the nucleus [42]. *STAT5B* performs various functions, mostly associated with hematopoiesis, mammary gland development, and immune systems [43–45].

In conclusion, we identified and described the genes expressed in long term in vitro cultures, belonging to the ontology groups "Regulation of steroid biosynthesis process" and "Regulation of steroids metabolic process". Analysis of the modes of their expression, together with their relation to each other, will broaden the understanding of the ways in which the cells of human granulosa are involved in the processes associated with steroid synthesis and metabolism, which may later be applied in both further research and, potentially, together with other research that improves the understanding of the molecular basis of this tissue's functioning, to clinical applications.

4. Material and Methods

4.1. Patients and Collection of Granulosa Cells

The granulosa cells (GCs) were derived from patients undergoing in vitro fertilization (IVF) procedures, who had given their informed consent to be included in this protocol. The study group consisted of eight patients, aged 18–40 years, with diagnosed infertility who were referred to the Division of Infertility and Reproductive Endocrinology, Poznan University of Medical Sciences, Poland. Patients underwent an IVF procedure based on controlled the ovarian hyperstimulation protocol, adjusted to the patient's initial infertility workup and ovarian response. Stimulation was performed with human recombinant FSH (Follicle-stimulating hormone; Gonal-F, Merck Serono, Darmstadt, Germany) and highly purified hMG-HP (Highly purified human menopausal gonadotropin; Menopur, Ferring, Saint-Prex, Switzerland). The injections with cetrorelix acetate (Cetrotide, Merck Serono,

Darmstadt, Germany) were administered in an adequate dose to suppress pituitary function. Ovulation triggering was based on subcutaneous injection of 6500 U of hCG (Human chorionic gonadotropin, Ovitrelle; Merck-Serono, Darmstadt, Germany). Follicular fluid, containing GCs, was collected during transvaginal ultrasound-guided oocyte pick-up, 36 h after human chorionic gonadotropin administration. The follicular content from follicles over 16 mm in diameter was given rapidly to the embryologist, who isolated the oocyte and pooled the fluids containing GCs together from each ovary. Fresh follicular fluid was centrifuged for 10 min at $200 \times g$, to separate and collect GCs. Patients with polycystic ovary syndrome (PCOS), endometriosis, and diminished ovarian reserve (serum antimüllerian hormone (AMH) less than 0.7 ng/mL and/or day 2–3 FSH serum level higher than 15 mU/mL and/or antral follicle count less than 9) were excluded from the study. This study was approved with resolution 558/17 by the Bioethical Committee at the Poznan University of Medical Sciences. All participants gave written informed consent for the research.

4.2. Primary Cell Culture

Collected cells were washed twice by centrifugation at $200 \times g$ for 10 min at room temperature with culture medium. Medium consisted of Dulbecco's Modified Eagle's Medium (DMEM/F12, Sigma-Aldrich, St. Louis, MO, USA), 2% fetal bovine serum FBS (FBS; Sigma-Aldrich Co., St. Louis, MO, USA), 200 mM L-glutamine (Invitrogen, Carlsbad, CA, USA), 10 mg/mL gentamicin (Invitrogen, USA), 10,000 units/mL penicillin, and 10,000 µg/mL streptomycin (Invitrogen, USA). Cells were cultivated at 37 °C under aerobic conditions (5% CO_2). Once adherent cells were more than 90% confluent, they were detached with 0.05% trypsin-EDTA (Invitrogen, USA) for 1–2 min and counted using a counting chamber "neubauer improved" (ISO LAB Laborgerate GmbH, Wertheim, Germany, DIN EN ISO CERTIFIED 9001). GCs were cultivated for 30 days. Medium was changed twice a week. Finally, total RNA was isolated from GCs after 24 h, 168 h, 15 days, and 30 days. The changes in cell morphology were presented in Figure 5.

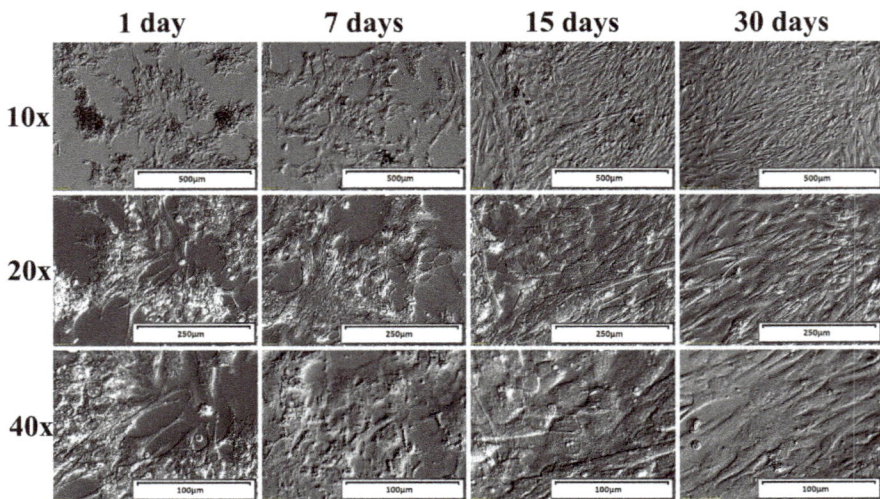

Figure 5. Morphology of human ovarian granulosa cells in long term in vitro culture shown using Nomarski phase/contrast images.

4.3. Total RNA Isolation

Total RNA was isolated at four time periods, after 24 h, 168 h, 15 days, and 30 days cultivation. For the isolation of total RNA, we used the improved Chomczyński-Sacchi method [46]. The GCs were suspended in 1 mL mixture of guanidine thiocyanate and phenol in monophasic solution (TRI Reagent®, Sigma-Aldrich, St. Luis, MO, USA). Then, chloroform was added and centrifuged to separate the mixture into three phases. RNA was located in the upper phase—an aqueous phase. The resulting RNA was intact with little or no contaminating DNA and protein. The last step was to strip the RNA with 2-propanol (Sigma-Aldrich, St. Luis, MO, catalog number I9516) per 1 mL of TRI-reagent and wash with 75% ethanol. Prepared RNA was used for further analysis.

4.4. Microarray Expression Analysis and Statistics

Total RNA (100 ng) from each pooled sample was subjected to two rounds of sense complementary DNA (cDNA) amplification (Ambion® WT Expression Kit; Ambion, Austin, TX, USA). The obtained cDNA was used for biotin labeling and fragmentation by Affymetrix GeneChip® WT Terminal Labeling and Hybridization (Affymetrix, Santa Clara, CA, USA). Biotin-labeled fragments of cDNA (5.5 µg) were hybridized to the Affymetrix® Human Genome U219 Array (48 °C/20 h; Affymetrix, Santa Clara, CA, USA). Microarrays were then washed and stained according to the technical protocol using the Affymetrix GeneAtlas Fluidics Station. The array strips were scanned employing the Imaging Station of the GeneAtlas System. Preliminary analysis of the scanned chips was performed using Affymetrix GeneAtlasTM Operating Software (Affymetrix, Santa Clara, CA, USA). The quality of gene expression data was confirmed according to the quality control criteria provided by the software. The obtained CEL files were imported into downstream data analysis software.

All of the presented analyses and graphs were performed using Bioconductor and R programming languages. Each CEL file was merged with a description file. In order to correct background, normalize, and summarize results, we used the Robust Multiarray Averaging (RMA) algorithm. To determine the statistical significance of the analyzed genes, moderated t-statistics from the empirical Bayes method were performed. The obtained *p*-value was corrected for multiple comparisons using Benjamini and Hochberg's false discovery rate. The selection of significantly altered genes was based on a *p*-value beneath 0.05 and expression higher than two-fold. The differentially expressed gene list (separated for up- and down-regulated genes) was uploaded to DAVID (Database for Annotation, Visualization and Integrated Discovery) software (Leidos Biomedical Research, Inc., National Cancer Institute, Frederick, MD, USA) [47].

Subsequently, sets of differentially expressed genes from selected GO BP terms were applied to STRING software (Search Tool for the Retrieval of Interacting Genes/Proteins; STRING Consortium) for interaction predictions. STRING is a huge database containing information of protein/gene interactions, including experimental data, computational prediction methods, and public text collections.

In order to further investigate the chosen gene sets, we investigated its mutual relations with the GOplot package [48]. Moreover, the GOplot package allowed us to calculate the z-score (the number of up- regulated genes minus the number of down- regulated genes divided by the square root of the count). Z-score analysis allowed us to compare the enrichment of selected GO BP terms.

4.5. Real-Time Quantitative Polymerase Chain Reaction (RT-qPCR) Analysis

Total RNA was isolated from granulosa cells after 24 h, 7 days, 15 days, and 30 days of culture. The RNA samples were re-suspended in 20 µL of RNase-free water and stored in liquid nitrogen. RNA samples were treated with DNase I and reverse-transcribed (RT) into cDNA. RQ-PCR was conducted in a LightCycler real-time PCR detection system (Roche Diagnostics GmbH, Mannheim, Germany) using SYBR® Green I as a detection dye, and target cDNA was quantified using the relative quantification method. For amplification, 2 µL of cDNA solution was added to 18 µL of QuantiTect® SYBR® Green PCR (Master Mix Qiagen GmbH, Hilden, Germany) and primers (Table 2).

Table 2. Oligonucleotide sequences of primers used for RT-qPCR analysis.

Gene	Gene Accession Number	Primer Sequence (5′-3′)	Product Size (bp)
STAR	NM_000349.2	GGCATCCTTAGCAACCAAGA TCTCCTTGACATTGGGGTTC	199
Il1B	NM_000576.2	GGGCCTCAAGGAAAAGAATC TTCTGCTTGAGAGGTGCTGA	205
POR	NM_000941.2	CACAAGGTCTACGTCCAGCA GCCACGATGTCGTAGAAGGT	143
GAPDH	NM_002046	TCAGCCGCATCTTCTTTTGC ACGACCAAATCCGTTGACTC	90
ACTB	NM_001101	AAAGACCTGTACGCCAACAC CTCAGGAGGAGCAATGATCTTG	132
HPRT	NM_000194	TGGCGTCGTGATTAGTGATG ACATCTCGAGCAAGACGTTC	141

One RNA sample of each preparation was processed without the RT-reaction to provide a negative control for subsequent PCR.

To quantify specific genes in the granulosa cells, expression levels of specific messenger RNAs (mRNAs) were calculated relative to *GAPDH* (Glyceraldehyde-3-Phosphate Dehydrogenase), *HPRT* (Hypoxanthine guanine phosphoribosyltransferase), and *ACTB* (Beta-actin). To ensure the integrity of these results, the additional housekeeping gene was used as an internal standard to demonstrate that *GAPDH*, *HPRT*, and *ACTB* mRNAs were not differentially regulated in the granulosa cells.

Acknowledgments: Publication of this article was made possible by grant number UMO-2012/07/N/NZ5/00069 from the Polish National Centre of Science and grant number 502-14-02227367-10694 from Poznan University of Medical Sciences.

Author Contributions: Wiesława Kranc—investigation (conducting research and investigation processes, specifically performing the experiments, or data/evidence collection), methodology (development of methodology), original draft preparation; Maciej Brązert—resources (provision of study materials and patients), design of experiment, methodology; Katarzyna Ożegowska—resources, wrote part of paper and completed revisions of the medical methodology; Mariusz J. Nawrocki—interpretation and description of microarray and RT-qPCR results; Joanna Budna—data curation (scrubbed data and maintained research data), review and editing, wrote the paper; Piotr Celichowski—software, creation and presentation of the published work, specifically writing the initial draft (including substantial translation), formal analysis, visualization; Marta Dyszkiewicz-Konwińska—revision of medical methodology; Maurycy Jankowski—writing (preparation and creation of the published work), editorial assistance, language corrections; Michal Jeseta—methodology (creation of models); Leszek Pawelczyk—supervision (oversight and leadership responsibility for the research activity planning and execution); Małgorzata Bruska—project administration; Michał Nowicki—supervision, provision of technical advice; Maciej Zabel—supervision, review and editing; Bartosz Kempisty—conceptualization, project administration, senior author, and major assistance.

Conflicts of Interest: The authors declare no conflict of interest.

References

1. Kranc, W.; Budna, J.; Kahan, R.; Chachuła, A.; Bryja, A.; Ciesiółka, S.; Borys, S.; Antosik, M.P.; Bukowska, D.; Brussow, K.P.; et al. Molecular Basis of Growth, Proliferation, and Differentiation of Mammalian Follicular Granulosa Cells. *J. Biol. Regul. Homeost. Agents* **2017**, *31*, 1–8. [PubMed]

2. Kranc, W.; Budna, J.; Dudek, M.; Bryja, A.; Chachuła, A.; Ciesiółka, S.; Borys, S.; Dyszkiewicz-Konwińska, M.; Jeseta, M.; Porowski, L.; et al. The Origin, in Vitro Differentiation, and Stemness Specificity of Progenitor Cells. *J. Biol. Regul. Homeost. Agents* **2017**, *31*, 365–369. [PubMed]

3. Budna, J.; Celichowski, P.; Karimi, P.; Kranc, W.; Bryja, A.; Ciesiółka, S.; Rybska, M.; Borys, S.; Jeseta, M.; Bukowska, D.; et al. Does Porcine Oocytes Maturation in Vitro Is Regulated by Genes Involved in Transforming Growth Factor Beta Receptor Signaling Pathway? *Adv. Cell Biol.* **2017**, *5*, 1–14. [CrossRef]

4. Kranc, W.; Celichowski, P.; Budna, J.; Khozmi, R.; Bryja, A.; Ciesiółka, S.; Rybska, M.; Borys, S.; Jeseta, M.; Bukowska, D.; et al. Positive Regulation Of Macromolecule Metabolic Process Belongs To The Main Mechanisms Crucial For Porcine Ooocytes Maturation. *Adv. Cell Biol.* **2017**, *5*, 15–31. [CrossRef]

5. Nawrocki, M.J.; Budna, J.; Celichowski, P.; Khozmi, R.; Bryja, A.; Kranc, W.; Borys, S.; Ciesiółka, S.; Knap, S.; Jeseta, M.; et al. Analysis of Fructose and Mannose—Regulatory Peptides Signaling Pathway in Porcine Epithelial Oviductal Cells (OECs) Primary Cultured Long-Term in Vitro. *Adv. Cell Biol.* **2017**, *5*, 129–135. [CrossRef]

6. Ciesiółka, S.; Budna, J.; Jopek, K.; Bryja, A.; Kranc, W.; Borys, S.; Jeseta, M.; Chachuła, A.; Ziółkowska, A.; Antosik, P.; et al. Time- and Dose-Dependent Effects of 17 Beta-Estradiol on Short-Term, Real-Time Proliferation and Gene Expression in Porcine Granulosa Cells. *Biomed. Res. Int.* **2017**, *2017*, 1–9. [CrossRef] [PubMed]

7. Wu, Y.-G.; Barad, D.H.; Kushnir, V.A.; Lazzaroni, E.; Wang, Q.; Albertini, D.F.; Gleicher, N. Aging-Related Premature Luteinization of Granulosa Cells Is Avoided by Early Oocyte Retrieval. *J. Endocrinol.* **2015**, *226*, 167–180. [CrossRef] [PubMed]

8. Borys, S.; Khozmi, R.; Kranc, W.; Bryja, A.; Dyszkiewicz-Konwińska, M.; Jeseta, M.; Kempisty, B. Recent Findings of the Types of Programmed Cell Death. *Adv. Cell Biol.* **2017**, *5*, 43–49. [CrossRef]

9. Ciesiółka, S.; Bryja, A.; Budna, J.; Kranc, W.; Chachuła, A.; Bukowska, D.; Piotrowska, H.; Porowski, L.; Antosik, P.; Bruska, M.; et al. Epithelialization and Stromalization of Porcine Follicular Granulosa Cells during Real-Time Proliferation—A Primary Cell Culture Approach. *J. Biol. Regul. Homeost. Agents* **2016**, *30*, 693–702. [PubMed]

10. Kossowska-Tomaszczuk, K.; De Geyter, C.; De Geyter, M.; Martin, I.; Holzgreve, W.; Scherberich, A.; Zhang, H. The Multipotency of Luteinizing Granulosa Cells Collected from Mature Ovarian Follicles. *Stem Cells* **2009**, *27*, 210–219. [CrossRef] [PubMed]

11. Borys, S.; Khozmi, R.; Kranc, W.; Bryja, A.; Jeseta, M.; Kempisty, B. Resveratrol and Its Analogues—Is It a New Strategy of Anticancer Therapy? *Adv. Cell Biol.* **2017**, *5*, 32–42. [CrossRef]

12. Alpy, F.; Tomasetto, C. Give Lipids a START: The StAR-Related Lipid Transfer (START) Domain in Mammals. *J. Cell Sci.* **2005**, *118*, 2791–2801. [CrossRef] [PubMed]

13. Stocco, D.M. Steroidogenic Acute Regulatory Protein. *Vitam. Horm.* **1998**, *55*, 399–441.

14. Clark, B.J.; Soo, S.C.; Caron, K.M.; Ikeda, Y.; Parker, K.L.; Stocco, D.M. Hormonal and Developmental Regulation of the Steroidogenic Acute Regulatory Protein. *Mol. Endocrinol.* **1995**, *9*, 1346–1355. [PubMed]

15. Caron, K.M.; Soo, S.C.; Wetsel, W.C.; Stocco, D.M.; Clark, B.J.; Parker, K.L. Targeted Disruption of the Mouse Gene Encoding Steroidogenic Acute Regulatory Protein Provides Insights into Congenital Lipoid Adrenal Hyperplasia. *Proc. Natl. Acad. Sci. USA* **1997**, *94*, 11540–11545. [CrossRef] [PubMed]

16. Bose, H.S.; Sugawara, T.; Strauss, J.F.; Miller, W.L. The Pathophysiology and Genetics of Congenital Lipoid Adrenal Hyperplasia. *N. Engl. J. Med.* **1996**, *335*, 1870–1879. [CrossRef] [PubMed]

17. Soccio, R.E.; Adams, R.M.; Romanowski, M.J.; Sehayek, E.; Burley, S.K.; Breslow, J.L. The Cholesterol-Regulated StarD4 Gene Encodes a StAR-Related Lipid Transfer Protein with Two Closely Related Homologues, StarD5 and StarD6. *Proc. Natl. Acad. Sci. USA* **2002**, *99*, 6943–6948. [CrossRef] [PubMed]

18. Rodriguez-Agudo, D.; Calderon-Dominguez, M.; Ren, S.; Marques, D.; Redford, K.; Medina-Torres, M.A.; Hylemon, P.; Gil, G.; Pandak, W.M. Subcellular Localization and Regulation of StarD4 Protein in Macrophages and Fibroblasts. *Biochim. Biophys. Acta* **2011**, *1811*, 597–606. [CrossRef] [PubMed]

19. Wang, X.; Sato, R.; Brown, M.S.; Hua, X.; Goldstein, J.L. SREBP-1, a Membrane-Bound Transcription Factor Released by Sterol-Regulated Proteolysis. *Cell* **1994**, *77*, 53–62. [CrossRef]

20. Brown, M.S.; Goldstein, J.L. The SREBP Pathway: Regulation of Cholesterol Metabolism by Proteolysis of a Membrane-Bound Transcription Factor. *Cell* **1997**, *89*, 331–340. [CrossRef]

21. Shimano, H.; Yahagi, N.; Amemiya-Kudo, M.; Hasty, A.H.; Osuga, J.; Tamura, Y.; Shionoiri, F.; Iizuka, Y.; Ohashi, K.; Harada, K.; et al. Sterol Regulatory Element-Binding Protein-1 as a Key Transcription Factor for Nutritional Induction of Lipogenic Enzyme Genes. *J. Biol. Chem.* **1999**, *274*, 35832–35839. [CrossRef] [PubMed]

22. Edwards, P.; Tabor, D.; Kast, H.R.; Venkateswaran, A. Regulation of Gene Expression by SREBP and SCAP. *Biochim. Biophys. Acta* **2000**, *1529*, 103–113. [CrossRef]

23. Nohturfft, A.; DeBose-Boyd, R.A.; Scheek, S.; Goldstein, J.L.; Brown, M.S. Sterols Regulate Cycling of SREBP Cleavage-Activating Protein (SCAP) between Endoplasmic Reticulum and Golgi. *Proc. Natl. Acad. Sci. USA* **1999**, *96*, 11235–11240. [CrossRef] [PubMed]

24. Shao, W.; Espenshade, P.J. Sterol Regulatory Element-Binding Protein (SREBP) Cleavage Regulates Golgi-to-Endoplasmic Reticulum Recycling of SREBP Cleavage-Activating Protein (SCAP). *J. Biol. Chem.* **2014**, *289*, 7547–7557. [CrossRef] [PubMed]

25. Yang, T.; Espenshade, P.J.; Wright, M.E.; Yabe, D.; Gong, Y.; Aebersold, R.; Goldstein, J.L.; Brown, M.S. Crucial Step in Cholesterol Homeostasis: Sterols Promote Binding of SCAP to INSIG-1, a Membrane Protein That Facilitates Retention of SREBPs in ER. *Cell* **2002**, *110*, 489–500. [CrossRef]

26. Huang, N.; Pandey, A.V.; Agrawal, V.; Reardon, W.; Lapunzina, P.D.; Mowat, D.; Jabs, E.W.; Vliet, G.; Van Sack, J.; Flück, C.E.; et al. Diversity and Function of Mutations in P450 Oxidoreductase in Patients with Antley-Bixler Syndrome and Disordered Steroidogenesis. *Am. J. Hum. Genet.* **2005**, *76*, 729–749. [CrossRef] [PubMed]

27. Hart, S.N.; Zhong, X. P450 Oxidoreductase: Genetic Polymorphisms and Implications for Drug Metabolism and Toxicity. *Expert Opin. Drug Metab. Toxicol.* **2008**, *4*, 439–452. [CrossRef] [PubMed]

28. Arlt, W.; Walker, E.A.; Draper, N.; Ivison, H.E.; Ride, J.P.; Hammer, F.; Chalder, S.M.; Borucka-Mankiewicz, M.; Hauffa, B.P.; Malunowicz, E.M.; et al. Congenital Adrenal Hyperplasia Caused by Mutant P450 Oxidoreductase and Human Androgen Synthesis: Analytical Study. *Lancet* **2004**, *363*, 2128–2135. [CrossRef]

29. Flück, C.E.; Tajima, T.; Pandey, A.V.; Arlt, W.; Okuhara, K.; Verge, C.F.; Jabs, E.W.; Mendonça, B.B.; Fujieda, K.; Miller, W.L. Mutant P450 Oxidoreductase Causes Disordered Steroidogenesis with and without Antley-Bixler Syndrome. *Nat. Genet.* **2004**, *36*, 228–230.

30. Waterham, H.; Wanders, R.J. Biochemical and Genetic Aspects of 7-Dehydrocholesterol Reductase and Smith-Lemli-Opitz Syndrome. *Biochim. Biophys. Acta* **2000**, *1529*, 340–356. [CrossRef]

31. Witsch-Baumgartner, M.; Löffler, J.; Utermann, G. Mutations in the Human DHCR7 Gene. *Hum. Mutat.* **2001**, *17*, 172–182. [CrossRef] [PubMed]

32. Saeed, M.; Andreo, U.; Chung, H.-Y.; Espiritu, C.; Branch, A.D.; Silva, J.M.; Rice, C.M. SEC14L2 Enables Pan-Genotype HCV Replication in Cell Culture. *Nature* **2015**, *524*, 471–475. [CrossRef] [PubMed]

33. Zingg, J.-M.; Libinaki, R.; Meydani, M.; Azzi, A. Modulation of Phosphorylation of Tocopherol and Phosphatidylinositol by hTAP1/SEC14L2-Mediated Lipid Exchange. *PLoS ONE* **2014**, *9*, e101550. [CrossRef] [PubMed]

34. Fechtner, S.; Singh, A.; Chourasia, M.; Ahmed, S. Molecular Insights into the Differences in Anti-Inflammatory Activities of Green Tea Catechins on IL-1β Signaling in Rheumatoid Arthritis Synovial Fibroblasts. *Toxicol. Appl. Pharmacol.* **2017**, *329*, 112–120. [CrossRef] [PubMed]

35. Miao, E.A.; Alpuche-Aranda, C.M.; Dors, M.; Clark, A.E.; Bader, M.W.; Miller, S.I.; Aderem, A. Cytoplasmic Flagellin Activates Caspase-1 and Secretion of Interleukin 1[beta] via Ipaf. *Nat. Immunol.* **2006**, *7*, 569–576. [CrossRef] [PubMed]

36. Ben-Sasson, S.Z.; Hu-Li, J.; Quiel, J.; Cauchetaux, S.; Ratner, M.; Shapira, I.; Dinarello, C.A.; Paul, W.E. IL-1 Acts Directly on CD4 T Cells to Enhance Their Antigen-Driven Expansion and Differentiation. *Proc. Natl. Acad. Sci. USA* **2009**, *106*, 7119–7124. [CrossRef] [PubMed]

37. Bartelmez, S.H.; Bradley, T.R.; Bertoncello, I.; Mochizuki, D.Y.; Tushinski, R.J.; Stanley, E.R.; Hapel, A.J.; Young, I.G.; Kriegler, A.B.; Hodgson, G.S. Interleukin 1 plus Interleukin 3 plus Colony-Stimulating Factor 1 Are Essential for Clonal Proliferation of Primitive Myeloid Bone Marrow Cells. *Exp. Hematol.* **1989**, *17*, 240–245. [PubMed]

38. Mangan, D.F.; Welch, G.R.; Wahl, S.M. Lipopolysaccharide, Tumor Necrosis Factor-Alpha, and IL-1 Beta Prevent Programmed Cell Death (Apoptosis) in Human Peripheral Blood Monocytes. *J. Immunol.* **1991**, *146*, 1541–1546. [PubMed]

39. Zhu, J.; Guo, L.; Min, B.; Watson, C.J.; Hu-Li, J.; Young, H.A.; Tsichlis, P.N.; Paul, W.E. Growth Factor Independent-1 Induced by IL-4 Regulates Th2 Cell Proliferation. *Immunity* **2002**, *16*, 733–744. [CrossRef]

40. Hock, H.; Hamblen, M.J.; Rooke, H.M.; Schindler, J.W.; Saleque, S.; Fujiwara, Y.; Orkin, S.H. Gfi-1 Restricts Proliferation and Preserves Functional Integrity of Haematopoietic Stem Cells. *Nature* **2004**, *431*, 1002–1007. [CrossRef] [PubMed]

41. Person, R.E.; Li, F.-Q.; Duan, Z.; Benson, K.F.; Wechsler, J.; Papadaki, H.A.; Eliopoulos, G.; Kaufman, C.; Bertolone, S.J.; Nakamoto, B.; et al. Mutations in Proto-Oncogene GFI1 Cause Human Neutropenia and Target ELA2. *Nat. Genet.* **2003**, *34*, 308–312. [CrossRef] [PubMed]

42. Wang, D.; Stravopodis, D.; Teglund, S.; Kitazawa, J.; Ihle, J.N. Naturally Occurring Dominant Negative Variants of Stat5. *Mol. Cell. Biol.* **1996**, *16*, 6141–6148. [CrossRef] [PubMed]

43. Ambrosioa, R.; Fimiania, G.; Monfregolaa, J.; Sanzaria, E.; Felicea, N.; De Salernob, M.C.; Pignatab, C.; D'Ursoa, M.; Valeria Ursini, M. The Structure of Human STAT5A and B Genes Reveals Two Regions of Nearly Identical Sequence and an Alternative Tissue Specific STAT5B Promoter. *Gene* **2002**, *285*, 311–318. [CrossRef]

44. Lin, J.X.; Mietz, J.; Modi, W.S.; John, S.; Leonard, W.J. Cloning of Human Stat5B. Reconstitution of Interleukin-2-Induced Stat5A and Stat5B DNA Binding Activity in COS-7 Cells. *J. Biol. Chem.* **1996**, *271*, 10738–10744. [CrossRef] [PubMed]

45. Miyoshi, K.; Shillingford, J.M.; Smith, G.H.; Grimm, S.L.; Wagner, K.U.; Oka, T.; Rosen, J.M.; Robinson, G.W.; Hennighausen, L. Signal Transducer and Activator of Transcription (Stat) 5 Controls the Proliferation and Differentiation of Mammary Alveolar Epithelium. *J. Cell Biol.* **2001**, *155*, 531–542. [CrossRef] [PubMed]

46. Chomczynski, P.; Sacchi, N. Single-Step Method of RNA Isolation by Acid Guanidinium Thiocyanate-Phenol-Chloroform Extraction. *Anal. Biochem.* **1987**, *162*, 156–159. [CrossRef]

47. Huang, D.W.; Sherman, B.T.; Tan, Q.; Kir, J.; Liu, D.; Bryant, D.; Guo, Y.; Stephens, R.; Baseler, M.W.; Lane, H.C.; et al. DAVID Bioinformatics Resources: Expanded Annotation Database and Novel Algorithms to Better Extract Biology from Large Gene Lists. *Nucleic Acids Res.* **2007**, *35*, W169–W175. [CrossRef] [PubMed]

48. Walter, W.; Sánchez-Cabo, F.; Ricote, M. GOplot: An R Package for Visually Combining Expression Data with Functional Analysis: Figure 1. *Bioinformatics* **2015**, *31*, 2912–2914. [CrossRef] [PubMed]

Article

Nicotine Alters Estrogen Receptor-Beta-Regulated Inflammasome Activity and Exacerbates Ischemic Brain Damage in Female Rats

Nathan D. d'Adesky [1], Juan Pablo de Rivero Vaccari [2], Pallab Bhattacharya [1], Marc Schatz [1], Miguel A. Perez-Pinzon [1], Helen M. Bramlett [2,3] and Ami P. Raval [1,*]

[1] Cerebral Vascular Disease Research Center, Department of Neurology and Neuroscience Program (D4-5), P.O. Box 016960, University of Miami School of Medicine, Miami, FL 33101, USA; nathandadesky@gmail.com (N.D.d.); pbhattachary@gmail.com (P.B.); marc.schatz@med.miami.edu (M.S.); perezpinzon@med.miami.edu (M.A.P.-P.)

[2] Department of Neurological Surgery, The Miami Project to Cure Paralysis, University of Miami School of Medicine, Miami, FL 33136, USA; jderivero@med.miami.edu (J.P.d.R.V.); hbramlett@med.miami.edu (H.M.B.)

[3] Bruce W. Carter Department of Veterans Affairs Medical Center, Miami, FL 33125, USA

*   Correspondence: araval@med.miami.edu; Tel.: +1-305-243-7491

Received: 18 April 2018; Accepted: 24 April 2018; Published: 30 April 2018

Abstract: Smoking is a preventable risk factor for stroke and smoking-derived nicotine exacerbates post-ischemic damage via inhibition of estrogen receptor beta (ER-β) signaling in the brain of female rats. ER-β regulates inflammasome activation in the brain. Therefore, we hypothesized that chronic nicotine exposure activates the inflammasome in the brain, thus exacerbating ischemic brain damage in female rats. To test this hypothesis, adult female Sprague-Dawley rats (6–7 months old) were exposed to nicotine (4.5 mg/kg/day) or saline for 16 days. Subsequently, brain tissue was collected for immunoblot analysis. In addition, another set of rats underwent transient middle cerebral artery occlusion (tMCAO; 90 min) with or without nicotine exposure. One month after tMCAO, histopathological analysis revealed a significant increase in infarct volume in the nicotine-treated group (64.24 ± 7.3 mm^3; mean ± SEM; $n = 6$) compared to the saline-treated group (37.12 ± 7.37 mm^3; $n = 7$, $p < 0.05$). Immunoblot analysis indicated that nicotine increased cortical protein levels of caspase-1, apoptosis-associated speck-like protein containing a CARD (ASC) and pro-inflammatory cytokines interleukin (IL)-1β by 88% ($p < 0.05$), 48% ($p < 0.05$) and 149% ($p < 0.05$), respectively, when compared to the saline-treated group. Next, using an in vitro model of ischemia in organotypic slice cultures, we tested the hypothesis that inhibition of nicotine-induced inflammasome activation improves post-ischemic neuronal survival. Accordingly, slices were exposed to nicotine (100 ng/mL; 14–16 days) or saline, followed by treatment with the inflammasome inhibitor isoliquiritigenin (ILG; 24 h) prior to oxygen-glucose deprivation (OGD; 45 min). Quantification of neuronal death demonstrated that inflammasome inhibition significantly decreased nicotine-induced ischemic neuronal death. Overall, this study shows that chronic nicotine exposure exacerbates ischemic brain damage via activation of the inflammasome in the brain of female rats.

Keywords: stroke; inflammasome; nicotine; estrogen; smoking; women's health

---

1. Introduction

Cigarette smoking is a preventable risk factor for stroke, and smoking-ingested nicotine exacerbates post-stroke brain damage [1,2]. Stroke disproportionately kills more women than men and remains one of the leading causes of death and disability in the U.S. Women have a higher risk of stroke, as well as a higher mortality rate associated with stroke, and a higher tendency for more frequent recurrent strokes than men [3–5].

Although we know that women are more susceptible to stroke, we have a limited understanding of the underlying mechanisms for increased stroke severity and possible sex-specific prevention and treatment of stroke. Sex differences in stroke are highly complicated, and sex-specific risk factors in part account for epidemiological findings in stroke incidence, prevalence, and mortality. As stated in a review by Girijala et al., there are a number of modifiable risk factors such as cardiac conditions, hypertension, diabetes mellitus type 2, metabolic syndrome, alcohol consumption, physical inactivity, and cigarette smoke exposure, which are common for both sexes [4,6,7]. At the same time, there are female specific risk factors, such as pregnancy, pre-eclampsia, gestational diabetes, migraine with aura during pregnancy, oral contraceptive (OC) use, menopause, and hormone replacement therapy (HRT), which make women more susceptible to stroke [4,8]. What is more striking is that even among women there are studies that clearly link more damaging effects of stroke in women who combine OC/HRT and cigarette smoking. This also points to the facts that: (1) OC/HRT and cigarette smoking/tobacco use have synergistic deleterious effects on a woman's brain and (2) a gender non-specific risk factor of cigarette smoking has unique effects on the female brain that need to be identified and targeted to reduce consequences of stroke in women. Cigarette smoking, however, is an important modifiable risk factor that an individual does have control over. Nicotine, the main active ingredient in tobacco, has been shown to aggravate ischemic brain damage [9], and the mechanism by which nicotine contributes to poor outcomes after cerebral ischemia is yet to be fully elucidated.

Smoking-attributed nicotine is known to inhibit aromatase enzyme activity, which catalyzes the conversion of androgens into estrogens [10]. Consequently, nicotine reduces circulating estrogen levels and leads to early onset of menopause in women [11–19]. In laboratory studies on female rats, we confirmed the aforementioned epidemiological findings that chronic nicotine exposure reduced endogenous 17β-estradiol (E$_2$; a potent estrogen) levels [9]. Estrogen-mediated neuroprotection requires activation of estrogen receptor-alpha (ER-α) and beta (ER-β). Silencing of hippocampal ER-β but not ER-α, abolishes E$_2$-induced ischemic protection, suggesting a key role of ER-α and/or ER-β-activation [20–22]. Our study demonstrated that ER-β activation regulates inflammasome activation. The inflammasome is an arm of the innate immune response involved in the activation of the pro-inflammatory cytokines interleukin (IL)-1β and IL-18 through the processing of caspase-1 [23]. The inflammasome is comprised of the signaling proteins caspase-1 and apoptosis-associated speck-like protein containing a CARD (ASC) [23]. In a published study, we demonstrated that silencing of ER-β attenuated E$_2$-mediated decrease in caspase-1, ASC and IL-1β [24]. On the other hand, ER-β agonist treatment reduces inflammasome activation and ischemic damage in reproductively senescent female rats. ER-β agonist treatment significantly decreased inflammasome activation and increased post-ischemic neuronal counts by 32% ($p < 0.05$), as compared to the vehicle-treated, reproductively senescent rats [24]. Studies from our laboratory also showed that chronic nicotine exposure decreased membrane-bound and mitochondrial ER-β, but not ER-α protein levels in the brain [9,25]. Therefore, in the current study, we hypothesized that chronic nicotine exposure activates the inflammasome in the brain, thus exacerbating ischemic brain damage in female rats.

2. Results

2.1. Nicotine Reduces ER-β Protein Levels in the Brain of Female Rats

Because our previous results demonstrated that nicotine reduced the level of membrane-bound and mitochondrial ER-β in the hippocampus, in this study, we investigated protein expression of ER-β in the cortex, the main brain area vulnerable after transient middle cerebral artery occlusion (tMCAO). Our results demonstrated that nicotine significantly reduced cortical ER-β protein levels as compared with the saline group (Figure 1). ER-β protein levels after nicotine showed a 30% ($n = 8$; $p < 0.05$)

and 31% ($n = 8$; $p < 0.05$) reduction in cortex and hippocampus, respectively, as compared with saline (100%; $n = 8$).

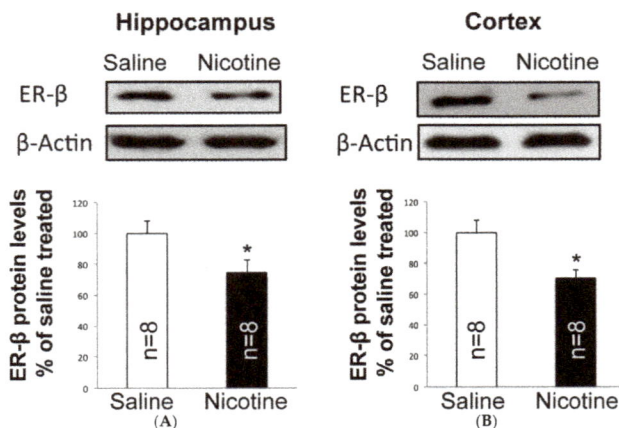

Figure 1. Nicotine decreases estrogen receptor beta (ER-β) protein expression in the hippocampus and cortex of female rats. Immunoblot analyses show significant reduction in the ER-β proteins of nicotine treated (**A**) hippocampus and (**B**) cortex when compared to the saline group. Data are presented mean ± SEM (* $p < 0.05$), $n = 8$.

2.2. Nicotine Increases Inflammasome Activation in the Brain of Female Rats

Since nicotine is an immunomodulatory agent and inflammasome activation plays a key role in ischemic brain damage, we then tested whether nicotine alters inflammasome protein expression. We obtained protein lysates from the cortex of female rats exposed to nicotine and resolved them by immunoblot analysis for the expression of active caspase-1, ASC and IL-1β. Our findings indicate a significant increase in these inflammasome proteins (Figure 2). Accordingly, nicotine increased protein levels of caspase-1, ASC and IL-1β by 88% ($p < 0.05$), 48% ($p < 0.05$) and 149% ($p < 0.05$) respectively in the cortex, as compared to the saline-treated group (Figure 2).

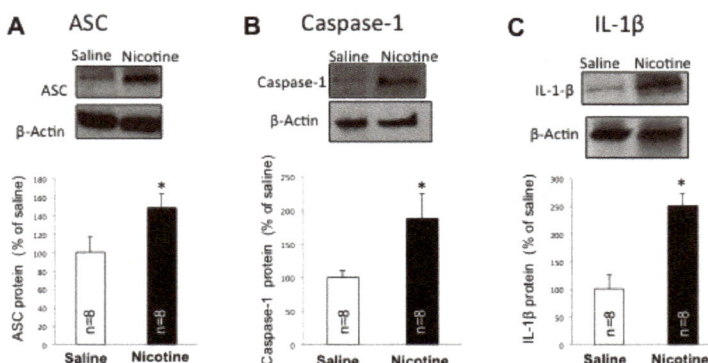

Figure 2. Nicotine increases inflammasome protein expression in the cortex of female rats. Immunoblot analyses show an increase in the inflammasome proteins (**A**) Caspase-1 (**B**) apoptosis-associated speck-like protein containing a CARD (ASC), and (**C**) IL-1β when compared to the saline group. Data are presented mean ± SEM (* $p < 0.05$), $n = 8$.

2.3. Nicotine Worsens Infarct Volume and Neurodeficit Score after tMCAO

Since inflammasome increase in the brain can exacerbate post-stroke outcomes, we tested the hypothesis that chronic nicotine exposure exacerbates stroke outcomes. Rats exposed to nicotine or saline for 16 days underwent tMCAO and were allowed to recover for 30 days before histological analysis. Our data demonstrated significantly higher mean infarct volume in the nicotine-treated group (64.24 ± 7.3 mm^3; $n = 6$) compared to the saline-treated group (37.12 ± 7.37 mm^3; Mean ± SEM; $n = 7$, $p < 0.05$) (Figure 3A). Histological analysis of nicotine- or saline-treated rat brains that underwent sham surgery did not show any infarct. Figure 3B shows that the neurodeficit score in each group was more than 10 when tested at 1 h after tMCAO. Neurodeficit scores 1 h, 1, 7, 15 and 30 days after tMCAO are shown in Figure 3B. The neurodeficit score remained unchanged in nicotine-treated animals. These results indicate that nicotine worsens outcome after stroke.

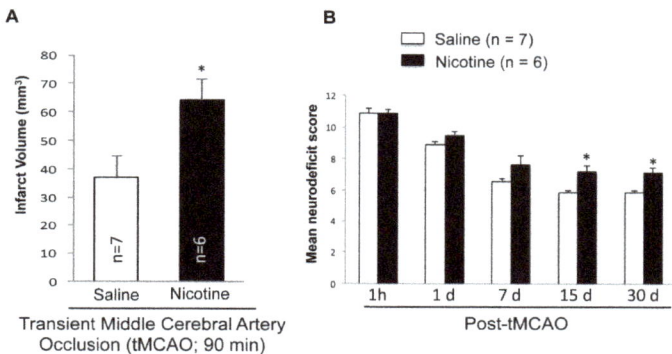

Figure 3. Nicotine increases infarct volume after tMCAO. (**A**) Infarct volume was measured by analyzing lesions after rats were subjected to 90 min of tMCAO. Volume was measured in mm^3. (**B**) Neurodeficit score was measured in rats post tMCAO. Higher scores represent a greater neurodeficit. Nicotine-treated rats showed significantly higher neurodeficit scores after ischemia when compared to control. Data presented are mean ± SEM (* $p < 0.05$)

2.4. Inhibition of Inflammasome Activation Decreases Neuronal Cell Death in an In Vitro Model of Stroke

Since chronic nicotine exposure increases inflammasome proteins in the brain of female rats, as a proof-of-principle, we tested the hypothesis that the inhibition of nicotine-induced inflammasome activation improves post-ischemic neuronal survival using our well-established in vitro model of cerebral ischemia. We tested this hypothesis in organotypic cultures by exposing slices to saline or nicotine for 14–19 days. We found that nicotine significantly increased ischemic neuronal death in the Cornu Ammonis area 1 (CA1) region of the hippocampus as compared to saline. The propidium iodide (PI) fluorescence values were 56 ± 5.5% ($n = 10$) and 70 ± 1.2% ($n = 10$) in the saline and nicotine + vehicle-control groups respectively ($p < 0.01$). Nicotine and saline exposed slices were treated with the inflammasome inhibitor isoliquiritigenin (ILG; 1, 10, and 40 µM). Inflammasome inhibition at 10 µM ILG concentration showed a significant decrease in nicotine-induced ischemic neuronal death when compared to the nicotine + vehicle group. The PI fluorescence values of the saline + ILG, nicotine + ILG, and vehicle treated nicotine groups were 44 ± 6.2% ($n = 8$), 27 ± 1.5% ($n = 8$), and 70 ± 1.2% ($n = 10$; $p < 0.05$), respectively. The PI fluorescence values of the nicotine + 1 µM ILG and nicotine + 40 µM ILG were 55 ± 5.3% ($n = 8$) and 61 ± 2.5% ($n = 8$), which were not significantly different as compared to PI values of saline + 1 µM ILG and saline + 40 µM ILG groups (Figure 4).

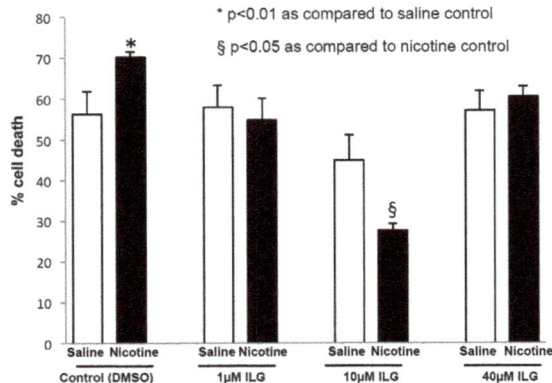

Figure 4. ILG decreases cell death in hippocampal organotypic slice cultures after oxygen-glucose deprivation (OGD). PI fluorescence, which represents cell death, was measured in organotypic brain slices exposed to nicotine or saline treatment, then ILG at varying concentrations 24 h before OGD. Data shown represent cell death normalized to the respective saline groups. To normalize data, the mean of the subgroup was divided by its respective control. PI fluorescence was significantly stronger in the nicotine group than the saline group. PI fluorescence significantly decreased nicotine induced cell death in the nicotine group treated with 10 μM ILG, as compared to the nicotine control. Data are presented mean ± SEM (* $p < 0.01$, § $p < 0.05$). $n = 4$ to 11 per group.

3. Discussion

This study revealed for the first time that chronic nicotine exposure increases inflammasome activation in the brain and exacerbates post-ischemic damage in the brain of female rats. Our study also demonstrated that inhibition of inflammasome activation specifically attenuates nicotine-induced ischemic cell death in an in vitro model of ischemia. Inhibition of inflammasome activation was achieved with ILG, which is a chalconoid found in licorice. Studies have found that ILG inhibits inflammasome activation via the inhibition of ASC oligomerization and the inhibition of nucleotide-binding oligomerization domain (NOD)-like receptor 3 (NLRP3) activation [26,27]. In our study, nicotine-exposed hippocampal slice-cultures were treated with ILG, one day prior to the induction of OGD. Therefore, the observed neuroprotective effect of ILG in our study suggests the suppression of nicotine-induced inflammasome activation in the brain. In vivo the half-life on ILG has been identified to be around 4 h [28]. In this study, ILG was added before and during OGD. In vitro, when ILG was applied to liver microsomes, the half-life was found to be 25.3 min. Since brain cells do not have the clearance capacity of liver cells, we estimate that the half-life of ILG in the brain slices is greater. Importantly, since OGD in our study was done in the presence of ILG for 45 min, we anticipate that for all or most of the duration of the experiment, ILG was present. At the low dose, the concentration of 1 μM ILG was not enough to prevent cell death. However, at the middle dose of 10 μM we found the concentration optimum at which ILG can prevent cell death in both the saline as well as the nicotine groups. The ILG treatment showed more rigorous neuroprotection in nicotine treated group as compared to saline group, which suggests that effects of ILG might wean off prior to post-OGD inflammasome activation and observed effects of ILG in nicotine group could be outcome of pre-OGD inhibition of ASC. Although the mechanism of action of how ILG inhibits inflammasome activation remains unknown, it has been shown that this inhibition occurs between 1 and 10 μM [29], and that at the higher dosages of even 30 μM ILG does not inhibit the inflammasome. Thus, it is possible that at the 40 μM the effects of ILG on inflammasome inhibition are absent, resulting in greater cell death in the 40 μM similar to the 1 μM group.

It has been shown that inflammasome activation in the brain is regulated by sex hormones [24]. In a previous study from our laboratory, we reported that ER-β attenuates inflammasome activation in the brain of female rats. Specifically, silencing of hippocampal ER-β increased caspase-1, ASC, and the pro-inflammatory cytokine interleukin-1β (IL-1β) in the hippocampus of ovariectomized rats. Conversely, periodic activation of ER-β significantly decreased inflammasome activation and reduced post-ischemic neuronal death in the hippocampus of reproductively senescent female rats. It has been shown that aging as well as long-term nicotine usage reduces ER-β availability in the brain [30,31]. Therefore, nicotine-induced inflammasome activation in the brain could be due to the direct immunomodulatory nature of nicotine or the loss of ER-β following long-term nicotine exposure in the brain. The latter scenario is consistent with females having a higher risk from nicotine toxicity. This notion is supported by a study showing that hippocampal neuronal damage is greater in female rats following chronic nicotine exposure [27].

Inflammasome activation occurs after the detection of pro-inflammatory molecules by pattern recognition receptors, such as a toll-like receptor or a NOD-like receptor (NLR) [32]. The mechanisms of inflammasome activation include the generation of mitochondrial reactive oxygen species (ROS) and the translocation of NLRP3 to the mitochondria [33]. Interestingly, the nicotinic acetylcholine receptor (7α-nAChR) has been found in neuronal mitochondria [34] and its physiological significance remains unknown. However, it is likely that it is involved in buffering cytoplasmic calcium [35,36]. It is through the mechanism of calcium buffering in which one study found that 7α-nAChRs located in the mitochondria regulates inflammasome activation in peritoneal mouse macrophages [37]. It is well documented that calcium overload can trigger mitochondrial dysfunction, promote the production of reactive oxygen species (ROS), and disrupt the mitochondrial permeability transition pore, eventually leading to neuronal death [38–40]. Therefore, nicotine-induced mitochondrial dysfunction, which has been described in several studies [41–43], may be responsible for the increase in stroke volume in female rats. However, studies are needed to understand the upstream mechanisms leading to mitochondrial dysfunction that increases the susceptibility to ischemia.

Finally, relinquishing the smoking habit reduces the risk for stroke; however, impact of smoking cessation on stroke outcome remains unknown. Because it is difficult to stop smoking, more tobacco users have switched to the "fashionable" e-Cigarette (electronic nicotine delivery systems) as an alternative to tobacco smoking or even as an aid for smoking cessation. However, the safety of e-Cigarettes remains questionable and as demonstrated in the current study, negative effects of nicotine on brain will persist, which makes current research in this area timely and of a high impact.

4. Materials and Methods

4.1. Animals

All animal procedures were carried out in accordance with the Guide for the Care and Use of Laboratory Animals published by the U.S. Public Health and procedures involving animal subjects and were approved by the Animal Care and Use Committee (#A-3224-01, effective 24 November 2015) of the University of Miami.

4.2. In Vivo

Adult female Sprague–Dawley rats (290 ± 20 g; 6–7 months old) were used for this study. The stages of estrous cycle were monitored as described [44] and only rats showing at least three consecutive normal (4 day) estrous cycles were used for our experiments.

4.3. Nicotine or Saline Treatment

To obtain sustained nicotine delivery, osmotic pumps (type 2ML2, Alzet Corp., Palo Alto, CA, USA) containing nicotine or saline were implanted in rats for 16–21 days as described [30]. The pump delivered a fixed and continuous dose of nicotine hydrogen tartrate (4.5 mg/kg/day, equivalent to

1.5 mg/kg/day free base) throughout the 16–21 days. Rats exposed to nicotine ($n = 21$) or saline ($n = 20$) were divided into two groups. One group of eight rats treated with either nicotine or saline was used for brain tissue collection, while the remaining rats in both treatment groups underwent transient middle cerebral artery occlusion (tMCAO) or sham surgery. Since we have previously demonstrated that the higher endogenous estrogen levels seen during proestrus protect CA1 neurons against cerebral ischemia [45], tMCAO or tissue collection was performed only when nicotine/saline treated rats were in proestrus stage. In those instances where rats were not in the proestrus stage on the last scheduled day of treatment, we extended the treatment by 1 or 2 days.

4.4. Transient Middle Cerebral Artery Occlusion (tMCAO) and Infarct Volume

Nicotine or saline treated rats were exposed to ischemic stroke (90 min) using an occluding intraluminal suture inserted past the internal carotid artery to occlude the middle cerebral artery as described previously [46,47]. In parallel, we also performed sham surgery on nicotine or saline treated rats. During this sham procedure rats were exposed to anesthesia for a period similar to that of the tMCAO group. During the surgical procedure of tMCAO or sham, physiological parameters including pCO_2, pO_2, pH, HCO_3 and arterial blood pressure were maintained within normal limits prior to and after tMCAO (data presented as Table 1). Body and head temperatures were maintained at 37 ± 0.2 °C throughout the experiment with assistance of lamps placed above the animal's body and head. One month after tMCAO surgery, rats were perfused with saline, then with FAM (a mixture of formaldehyde, glacial acetic acid, and methanol; 10:10:80), and brains were prepared, sectioned, and stained to obtain infarct volumes. The sample preparations and procedures are described in more detail in our previous publication [48]. Brains were then embedded with paraffin and 10 µm thick sections were obtained. Infarct volume measurements were adapted from previous publications [49,50]. Briefly, the same nine coronal sections were selected for every animal (Bregma levels 5.2, 2.7, 1.2, −0.3, −1.3, −1.8, −3.8, −5, −7.3). Infarcted area was then measured on each section using MCID software (version, Manufacturer, City, US State abbrev. if applicable, Country). Infarct volume was then determined similar to previously used methods [49,50].

Table 1. Physiological variable. Physiological variables were recorded during the period when the animal was under anesthesia. Rows with bold background denote values after tMCAO. Mean arterial blood pressure (MABP).

Groups	Glucose	pH	pCO₂	pO₂	MABP	Post-tMCAO Mortality
Saline + Sham ($n = 4$)	131.0 ± 7.4	7.39 ± 0.5 **7.39 ± 0.4**	36.3 ± 2.3 **36.0 ± 4.1**	135 ± 35 **133 ± 15**	132.4 ± 6 **134 ± 11.7**	0
Nicotine + Sham ($n = 5$)	126.4 ± 2.8	7.36 ± 0.4 **7.37 ± 0.04**	34 ± 7.2 **38.0 ± 3.4**	133 ± 37 **127 ± 28**	135 ± 3 **135 ± 11.4**	0
Saline + tMCAO ($n = 8$)	130.3 ± 8.7	7.36 ± 0.4 **7.37 ± 0.05**	35.3 ± 2.3 **34.3 ± 5.2**	134 ± 39 **140 ± 15**	131 ± 7.7 **134 ± 9.3**	1
Nicotine + tMCAO ($n = 8$)	127.4 ± 6.8	7.4 ± 0.2 **7.39 ± 0.04**	38.08 ± 4.5 **35.8 ± 3.9**	130 ± 35 **138 ± 25**	129 ± 7.8 **132 ± 5.3**	2

4.5. Neurodeficit Scoring

The neurological score was monitored at an hour, 1, 7, 15 and 30 days after tMCAO. A standardized neurobehavioral test battery was conducted as described previously [51], which includes tests for postural reflex, sensorimotor integration, and proprioception. Total neurological score ranged from a normal score of 0 to a maximal possible score of 12.

4.6. Western Blotting

Rats exposed to nicotine (16–21 days) or saline were anesthetized using 5% isoflurane, decapitated, and the hippocampal and cortical tissues were collected, flash frozen, and stored at −80 °C. At the time of immunoblotting, hippocampal and cortical tissues were homogenized, protein content was analyzed, and proteins were separated by 12% SDS-PAGE as described [47]. Proteins were transferred to Immobilon-P (Millipore, Burlington, MA, USA) membrane and incubated with primary antibodies against rabbit polyclonal anti-ER-β (1:500; Santa Cruz Biotechnology, Dallas, TX, USA), IL-1β (1:1000; Cell Signaling, Danvers, MA, USA), ASC (1:1000; Santa Cruz Biotechnology), Caspase-1 (1:1000; Novus Biologicals, Littleton, CO, USA), and β-Actin (1:5000; Sigma, St. Louis, MO, USA). All data were normalized to β-Actin (monoclonal; 1:1000; Sigma). Immunoblot images were digitized and subjected to densitometric analysis [45].

4.7. In Vitro Organotypic Slice Cultures and Oxygen-Glucose Deprivation

Hippocampal organotypic slice cultures were prepared from female neonatal (9–11 days old) Sprague–Dawley rats and the details of slice culture are as described previously [46,52–54]. Briefly, hippocampal slices were cultured for 14–20 days followed by exposure to nicotine (100 ng/mL) or saline (vehicle) for 14–16 days. At the end of the treatment period (~29–37 days), a subgroup of slices (either treated with nicotine or saline) were exposed to the inflammasome inhibitor isoliquiritigenin (ILG; Invivogen, San Diego, CA, USA) at different concentrations (1, 10, or 40 μM) dissolved in dimethyl sulfoxide (DMSO; Sigma-Aldrich, St. Louis, MO, USA) or vehicle control DMSO (1 μL/mL of medium) for 24 h. After inhibitor/control treatment, each subgroup was exposed to OGD (45 min) as described [46,52–54]. To determine the extent of neuronal damage following OGD, we used the propidium iodide (PI) method [46,52–54]. Briefly, slices were incubated in culture medium supplemented with 2 μg/mL PI (Sigma, St. Louis, MO, USA) for 1 h. Images were taken using an inverted fluorescence microscope. Images of the cultured slices were taken (1) at baseline prior to the 'test' ischemia procedure; (2) 24 h after the 'test' ischemic insult to assess ischemic damage; and (3) 24 h after *N*-methyl-D-aspartate (NMDA) treatment to assess maximum damage to neuronal cells. The hippocampal CA1 subfield was chosen as the region of interest, and quantification was performed using Scion Image software [46,52–54]. The percentage of relative optical intensity (ROI) served as an index of neuronal cell death [46,52–54].

4.8. Statistical Analysis

The data are presented as mean value ± SEM and results were analyzed by a two-tailed Student's *t* test and $p < 0.05$ was considered statistically significant.

Author Contributions: A.P.R. and J.P.d.R.V. conceived and designed the experiments; N.D.d., M.S. and P.B. performed the experiments; A.P.R., J.P.d.R.V. and H.M.B. analyzed the data; A.P.R., J.P.d.R.V, N.D.d. and M.S. wrote the paper; and H.M.B. and M.A.P.-P. edited the paper.

Acknowledgments: This work was supported by an Endowment from Chantal Scheinberg and Peritz Scheinberg (A.P.R.), Florida Department of Heath #7JK01 funds (H.M.B. & A.P.R.), the American Heart Association grants Grant-in-aid #16GRNT31300011 (A.P.R.) and 12SDG11970010 (J.P.d.R.V.) as well as The Miami Project to Cure Paralysis (J.P.d.R.V., H.M.B.). J.P.d.R.V. and H.M.B. are co-founders and managing members of InflamaCORE, LLC, a company dedicated to developing therapies and diagnostic tools focusing on the inflammasome.

Conflicts of Interest: J.P.d.R.V. and H.M.B. are co-founders and managing members of InflamaCORE, LLC, a company dedicated to developing therapies and diagnostic tools focusing on the inflammasome.

Abbreviations

ASC	Apoptosis associated speck-like protein containing a caspase recruitment domain
BBB	Blood brain barrier
DMSO	Dimethyl sulfoxide
e-Cigarette	Electronic cigarettes

ER-β	Estrogen receptor subtype beta
FAM	Formaldehyde, glacial acetic acid, and methanol
IL-1β	Interleukin-1β
ILG	Isoliquiritigenin
nAChR	Nicotinic acetylcholine receptor
NLR	NOD-like receptor
NMDA	*N*-methyl-D-aspartate
OGD	Oxygen-glucose deprivation
PI	Propidium iodide
tMCAO	Transient middle cerebral artery occlusion
TNF	Tumor necrosis factor

References

1. Boehme, A.K.; Esenwa, C.; Elkind, M.S. Stroke Risk Factors, Genetics, and Prevention. *Circ. Res.* **2017**, *120*, 472–495. [CrossRef] [PubMed]
2. Peters, S.A.; Huxley, R.R.; Woodward, M. Smoking as a risk factor for stroke in women compared with men: A systematic review and meta-analysis of 81 cohorts, including 3,980,359 individuals and 42,401 strokes. *Stroke* **2013**, *44*, 2821–2828. [CrossRef] [PubMed]
3. Loraine, A.; West, S.C.; Goodkind, D.; He, W. 65+ in the United States: 2010. *Curr. Popul. Rep.* **2014**, 23–212.
4. Girijala, R.L.; Sohrabji, F.; Bush, R.L. Sex differences in stroke: Review of current knowledge and evidence. *Vasc. Med.* **2017**, *22*, 135–145. [CrossRef] [PubMed]
5. Mozaffarian, D.; Benjamin, E.J.; Go, A.S.; Arnett, D.K.; Blaha, M.J.; Cushman, M.; Das, S.R.; de Ferranti, S.; Després, J.P.; Fullerton, H.J.; et al. Heart Disease and Stroke Statistics-2016 Update: A Report From the American Heart Association. *Circulation* **2016**, *133*, e38–e360. [CrossRef] [PubMed]
6. Romero, J.R.; Morris, J.; Pikula, A. Stroke prevention: Modifying risk factors. *Ther. Adv. Cardiovasc. Dis.* **2008**, *2*, 287–303. [CrossRef] [PubMed]
7. Aoki, J.; Uchino, K. Treatment of risk factors to prevent stroke. *Neurotherapeutics* **2011**, *8*, 463–474. [CrossRef] [PubMed]
8. Di Carlo, A.; Lamassa, M.; Baldereschi, M.; Pracucci, G.; Basile, A.M.; Wolfe, C.D.; Giroud, M.; Rudd, A.; Ghetti, A.; Inzitari, D. Sex differences in the clinical presentation, resource use, and 3-month outcome of acute stroke in Europe: Data from a multicenter multinational hospital-based registry. *Stroke* **2003**, *34*, 1114–1119. [CrossRef] [PubMed]
9. Raval, A.P.; Hirsch, N.; Dave, K.R.; Yavagal, D.R.; Bramlett, H.; Saul, I. Nicotine and estrogen synergistically exacerbate cerebral ischemic injury. *Neuroscience* **2011**, *181*, 216–225. [CrossRef] [PubMed]
10. Rune, G.M.; Frotscher, M. Neurosteroid synthesis in the hippocampus: Role in synaptic plasticity. *Neuroscience* **2005**, *136*, 833–842. [CrossRef] [PubMed]
11. Barbieri, R.L.; Gochberg, J.; Ryan, K.J. Nicotine, cotinine, and anabasine inhibit aromatase in human trophoblast in vitro. *J. Clin. Investig.* **1986**, *77*, 1727–1733. [CrossRef] [PubMed]
12. Cassidenti, D.L.; Vijod, A.G.; Vijod, M.A.; Stanczyk, F.Z.; Lobo, R.A. Short-term effects of smoking on the pharmacokinetic profiles of micronized estradiol in postmenopausal women. *Am. J. Obstet. Gynecol.* **1990**, *163*, 1953–1960. [CrossRef]
13. Cramer, D.W.; Harlow, B.L.; Xu, H.; Fraer, C.; Barbieri, R. Cross-sectional and case-controlled analyses of the association between smoking and early menopause. *Maturitas* **1995**, *22*, 79–87. [CrossRef]
14. Grainge, M.J.; Coupland, C.A.C.; Cliffe, S.J.; Chilvers, C.E.D.; Hosking, D.J.; Nottingham EPIC Study Group. Cigarette smoking, alcohol and caffeine consumption, and bone mineral density in postmenopausal women. *Osteoporos. Int.* **1998**, *8*, 355–363. [CrossRef] [PubMed]
15. Greenberg, G.; Thompson, S.G.; Meade, T.W. Relation between cigarette smoking and use of hormonal replacement therapy for menopausal symptoms. *J. Epidemiol. Community Health* **1987**, *41*, 26–29. [CrossRef] [PubMed]
16. Jensen, J.; Christiansen, C.; Rodbro, P. Cigarette smoking, serum estrogens, and bone loss during hormone-replacement therapy early after menopause. *N. Engl. J. Med.* **1985**, *313*, 973–975. [CrossRef] [PubMed]
17. Michnovicz, J.J.; Naganuma, H.; Hershcopf, R.J.; Bradlow, H.L.; Fishman, J. Increased urinary catechol estrogen excretion in female smokers. *Steroids* **1988**, *52*, 69–83. [CrossRef]

18. Mueck, A.O.; Seeger, H. Smoking, estradiol metabolism and hormone replacement therapy. *Curr. Med. Chem. Cardiovasc. Hematol. Agents* **2005**, *3*, 45–54. [CrossRef] [PubMed]
19. Windham, G.C.; Elkin, E.P.; Swan, S.H.; Waller, K.O.; Fenster, L. Cigarette smoking and effects on menstrual function. *Obstet. Gynecol.* **1999**, *93*, 59–65. [PubMed]
20. Zhang, Q.G.; Raz, L.; Wang, R.; Han, D.; De Sevilla, L.; Yang, F.; Vadlamudi, R.K.; Brann, D.W. Estrogen attenuates ischemic oxidative damage via an estrogen receptor α-mediated inhibition of NADPH oxidase activation. *J. Neurosci.* **2009**, *29*, 13823–13836. [CrossRef] [PubMed]
21. Dubal, D.B.; Rau, S.W.; Shughrue, P.J.; Zhu, H.; Yu, J.; Cashion, A.B.; Suzuki, S.; Gerhold, L.M.; Bottner, M.B.; Dubal, S.B. Differential modulation of estrogen receptors (ERs) in ischemic brain injury: A role for ER-α in estradiol-mediated protection against delayed cell death. *Endocrinology* **2006**, *147*, 3076–3084. [CrossRef] [PubMed]
22. Lebesgue, D.; Chevaleyre, V.; Zukin, R.S.; Etgen, A.M. Estradiol rescues neurons from global ischemia-induced cell death: Multiple cellular pathways of neuroprotection. *Steroids* **2009**, *74*, 555–561. [CrossRef] [PubMed]
23. De Rivero Vaccari, J.P.; Dietrich, W.D.; Keane, R.W. Activation and regulation of cellular inflammasomes: Gaps in our knowledge for central nervous system injury. *J. Cereb. Blood Flow Metab.* **2014**, *34*, 369–375. [CrossRef] [PubMed]
24. de Rivero Vaccari, J.P.; Patel, H.H.; Brand, F.J.; Perez-Pinzon, M.A.; Bramlett, H.M.; Raval, A.P. Estrogen receptor β signaling alters cellular inflammasomes activity after global cerebral ischemia in reproductively senescence female rats. *J. Neurochem.* **2016**, *136*, 492–496. [CrossRef] [PubMed]
25. Raval, A.P.; Dave, K.R.; Saul, I.; Gonzalez, G.J.; Diaz, F. Synergistic inhibitory effect of nicotine plus oral contraceptive on mitochondrial complex-IV is mediated by estrogen receptor β in female rats. *J. Neurochem.* **2012**, *121*, 157–167. [CrossRef] [PubMed]
26. Zeng, J.; Chen, Y.; Ding, R.; Feng, L.; Fu, Z.; Yang, S.; Deng, X.; Xie, Z.; Zheng, S. Isoliquiritigenin alleviates early brain injury after experimental intracerebral hemorrhage via suppressing ROS- and/or NF-κB-mediated NLRP3 inflammasome activation by promoting Nrf2 antioxidant pathway. *J. Neuroinflammation* **2017**, *14*, 119. [CrossRef] [PubMed]
27. Gomes, P.X.; de Oliveira, G.V.; de Araújo, F.Y.R.; de Barros Viana, G.S.; de Sousa, F.C.F.; Hyphantis, T.N.; Grunberg, N.E.; Carvalho, A.F.; Macêdo, D.S. Differences in vulnerability to nicotine-induced kindling between female and male periadolescent rats. *Psychopharmacology* **2013**, *225*, 115–126. [CrossRef] [PubMed]
28. Qiao, H.; Zhang, X.; Wang, T.; Liang, L.; Chang, W.; Xia, H. Pharmacokinetics, biodistribution and bioavailability of isoliquiritigenin after intravenous and oral administration. *Pharm. Biol.* **2014**, *52*, 228–236. [CrossRef] [PubMed]
29. Honda, H.; Nagai, Y.; Matsunaga, T.; Okamoto, N.; Watanabe, Y.; Tsuneyama, K.; Hayashi, H.; Fujii, I.; Ikutani, M.; Hirai, Y.; et al. Isoliquiritigenin is a potent inhibitor of NLRP3 inflammasome activation and diet-induced adipose tissue inflammation. *J. Leukoc. Biol.* **2014**, *96*, 1087–1100. [CrossRef] [PubMed]
30. Raval, A.P.; Sick, J.T.; Gonzalez, G.J.; DeFazio, R.A.; Dong, C.; Sick, T.J. Chronic nicotine exposure inhibits estrogen-mediated synaptic functions in hippocampus of female rats. *Neurosci. Lett.* **2012**, *517*, 41–46. [CrossRef] [PubMed]
31. Waters, E.M.; Yildirim, M.; Janssen, W.G.; Lou, W.W.; McEwen, B.S.; Morrison, J.H.; Milner, T.A. Estrogen and aging affect the synaptic distribution of estrogen receptor beta-immunoreactivity in the CA1 region of female rat hippocampus. *Brain Res.* **2011**, *1379*, 86–97. [CrossRef] [PubMed]
32. Hauenstein, A.V.; Zhang, L.; Wu, H. The hierarchical structural architecture of inflammasomes, supramolecular inflammatory machines. *Curr. Opin. Struct. Biol.* **2015**, *31*, 75–83. [CrossRef] [PubMed]
33. Yang, C.S.; Kim, J.J.; Kim, T.S.; Lee, P.Y.; Kim, S.Y.; Lee, H.M.; Shin, D.M.; Nguyen, L.T.; Lee, M.S.; Jin, H.S.; et al. Small heterodimer partner interacts with NLRP3 and negatively regulates activation of the NLRP3 inflammasome. *Nat. Commun.* **2015**, *6*, 6115. [CrossRef] [PubMed]
34. Gergalova, G.; Lykhmus, O.; Kalashnyk, O.; Koval, L.; Chernyshov, V.; Kryukova, E.; Tsetlin, V.; Komisarenko, S.; Skok, M. Mitochondria express α7 nicotinic acetylcholine receptors to regulate Ca^{2+} accumulation and cytochrome c release: Study on isolated mitochondria. *PLoS ONE* **2012**, *7*, e31361. [CrossRef] [PubMed]
35. Picciotto, M.R.; Zoli, M. Neuroprotection via nAChRs: The role of nAChRs in neurodegenerative disorders such as Alzheimer's and Parkinson's disease. *Front. Biosci.* **2008**, *13*, 492–504. [CrossRef] [PubMed]
36. Slotkin, T.A. Nicotine and the adolescent brain: Insights from an animal model. *Neurotoxicol. Teratol.* **2002**, *24*, 369–384. [CrossRef]

37. Lu, B.; Kwan, K.; Levine, Y.A.; Olofsson, P.S.; Yang, H.; Li, J.; Joshi, S.; Wang, H.; Andersson, U.; Chavan, S.S.; et al. α7 nicotinic acetylcholine receptor signaling inhibits inflammasome activation by preventing mitochondrial DNA release. *Mol. Med.* **2014**, *20*, 350–358. [CrossRef] [PubMed]

38. Duchen, M.R. Mitochondria, calcium-dependent neuronal death and neurodegenerative disease. *Pflügers Arch. Eur. J. Physiol.* **2012**, *464*, 111–121. [CrossRef] [PubMed]

39. Barsukova, A.G.; Bourdette, D.; Forte, M. Mitochondrial calcium and its regulation in neurodegeneration induced by oxidative stress. *Eur. J. Neurosci.* **2011**, *34*, 437–447. [CrossRef] [PubMed]

40. Toman, J.; Fiskum, G. Influence of aging on membrane permeability transition in brain mitochondria. *J. Bioenerg. Biomembr.* **2011**, *43*, 3–10. [CrossRef] [PubMed]

41. Zanetti, F.; Giacomello, M.; Donati, Y.; Carnesecchi, S.; Frieden, M.; Barazzone-Argiroffo, C. Nicotine mediates oxidative stress and apoptosis through cross talk between NOX1 and Bcl-2 in lung epithelial cells. *Free Radic. Biol. Med.* **2014**, *76*, 173–184. [CrossRef] [PubMed]

42. Arany, I.; Clark, J.; Reed, D.K.; Juncos, L.A. Chronic nicotine exposure augments renal oxidative stress and injury through transcriptional activation of p66shc. *Nephrol. Dial. Transplant.* **2013**, *28*, 1417–1425. [CrossRef] [PubMed]

43. Bhagwat, S.V.; Vijayasarathy, C.; Raza, H.; Mullick, J.; Avadhani, N.G. Preferential effects of nicotine and 4-(N-methyl-N-nitrosamine)-1-(3-pyridyl)-1-butanone on mitochondrial glutathione S-transferase A4-4 induction and increased oxidative stress in the rat brain. *Biochem. Pharmacol.* **1998**, *56*, 831–839. [CrossRef]

44. Marcondes, F.K.; Bianchi, F.J.; Tanno, A.P. Determination of the estrous cycle phases of rats: Some helpful considerations. *Braz. J. Biol.* **2002**, *62*, 609–614. [CrossRef] [PubMed]

45. Raval, A.P.; Saul, I.; Dave, K.R.; DeFazio, R.A.; Perez-Pinzon, M.A.; Bramlett, H. Pretreatment with a single estradiol-17β bolus activates cyclic-AMP response element binding protein and protects CA1 neurons against global cerebral ischemia. *Neuroscience* **2009**, *160*, 307–318. [CrossRef] [PubMed]

46. Bright, R.; Raval, A.P.; Dembner, J.M.; Pérez-Pinzón, M.A.; Steinberg, G.K.; Yenari, M.A.; Mochly-Rosen, D. Protein kinase C delta mediates cerebral reperfusion injury in vivo. *J. Neurosci.* **2004**, *24*, 6880–6888. [CrossRef] [PubMed]

47. Maier, C.M.; Sun, G.H.; Kunis, D.; Yenari, M.A.; Steinberg, G.K. Delayed induction and long-term effects of mild hypothermia in a focal model of transient cerebral ischemia: Neurological outcome and infarct size. *J. Neurosurg.* **2001**, *94*, 90–96. [CrossRef] [PubMed]

48. Lin, B.; Ginsberg, M.D. Quantitative assessment of the normal cerebral microvasculature by endothelial barrier antigen (EBA) immunohistochemistry: Application to focal cerebral ischemia. *Brain Res.* **2000**, *865*, 237–244. [CrossRef]

49. Zhang, Y.; Belayev, L.; Zhao, W.; Irving, E.A.; Busto, R.; Ginsberg, M.D. A selective endothelin ETA receptor antagonist, SB 234551, improves cerebral perfusion following permanent focal cerebral ischemia in rats. *Brain Res.* **2005**, *1045*, 150–156. [CrossRef] [PubMed]

50. Belayev, L.; Khoutorova, L.; Zhao, W.; Vigdorchik, A.; Belayev, A.; Busto, R.; Magal, E.; Ginsberg, M.D. Neuroprotective effect of darbepoetin alfa, a novel recombinant erythropoietic protein, in focal cerebral ischemia in rats. *Stroke* **2005**, *36*, 1071–1076. [CrossRef] [PubMed]

51. Ley, J.J.; Vigdorchik, A.; Belayev, L.; Zhao, W.; Busto, R.; Khoutorova, L.; Becker, D.A.; Ginsberg, M.D. Stilbazulenyl nitrone, a second-generation azulenyl nitrone antioxidant, confers enduring neuroprotection in experimental focal cerebral ischemia in the rat: Neurobehavior, histopathology, and pharmacokinetics. *J. Pharmacol. Exp. Ther.* **2005**, *313*, 1090–1100. [CrossRef] [PubMed]

52. Xu, G.P.; Dave, K.R.; Vivero, R.; Schmidt-Kastner, R.; Sick, T.J.; Pérez-Pinzón, M.A. Improvement in neuronal survival after ischemic preconditioning in hippocampal slice cultures. *Brain Res.* **2002**, *952*, 153–158. [CrossRef]

53. Raval, A.P.; Dave, K.R.; Mochly-Rosen, D.; Sick, T.J.; Pérez-Pinzón, M.A. Epsilon PKC is required for the induction of tolerance by ischemic and NMDA-mediated preconditioning in the organotypic hippocampal slice. *J. Neurosci.* **2003**, *23*, 384–391. [CrossRef] [PubMed]

54. Lange-Asschenfeldt, C.; Raval, A.P.; Dave, K.R.; Mochly-Rosen, D.; Sick, T.J.; Pérez-Pinzón, M.A. Epsilon protein kinase C mediated ischemic tolerance requires activation of the extracellular regulated kinase pathway in the organotypic hippocampal slice. *J. Cereb. Blood Flow Metab.* **2004**, *24*, 636–645. [CrossRef] [PubMed]

International Journal of
Molecular Sciences

MDPI

Article

Immune-Specific Expression and Estrogenic Regulation of the Four Estrogen Receptor Isoforms in Female Rainbow Trout (*Oncorhynchus mykiss*)

Ayako Casanova-Nakayama [1,2], **Elena Wernicke von Siebenthal** [1,2], **Christian Kropf** [1,2], **Elisabeth Oldenberg** [1,2] **and Helmut Segner** [1,2,*]

[1]   Centre for Fish and Wildlife Health, 3012 Bern, Switzerland; ayako0109@yahoo.com (A.C.-N);
      elena.wernicke@vetsuisse.unibe.ch (E.W.v.S.); christian.kropf@vetsuisse.unibe.ch (C.K.);
      Elisabeth.oldenberg@sunrise.ch (E.O.)
[2]   Department of Infectious Diseases and Pathobiology, Vetsuisse Faculty, University of Bern,
      Länggassstrasse 122, 3012 Bern, Switzerland
*   Correspondence: helmut.segner@vetsuisse.unibe.ch; Tel: +41-316-312-441

Received: 2 February 2018; Accepted: 7 March 2018; Published: 21 March 2018

Abstract: Genomic actions of estrogens in vertebrates are exerted via two intracellular estrogen receptor (ER) subtypes, ERα and ERβ, which show cell- and tissue-specific expression profiles. Mammalian immune cells express ERs and are responsive to estrogens. More recently, evidence became available that ERs are also present in the immune organs and cells of teleost fish, suggesting that the immunomodulatory function of estrogens has been conserved throughout vertebrate evolution. For a better understanding of the sensitivity and the responsiveness of the fish immune system to estrogens, more insight is needed on the abundance of ERs in the fish immune system, the cellular ratios of the ER subtypes, and their autoregulation by estrogens. Consequently, the aims of the present study were (i) to determine the absolute mRNA copy numbers of the four *ER* isoforms in the immune organs and cells of rainbow trout, *Oncorhynchus mykiss*, and to compare them to the hepatic *ER* numbers; (ii) to analyse the *ER* mRNA isoform ratios in the immune system; and, (iii) finally, to examine the alterations of immune *ER* mRNA expression levels in sexually immature trout exposed to 17β-estradiol (E2), as well as the alterations of immune *ER* mRNA expression levels in sexually mature trout during the reproductive cycle. All four ER isoforms were present in immune organs—head kidney, spleen-and immune cells from head kidney and blood of rainbow trout, but their mRNA levels were substantially lower than in the liver. The ER isoform ratios were tissue- and cell-specific, both within the immune system, but also between the immune system and the liver. Short-term administration of E2 to juvenile female trout altered the *ER* mRNA levels in the liver, but the ERs of the immune organs and cells were not responsive. Changes of *ER* gene transcript numbers in immune organs and cells occurred during the reproductive cycle of mature female trout, but the changes in the immune ER profiles differed from those in the liver and gonads. The correlation between *ER* gene transcript numbers and serum E2 concentrations was only moderate to low. In conclusion, the low mRNA numbers of nuclear ER in the trout immune system, together with their limited estrogen-responsiveness, suggest that the known estrogen actions on trout immunity may be not primarily mediated through genomic actions, but may involve other mechanisms, such as non-genomic pathways or indirect effects.

Keywords: estrogen receptor; isoforms; rainbow trout; immune system; reproductive cycle

---

1. Introduction

The main physiological function of estrogens in vertebrates is to regulate sexual development and reproduction. However, estrogens have pleiotropic functions and beyond the "classical" function in the reproductive axis, estrogens target a number of other physiological systems including the immune system [1]. In fact, for mammals it is well documented that estrogens like 17β-estradiol (E2) modulate the development, differentiation, life span, activation, and functioning of immune cells, and can have both immunostimulating and immunosuppressive actions [2–5]. The immunomodulatory activity of estrogens is a key proximate mechanism contributing to the known sexual dimorphism of mammalian immunity [6,7]. The primary effects of estrogens on the immune cells are mediated via rapid non-genomic signaling pathways as well as via the two nuclear estrogen receptor (ER) subtypes of mammals, ERα and ERβ [4]. Nuclear ER can either directly bind to estrogen response elements in gene promoters or serve as cofactors with other transcription factors such as nuclear factor-kappa beta (NFκB) [8]. ERα and ERβ are expressed in most cells of the myeloid and lymphoid cell lineages and in many hematopoietic progenitor cells [4,9–11]. The ratios of the two ER subtypes differ between immune tissues and cells, what has relevance for the diverse immunological effects of estrogens [12–14].

The immunomodulatory actions of estrogens in mammals vary with respect to target cell type, physiological condition of the organism or estrogen concentrations [2,3,15–17]. In particular, the female reproductive status and the associated changes of estrogen and ER levels have a major influence on the immune system response to estrogens [2]. With the evolution of internal fertilization and viviparity, mammals had to master a delicate balance between immunological protection of the mother against pathogens that are transmitted with fertilization, the prevention of immune responses against the spermatozoa, and immunological tolerance against the implantation of the semi-allogeneic embryos and the developing foetus [18–20]. In contrast to mammals, the reproductive strategy of lower vertebrates, such as teleost fish, relies on external fertilization and ovipary. Despite this difference, estrogens appear to have immunomodulatory actions in teleosts as well. A number of studies could show that immune parameters and immunocompetence of fish are influenced by estrogens, both by endogenous estrogens and by environmental (xeno) estrogens [21–23]. Moreover, recent research provided evidence that both membrane and nuclear ERs are expressed in immune organs and cells of teleosts [24–31]. In fact, the available evidence suggest that the immunomodulatory function of estrogens has been conserved throughout vertebrate evolution, despite the differences of reproductive strategies between oviparous and viviparous vertebrates [23].

The responsiveness of target cells to estrogens depends in large part on the cellular ratios of the various ER isoforms, their numbers and stability, and the regulation of ER activity and stability by the hormone signal, as well as by co-regulators and cross-talk with other signaling pathways [32–35]. While research during recent years has greatly advanced our understanding of the regulation of ER activity and turnover in mammalian cells [36–38] and how this drives the responsiveness of distinct cell types to estrogens, the current knowledge for teleost fish of the factors regulating ER activity and cell type-specific estrogen responsiveness is rather limited. With respect to the immune system of fish, information on absolute gene copy numbers of the ER in the immune organs and cells is lacking. Also, it is not clear yet whether piscine immune cells express all nuclear ER isoforms. Particularly for ERβ isoforms, there have been reports that they are not ubiquitously expressed in immune cells and organs [25,27,31,39]. Finally, while we have a reasonably good understanding of the autoregulation of the hepatic ERs in fish [35], no such database exists with respect to the estrogenic regulation of the ERs in the immune system. Given these knowledge gaps, the aims of the present study were to determine the absolute numbers of ER in immune organs and cells, and to compare them to the hepatic ER numbers, to analyse ER subtype ratios in the immune organs and cells, and to examine the alteration of immune ER expression levels in response to exogenous E2 and in association with the reproductive cycle. As experimental species, the rainbow trout, *Oncorhynchus mykiss,* was used. This species possesses four nuclear ER isoforms *ERα1, ERα2, ERβ1,* and *ERβ2,* which share a high degree of similarity of their amino acid sequences, particularly in the C-domain/zinc finger motif, in the activation function 1 (AF1) and AF2 domains [40]. In a first step, absolute gene copy numbers of the four *ER* were determined in the head kidney, the spleen, as well as in leukocytes that were isolated from the head kidney and from the blood of juvenile trout. In a next step, we aimed to gain insight into the regulation of the four ER subtypes in the immune system and examined the influence of exogenous E2 exposure on immune-specific ER profiles of juvenile rainbow trout, and we evaluated the immune *ER* mRNA profiles variation during the reproductive cycle and the associated fluctuations of endogenous levels of circulating E2 in mature female trout.

2. Results

2.1. Absolute Gene Transcript Levels of Erα1, α2, β1, and β2 in Immune Organs and Cells of Juvenile Rainbow Trout in Comparison to Liver ER Gene Transcript Levels

In juvenile rainbow trout, there exist distinct differences of the ER subtype ratios and profiles between the various organs and cells (Figure 1). Generally, the liver has significantly higher *ER* gene transcript levels than the immune tissues (except for *ERα2*). This applies particularly for *ERβ2*, where the liver gene transcripts are about 18 times higher than in the spleen, 55 times higher than in the head kidney and more than 1000 times higher than in the isolated leukocytes. Similar differences are observed for *ERα1*, with hepatic gene transcript levels being 10 times higher than in spleen and blood leukocytes, 160 times higher than in the head kidney, and 90 times higher than in the head kidney leukocytes. For *ERβ1* mRNA, expression levels in the liver are about 1.5 times higher than in spleen, six times higher than in head kidney, and about 80 times higher than in the isolated leukocytes, regardless whether they originate from the head kidney or the blood. In general, the mRNA lowest levels were found in the isolated immune cells (with the exception of *ERβ2*).

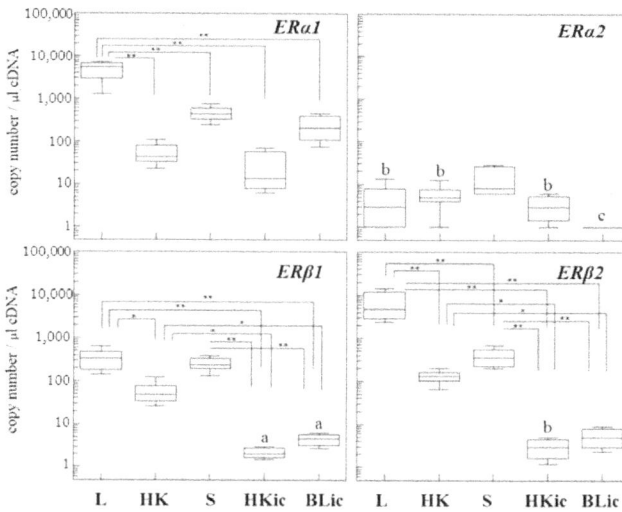

Figure 1. Absolute mRNA quantification of the four estrogen receptor (ER) isoforms in liver (L), head kidney (HK), spleen (S) and immune cells isolated from either head kidney (HKic) or blood (BLic) of 6-month-old female rainbow trout. The gene copy number of each isoform per 1 μL cDNA is presented by Box-Whisker plots (*n* = 5 individuals). Note logarithmic scale of y-axis. * *p* < 0.05, ** *p* < 0.01. a: under detection limit. b: Part of the sample was not detectable or under detection limit. c: not detected.

When considering the mRNA profiles of ER isoforms for the various tissues and cells using *ERα1* mRNA as a reference point (Table 1), the *ERα1* isoform has slightly lower expression levels than *ERβ2* in liver and head kidney, equal levels in the spleen, and 10 to 110 times higher levels in the leukocytes. *ERβ1* mRNA levels have the greatest difference to *ERα1* mRNA in the blood leukocytes and the smallest in head kidney and spleen. *ERα2* is the isoform with the lowest mRNA expression levels, relative to *ERα1* mRNA, in all of the organs and cells of control animals. Thus, each organ and cell has a specific profile of the ER isoforms.

Table 1. The mRNA ratios of the four ER isoforms in liver, head kidney, spleen, head kidney leukocytes and blood leukocytes.

Organ	Liver	Head Kidney	Spleen	HK Leukocytes	Blood Leukocytes
Ratio	Ratio *ERα1* mRNA to Other Isoforms	Ratio *ERα1* mRNA to Other Isoforms	Ratio *ERα1* mRNA to Other Isoforms	Ratio *ERα1* mRNA to Other Isoforms	Ratio *ERα1* mRNA to Other Isoforms
ERα1 mRNA	1	1	1	1	1
ERα2 mRNA	994	10	30	10	450
ERβ1 mRNA	14	1	2	10	110
ERβ2 mRNA	0.7	0.5	1.1	10	110

The ratios are calculated by dividing the absolute gene copy number (mean value) of *ERα1* in the respective organ or cell type by the absolute gene copy numbers (mean values) of the other isoforms. For instance, a value like "*ERα2* mRNA = 994" indicates that in this organ there are 994 times more gene copy numbers of *ERα1* than of *ERα2*.

2.2. Changes of ER Gene Transcript Levels in Sexually Immature Juvenile Rainbow Trout Exposed to Exogenous E2

Short-term (five days) exposure of sexually immature rainbow trout to E2 (via the diet) resulted in a significant elevation of plasma E2 concentrations and hepatic *VTG* gene transcript levels (Figure 2A), indicating that the treatment indeed induced an "estrogenic condition" in the animals.

The E2 treatment also affected the hepatic gene transcript levels of the two *ERα* isoforms: *ERα1* mRNA levels were significantly upregulated (4-fold) and those of *ERα2* mRNA even 17-fold (Figure 2B). In contrast, *ERβ2* was significantly downregulated, while *ERβ1* gene copy numbers showed no significant change. Interestingly, it was the *ERα2* isoform that showed the strongest E2 response among the hepatic ERs isoforms. In head kidney the E2 treatment remained without significant effects on the *ER* gene transcript levels, although there was a trend for elevated values, particularly for *ERα1*. Also, in the isolated immune cells, the estrogenic condition showed no significant effect on the ER expression levels, regardless whether the cells originated from the head kidney or the blood. Thus, the estrogenic condition had a prominent effect on the ER expression levels in the liver but did not clearly modulate ER expression in the immune system.

By means of *in situ* hybridization (ISH), we tried to visualize the cellular localization of the ERs mRNA in the immune organs of control and E2-treated fish. Liver tissue was used as control. We obtained a weak positive staining in the liver of control rainbow trout, and a very strong staining in the liver of E2-exposed trout (Figure 3). This finding is well in agreement with the RT-PCR results. In the immune organs, head kidney, and spleen, we did not obtain a positive staining result. Apparently, the sensitivity of the ISH was not sufficient to stain the low mRNA numbers of ERs in the immune organs.

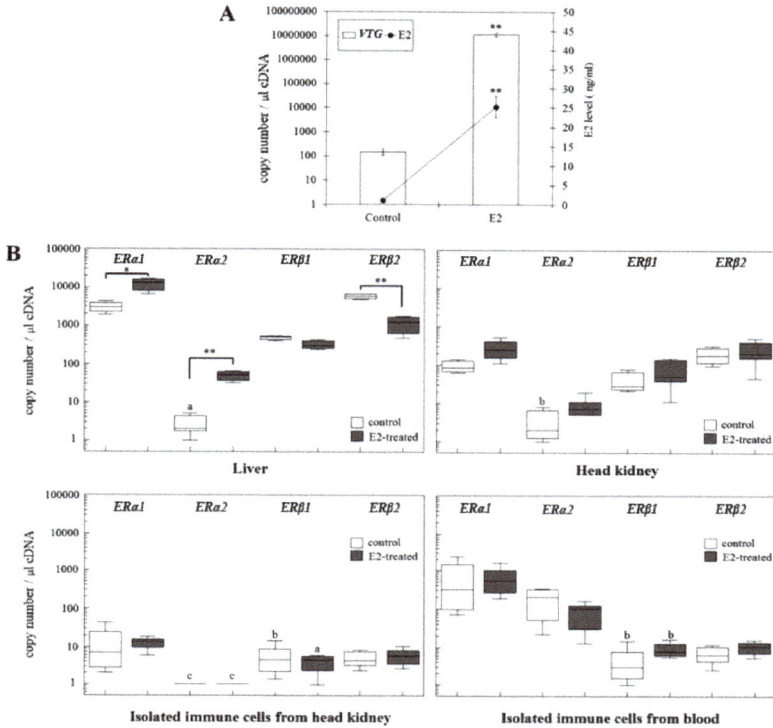

Figure 2. Response to exogenous 17β-estradiol treatment in 6-month-old juvenile female trout. Fish were fed with E2 containing pellets for five days and pellets prepared with only vehicle (ethanol) were used as control diets. (**A**) Absolute quantification of vitellogenin (VTG) mRNA in the liver and E2 levels in serum of the control (**C**) and E2-treatment (E2) groups. The absolute *VTG* gene copy number per 1 μL cDNA in the liver is shown as mean ± SE (n = 5 individuals). (**B**) Absolute mRNA quantification of the four ER isoforms in liver, head kidney, immune cells isolated from head kidney and blood of 6-month-old rainbow trout treated with E2. The gene copy number of each isoform per 1 μL cDNA is presented by Box-Whisker plots (n = 5 individuals). Control and E2-treated group were compared for statistical analysis. The asterisks denote statistically significant differences between control and E2-treated groups. * $p < 0.05$, ** $p < 0.01$. a: under detection limit. b: Part of the sample was not detectable or under detection limit. c: not detected.

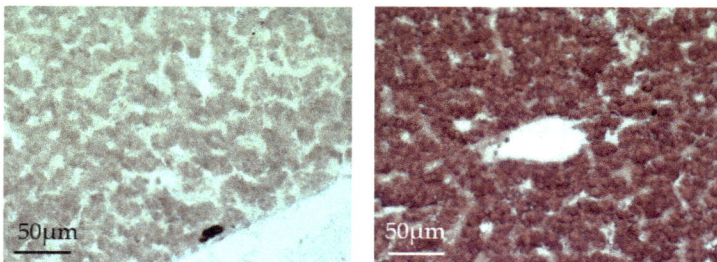

Figure 3. In situ hybridization of the *ERα1* mRNA in the liver of juvenile rainbow trout. Detection of the hybridization product was done using *ERα1* probes on liver sections of control (**left**) and E2-exposed (**right**) juvenile rainbow trout and detected with NBT-BCIP (dark-purple).

2.3. Changes of ER Gene Transcript Levels in Sexually Mature Adult Rainbow Trout Females during the Reproductive Cycle

Changes of hepatic and immune *ER* gene transcript levels were studied in female rainbow trout over a full spawning cycle. The reproductive status of the fishes was assessed by measuring liver-somatic index (LSI), mRNA levels of hepatic vitellogenin (Figure 4), plasma E2 concentrations (Figure 4), and gonadosomatic index (GSI). Additionally, the ovaries were examined by histology to assess the maturation status of the oocytes. Based on these criteria, fish were categorised into four stages: Stage A—fish at the beginning of the reproductive cycle, with low LSI, a GSI less than 1, low hepatic vitellogenin mRNA levels, low serum E2 levels and immature and partly cortico-alveoloar oocytes; Stage B—vitellogenic fish, with enlarged liver (LSI > 1.5), increased ovaries (GSI 12–18), significantly elevated hepatic vitellogenin mRNA and serum E2 levels, and vitellogenic oocytes; Stage C—spawning fish, with high LSI, high GSI, significantly reduced serum E2 and hepatic vitellogenin mRNA levels, and mature oocytes; Stage D—post-spawning fish, with reduced LSI, low GSI (close to stage A), low vitellogenin mRNA, and low E2 levels, similar to stage A. The ovaries of stage D fish display spent follicles.

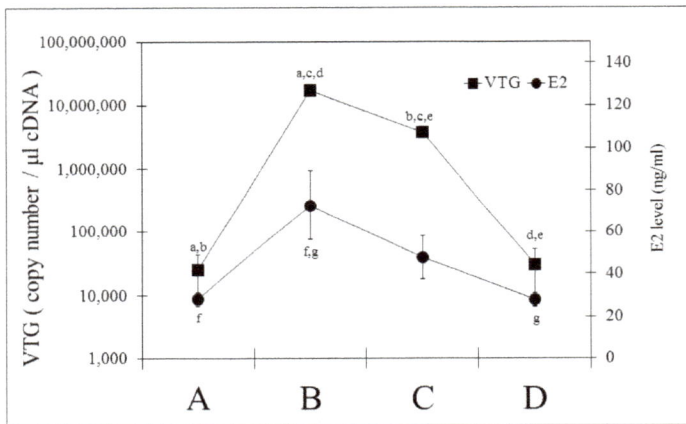

Figure 4. Physiological changes during the reproductive cycle of mature female rainbow trout from September to January: Alterations of the hepatic vitellogenin (*VTG*) mRNA levels of the liver and the serum 17β-estradiol (E2). Categorization of the fishes into maturation stages was done based on the gonadosomatic index (GSI) and the histological appearance of the oocytes; Stage A: fish at the beginning of ovarian development (GSI < 1), with primary follicles and partly cortical alveolar oocytes (*n* = 4), Stage B: fish with enlarged ovaries (GSI 12 < 18) and vitellogenic oocytes; additionally they possess an enlarged liver (liver somatic index LSI > 1.5) (*n* = 5), Stage C: Spawning fish with large ovaries (GSI > 18), mature oocytes and reduced liver size (*n* = 3), Stage D: Post-Spawning fish, with low (GSI < 5), spent follicles and a LSI close to 1 (*n* = 4). Statistically significant differences between groups are indicated by the same letter (a–g). The absolute gene copy number of *VTG* per 1 μL cDNA in the liver and E2 concentrations are shown as mean ± SE.

Figure 5 reports the mRNA changes of the four ER subtypes in the liver, gonads, and immune organs and cells of mature rainbow trout over the reproductive cycle, i.e., from stage A to stage D. In the liver, *ERα1* mRNA showed a slight tendency for increasing values with maturation and a decrease towards the post-spawning stage; however, the differences are not statistically significant. In contrast, hepatic *ERα2* mRNA experienced strong and significant changes during the reproductive cycle. With *ERβ1*, we observed a significant downregulation in the liver with increasing maturation of the fishes, and a partial recovery during the post-spawning stage. For the hepatic *ERβ* isoforms,

Int. J. Mol. Sci. **2018**, *19*, 932

alterations took place from stage A to C, with significant downregulation in the case of *ERβ1* and significant upregulation in the case of *ERβ2*. Thus, each of the four ER subtypes in the liver showed an individual pattern over the reproductive cycle, and the pattern was partly different to the changes of the hepatic ER profile of juvenile trout under E2 exposure.

In the head kidney, mRNA levels of the *ERα2* isoform varied over the reproductive cycle similar to the behaviour of the *ERα2* isoform in the liver. In contrast to the liver, however, *ERα1* gene transcript levels experienced significant variations in the head kidney, whereas the *ERβ2* isoform remained unchanged.

The reproductive cycle was associated with alterations of ER expression levels in the leukocytes. A significant mRNA upregulation of *ERα2* and *ERβ1* was observed in the head kidney leukocytes of post-spawning females, and also in the blood leukocytes, *ERβ2* gene copy numbers increased in the post-spawning females. *ERα1* gene transcript levels of blood leukocytes, however, decreased towards the post-spawning stage, after they had increased from stage A to C.

In the gonads (Figure 5B), the most prominent response of the ER expression patterns during the reproductive cycle was the strong mRNA downregulation of the two *ERα* isoforms in reproductive stage C. The gene transcripts levels of the *ERβ* isoforms in the gonads showed limited variation during the reproductive cycle.

A regression analysis between the changes of serum E2 concentrations and tissue *ER* mRNA levels in mature rainbow trout yielded overall moderate to low correlation coefficients (Table 2). The strongest correlations to E2 were observed for *ERα2* and *ERβ1*. The *ERα* isoforms usually showed a positive correlation, i.e., mRNA increased with increasing E2 concentrations, whereas with the *ERβ* isoforms, also negative correlations were found. In general, the poorest ER-E2 correlation existed for the blood leukocytes. This could be due to a low estrogen sensitivity of the cells or to alterations in the cellular composition of the blood leukocyte population [41].

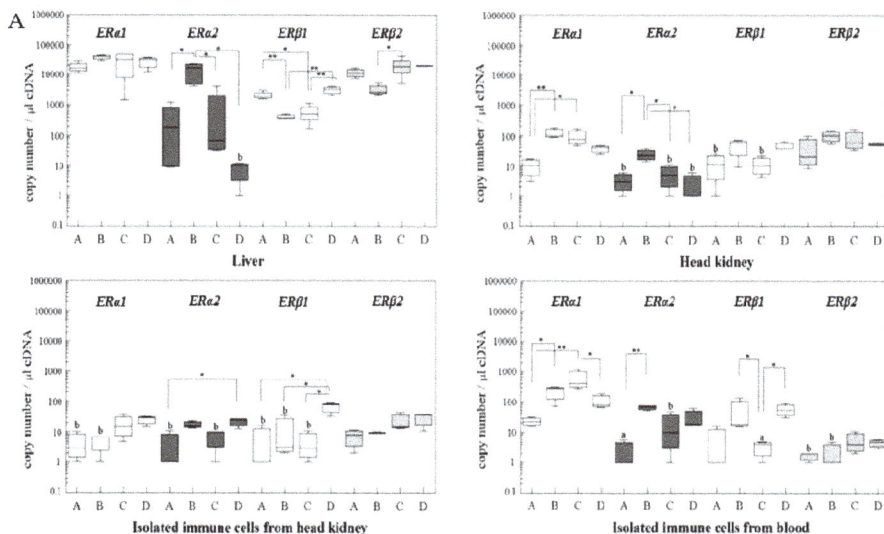

Figure 5. *Cont.*

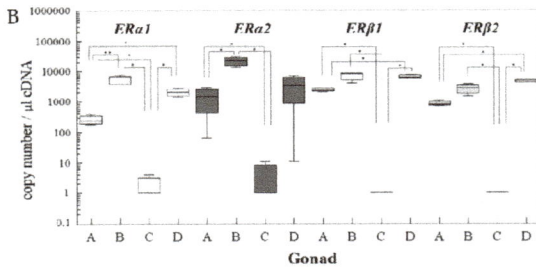

Figure 5. (**A**): The mRNA expression levels of the four ER isoforms in the liver, head kidney (HK), immune cells isolated from head kidney or from blood during the reproductive cycle. The reproductive cycle was subdivided into four stages: stage A = start of reproductive cycle, stage B = vitellogenic stage, stage C = spawning stage, stage D = post-spawning stage (see Figure 4). The gene copy number of each isoform gene per 1 μL cDNA is presented by Box-Whisker plots (group A: $n = 4$, B: $n = 5$, C: $n = 3$, D: $n = 4$). Note logarithmic scale of y-axis. * $p < 0.05$, ** $p < 0.01$. (**B**): The mRNA expression levels of the four ER isoforms in the ovaries during the reproductive cycle.

Table 2. Correlation coefficients (r^2) between serum E2 concentrations and *ER* mRNA abundance in liver and leukocytes of female rainbow trout over the reproductive cycle.

ER Isoforms	Liver	Head Kidney Leukocytes	Blood Leukocytes
ERα1	0.138 ↑	0.039 ↑	0.044 ↑
ERα2	0.241 ↑	0.203 ↑	0.014 ↑
ERβ1	0.282 ↓	0.282 ↑	0.009 ↓
ERβ2	0.019 ↓	0.054 ↓	0.041 ↓

Linear regressions were calculated between serum E2 concentrations and mRNA numbers of the four *ER* isoforms in liver and leukocytes of adult rainbow trout from different stages of the reproductive cycle. ↑ positive correlation (*ER* gene transcript levels increase with increasing E2 concentrations); ↓ negative correlation (*ER* gene transcript levels decrease with increasing E2 concentrations).

3. Discussion

To provide a baseline for understanding the physiological role of estrogens in the immune system of teleost fish, this study (1) characterized the mRNA expression levels and ratios of the four ER isoforms [40] in immune organs and cells of rainbow trout, (2) examined their response to exogenous or endogenous variations of estrogen concentrations, and (3) compared the mRNA levels of the ER isoforms in the immune system to that of the hepatic, and partly also, gonadal ERs. A first finding of this study is that the immune organs and immune cells of rainbow trout express all four ER isoforms, namely ERα1, α2, β1 and β2. Expression of nuclear ERs in immune cells is well documented for mammals, where both nuclear ER subtypes, ERα, and ERβ, are present in most immune cells and hematopoietic progenitor cells [4,9,10,42,43]. The differential expression of the ER subtypes in the immune cells influences gene regulation and appears to be important to balance the multiple effects of estrogens in the mammalian immune system [13,44,45]. Generally, the ERα subtype appears to have a more prominent expression and distribution in mammalian immune cells than ERβ [5]. Also, in the trout immune system, ERα is prominently expressed, but at least in the immune organs, head kidney and spleen, the *ERβ2* gene copy numbers are in the same range as those of *ERα*, pointing to an important role of this ER isoform in teleostean immune organs.

Presence of nuclear ERs in isolated immune cells has been assessed by means of relative mRNA quantification for a number of teleost species other than the rainbow trout: For seabream (*Sparus aurata*), Liarte et al. [27] reported no presence of nuclear *ER* gene transcripts in the testicular and head kidney acidophilic granulocytes, whereas macrophages and lymphocytes isolated from the head kidney contained *ERα* mRNA, but not *ERβ1* or *ERβ2* mRNA. For channel catfish (*Ictalurus punctatus*),

Iwanowicz et al. [39] described expression of *ERα* and *ERβ* mRNA in primary leukocytes from head kidney and spleen, while only *ERα* was detected in peripheral blood leukocytes. For carp (*Cyprinus carpio*), Szwejser et al. [31] found high mRNA levels of *ERα*, but no *ERβ* gene transcripts in peripheral blood leukocytes. In leukocytes that were isolated from the head kidney of carp, *ERβ* could be detected although at very low levels. Thus, in all three species the tissue leukocytes displayed higher gene transcript levels of *ERα* than of *ERβ*, and the later was completely absent from peripheral blood leukocytes. In contrast to these studies, we detected both *ERβ* subtypes in the peripheral blood leukocytes of rainbow trout. Interestingly, however, while in the intact head kidney and spleen, the *ERβ2* mRNA numbers equalled those of *ERα1*, the isolated leukocytes displayed 10–100 times lower mRNA numbers of *ERβ2* than of *ERα1*. We found the two ER isoforms not only in blood leukocytes but also in head kidney leukocytes, together with *ERα1* and *ERα2*. Also Shelley et al. [30] reported the presence of mRNA of all four ER subtypes in head kidney leukocytes of rainbow trout. Thus, the overall picture arising from the various studies on nuclear ER in the immune system of diverse teleost species point to ERα/ERα1 being the dominant nuclear ER isoform in the immune cells, but not necessarily in the immune organs. The expression of ERβ in fish immune cells appears to vary with the origin of the cells and across species.

Expression levels of *ERα1* in the immune organs of juvenile rainbow trout were significantly lower than in the liver. Also for *ERβ1*, the head kidney and the isolated leukocytes (but not spleen) displayed significantly lower mRNA levels than the liver, while no significant tissue differences existed for *ERα2*. Our findings agree with those of Nagler et al. [40] who identified the liver of rainbow trout to be the organ with the highest gene transcript levels of *ERα1* and *ERβ2* and clearly lower levels in immune organs. Similarly, Massart et al. [29] observed much higher *ERα1* mRNA levels in the liver of rainbow trout than in head kidney and spleen. However, at the protein level, Massart et al. [29] found no clear difference of the ERα expression in liver compared to head kidney and spleen. Discrepancies between ER levels at the protein and mRNA levels have been observed also in other studies, for instance, Pinto et al. [46] found no measurable *ERα* mRNA in the scales of sea bream scale, whereas ERα protein was well detectable. In this context it is important to keep in mind the complexity of ER regulation as it has been highlighted from recent studies with mammals [36–38,47]. The "classical" view of estrogen receptor activity is that, after binding of E2, ER dimerizes, and translocates into the nucleus where it binds to Estrogen-Response Elements (ERE) on target gene promotors to activate or repress transcription. However, there are a number of different regulation processes involved, including the cell-specific availability of co-repressors and co-activators, ER stability or proteolysis as well as post-translational modifications, such as ER phosphorylation. In addition, cross-talks with other signaling pathways such as the insulin-like growth factor 1 receptor pathway modulate the dynamics of ER-mediated gene regulation. Vice versa, both liganded and unliganded ERs are able to influence other signaling pathways. Altogether, these diverse processes of ER regulation and activity largely drive the target cell-specific estrogen actions. ER sequences influence isoform conformation, turnover rates and also the regulation by co-regulators, and thus can provide a basis to understand the E2 dependence of ER expression. Here, the information that the four ER isoforms of rainbow trout show similarity of their amino acid sequences, particularly in the AF1 and AF2 domains [40], is an important starting point for unravelling the mechanisms of ER functions in the trout immune cells.

A striking difference in absolute ER gene copy numbers that was observed in this study existed between intact immune organs and the pure leukocyte preparations. *ERβ1* and *ERβ2* gene transcript numbers were significantly lower in the isolated leukocytes. Only for *ERα1*, the blood leukocytes had higher gene copy number levels than the head kidney and as a high levels as the spleen. Similar results have been reported by Iwanowicz et al. [39] for channel catfish. This suggests that the ERβ isoforms have a prominent function in the leukocytes. Blood and head kidney leukocytes differed in their ER expression profiles in what is likely to reflect a different cellular composition [31].

Taken together, the findings from this study provide evidence that immune organs and cells of rainbow trout express all four ER isoforms, although mostly at low levels, and that ER profiles of the immune organs and cells differ strongly to each other. With a relatively strong expression of ERβ2, the immune organs are more similar to the liver than to the leukocytes, which show a dominance of ERα1.

A second aim of this study was to evaluate how *ER* mRNA levels in immune organs and cells of rainbow trout respond to changing E2 concentrations under different physiological conditions. This was investigated on one hand by exposing sexually immature juvenile trout to exogenous E2. At this life stage, the gonads of salmonids are already differentiated into ovaries and testes but endogenous sex steroid production is still negligible or very low [48,49]. Thus, elevating the estrogen concentrations of these animals by exposure to exogenous E2 was considered to represent a non-physiological situation. On the other hand, we examined mature female rainbow trout over a full reproductive cycle. In this situation, the endogenous alterations of E2 levels are embedded in a number of additional physiological changes, and thus, E2 is not acting in isolation, as in the juvenile fish, but in concert with other factors. We were interested to compare these two situations since differences of the physiological states can strongly influence the estrogenic regulation of ER expression [35,50,51].

The induction or suppression of the number of nuclear ER by E2 (autoregulation) is a way by which a target organ or cell can modulate its sensitivity to estrogens [34,35,52]. In mammals, ER autoinduction has been demonstrated for the liver and for reproductive tissues, as well as for immune cells. Molero et al. [53] showed that an increases of plasma E2 concentrations during the menstrual cycle of women are accompanied by an elevation of ERα and ERβ expression in the neutrophils. In contrast, in isolated neutrophils of males, E2 upregulated only ERα, but not ERβ. In human macrophages, E2 upregulated the expression of the ER splice variant, ERα46 [33]. In teleost fish, ER autoinduction has been described to date mainly for the liver [35]. For instance, Menuet et al. [54] reported that short-term exposure of mature zebrafish with E2 resulted in a strong upregulation of hepatic *ERα*, a marked reduction of the mRNA levels of hepatic *ERβ1* and virtually no change of *ERβ2*. Injection of male largemouth bass (*Micropterus salmoides*) with E2 led to a dose-dependent upregulation of hepatic *ERα*, but had no clear effect on the hepatic *ERβ* isoforms [55]. Comparable findings were reported from *in vivo* studies with fathead minnow (*Pimephales promelas*) [56], and from *in vitro* studies with isolated trout hepatocytes [57]. In the liver of male goldfish receiving E2 implants, *ERα* was highly upregulated, *ERβ1* was significantly downregulated and *ERβ2* did not change [58]. As summarized by Nelson and Habibi [35], estrogen-dependent upregulation of hepatic *ERα* appears to be fairly ubiquitous across species, whereas the estrogenic regulation of the hepatic *ERβ* isoforms varies strongly with species and experimental/physiological conditions. This is confirmed by the results of the present study: E2 exposure of juvenile trout led to significant mRNA upregulation of the two *ERα* isoforms but had no effect on *ERβ1* mRNA and significantly downregulated *ERβ2* mRNA.

Tissue differences in the response of the nuclear ER to estrogens are prominent. This has been demonstrated for mammals [59] and for fish as well [60]. Here, we focused on the regulation of the ERs in juvenile trout immune organs and cells by short-term (five days) exogenous E2 administration. The key finding is that exposure of sexually immature female rainbow trout to exogenous E2 concentrations that were sufficiently high to cause a significant vitellogenin mRNA induction did not lead to significant changes in the mRNA levels of all four ER isoforms, in the head kidney organ, in the head kidney leukocytes, or in the blood leukocytes. This behaviour is in contrast to the prominent responses of the hepatic ER. In another study with *in vivo* exposure of rainbow trout to E2, Shelley et al. [61] found an upregulation of *ERα1* mRNA in leukocytes from head kidney and blood, an upregulation of *ERα2* mRNA in head kidney leukocytes, but a downregulation in blood leukocytes, and no change of the gene transcript levels of the *ERβ* isoforms. Interestingly, in vitro exposure of rainbow trout blood leukocytes had no effect on the gene transcript levels of the four ER isoforms [61]. Developmental exposure of tilapia (*Oreochromis niloticus*) to ethinylestradiol was associated with elevated *ERα* gene transcript levels in the spleen, but not in the head kidney [26]. Finally, Liarte et al. [27] found an upregulation of *ERα* and *ERβ2* mRNA after in vitro treatment of specific macrophage cultures with E2. Given the variations of experimental conditions between the cited studies, as well as the species differences, it appears to be too pre-mature to come up with a general statement on whether ER autoregulation does exist in the immune system of fish or not.

In the third part of the present study we examined how immune *ER* mRNA levels of mature female rainbow trout change with the reproductive cycle and the associated fluctuations of plasma E2 concentrations. In contrast to sexually immature fish, the immune ERs of mature fish experienced changes of their mRNA expression levels. This may indicate that the effect of E2 in the immune system is not a simple function of estrogen concentration, but depends on the overall physiological context [35,50,51]. One key finding from the analysis of the *ER* mRNA expression levels in the immune system of mature female rainbow trout is that the reproduction-related changes of ER isoform profiles in the immune tissues and cells are clearly different to the corresponding changes of ER profiles in liver and gonads. Even within the immune system, there exist distinct differences between the leukocytes from head kidney and those from blood. A second key finding that the reproduction-related changes of nuclear ER expression in the immune system are mainly restricted to the *ERα* isoforms, whereas the *ERβ* isoforms are less responsive. Also, while the *ERα* isoforms tend to increase with increasing E2 concentrations, *ERβ* isoforms tend to decrease if they respond at all. Finally, a third important observation is that the correlation between the plasma E2 concentrations and the immune *ER* gene transcript levels is overall moderate to low.

The organ differences of the ER changes highlight again the importance of the specific cell and tissue environment for shaping expression and activity of the nuclear ERs [4,59]. The differences between the leukocyte populations of head kidney and blood are likely to reflect differences in their cellular composition. The head kidney population, in addition to differentiated immune cells, contains also diverse developmental stages of immune cells. Estrogens are master regulators of cell proliferation and differentiation and in line with this, ER are well expressed in developing immune cells of mammals. Importantly, the ER isoform profile of mammalian immune progenitor cells differs from that of mature immune cells [4,9]. If the situation is similar in fish, this may explain our finding of contrasting ER profiles between head kidney leukocytes and blood leukocytes of trout.

The functional interpretation of the reproduction-related changing the *ER* mRNA profiles of the trout immune cells is difficult if not impossible at the current state of knowledge on the immune functional roles of the four isoforms. In mammals there exists evidence that the ERα subtype mediates anti-inflammatory actions in the immune system, [13,62], and the upregulation of this subtype by the elevated E2 levels during pregnancy is considered as one mechanism of the pregnancy-associated lowering of the immune activity in women. Likewise, the increase of immune ERα isoforms in trout with progressing ovarian maturation may represent an immunosuppressive mechanism as well. However, different to mammals, the purpose of this mechanism in oviparous fish could not be the protection of the embryos, but should have an alternative function, for instance, it may be speculated that it is mediating resource trade-offs between the immune and reproductive systems [63].

When initiating this study, we expected a rather close correlation between nuclear ERs in the immune system of rainbow trout and E2 levels, and we expected relatively high gene copy numbers of the *ER*s in the immune cells since E2 has prominent immunomodulatory actions in fish [21]. Our results prove the opposite to our expectations—the correlation between E2 levels and nuclear ER mRNA levels is moderate at its best, and the *ER* mRNA numbers in immune organs and cells are very low. The discrepancy between the pronounced immunomodulatory activity of estrogens in trout and low nuclear ER numbers and the limited estrogen-responsiveness suggests that the estrogen actions on the trout immune system involve, in addition to genomic signaling, alternative mechanisms. These could include membrane estrogen receptors [28,31], or indirect effects via interaction with other endocrine systems. Such indirect effects are well documented for the immune effects of estrogens in mammals [64–67], and may be of particular importance to mediate the resource trade-offs between the immune system and other fitness-relevant traits.

In conclusion, the results from this study provide insight into the tissue-specific and physiological status-related expression and estrogenic regulation of the four nuclear ER isoforms in rainbow trout.

While all four nuclear ER isoforms are present in the immune organs and immune cells of rainbow trout, their expression levels, ratios, as well as their autoregulation by E2, show distinct differences to liver or gonads. This data provides important baseline information for the immunomodulatory role of estrogens in fish, but to advance our understanding we need more insight into the functional role of the ER isoforms in the immune system, as well as an on the relative importance of genomic estrogenic signaling versus non-genomic and/or indirect pathways of estrogen action.

4. Materials and Methods

4.1. Animal Experiments

4.1.1. Juvenile Rainbow Trout

Juvenile all-female rainbow trout (*Oncorhynchus mykiss*) of an average weight of three grams were bought at DSM SA (Village Neuf, France) and were reared at the Centre for Fish and Wildlife Health, University of Berne, Switzerland. Fish were kept at 11.3–11.8 °C, in 130 L flow-through glass tanks supplied with tap-water (approx. 1 L/m), constant aeration, and artificial light (12 h light to 12 h dark). On arrival, ten fish were randomly sampled and were screened for the presence of pathogens. No infectious agents were found. Any mortalities were recorded, and necropsied and investigated for the presence of parasites and other infectious agents. The fish were fed with a commercial dry pellet (Hokovit, Bützberg, Switzerland) with 1.5% body weight per day.

When the fish were six months old and had achieved an average weight of 50 g samples, the fishes were split into two groups: a control group that received the commercial diet and a 17β-estradiol (E2)-exposed group that received the commercial diet enriched with 20 mg E2/kg diet: this concentration was found to be sufficient to induce an estrogenic condition of juvenile trout in previous studies [11]. The feeding with the E2-enriched diet lasted for five days; the feeding level was 1% body weight per day both in the control and in the E2-exposed groups.

4.1.2. Adult rainbow trout

Two-year-old rainbow trout of the breeding stock of the Centre for Fish and Wildlife Health were maintained in 1500 L tanks under flow-through conditions and light/dark cycles of Berne, Switzerland from September 2012 to January 2013. Water temperatures varied between 11 °C and 15 °C. The period from September to January covered the reproductive cycle of the fish, form the onset of ovarian maturation through the vitellogenic and spawning stage to the post-spawning stage (see Results). The fish were fed with the commercial diet at 0.5% body weight/day.

4.2. Preparation of Samples and Immune Cell Isolation

Trout were euthanized in neutralized MS222, and liver, head kidney, spleen, ovary, and blood were sampled. All procedures were carried out according to the Swiss legislation for animal experimentation guidelines (Ethics Comitee Bern, approval date 31 August 2017, approval No. BE84/11). The blood was taken from the caudal vein. In addition to the tissue sampling, leukocytes were prepared from blood and head kidney. A thousand-fold dilution from blood or head kidney cell preparations was used to count the number of leukocytes using a Neubauer chamber. Moreover, serum was collected to determine plasma E2 concentrations by means of competitive enzyme-linked immunosorbent assay (ELISA).

For the immune cell isolation from the head kidney, the tissue was mechanically disrupted and passed through nylon nets with 250 μm and 125 μm nylon mesh, and the cells were collected in L-15 medium (Gibco) containing 10 IU/mL heparin. For the immune cell preparation that was isolated from the blood, the blood was diluted 10 times with L-15 medium containing 10 IU/mL heparin. The resulting cell suspensions from blood or head kidney were layered onto a Ficoll solution (Biochrom AG, Berlin, Germany) and were centrifuged at 400× *g*, 4 °C for 40 min. The immune cell fractions were collected in L-15 medium, washed repeatedly, and then adjusted to the appropriate different concentrations.

4.3. RNA Extraction and Gene Expression Analysis

Isolated immune cells adjusted to 10^7 cells were stored in 1 mL of TRIzol reagent (Sigma-Aldrich, St. Louis, MO, USA), homogenized. After adding 200 µL of bromochloropropane (Sigma-Aldrich, Buch, Switzerland), cell sample was mixed and centrifuged at $10,000 \times g$ for 15 min at room temperature. An aqueous phase of each cell sample was replaced by 500 µL of isopropanol and samples were stored at -80 °C until use. Tissue samples (approximately $5 \times 5 \times 5$ mm) were kept in RNAlater (Sigma-Aldrich) at 4 °C overnight and were then stored -20 °C before use. Tissues were replaced in TRIzol reagent and homogenized, followed by the phase separation with bromochloropropane. The RNA precipitation with isopropanol and ethanol wash for both cell and tissue samples were performed and the resulting RNA was dissolved in nuclease-free water. After the digestion of resting DNA with RQ1 RNase-Free DNase (Promega AG, Dübendorf, Switzerland), 500 ng of RNA were reverse-transcribed to cDNA using GoScriptTM reverse transcriptase containing random primers, and dNTP as described in the manufacturer's protocol (Promega AG) and total volume of cDNA was adjust to 25 µL. The TaqMan®-based real-time RT-PCR was carried out in triplicate for each sample mixture of total volume (12.5 µL) with 1 µL of cDNA template, 0.5 µM of each forward and reverse primer, 0.2 µM of the probe and TaqMan® Gene Expression Master Mix (Applied Biosystems, Foster City, CA, USA) using a 7500 Fast Real-time PCR System (Applied Biosystems). The used primer and probe sequences were listed in Table 3. Expression of each ER isoform was calculated by absolute quantification using each plasmid DNA that prepared with a pGEM-T Easy Vector System I (for *ERα1* with fwd: 5′-CGGCCCCTCTCTATTACTCC-3′, rev: 5′-TGTACGACTGCTGCCTATCG-3′, for *ERα2* with fwd: 5′-TGCTGGTGACAACAGTGTCC-3′, rev: 5′-GGCCCAACTGCTGACTAGAA-3′, for *ERβ1* with fwd: 5′-CAGCTACCGGGGTCATAAAC-3′, rev: 5′-ACAGGCACAGGTCCACAAAT-3′, for *ERβ2* with fwd: 5′-TCATTCCAGCAGCAGTCATC-3′, rev: 5′-CTGAGGTACACATCTCCCCTCT-3′), and expressed mean of copy number per 1 µL cDNA ± standard error. In accordance with our PCR-system, the detection limit of *ERα1*, *α2*, *β1*, and *β2* was 1, 5, 10, and 1 copy/µL cDNA, respectively. As an endogenous reference, 18S rRNA (Applied Biosystems, Foster City, CA, USA) was measured for the quality check of reverse-transcription of each cDNA. The gene expression level of liver-vitellogenin (VTG, Hamburg, Germany) [68] was utilized as an indicator for E2 response.

Table 3. Primer sequences used for the gene expression analysis and related accession numbers.

Gene		Sequence (5′-3′)	Accession No.
ERα1	Forward	CCCCCCAAGCCACCAT	AJ242741
	Reverse	TGATTGGTTACCACACTCGACCTATAT	
	Probe	CATACTACCTGGAGACCTCGTCCACACCC	
ERα2	Forward	TCCTGGAGCACAGCAAAGC	DQ177438
	Reverse	TGATCTTGAGACGCCCTTCTC	
	Probe	CCTCAGGACAGTAGCAAGAACAGCAGCTTC	
ERβ1	Forward	GGAGCGAGCCAATCAAGGA	DQ177439
	Reverse	GCCATGATCCGGCCAAT	
	Probe	TCTGCCCCACAGTATTAACCCCGGA	
ERβ2	Forward	CAGCTCCTGCTGTAGACACTCAGT	DQ248229
	Reverse	GGATGTACTAATGCTCTCGAGTGTTT	
	Probe	TGCTAACATTCCAAAACCCAGAGGAGAGC	

4.4. In Situ Hybridization

Plasmid DNA of ERs (*ERα1* with fwd: 5′-CTCTCCCCAGCCAGTCATAC-3′ and rev: 5′-CCTCCACCACCATTGAGACT-3′, *ERβ1*, and *β2*, as described above) was cloned in pGEM-T

Easy Vector System I. Following digestion with NdeI and NcoI (Promega, Medison, MI, USA), linearized plasmid DNA was transcribed with T7 and SP6 polymerases (Roche Diagnostics AG, Rotkreuz, Switzerland), respectively, and labelled with digoxigenin (DIG) (Roche Diagnostics AG), as described in the manufacturer's protocol. Synthesized labelled probes were stored at -20 °C in 50/50 (*v/v*) nuclease-free water/formamide buffer before use.

Dissected organs, liver, and head kidney were placed immediately into cold Histochoice MB (Electron Microscopy Sciences, Hatfield, PA, USA) and were fixed at 4 °C for 3 h. Fixed organs were dehydrated in a graded ethanol series at 4 °C. For paraffin-embedding, the tissue were infiltrated with Histoclear (National Diagnostic, Chemie Brunschwig, Lausanne, Switzerland) for 60 min at room temperature, followed by Histoclear/Paraplast (50/50, *v/v*) for 60 min at 65 °C twice. After repeated cleaning in 100% of Paraplast for 60 min at 65 °C, tissues were incubated in 100% of Paraplast for overnight at 65 °C. The tissues were embedded in the fresh prepared Paraplast and stored at 4 °C before sectioning.

Tissues were deparaffinised and washed in diethyl pyrocarbonate (DEPC)-treated water. The acetylation of sections was performed in a buffer containing 100 mM of triethanolamine (pH 8.0) and 0.25% of acetic anhydride by shaking for 10 min. After repeated washing, hybridization was done using an antisense RNA- digoxigenin (DIG) probe in a hybridization buffer that was mixed with 50% deionized formamide, 4 × saline-sodium citrate (SSC), 10% dextran sulfate, 1 × Denhardt's and 1 mg/mL ribonucleic acid from torula yeast for 16 h at 50 °C in a humid box. Sense RNA-DIG probe was applied in the same hybridization buffer as negative control. For post-hybridization, the slides were washed in tris-buffered saline with Tween20 (TTBS) (0.5 M NaCl, 0.1 M Tris-HCl (pH 8.0), 0.1% Tween-20). Following blocking with 6% milk powder that was diluted in TTBS for 1 h and bovine serum albumin (BSA)-Triton X-100 buffer containing 0.1 M Tris-HCl (pH 7.5), 0.15 M NaCl, 1% BSA and 0.3% Triton X-100 for 1 h, the specimens were incubated with a sheep anti-DIG antibody-alkaline phosphatase (AP) (Roche Diagnostics AG, Basel, Switzerland) diluted to 1:500 in the BSA-Triton X-100 buffer for 2 h at room temperature. The slides were then washed in the BSA-Triton X-100 buffer three times for 20 min. To equilibrate the slide, a buffer containing 0.1 M Tris-HCl (pH 9.5), 0.05 M $MgCl_2$ and 0.1 M NaCl was used for 15 min, then the nitro blue tetrazolium (NBT)/5-bromo-4-chloro-3-indolyl-phosphate (BCIP) was applied on the slide for the development. The reaction was stopped by Tris-EDTA (TE)-buffer containing 0.01 M Tris-HCl (pH 7.5) and 1 mM EDTA (pH 8.0). For the head kidney, the same procedure as described for liver until post-hybridization was done; then, an additional endogenous peroxidase-blocking step with 1% of hydrogen peroxide was performed to account for the high endogenous alkaline phosphatase in the head kidney, Afterwards, the visualization was done as follows: The sections were blocked using 5% normal donkey serum (Jackson ImmunoResearch, West Grove, PA, USA) diluted in TTBS; this was followed by 30 min incubation with a sheep anti-DIG antibody diluted to 1:1000 in TTBS. Then, the sections were incubated with a donkey anti-sheep antibody (Jackson ImmunoResearch) diluted to 1:100 in TTBS, and after repeated washing a sheep peroxidase anti-peroxidase (PAP) soluble complex diluted to 1:100 with TTBS was applied. NBT-BCIP was used for visualization.

4.5. Competitive Enzyme-Linked Immunosorbent Assay (Celisa) to Determine 17β-Estradiol Concentrations in Serum

The blood samples were centrifuged at 3000× *g* for 15 min at 4 °C. 200 µL of serum were diluted in 300 µL of PBS (pH 7.4) and then extracted by adding 3 mL of diethyl ether, vortexing for 10 s 6 times, and centrifuging at 1800× *g* for 10 min at 20 °C. After the samples were frozen at -80 °C for 20 min, the organic phase was transferred into a new glass tube and were completely dried in a heat block at 30 °C for overnight prior to be resuspended in 200 µL of PBS.

A high binding ELISA-plate (Greiner bio-one, Frickenhausen, Germany) was coated with a mouse anti-rabbit antibody (Sigma-Aldrich, 1:2000 diluted in PBS) for 24 h at 4 °C. After repeated washes with PBST (0.05% Tween-20), the plate was blocked with 1% of BSA-PBST for 12 h at 4 °C. Fifty µl of the sample, 50 μL of the estradiol- horseradish peroxidase (HRP) (Cal Bioreagents, San Mateo, CA, USA, 1:10,000 diluted in PBS) and 50 μL of a rabbit anti-estradiol antibody (Cal Bioreagents, 1:2500 diluted in PBS) were mixed and incubated for 2 h at room temperature. For the standard, first 17β-Estradiol (Sigma-Aldrich) was dissolved in ethanol, and then the same volume of 17β-Estradiol instead of the sample ranging from 0.36 to 40 ng/mL diluted in PBS was used. Following five washes with PBST for 5 min each, the ABTS® Peroxidase Substrate (Kirkegaard & Perry Laboratories, Maryland, USA) was applied for the color development. The plate was measured at 405 nm by an EnSpire 2300 Multimode Plate Reader (Perkin Elmer, Waltham, MA, USA).

4.6. Statistical Analysis

Normal distribution and homogeneity of variances of qRT-PCR data from control and E2-treatment group (Figure 2A,B) were first individually estimated. For statistical analysis between control and E2-treatment group within the same gene expression analysis, Student's *t*-test or Mann-Whitney's U test were applied. Multiple comparisons between different maturation stages were performed by Kruskal-Wallis test, followed by Sheffè multiple comparison test. Results were considered statistically significant when $p < 0.05$.

Acknowledgments: This study was financially supported by the grants 31003A_153427 and 31003A_130640 by the Swiss National Science Foundation to Helmut Segner.

Author Contributions: Ayako Casanova-Nakayama led the experimental work and contributed to the writing of the manuscript. Elena Wernicke von Siebenthal, Christian Kropf and Elisabeth Oldenberg contributed to the experimental work. Helmut Segner, together with Ayako Casanova, designed the study, drafted the manuscript, and together with Elena Wernicke von Siebenthal finalized the manuscript.

Conflicts of Interest: The authors declare no conflict of interest.

References

1. Hall, J.M.; Couse, J.F.; Korach, K.S. The multifaceted mechanisms of estradiol and estrogen receptor signaling. *J. Biol. Chem.* **2001**, *276*, 36869–36872. [CrossRef] [PubMed]
2. Straub, R.H. The complex role of estrogens in inflammation. *Endocr. Rev.* **2007**, *28*, 521–574. [CrossRef] [PubMed]
3. Nadkarni, S.; McArthur, S. Oestrogen and immunomodulation: New mechanisms that impact on peripheral and central immunity. *Curr. Opin. Pharmacol.* **2013**, *13*, 576–581. [CrossRef] [PubMed]
4. Kovats, S. Estrogen receptors regulate innate immune cells and signaling pathways. *Cell. Immunol.* **2015**, *294*, 63–69. [CrossRef] [PubMed]
5. Khan, D.; Ansar Ahmed, S. The immune system is a natural target for estrogen action: Opposing effects of estrogen in two prototypical autoimmune diseases. *Front. Immunol.* **2016**, *6*, 635. [CrossRef] [PubMed]
6. Klein, S. Hormonal and immunological mechanisms mediating sex differences in parasite infection. *Parasite Immunol.* **2004**, *26*, 247–264. [CrossRef] [PubMed]
7. Nunn, C.L.; Lindenfors, P.; Pursall, E.R.; Rolff, J. On sexual dimorphism in immune function. *Philos. Trans. R. Soc. B Biol. Sci.* **2009**, *364*, 61–69. [CrossRef] [PubMed]
8. Cunningham, M.; Gilkeson, G. Estrogen receptors in immunity and autoimmunity. *Clin. Rev. Allergy Immunol.* **2011**, *40*, 66–73. [CrossRef] [PubMed]
9. Igarashi, H.; Kouro, T.; Yokota, T.; Kincade, P.W. Age and stage dependency of estrogen receptor expression by lymphocyte precursors. *Proc. Natl. Acad. Sci. USA* **2001**, *98*, 15131–15136. [CrossRef] [PubMed]
10. Phiel, K.L.; Henderson, R.A.; Adelman, S.J.; Elloso, M.M. Differential estrogen receptor gene expression in human peripheral blood mononuclear cell populations. *Immunol. Lett.* **2005**, *97*, 107–113. [CrossRef] [PubMed]
11. Stygar, D.; Masironi, B.; Eriksson, H.; Sahlin, L. Studies on estrogen receptor (ER) α and β responses on gene regulation in peripheral blood leukocytes in vivo using selective ER agonists. *J. Endocrinol.* **2007**, *194*, 101–119. [CrossRef] [PubMed]

12. Lambert, K.C.; Curran, E.M.; Judy, B.M.; Milligan, G.N.; Lubahn, D.B.; Estes, D.M. Estrogen receptor α (ERα) deficiency in macrophages results in increased stimulation of CD4+ T cells while 17β-estradiol acts through ERα to increase IL-4 and GATA-3 expression in CD4+ T cells independent of antigen presentation. *J. Immunol.* **2005**, *175*, 5716–5723. [CrossRef] [PubMed]

13. Tiwari-Woodruff, S.; Morales, L.B.J.; Lee, R.; Voskuhl, R.R. Differential neuroprotective and antiinflammatory effects of estrogen receptor (ER) α and ERβ ligand treatment. *Proc. Natl. Acad. Sci. USA* **2007**, *104*, 14813–14818. [CrossRef] [PubMed]

14. Cvoro, A.; Tatomer, D.; Tee, M.-K.; Zogovic, T.; Harris, H.A.; Leitman, D.C. Selective estrogen receptor-β agonists repress transcription of proinflammatory genes. *J. Immunol.* **2008**, *180*, 630–636. [CrossRef] [PubMed]

15. Olsen, N.J.; Kovacs, W.J. Gonadal steroids and immunity. *Endocr. Rev.* **1996**, *17*, 369–384. [PubMed]

16. Fish, E.N. The X-files in immunity: Sex-based differences predispose immune responses. *Nat. Rev. Immunol.* **2008**, *8*, 737–744. [CrossRef] [PubMed]

17. Gilliver, S.C.; Emmerson, E.; Campbell, L.; Chambon, P.; Hardman, M.J.; Ashcroft, G.S. 17β-Estradiol inhibits wound healing in male mice via estrogen receptor-α. *Am. J. Pathol.* **2010**, *176*, 2707–2721. [CrossRef] [PubMed]

18. Beagley, K.W.; Gockel, C.M. Regulation of innate and adaptive immunity by the female sex hormones oestradiol and progesterone. *Pathog. Dis.* **2003**, *38*, 13–22. [CrossRef]

19. Abrams, E.T.; Miller, E.M. The roles of the immune system in women's reproduction: Evolutionary constraints and life history trade-offs. *Am. J. Phys. Anthropol.* **2011**, *146*, 134–154. [CrossRef] [PubMed]

20. Wira, C.R.; Rodriguez-Garcia, M.; Patel, M.V. The role of sex hormones in immune protection of the female reproductive tract. *Nat. Rev. Immunol.* **2015**, *15*, 217–230. [CrossRef] [PubMed]

21. Milla, S.; Depiereux, S.; Kestemont, P. The effects of estrogenic and androgenic endocrine disruptors on the immune system of fish: A review. *Ecotoxicology* **2011**, *20*, 305–319. [CrossRef] [PubMed]

22. Chaves-Pozo, E.; Cabas, I.; García-Ayala, A. Sex steroids modulate fish immune response. In *Sex Steroids*; InTech: London, UK, 2012.

23. Segner, H.; Casanova-Nakayama, A.; Kase, R.; Tyler, C.R. Impact of environmental estrogens on fish considering the diversity of estrogen signaling. *Gen. Comp. Endocrinol.* **2013**, *191*, 190–201. [CrossRef] [PubMed]

24. Casanova-Nakayama, A.; Wenger, M.; Burki, R.; Eppler, E.; Krasnov, A.; Segner, H. Endocrine disrupting compounds: Can they target the immune system of fish? *Mar. Pollut. Bull.* **2011**, *63*, 412–416. [CrossRef] [PubMed]

25. Iwanowicz, L.R.; Ottinger, C.A. Estrogens, estrogen receptors and their role as immunoregulators in fish. *Fish Def.* **2009**, *1*, 277–322.

26. Shved, N.; Berishvili, G.; Häusermann, E.; D'cotta, H.; Baroiller, J.-F.; Eppler, E. Challenge with 17α-ethinylestradiol (EE2) during early development persistently impairs growth, differentiation, and local expression of IGF-I and IGF-II in immune organs of tilapia. *Fish Shellfish Immunol.* **2009**, *26*, 524–530. [CrossRef] [PubMed]

27. Liarte, S.; Chaves-Pozo, E.; Abellán, E.; Meseguer, J.; Mulero, V.; García-Ayala, A. 17β-Estradiol regulates gilthead seabream professional phagocyte responses through macrophage activation. *Dev. Comp. Immunol.* **2011**, *35*, 19–27. [CrossRef] [PubMed]

28. Cabas, I.; Rodenas, M.C.; Abellán, E.; Meseguer, J.; Mulero, V.; García-Ayala, A. Estrogen Signaling through the G Protein–Coupled Estrogen Receptor Regulates Granulocyte Activation in Fish. *J. Immunol.* **2013**, *191*, 4628–4639. [CrossRef] [PubMed]

29. Massart, S.; Milla, S.; Kestemont, P. Expression of gene, protein and immunohistochemical localization of the estrogen receptor isoform ERα1 in male rainbow trout lymphoid organs; indication of the role of estrogens in the regulation of immune mechanisms. *Comp. Biochem. Physiol. Part B Biochem. Mol. Biol.* **2014**, *174*, 53–61. [CrossRef] [PubMed]

30. Shelley, L.K.; Osachoff, H.L.; van Aggelen, G.C.; Ross, P.S.; Kennedy, C.J. Alteration of immune function endpoints and differential expression of estrogen receptor isoforms in leukocytes from 17β-estradiol exposed rainbow trout (*Oncorhynchus mykiss*). *Gen. Comp. Endocrinol.* **2013**, *180*, 24–32. [CrossRef] [PubMed]

31. Szwejser, E.; Maciuszek, M.; Casanova-Nakayama, A.; Segner, H.; Verburg-van Kemenade, B.L.; Chadzinska, M. A role for multiple estrogen receptors in immune regulation of common carp. *Dev. Comp. Immunol.* **2017**, *66*, 61–72. [CrossRef] [PubMed]

32. Moggs, J.G.; Orphanides, G. Estrogen receptors: Orchestrators of pleiotropic cellular responses. *EMBO Rep.* **2001**, *2*, 775–781. [CrossRef] [PubMed]

33. Murphy, A.J.; Guyre, P.M.; Wira, C.R.; Pioli, P.A. Estradiol regulates expression of estrogen receptor ERα46 in human macrophages. *PLoS ONE* **2009**, *4*, e5539. [CrossRef] [PubMed]

34. Bagamasbad, P.; Denver, R.J. Mechanisms and significance of nuclear receptor auto-and cross-regulation. *Gen. Comp. Endocrinol.* **2011**, *170*, 3–17. [CrossRef] [PubMed]

35. Nelson, E.R.; Habibi, H.R. Estrogen receptor function and regulation in fish and other vertebrates. *Gen. Comp. Endocrinol.* **2013**, *192*, 15–24. [CrossRef] [PubMed]

36. Tecalco-Cruz, A.C.; Ramírez-Jarquín, J.O. Mechanisms that increase stability of estrogen receptor alpha in breast cancer. *Clin. Breast Cancer* **2017**, *17*, 1–10. [CrossRef] [PubMed]

37. Zhou, W.; Slingerland, J.M. Links between oestrogen receptor activation and proteolysis: Relevance to hormone-regulated cancer therapy. *Nat. Rev. Cancer* **2014**, *14*, 26. [CrossRef] [PubMed]

38. Leclercq, G.; Lacroix, M.; Laïos, I.; Laurent, G. Estrogen receptor alpha: Impact of ligands on intracellular shuttling and turnover rate in breast cancer cells. *Curr. Cancer Drug Targets* **2006**, *6*, 39–64. [CrossRef] [PubMed]

39. Iwanowicz, L.R.; Stafford, J.L.; Patiño, R.; Bengten, E.; Miller, N.W.; Blazer, V.S. Channel catfish (*Ictalurus punctatus*) leukocytes express estrogen receptor isoforms ERα and ERβ2 and are functionally modulated by estrogens. *Fish Shellfish Immunol.* **2014**, *40*, 109–119. [CrossRef] [PubMed]

40. Nagler, J.J.; Cavileer, T.; Sullivan, J.; Cyr, D.G.; Rexroad, C. The complete nuclear estrogen receptor family in the rainbow trout: Discovery of the novel ERα2 and both ERβ isoforms. *Gene* **2007**, *392*, 164–173. [CrossRef] [PubMed]

41. Szwejser, E.; Pijanowski, L.; Maciuszek, M.; Ptak, A.; Wartalski, K.; Duda, M.; Segner, H.; Verburg-van Kemenade, B.L.; Chadzinska, M. Stress differentially affects the systemic and leukocyte estrogen network in common carp. *Fish Shellfish Immunol.* **2017**, *68*, 190–201. [CrossRef] [PubMed]

42. Stygar, D.; Westlund, P.; Eriksson, H.; Sahlin, L. Identification of wild type and variants of oestrogen receptors in polymorphonuclear and mononuclear leucocytes. *Clin. Endocrinol.* **2006**, *64*, 74–81. [CrossRef] [PubMed]

43. Yakimchuk, K.; Jondal, M.; Okret, S. Estrogen receptor α and β in the normal immune system and in lymphoid malignancies. *Mol. Cell. Endocrinol.* **2013**, *375*, 121–129. [CrossRef] [PubMed]

44. Li, J.; McMurray, R.W. Effects of estrogen receptor subtype-selective agonists on immune functions in ovariectomized mice. *Int. Immunopharmacol.* **2006**, *6*, 1413–1423. [CrossRef] [PubMed]

45. Suzuki, S.; Gerhold, L.M.; Böttner, M.; Rau, S.W.; Dela Cruz, C.; Yang, E.; Zhu, H.; Yu, J.; Cashion, A.B.; Kindy, M.S. Estradiol enhances neurogenesis following ischemic stroke through estrogen receptors α and β. *J. Comp. Neurol.* **2007**, *500*, 1064–1075. [CrossRef] [PubMed]

46. Pinto, P.I.S.; Estevao, M.D.; Redruello, B.; Socorro, S.M.; Canario, A.V.M.; Power, D.M. Immunohistochemical detection of estrogen receptors in fish scales. *Gen. Comp. Endocrinol.* **2009**, *160*, 19–29. [CrossRef] [PubMed]

47. Stellato, C.; Porreca, I.; Cuomo, D.; Tarallo, R.; Nassa, G.; Ambrosino, C. The "busy life" of unliganded estrogen receptors. *Proteomics* **2016**, *16*, 288–300. [CrossRef] [PubMed]

48. Van den Hurk, R.; Lambert, J. Temperature and steroid effects on gonadal sex differentiation in rainbow trout. In Proceedings of the International Symposium on Reproductive Physiology of Fish, Wageningen, The Netherlands, 2–6 August 1982; pp. 69–72.

49. Feist, G.; Schreck, C.B.; Fitzpatrick, M.S.; Redding, J.M. Sex steroid profiles of coho salmon (*Oncorhynchus kisutch*) during early development and sexual differentiation. *Gen. Comp. Endocrinol.* **1990**, *80*, 299–313. [CrossRef]

50. Marlatt, V.L.; Lakoff, J.; Crump, K.; Martyniuk, C.J.; Watt, J.; Jewell, L.; Atkinson, S.; Blais, J.M.; Sherry, J.; Moon, T.W. Sex-and tissue-specific effects of waterborne estrogen on estrogen receptor subtypes and E2-mediated gene expression in the reproductive axis of goldfish. *Comp. Biochem. Physiol. Part A Mol. Integr. Physiol.* **2010**, *156*, 92–101. [CrossRef] [PubMed]

51. Szwejser, E.; Verburg-van Kemenade, B.L.; Maciuszek, M.; Chadzinska, M. Estrogen-dependent seasonal adaptations in the immune response of fish. *Horm. Behav.* **2017**, *88*, 15–24. [CrossRef] [PubMed]

52. Pakdel, F.; Le Guellec, C.; Vaillant, C.; Le Roux, M.G.; Valotaire, Y. Identification and estrogen induction of two estrogen receptors (ER) messenger ribonucleic acids in the rainbow trout liver: Sequence homology with other ERs. *Mol. Endocrinol.* **1989**, *3*, 44–51. [CrossRef] [PubMed]

53. Molero, L.; García-Durán, M.; Diaz-Recasens, J.; Rico, L.; Casado, S.; López-Farré, A. Expression of estrogen receptor subtypes and neuronal nitric oxide synthase in neutrophils from women and men: Regulation by estrogen. *Cardiovasc. Res.* **2002**, *56*, 43–51. [CrossRef]

54. Menuet, A.; Le Page, Y.; Torres, O.; Kern, L.; Kah, O.; Pakdel, F. Analysis of the estrogen regulation of the zebrafish estrogen receptor (ER) reveals distinct effects of ERα, ERβ1 and ERβ2. *J. Mol. Endocrinol.* **2004**, *32*, 975–986. [CrossRef] [PubMed]

55. Sabo-Attwood, T.; Kroll, K.J.; Denslow, N.D. Differential expression of largemouth bass (*Micropterus salmoides*) estrogen receptor isotypes alpha, beta, and gamma by estradiol. *Mol. Cell. Endocrinol.* **2004**, *218*, 107–118. [CrossRef] [PubMed]

56. Filby, A.; Tyler, C. Molecular characterization of estrogen receptors 1, 2a, and 2b and their tissue and ontogenic expression profiles in fathead minnow (*Pimephales promelas*). *Biol. Reprod.* **2005**, *73*, 648–662. [CrossRef] [PubMed]

57. Boyce-Derricott, J.; Nagler, J.J.; Cloud, J.G. Variation among Rainbow Trout (Oncorhynchus mykiss) Estrogen Receptor Isoform 3′ Untranslated Regions and the Effect of 17β-Estradiol on mRNA Stability in Hepatocyte Culture. *DNA Cell Biol.* **2010**, *29*, 229–234. [CrossRef] [PubMed]

58. Marlatt, V.; Martyniuk, C.; Zhang, D.; Xiong, H.; Watt, J.; Xia, X.; Moon, T.; Trudeau, V. Auto-regulation of estrogen receptor subtypes and gene expression profiling of 17β-estradiol action in the neuroendocrine axis of male goldfish. *Mol. Cell. Endocrinol.* **2008**, *283*, 38–48. [CrossRef] [PubMed]

59. Grčević, M.; Kralik, Z.; Kralik, G.; Galović, D.; Pavić, M. The effect of lutein additives on biochemical parameters in blood of laying hens. *Poljoprivreda* **2016**, *22*, 34–38. [CrossRef]

60. Chandrasekar, G.; Archer, A.; Gustafsson, J.-Å.; Lendahl, M.A. Levels of 17β-estradiol receptors expressed in embryonic and adult zebrafish following in vivo treatment of natural or synthetic ligands. *PLoS ONE* **2010**, *5*, e9678. [CrossRef] [PubMed]

61. Shelley, L.K.; Ross, P.S.; Kennedy, C.J. The effects of an in vitro exposure to 17β-estradiol and nonylphenol on rainbow trout (*Oncorhynchus mykiss*) peripheral blood leukocytes. *Comp. Biochem. Physiol. Part C Toxicol. Pharmacol.* **2012**, *155*, 440–446. [CrossRef] [PubMed]

62. Dulos, J.; Vijn, P.; van Doorn, C.; Hofstra, C.L.; Veening-Griffioen, D.; de Graaf, J.; Dijcks, F.A.; Boots, A.M. Suppression of the inflammatory response in experimental arthritis is mediated via estrogen receptor α but not estrogen receptor β. *Arthritis Res. Ther.* **2010**, *12*, R101. [CrossRef] [PubMed]

63. Segner, H.; Verburg-van Kemenade, B.L.; Chadzinska, M. The immunomodulatory role of the hypothalamus-pituitary-gonad axis: Proximate mechanism for reproduction-immune trade offs? *Dev. Comp. Immunol.* **2017**, *66*, 43–60. [CrossRef] [PubMed]

64. Myers, M.J.; Butler, L.D.; Petersen, B.H. Estradiol-induced alteration in the immune system. II. Suppression of cellular immunity in the rat is not the result of direct estrogenic action. *Immunopharmacology* **1986**, *11*, 47–55. [CrossRef]

65. McMurray, R.W. Estrogen, prolactin, and autoimmunity: Actions and interactions. *Int. Immunopharmacol.* **2001**, *1*, 995–1008. [CrossRef]

66. Lang, T.J. Estrogen as an immunomodulator. *Clin. Immunol.* **2004**, *113*, 224–230. [CrossRef] [PubMed]

67. Giefing-Kröll, C.; Berger, P.; Lepperdinger, G.; Grubeck-Loebenstein, B. How sex and age affect immune responses, susceptibility to infections, and response to vaccination. *Aging Cell* **2015**, *14*, 309–321. [CrossRef] [PubMed]

68. Burki, R.; Krasnov, A.; Bettge, K.; Rexroad, C.E.; Afanasyev, S.; Antikainen, M.; Burkhardt-Holm, P.; Wahli, T.; Segner, H. Pathogenic infection confounds induction of the estrogenic biomarker vitellogenin in rainbow trout. *Environ. Toxicol. Chem.* **2012**, *31*, 2318–2323. [CrossRef] [PubMed]

International Journal of
Molecular Sciences

MDPI

Article

Understanding the Inguinal Sinus in Sheep (*Ovis aries*)—Morphology, Secretion, and Expression of Progesterone, Estrogens, and Prolactin Receptors

Graça Alexandre-Pires [1,*], Catarina Martins [2], António M. Galvão [3], Margarida Miranda [4], Olga Silva [4], Dário Ligeiro [5], Telmo Nunes [6] and Graça Ferreira-Dias [7]

[1] CIISA-Faculty of Veterinary Medicine (FMV), Universidade de Lisboa, Av. Universidade Técnica, 1300-477 Lisboa, Portugal
[2] CEDOC-Chronic Diseases Research Center, Immunology, NOVA Medical School, Universidade Nova de Lisboa, Rua Câmara Pestana n° 6, 6-A Edifício CEDOC II, 1150-082 Lisboa, Portugal; catarina.martins@fcm.unl.pt
[3] CIISA-Faculty of Veterinary Medicine (FMV), Universidade de Lisboa, Av. Universidade Técnica, 1300-477 Lisboa, Portugal; agalvao@fmv.ulisboa.pt
[4] iMed.ULisboa, Pharmacological and Regulatory Sciences Group, Faculty of Pharmacy, Universidade de Lisboa, Av. Gama Pinto, 1649-003 Lisbon, Portugal; mimiranda@live.com.pt (M.M.); odsilva@campus.ul.pt (O.S.)
[5] Centro de Sangue e Transplantação de Lisboa, IPST,IP Alameda das Linhas de Torres 117, 1749-005 Lisbon, Portugal; dario@ipst.min-saude.pt
[6] Microscopy Center, Faculty of Sciences, Campo Grande, 1749-016 Lisbon, Portugal; telmonunes@hotmail.com
[7] CIISA-Faculty of Veterinary Medicine (FMV), Universidade de Lisboa, Av. Universidade Técnica, 1300-477 Lisboa, Portugal; gmlfdias@fmv.ulisboa.pt
*   Correspondence: gpires@fmv.ulisboa.pt; Tel.: +351-21-365-2800 (ext. 58)

Received: 1 May 2017; Accepted: 10 July 2017; Published: 13 July 2017

Abstract: Post-parturient behavior of mammalian females is essential for early parent–offspring contact. After delivery, lambs need to ingest colostrum for obtaining the related immunological protection, and early interactions between the mother and the lamb are crucial. Despite visual and auditory cues, olfactory cues are decisive in lamb orientation to the mammary gland. In sheep, the inguinal sinus is located bilaterally near the mammary gland as a skin pouch (IGS) that presents a gland that secretes a strong-smelling wax. Sheep IGS gland functions have many aspects under evaluation. The objective of the present study was to evaluate sheep IGS gland functional aspects and mRNA transcription and the protein expression of several hormone receptors, such as progesterone receptor (PGR), estrogen receptor 1 (ESR1), and 2 (ESR2) and prolactin receptor (PRLR) present. In addition, another aim was to achieve information about IGS ultrastructure and chemical compounds produced in this gland. All hormone receptors evaluated show expression in IGS during the estrous cycle (follicular/luteal phases), pregnancy, and the post-partum period. IGS secretion is rich in triterpenoids that totally differ from the surrounding skin. They might be essential substances for the development of an olfactory preference of newborns to their mothers.

Keywords: inguinal sinus; morphology; transcription; ESR1; ESR2; PGR; PRLR; chemical compounds; triterpenoids

1. Introduction

Shortly after birth, the mammalian neonate and its mother interact in order to favor parent–offspring contact. Under evolutionary pressure to meet the needs of their neonates, selected behaviors of post-parturient females occur, with nursing patterns being very broad across species concerning duration or frequency [1]. In fact, nursing patterns are different depending on the species. The amount of parental nursing and milk investment determines not only the number of young the parent can produce, but also affects the offspring's fitness. As a an example, while penguins are known to sacrifice their own health and wellbeing in exchange for the survival of their young, Sprague–Dawley rats eject milk to feed their pups only when the mother is asleep and therefore without any nursing care burden [2,3].

In eutherian mammals, the mammary gland suffers modifications due to the presence of a prehensile nipple that facilitates the milk intake. The neonate, especially in a precocious mammal, is highly aroused by the stimulatory process caused by birth and is tuned towards the sensory cues that facilitate localization of the mother's nipples. Since lambs are born with fully functional sensory systems, they can use multisensory cues to find the teat. In addition, ewes lick their lambs after birth and make a low-pitched moaning, which is an important behavior for the development of ewe–lamb binding and directs the lamb towards the inguinal region. That area is wool-free and has a higher temperature, eliciting lambs' nosing and directing them towards IGS. When considering lambing, the desired outcome is to have more lambs born alive and allowing them to get adequate colostrum so they stay alive [4,5].

Newly born mammals have to reach the source of milk as promptly as possible to ensure uninterrupted mother-to-offspring transfer of hydration and nutrients. Colostrum intake guards against immediate exposure to micronutrients and antioxidants, passive immunization, innocuous bacterial strains, growth factors, and a range of bioactive peptides that control conservative behavioral function [4,5].

While some species give birth to altricial neonates, other species bear precocial or semiprecocial newborns whose behavior is under the control of all their senses (e.g., ungulates, some rodents). We do not fully understand how mammalian neonates perceive and analyze chemosignals and the olfactory scene that is associated with the mammary structure. The fact is that, by staying alive, the majority of them prove their competence to position themselves adequately from the very first exposure to a nipple. Suckling is the most intimate form of contact with the maternal body, which strengthens the relationship with the mother [6]. Besides maintaining the newborn's warmth, nursing/suckling interactions also facilitate olfactory learning in newborns, as long as suckling and the olfactory stimulus temporally overlap [7].

Chemical communication plays a major role in mammalian behavior and starts by understanding olfactory communication and acquiring knowledge on body sources of odors and the behaviors associated with their deposition. This is the starting point for knowledge concerning putative chemosignals. Some insights concerning the mechanisms used by the newborn for detection and transduction of chemical stimuli are also needed. Odors refer to any chemical released by an individual, that is potentially detectable by another individual, and a pheromone is an odor that elicits a predictable and stereotypical behavior or physiological response, provides specific information, or modulates responses in the receiving individual [8]. Some of those chemosignals are of skin and gland origin (examples: the tarsal gland of the black-tailed deer, the inguinal and chin glands of rabbits for individual identification, the preputial glands in mature male pigs, and submaxillary glands for mating stance in estrus gilts) [9–11].

Across very different phylogenies, the mechanisms for pheromones and odor learning have much in common. The medial amygdala appears to be involved in both the recognition of social odors and their association with chemosensory information sensed by the vomeronasal system and sensory neurons and GABAergic interneurons in the olfactory bulb, which are continuously replaced. In fact, this contradicts the idea that neurogenesis is purely restorative in the adult stage since it develops the necessary plasticity for induction and maintenance of learned chemosensory responses [12,13].

In the ungulate family (*Caprinae, Cephalophinae, Antilopinae*), circumscribed scent glands are located bilaterally as pouches or pockets near the exterior base of the udder. Several morphological features that present a gland, such as the infraorbital sinus, the interdigital sinus, and the inguinal sinus, characterize the genus *Ovis aries*. Therefore, we can find the inguinal sinus in the ram as well as in the ewe. Different research groups have pointed out different responsibilities in the secretion of these three different glands. In some cases, they function as trail glands, while in others they have a role in the known "male effect". By producing pheromones in the male, they elicit female reproductive physiologic responses [14,15].

After giving birth, mothers position their bodies so that the neonate finds the mammary zone. The suckling/nursing relationship becomes the core of their physiological and behavioral relationship and involves the blocking of the estrous cycle, the stimulation of lactation, and the development of a bond with repercussions for the future of the neonate, as this contact stimulates the mother's care and the infant's willing to find the nipple and ingest milk. Since lambs are born with fully functional sensory modalities, they can use multisensory cues to find the teat. In return, ewes lick their lambs after birth and present low-pitched bleating, an important behavior for the development of ewe–lamb attachment and direct the lamb towards the inguinal region, where areas free of wool have a higher temperature, eliciting the lambs' nosing and directing them towards IGS. When considering lambing, the pressure is to have as many lambs born alive as possible and allow them to get adequate colostrum so that they stay alive. The parental investment in producing a new young life (or two) per reproductive cycle is very high and the behavior of the young and its likelihood to survive is under the responsibility of its mother at the beginning of its life throughout lactation. This mother–young unit is a major part of the welfare of the neonate and inadequate maternal care invariably leads to early death [6,16].

In the literature, references to inguinal gland wax production describe a strong-smelling substance that seems to activate udder-seeking behavior, in combination with vocal and tactile stimuli [17,18]. Records of respiration and heart rates demonstrated that non-suckling lambs respond to the smell of mother's inguinal sinus production in a more reactive way than the response found when odors of wool or milk from an unfamiliar ewe are available. In addition, newborns can discriminate between the smell of their mothers and that of an alien ewe and it was shown that when lambs are made anosmic by applying lidocaine to their nostrils, localization of the teat was delayed [19–21]. Initiation of suckling is dependent on variable blends of maternal "signature odors" that are learned and recognized prior to first suckling. In rabbits, the pheromone responsible for initiating suckling has already been identified and newborn mice require maternal olfactory or "signature odors" cues to start suckling since this blend of volatile odors (not necessarily a classical pheromone) produced from the mother elicits the behavior [22]. In sheep, maternal behavior (low-pitched bleats, licking, and nursing) is triggered by changes of plasma progesterone and estradiol around parturition, and the release of oxytocin in the brain [17]. Thus, the hypothesis we presented was that sheep IGS might play a role as a chemosensory clue to the newborn lamb.

As IGS gland functions are in many aspects under evaluation, the objective of the present study was to evaluate: (i) IGS morphology, histology and ultrastructure; (ii) mRNA transcription and protein expression of progesterone (PGR), estrogen receptors 1 (ESR1) and 2 (ESR2) and prolactin receptor (PRLR). Also, as these hormones show associations with sexual and nursing behavior in many aspects, achieving information about putative changes in the chemical compounds produced in this body sinus along the estrous cycle was also one of the goals.

2. Results

Near the mammary gland, one can observe by abduction of the hind limb the inguinal sinus located bilaterally as pockets at the external base of this gland. When exposed, a yellowish substance with a wax appearance seems to spread downward to the teats (Figure 1). The histology of the inguinal sinus demonstrates invaginated skin presenting sebaceous and acinar glandular fields, with collagen sheath fibers sustaining the secretory epithelium (Figure 2A–C). The acinar glands appear with different patterns of secretion, from a resting phase to the development of huge cellular protrusions of apocrine secretions towards the glandular lumen (for example comparing Figure 2C,F,G,H). According to dye affinities, the parenchyma is shown to be rich in glycogen (PAS; Figure 2E,F) and lipid granules (Sudan Black stain, 2 G), and mucin content as well (Alcian blue-2H-I). Myoepithelial cells can be depicted (Figure 2E). Secretion production appears with an uneven distribution in different areas of the inguinal pouch and in different acinar units, as shown in different plates of Figure 2.

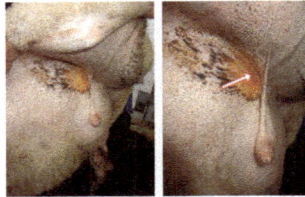

Figure 1. By abduction of the hind-left leg one can observe the skin pouch (IGS) over the mammary gland. On the right, the arrow points out the IGS and its yellowish secretion.

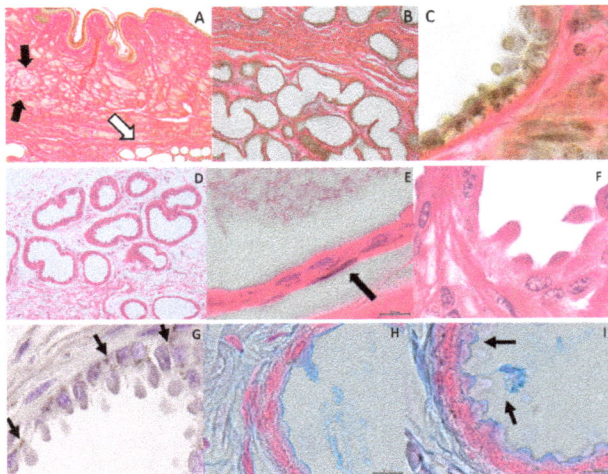

Figure 2. Histology-Bar = 100 μm. *Van Gieson's* Stain. (**A**) Organization of the inguinal sinus presenting sebaceous (black arrows) and acinar glandular fields (white arrow); (**B**) collagen fibers sustain the secretory epithelium of the acinar glands that presents; (**C**) cellular protrusions towards the glandular lumen (apocrine secretion). Parenchyma rich in glycogen can be observed and its amount varies in different areas of the gland being the secretory cells in different stages of secretion production in different areas of the gland–PAS; (**D–F**); In (**E**) the arrow points out a myoepithelial cell. (**D**) = 40× magnification; (**E**) = 1000×. Visualizations of lipid granules with Sudan black stain; (**G**) Alcian blue stain depicts mucin content; (**H,I**) Magnification = 1000×.

Ultrastructural observations of scanning electronic microscopy (SEM) (Figure 3) showed that the apocrine glandular units are of alveolar type but not tubuloalveolar and that these secretion units appear in clusters (Figure 3A,B). Luminal surface secretion presents a paved appearance, with secretory cells in diverse stages of differentiation (Figure 3C). In fact, some acini at different stages depict fragments of secretion being "pinched off", exhibiting secretory vesicles or secretion blebs (Figure 3D,E), although cells preserve a clear demarcation with neighboring cells by means of rows of microvilli. Apical end-pieces show a progressive filling process that upsurges as bulge-like structures. In other acini, cells appear to be in a transitional process where the clear demarcation with surrounding cells is no longer as visible, resulting from the development of apical protrusions that in a final stage denote a smooth plasma membrane devoid of microvilli and covering the protrusions (Figure 3F,H).

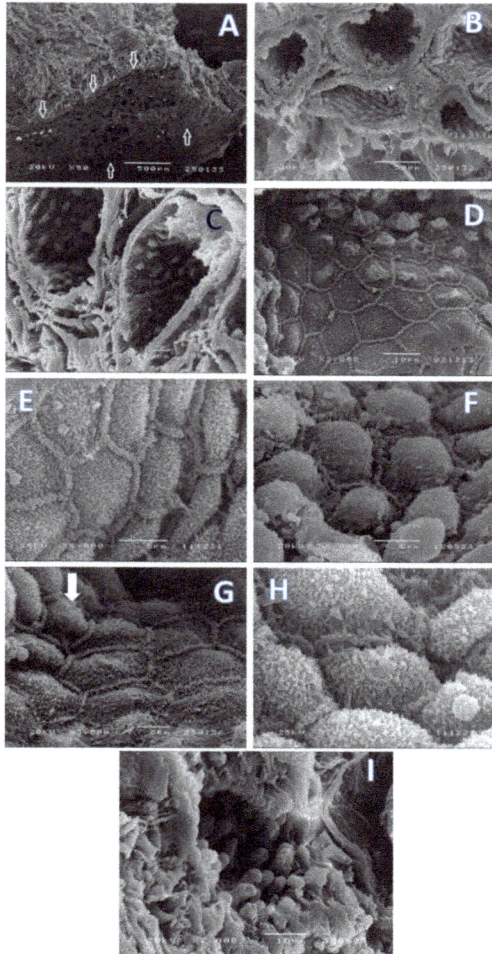

Figure 3. Scanning electronic images of IGS. In (**A**) (bar = 500 μm); (**B**) (bar = 50 μm) and (**C**) (bar = 10 μm) it is clear that apocrine glandular structures appear in clusters inside the IGS. Luminal surface can show a paved appearance; (**D**) bar = 10 μm or an irregular one resulting from the secretory process. Secretory cells appear in different stages of differentiation, where fragments of secretion are being "pinched off" exhibiting secretory vesicles (secretion blebs), while cells maintain a clear demarcation with neighboring cells by means of rows of microvilli; (**D**) (bar = 10 μm); (**E**) (bar = 5 μm) and (**F**) (bar = 5 μm). A progressive gland filling process results on the upsurge of bulge-like structure; (**G**) (bar = 5 μm) and (**H**) (bar = 5 μm). Some cells appear to be in a transitional process where demarcation with surrounding cells is no longer as visible resulting from the development of apical protrusions that in a final stage denote a smooth plasma membrane devoid of microvilli and covering the protrusions; (**I**) bar = 10 μm.

According to specific primer sequences used for quantitative real-time PCR (reported in Section 4), it was possible to show transcription of mRNA for *ESR1*, *ESR2*, *PGR* and *PRLR* during the follicular and mid-luteal phase of the estrous cycle (Figure 4).

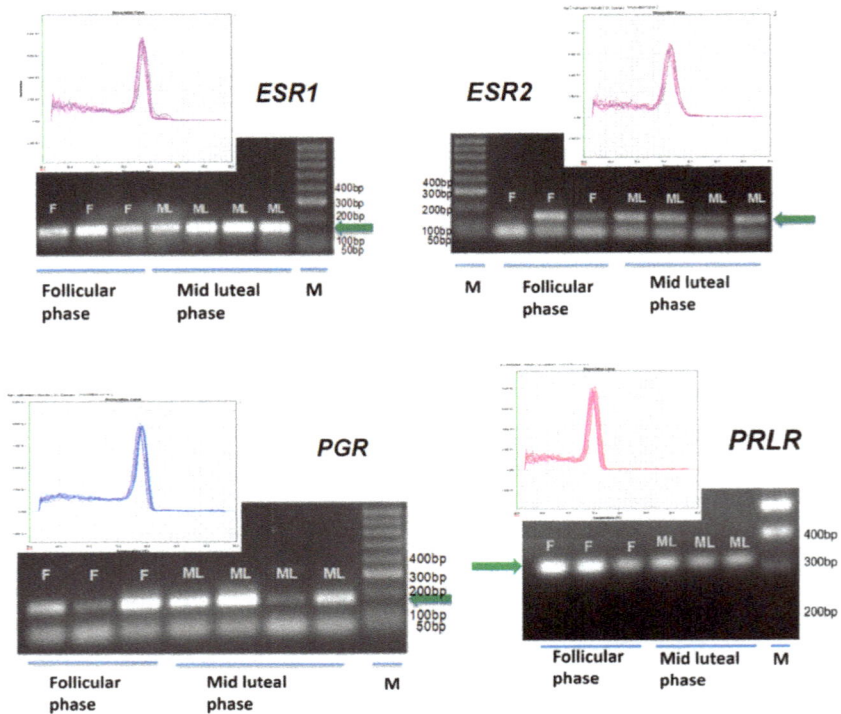

Figure 4. Qualitative PCR electrophoresis gel and dissociation curves of real-time PCR confirming estrogen receptor 1 (*ESR1*), and 2 *ESR2*, progesterone receptor (*PGR*) and prolactin receptor (*PRLR*) gene transcription in the IGS in different phase of the estrous cycle. Green arrow indicates the specific gene band. (F) follicular phase; (ML) mid luteal phase; (M) DNA marker; bp (base pairs). All primers validated for 80 nM in the real time PCR run. Single product confirmation with the single peak in the dissociation curve.

Confocal scanning microscopy demonstrated immunoreactivity towards ESR1, ESR2, PGR and PLRL receptors regardless of the estrous phase or the differentiation of acini cells concerning secretion process. Evidence was found of non-nuclear estrogen receptor and PGR in a clear border basilar position (Figure 5).

Flow cytometry analysis of cell suspensions showed cells with distinct auto fluorescence levels and different behavior towards ESR1, ESR2, PRLR and PGR); (Figure 6). Along with the different estrous cycle phases studied, PRLR and ESR2 positive cell populations showed always a higher fluorescence intensity compared to PGR and ESR1 positive cells ($p < 0.05$). At pregnancy, PRLR also showed a higher expression ($p < 0.01$) in comparison with other fluorescence intensities.

Thin-layer chromatography was performed on the available set of samples ($n = 28$), applied semi-quantitatively. The conditions allowed for the identification of a chromatographic profile (retention factor, fluorescence) characteristic of the presence of triterpenoids, with three major bands (Figure 7, compounds 1–3) in post-partum ewes (PP), non-pregnant (NP), and pregnant ewes IGS (P). The relative intensities of these three bands in these groups of samples made it possible to infer the existence of a higher relative content of these triterpenoids in P samples, as also demonstrated in Supplementary Data.

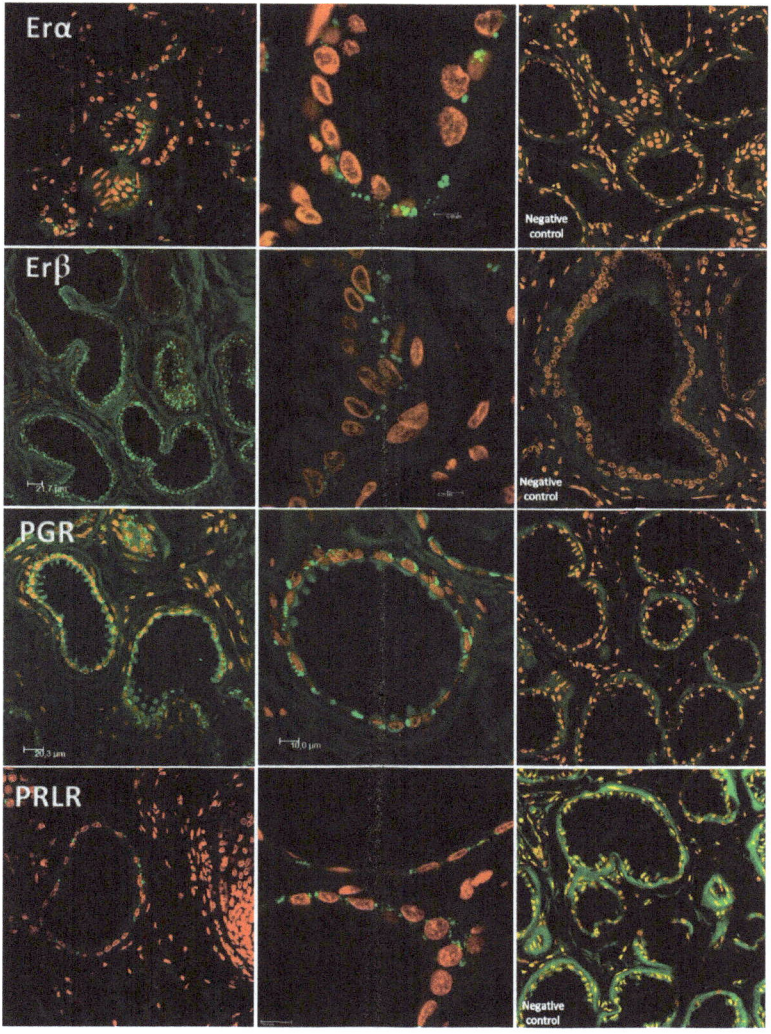

Figure 5. Examples of laser-scanning confocal fluorescence (LSC—lens 63.03 oil) images of IGS. One can observe immunoreactivity towards ESR1, ESR2, PGR and PRLR in cells of the apocrine glands labeled with PE and stained for the different receptors (fluorescence in green). Use of To-Pro-3 iodide for nuclear counterstaining (fluorescence in red).

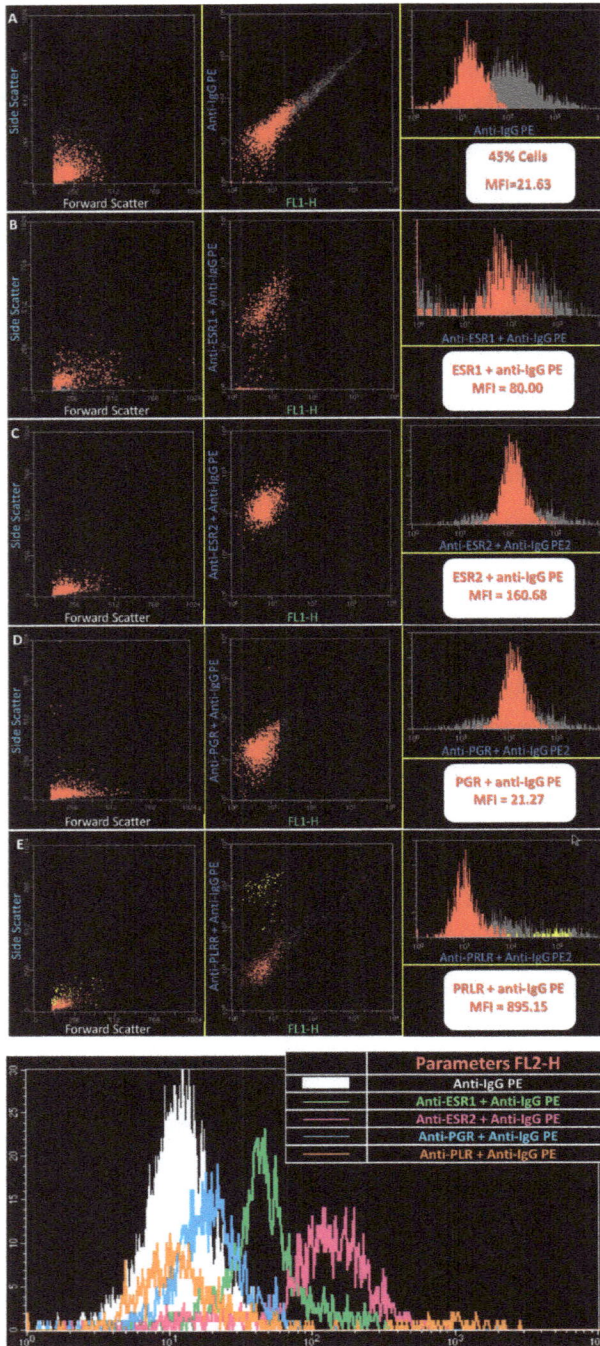

Figure 6. Flow cytometry analysis of gated cells of IGS. Examples of gated and dot plots and histograms showing the expression of ESR1, ESR2, PLRL and PGR. Shown flow cytometry data depict a positive cell expression towards the receptors under evaluation.

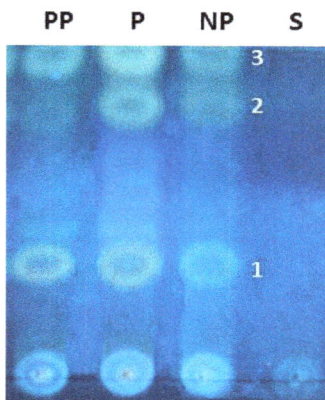

Figure 7. Chromatogram analysis revealed the presence of triterpenoids as marker chemical classes of post-partum ewes (PP), pregnant ewes (P) and non-pregnant (NP) samples, and a chemical profile that clearly portrait three major bands: 1, 2 and 3. These compounds were absent on S, the surrounding skin sample used as a negative control. The background color of the chromatogram is not uniform, which is a normal and recognized situation and does not interfere with the interpretation of the results.

3. Discussion

Several studies suggest that mammalian species have evolved multiple strategies to release olfactory signals to their offspring and ensure the onset of suckling as a critical behavior [23]. These tactics can be a pheromone-mediated behavior, or a reply to signature odors. Both appear to elicit innate behavior [22]. In the domestic pig, for example, after washing the sow's abdomen with organic solvents, the piglets lost teat localization [24]. As reported, mammalian females are known to use odor cues to control infant state, attention and directional responses, to delay distress responses, to stimulate breathing and positive oral actions, and finally they can boost learning. Female–offspring odor communication in European rabbits and humans is representatives of evolutionary extremes in terms of the structure and dynamics of mother–infant relationships, and levels of neonatal autonomy. In fact, both species have evolved mammary structures and chemosignal sources under greasy fixatives [24]. These features confer on them a chemo-communicative function and promote the success of the offspring's approach and exploration of the maternal body surface, resulting in effective initial feeds and rapid learning of maternal identity [25]. In women, for example, the literature reports several volatile compounds in the nipple/areola region during pregnancy and after childbirth that are not present in other phases [26]. Human neonatal reactivity to these areolar odors tested against several reference stimuli (e.g., human milk or sebum, solvent, vanilla, fresh cow's milk, cow's-milk-based formula) showed that pure Montgomery gland secretion elicits more orofacial activity. Nevertheless, further investigation is necessary to pin down the volatile compounds that can then be evaluated under the nose of human newborns in repeatable bioassays. So far, no evidence for any chemo-stimulus that would qualify as a pheromone is at hand in primates, including human mother-to-infant communication [23,27].

Lactating rabbits emit in their milk a volatile aldehyde, 2-methylbut-2-enal, that provokes searching and grasping behaviors in neonates [28,29]; newborn rabbits' survival depends on the perception of this odor signal emitted from the mother's *ventrum* area that allows them to locate the mother's nipples and suckle. Emission of nipple pheromones in the rabbit is induced during pregnancy by the combined action of estrogen and progesterone [11,30].

It was recently demonstrated that peripheral olfactory neurogenesis driven by estrogen occurs in the vomeronasal organ. A fraction of those cells are able to extend their dendrites to contact the vomeronasal lumen and detect proteins that show pheromone activity, while concomitant differences in gene expression in the vomeronasal transcriptomes of pregnant female mice also occur [11]. In sheep, endocrine changes at parturition, together with interactions with the newborn, modulate cell proliferation and neurogenesis in the sub-ventricular zone, the main olfactory bulb, and the dentate gyrus. Also at parturition, with or without interactions with the lamb for two days, a downregulation of the number of newly born cells in those neurological areas occurs in comparison to non-pregnant sheep. Therefore, in sheep it was postulated that the downregulation of cell proliferation observed in the early post-partum period could facilitate the olfactory perceptual and memory demands associated with maternal behavior, by favoring the survival and integration of neurons born earlier [31]. This situation contrasts with the occurrence of peripheral olfactory neurogenesis driven by estrogen and prolactin, which induce an increase in neural progenitors in the sub-ventricular zone of the lateral ventricle of the brain in other species [13]. A huge field of possibilities must be tested since these facts suggest the likelihood that lamb post-partum survival might depend on an olfactory perception fulfilled by the production of substances present in the IGS of the mother.

A common factor that characterizes most post-partum lamb deaths is the disturbance of bond formation between ewes and their offspring [32]. The glandular area near the mammary gland produces a substance with a yellowish aspect that spreads down the teats. This glandular substrate can dislodge itself while the animal walks or licks the area, making lambs reactive when exposed to it by means of directional movements of the head resulting in an increase in respiratory and cardiac rates [33]. Moreover, the odor of the IGS secretion elicits a more major response than the odor of wool or milk. It should be underlined that the maternal "wax production" by the IGS is more reactogenic than a similar product from an alien ewe. Moreover, newly born lambs showed a reaction to ovine milk odor even if impregnated in a cloth, a situation that did not occur when a scentless and humid cloth was used instead, and it was shown that when lambs are made anosmic by having lidocaine applied to their nostrils, localization of the teat was delayed [1]. Altogether, data from different researchers call attention to IGS production as a potential candidate for a strong scent effect. Although no studies have pointed out the exact constitutive compounds of IGS, our data have shown that when comparing IGS secretions it is clear that they show differences compared to the surrounding skin secretions and change from early motherhood. Although it was outside the scope of this work, by looking at the possible neurogenesis associated with olfactory neurogenic repercussion of this IGS secreted substance or how this complex material will behave under down/upregulation of steroid hormones or prolactin, we gained an overall picture of the complexity of the mechanisms involved, considering the putative hormonal influence in olfactory signaling . As a unique feature, not present in the majority of other animals, evaluation of IGS and its production in our work aimed to find out their particularities with respect to normal skin and the nature of this substance in different reproductive phases of the ewe life cycle (estrous cycle, pregnancy and post-partum). The differences we have found might have an impact on sheep herd management.

Thin-layer chromatography (TLC) is a method for the chemical screening of compounds, considered one of the first steps to establishment of the chemical profile of a sample. The thin-layer chromatography conditions used allowed us to identify a chromatographic profile characteristic of triterpenoids [34] present on PP, NP and P. The relative intensities of these three bands on these groups of samples made it possible to infer the existence of a higher relative content of these triterpenoids in P samples. This technique was applied semi-quantitatively to the available samples. It is expected that after structural identification of the three detectable marker compounds, they will be quantified by means of a suitable method specifically validated for that purpose. Nevertheless, the NP group's lowest relative content on these compounds calls attention to their putative involvement in chemo-communication and, as mentioned, further studies will be carried out in order to confirm the involvement of those triterpenoids in the development of an olfactory preference of the newborn lambs towards the mother's mammary gland.

In sheep and goats the beginning of maternal behavior at parturition is under the control of hormonal changes and fetus expulsion [35,36]. Thus, our interest concerning the presence of progesterone, estrogens, and prolactin receptors in IGS in the ewe is justified beyond a down- or upregulation involved in peripheral olfactory neurogenesis, previously discussed. Prolactin (PRL) is a versatile hormone in mammals with effects in reproductive, sexual, metabolic, and immune functions, among others [37]. The distribution of PRLR and the awareness of several extra-pituitary PRL-expressing tissues, has called attention to the range of PRL actions beyond mammary gland function [38–40]. Rabbit maternal behavior consists of building an underlayer of fur during late pregnancy and displaying, with circadian periodicity, a single 3-min nursing bout/day across lactation. It is synthesized in multiple tissues and its biological actions are not limited solely to reproduction, as it has been shown to control a variety of behaviors [41]. Synthesis of female-attracting pheromones in amphibians is regulated by prolactin (PRL) and the responsiveness of the female vomeronasal epithelium is enhanced by PRL and estrogen [42]. Estrogen, androgen, progesterone, and prolactin regulate specific aspects of nest-building and promote the onset of maternal responsiveness. However, the maintenance of this behavior relies on stimuli from the litter. By preventing mother/young contact at parturition or during early lactation, maternal responsiveness changes or became abolished. In the rabbit, brain areas controlling the expression of nest- building and nursing were under investigation by implanting estradiol and locating the distribution of estrogen and prolactin receptors in the forebrain [43]. That work showed that ESR1 present in the preoptic region may mediate the stimulation of nest-building by estradiol and that prolactin-binding sites, located mainly in periventricular structures, are more abundant in late pregnancy and early lactation [34]. Moreover, in the rabbit, scent emission is responsible for nipple search and is depressed following ovariectomy but further stimulated by estradiol administration [43].

These aspects legitimate our findings of ESR1 and ESR2 receptors in the IGS, as these estrogen receptors might be involved in the process of signaling pathways that may regulate conspecific chemical messages attributed to IGS as binding steroid hormones to specific receptors. Therefore, generating changes in the rates of nucleic acids and proteins synthesis might result in chemosensation. We should underline that there are subtypes of cytoplasmic estrogen receptors, as they perform their biological actions in the cytosol at a fast rate to be compatible with transcriptional mechanisms [44–52]. The canonical model for ER-mediated regulation of gene expression involves the direct binding of dimeric ER to DNA sequences known as estrogen response elements, which are specific and inverted palindromic sequences [53]. ER can associate indirectly with promoters through protein–protein interactions with other DNA-binding transcription factors and interaction of ERs with E2 leads to transcriptional activation of the associated genes via the recruitment of coactivators and components of the basal transcriptional machinery [54,55]. The "genomic action" of steroid hormones occurs after a time lag of at least 2 h after E2 stimulation and explains some of the hormone functions in physiological and pathological situations. However, some effects are too rapid to account for genomic action [56]. In fact, cytosol receptors can be turned on and off, suggesting different roles in physiological functions and pathogenesis [51,52]. These findings point out the evidence for an important role of the non-nuclear estrogen receptor in a fast non-transcriptional response of cells to estrogen. The way they contribute to the signaling of the mammary gland has been ascertained and already demonstrated in the interdigital sinus of the ewe as well [57].

In the present study, we demonstrated that the IGS appears as an invaginated skin fold presenting sebaceous and acinar glandular areas sustained by collagen sheath fibers. Since the apocrine unit is alveolar, but not tubuloalveolar, the spread of the wax is probably a result of the compression of the medial face of the hind limb against the upper area of the mammary gland where the IGS is located. In addition, myoepithelial cells contribute to the expelling of the apocrine secretion toward the lumen of the acinar units. Acinar cells present lipid granules, which we should link with the greasy aspect of the IGS production. Mucin content is present and seen in the matrix. It might eventually contribute to the capacity to resist proteolysis to maintain the "scent" production characteristics of this gland for a longer period.

According to our data, we can pinpoint the same acini cells where exocytosis of the glandular content release occurs via a mechanism of non-protrusion [58–60]. In addition, a process of gradual accumulation of secretory products that form balloon-like swellings protruding into the lumen was also depicted [61]. In accordance with the presence of a mixed population of epithelial cells in the acini, a dual pattern secretion can result in the same alveolar unit, as demonstrated in the sweat glands of Karagouniko sheep [62], although the signaling pathway that triggers vesicular sorting in still under discussion [60–63]. The parenchyma is rich in glycogen, whose amount varies in different areas of the gland.

PRL simultaneously embraces a high diversity of physiological actions beyond mammary gland development or milk production. It is synthesized in multiple tissues and its biological actions are not limited solely to reproduction, as it has been shown to control a variety of behaviors [41,42]. This hormone appears in the background of pathological skin conditions and skin derivatives such as disruption of time regulation of hair growth cycles in mice [64,65], alopecia, and psoriasis [40].

Indeed, this fact raises questions about the primitive function of PRL. This would explain its maintenance during pre-mammalian evolution, as some of those functions attributed to PRL are associated with the post-mating phase of reproductive cycles in different reproductive strategies. These fit with seasonal gonadal suppression or behavioral changes, such as inhibition of aggression [66,67]. Seeing as the integument and its appendages (feathers, hair, glands, and the mammary gland itself) are the focal point of numerous PRL actions, even in non-mammalian vertebrates [40,68–71], and as it has already been revealed that PRL controls lobule-alveolar proliferation and differentiation of secretory epithelium [72,73], it is important to find its receptor's presence in the IGS. Moreover, since PRL shares the signal transduction pathway used by a variety of cytokines and growth factors [74,75], this hormone might therefore be important in signaling the onset of lactation for lambs. Epithelial growth is known to result from combined effects of P_4 and PRL, both triggering a juxtacrine RANKL signal, which induces alveolar growth [76–80]. The simultaneous presence of PGR and PRLR in the inguinal sinus gland can also suggest the role of PG and PRL in IGS function in sheep. Synthesis of female-attracting pheromones in some species is regulated by prolactin (PRL) and responsiveness of the female vomeronasal epithelium is enhanced by PRL and estrogen [42]. Overall, the evaluation of the chosen receptors under research in our study seems to be of interest since these receptors are needed for the related hormone action, which may work together in the putative mechanisms involved in IGS function.

Data found in the present work agree with previous reports and contribute to the body of knowledge. At this point, it is not possible to state that there is a direct cause/effect action between plasma E_2, P_4 and prolactin on IGS function. Nevertheless, the receptors of these hormones found in IGS present variations in the different phases of the estrous cycle and pregnancy. Thus, we do suggest the involvement of those hormones in IGS function in ewes. Their presence constitutes a hallmark and an important point to start to manipulate responses [81]. Further studies investigating these hormones' regulation will be mandatory since estrogen and prolactin up- or downregulation behave differently among species when considering the generation of neural progenitors in the brain, as mentioned earlier. By widening this field of research, which will contribute to an understanding and further development of odor-specific products, we may be able to promote newborn lambs' survival.

4. Materials and Methods

For the present work, IGS ($n = 92$) were collected post mortem from adult merino ewes for different evaluations, as described below. As the reproductive history of the ewes was unknown, their estrous cycle phases were determined based on ovarian structures and plasma progesterone (P_4) concentrations. Therefore, when a pre-ovulatory follicle was present in the ovary, in the absence of a corpus luteum (CL), and plasma P_4 concentration was below 1 ng/mL, the ewes were considered as being in the follicular phase. Nevertheless, the presence of a CL in the ovary, and plasma P_4 concentration above 1 ng/mL, indicated the ewe was in the luteal phase. Right after collection, IGS samples were immersed in (i) RNAlater (AM7020, Ambion, Applied Biosystems, Carlsbad, CA, USA) for mRNA transcription quantification; (ii) 4% buffered formaldehyde, for histology, immunohistochemistry, and confocal microscopy; (iii) Karnovsky's solution for ultrastructure studies; or (iv) a sterile tube with RPMI 1640 (Gibco-Brl, Gaithersburg, MD, USA) for flow cytometry studies. In addition, the contents of IGS were collected into sterile tubes for biochemical quantifications. Blood samples were drawn into heparinized tubes at the time of exsanguination (MonovettesVR-Sarstedt, Numbrecht, Germany) for further estrous cycle confirmation. Furthermore, from 28 sheep (*n* = 8 from follicular phase, *n* = 10 pregnant, and *n* = 10 post-partum period (1–3 days after delivery)), the content of IGS obtained post mortem as a byproduct from animals used for other research purposes was collected to evaluate putative variations in secretions.

Competent veterinary authorities monitored the experiments. The ethical committee of the Faculty of Veterinary Medicine (Lisbon, Portugal) approved these. Several authors are holders of Federation of European Laboratory Animal Science Associations (FELASA) grade C certificate, which permits designing and conducting laboratory animal experimentation in the European Union.

4.1. Histology Evaluation

Ovine IGS samples (follicular phase, *n* = 5; luteal phase, *n* = 5) were cut into small pieces, fixed in buffered formaldehyde for 24 h, and processed for light microscopic study. Tissue serial sections were cut (5 mM thick—Microtome Leica SM2000R, Berlin, Germany) and stained with Weigert Van Gieson for collagen detection, Periodic Acid Schiff to assess glycogen content, Alcian Blue for mucin detection, and Black Sudan for detection of lipid production [82].

4.2. Scanning Electron Microscopy

Scanning electron microscopy (SEM) evaluation of intact IGS tissue (follicular phase, *n* = 5; luteal phase, *n* = 5) was performed. Immersion of the intact IGS tissue in Karnovsky's solution (Sigma-Aldrich, Lisboa, Portugal), rinsed in cacodylate buffer, and post-fixed in a 2% osmium tetroxide solution for 1h. Rinsed once again with cacodylate buffer and subsequently dehydrated in a graded ethanol series. Samples dried using the critical point drying method and coated with gold palladium. IS were mounted on stubs, observed in a scanning electronic microscope (JEOL 5200-LV, Tokyo, Japan), and photographed.

4.3. Flow Cytometry Analysis

Flow cytometry analysis of IGS was carried out to quantify the expression of ESR1, ESR2, PGR and PRLR proteins. Ewe IGS (estrus *n* = 14; diestrus *n* = 14) were removed with a surgical blade and collected in a sterile tube with 1 mL of RPMI 1640 (Gibco-Brl). After disaggregation of the tissue with a surgical blade, samples of whole IGS were centrifuged at 190 *g* for 10 min. Then, they were resuspended in phosphate-buffered saline solution (PBS-P3813 Sigma). Fixation and permeabilization of cell suspensions with FIX & PERM VR-Fixation and Permeabilization Kit (Invitrogen Laboratories, Life Technologies, Waltham, MA, USA) were performed for 15 min longer in the dark at room temperature. After a final washing step, the pellet was suspended once again in 500 mL of BD FACS Flow for no longer than 15 min in the dark, at room temperature. After a new washing and centrifugation step, RPE-conjugated secondary antibody (10 mL) was added and cells were incubated 15 min longer in the dark at room temperature. A final washing step was necessary, and the resulting pellet was suspended once again in 500 μL of BD FACS Flow (BD Biosciences, San Jose, CA, USA). Cell acquisition was performed on a BD FACS Calibur flow cytometer (BD Biosciences) and data were analyzed using Paint-A-Gate Pro and Cell-Quest Pro software (BD Biosciences). In each experiment, incubation of cells was done according to the above protocol but with the secondary antibody only. This control tube was performed in order to assess the level of unspecific fluorescence signal of the secondary antibody. Selection of primary antibodies' dilutions was as follows: 1. Mouse monoclonal anti progesterone receptor (77201704 AbD Serotec, Kidlington, UK), diluted at 1:10 in PBS; 2. Mouse anti-human monoclonal antibody ESR1 (ref. 41700, Invitrogen, Dorset, UK), diluted at 1:10 in PBS; 3. Mouse anti-human polyclonal ESR2 (MCA2279S, AbD Serotec), diluted at 1:10 in PBS, 4. Mouse monoclonal (U5) to prolactin receptor (abcam 2772, Cambridge, UK), diluted at 1:10 in PBS. The secondary antibody used was R-phycoerythrin F(ab')2 frag. of goat anti-mouse (F2653 Sigma).

4.4. Laser-Scanning Confocal Microscopy

The locations of ESR1, ESR2, PGR and PRLR protein in IGS were assessed using laser-scanning confocal microscopy (Leica TCS SP2, Leica Microsystems; Berlin, Germany). The same antibodies used in flow cytometry evaluation were employed in this study (n = 6). Incubation of antibodies was performed overnight with the following dilutions: (1) Mouse monoclonal anti-progesterone receptor diluted at 1:50; (2) estrogen receptor (ESR1) diluted at 1:50 and rabbit anti-human estrogen receptor (ESR2) diluted at 1:50. Again, the FIX & PERM VR Fixation and Permeabilization Kit (Invitrogen Laboratories, Life Technologies, Waltham, MA, USA) was used. Briefly, reagent A was added for no longer than 15 min in the dark at room temperature before the addition of the primary antibody, which was incubated for an hour. For another 15 min, solution B was added followed by the addition of the second antibody.

To-Pro-3 iodide 1 mM solution (Invitrogen Molecular Probes, Eugene, OR, USA) was used for nuclear counterstaining. Negative controls were performed by replacing the primary antibody with either rabbit polyclonal IgG (ab27478, Abcam), for antibodies developed in a rabbit, or mouse IgG (550878, BD Biosciences, San Jose, CA, USA) for antibodies developed in a mouse, with the same dilution and incubation times as the primary antibody, followed by To-Pro-3 iodide for nuclear counterstaining. Selected sections were photographed with confocal laser microscopy, Leica TCS SP2.

4.5. Genomic Analysis

Conventional polymerase chain reaction (PCR) was used to assess mRNA gene expression of *PGR*, *ESR1*, *ESR1* and *PRLR* in sheep's inguinal glands (follicular phase, n = 5; luteal phase, n = 5) Specific primers for *PGR*, *ESR1*, *ESR2* and *PRLR* were designed (Table 1), as follows:

Table 1. Specific primers were designed—sequences used for quantitative real-time PCR (bp = base pair).

Gene (Acession Number)	Sequence 5'–3' Amplicon (Base Pairs)
ESR1 (XM_015097472.1)	Forward: CCATGGAATCTGCCAAGGAG (167 bp) Reverse: ATCAATTGTGCACTGGTTGGT
ESR2 (NM_001009737.1)	Forward: TGGAGTCTGGTCATGTGAAGGA (150 bp) Reverse: TCATAGCACTTCCGCAGTCG
PGR (XM_015100878.1)	Forward: CAGCCAGAGCCCACAGTACA (176 bp) Reverse: TGCAATCGTTTCTTCCAGCA
PRLR (NM_001009204.1)	Forward: GTCTCCACCCACCCTGACTG (320 bp) Reverse: AAGCCACTGCCCAGACCATA

RNA was extracted from IGS tissue (Qiagen's Kit for Total RNA Extraction and Purification; ref. 28704, Qiagen, Hilden, Germany) and DNA digested (RNase-free DNase Set; ref. 50979254, Qiagen), according to the manufacturer's instructions. By the use of a spectrophotometer, RNA concentration was determined (260 and 280 nm) and RNA quality assessed by visualization of 28S and 18S rRNA bands, after electrophoresis through a 1.5% gel agarose and ethidium bromide staining. Reverse transcription was carried out using Reverse Transcriptase Superscript III enzyme (ref. 18080093, Invitrogene, Gibco, Carlsbad, CA, USA), from 1 mg total RNA in a 20 mL reaction volume, using oligo (dT) primer (27–7858-01, GE Healthcare, Buckinghamshire, UK). Different Internet-based interfaces, such as Primer-3 (Untergasser et al. 2012) and Primer Premier Software (Premier Biosoft Int., Palo Alto, CA, USA) were used for specific primers for target genes. Several conventional PCR reactions were carried out using a default thermocycler (Applied Biosystems, Foster City, CA, USA) as follows:

2 min at 94 °C for denaturation; 35 cycles of 15 s at 94 °C for enzyme activation, 45 s at 57–60 °C for annealing (depending on the gene-*PGR*-57.8 °C; *ESR2*-58.5 °C and *ESR1*-60 °C) and 45 s at 68 °C for extension; and 5 min at 68 °C for finalization. The design of all primers for two different exons followed specific guidelines in order to avoid genomic DNA amplification (www.qiagen.com). All reactions were carried out in duplicate in 0.2-mL PCR tubes (PCR-0.2-C, Axygen 321-02-051, Corning, CA, USA) in a 25 μL reaction volume: 8.5 μL water; 1 μL forward primer (10 pmol/μL); 1 μL reverse primer (10 pmol/μL); 12.5 μL using FideliTaq DNA polymerase master mix (71180, USB, Cleveland, OH, USA), and 2 μL of cDNA. All Agarose (2%) (BIO-41025, Bioline, Luckenwalde, Germany) electrophoresis gel and ethidium bromide (17896, Thermo, Waltham, MA, USA) staining showed a specific and single product. For dissociation curve analysis, cDNA was amplified with real-time PCR, as describe before. All primers were validated and used at 80 nM.

4.6. Progesterone Analysis

Evaluation of progesterone concentration was in plasma using a solid-phase radioimmunoassay (Coat-a-Count Progesterone, Diagnostic Product Corp., Los Angeles, CA, USA). Intra-assay coefficient was 6.4% for the level of 3.2 nmol/L (1 ng/mL) and 4.2% for the level of 15.9 nmol/L (5 ng/mL).

4.7. Chemical Studies

The total content of the inguinal sinus was collected from non-pregnant (NP), pregnant (P), and post-parturient (PP) ewes ($n = 28$) and extracted with ethyl acetate using an ultrasonic bath (1 mg/mL; 5 min). The total content of the inguinal sinus was collected from non-pregnant (NP), pregnant (P) and post parturient (PP) ewes ($n = 28$) and extracted with ethyl acetate using an ultrasonic bath (1 mg/mL; 5min). After, centrifugation (2000 rpm/15min), the supernatant was evaporated to dryness under vacuum at a temperature below 40 °C. The obtained residue was dissolved in ethyl acetate (1 mL) and then analyzed by thin layer chromatography (TLC) using different chromatographic and derivatization systems, including silica gel F254 Merck as stationary phase, and toluene: ethyl acetate (9:1 v/v) as mobile phase and Liebermann and Dragendorff as spraying reagents [83]. After derivatization, visible and UV light at 365 nm used for data acquisition. Samples of the surrounding skin prepared in the same way used as negative control.

4.8. Statistical Analysis

Flow cytometry data of ERSR1, ESR2, PRLR, and PGR proteins in IGS from ewes in the follicular, luteal, pregnancy, and post-partum phases, were subjected to a one-way analysis of variance (ANOVA). Significance was defined as values of $p < 0.05$. For statistically different results, the means were further analyzed by post hoc comparison test, such as LSD (least significant differences) and Scheffé tests (probabilities for post hoc tests).

5. Conclusions

Our data, to the best of our knowledge, have not been reported elsewhere. These findings concern the expression of steroid hormones and prolactin receptors in sheep IGS and point out that a particular triterpenoid-rich secretion and organic nitrogen compounds that totally differ from the surrounding skin are present in IGS. These secretions show modifications, particularly during pregnancy and post-partum. Altogether, IGS secretion in the ewe might contribute to the putative involvement of this gland in the signaling cues that direct the lamb to the mammary gland.

Since impairment of bond formation between ewes and their offspring is a common cause of offspring death, this research will potentially benefit farmers by contributing to the increased survival of lambs due to understanding and using these odor products. As stated, IGS fluctuations in the expression of ERS1, ERS2, PGR, and PRLR are present and might be linked to putative chemosignals. Further research by our team will give rise to new data regarding a direct hormonal cause/effect on IGS, in agreement with other researchers who claim to have achieved methods of manipulating reproduction in domestic ungulates by using specific odors.

Supplementary Materials: Supplementary materials can be found at www.mdpi.com/1422-0067/18/7/1516/s1.

Acknowledgments: Funding for this work was provided by the Portuguese Foundation for Science and Technology (FCT) with co-participation of European Union Fund (FEDER) through the research project UID/CVT/00276/2013. The authors present special thanks to Sandra Carvalho and Rosário Luís (Histology Dept., FMV) and Gloria Nunes (Immunology Dept., FCM) for technical assistance.

Author Contributions: Graça Alexandre-Pires designed the study and performed the histology, laser scanning confocal, and scanning electronic studies. Catarina Martins performed the cytometry analysis. António M. Galvão designed the primers and performed the genomic analysis. Margarida Miranda and Olga Silva were responsible for the chemical studies. Dário Ligeiro was involved in laser scanning confocal fluorescence studies. Telmo Nunes was responsible for the preparation of SEM material. Graça Ferreira-Dias co-designed the study, performed the statistical analysis, and revised the manuscript. All authors approved the manuscript.

Conflicts of Interest: The authors declare no conflict of interest.

References

1. Brockman, J.H.; Snowdon, C.; Roper, T.; Naguib, M.; Wynne-Eduards, K. *Advances in the Study of Behavior*; Elsevier Inc.: San Diego, CA, USA, 2009; Volume 36, ISBN 978-0-12-374474-6.

2. Fuchs, S. Optimality of parental investment: The influence of nursing on reproductive success of mother and female young house mice. *Behav. Ecol. Sociobiol.* **1982**, *10*, 39–51. [CrossRef]

3. Voloschin, L.M.; Tramezzani, J.H. Milk Ejection Reflex Linked to Slow Wave Sleep in Nursing Rats. *Endocrinology* **1979**, *105*, 1202–1207. [CrossRef] [PubMed]

4. Korhonen, H.; Marnila, P.; Gill, H.S. Milk immunoglobulins and complement factors. *Br. J. Nutr.* **2000**, *84*, 75–80. [CrossRef]

5. Schaal, B. *Pheromones for Newborns*; Mucignat-Caretta, C., Raton, B., Eds.; Taylor & Francis: Abingdon, UK, 2014; Chapter 17; ISBN-13: 978-1-4665-5341.

6. Nowak, R. Suckling, Milk, and the Development of Preferences toward Maternal Cues by Neonates: From Early Learning to Filial Attachment? *Adv. Study Behav.* **2006**, *36*, 1–58. [CrossRef]

7. Serra, J.; Ferreira, G.; Mirabito, L.; Lévy, F.; Nowak, R. Post-oral and Perioral Stimulations during Nursing Enhance Appetitive Olfactory Memory in Neonatal rabbits. *Chem. Senses* **2009**, *34*, 405–413. [CrossRef] [PubMed]

8. Litwack, G. *Vitamines and Hormones in: Pheromones*; Elsevier: San Diego, CA, USA, 2010; ISBN 978-0-12-381516-3.

9. Mykytowycz, R. Skin Glands as Organs of Communication in Mammal. *J. Investig. Dermatol.* **1974**, *62*, 124–131. [CrossRef] [PubMed]

10. Hudson, R.; González-Mariscal, G.; Beyer, C. Chin marking behavior, sexual receptivity, and pheromone emission in steroid-treated, ovariectomized rabbits. *Horm. Behav.* **1990**, *24*, 1–13. [CrossRef]

11. Gonzalez-Mariscal, G.; Chirino, R.; Hudson, R. Prolactin Stimulates Emission of Nipple Pheromone in Ovariectomized New Zealand White Rabbits. *Biol. Reprod.* **1994**, *50*, 373–376. [CrossRef] [PubMed]

12. Brennan, P.; Keverne, E.B. Biological complexity and adaptability of simple mammalian olfactory memory systems. *Neurosci. Biobehav. Rev.* **2015**, *50*, 29–40. [CrossRef] [PubMed]

13. Oboti, L.; Ibarra-Soria, X.; Pérez-Gómez, A.; Schmid, A.; Pyrski, M.; Paschek, N.; Kircher, S.; Logan, D.W.; Leinders-Zufall, T.; Zufall, F.; et al. Pregnancy and estrogen enhance neural progenitor-cell proliferation in the vomeronasal sensory epithelium. *BMC Biol.* **2015**, *13*, 104. [CrossRef] [PubMed]

14. Rekwot, P.I.; Ogwu, D.; Oyedipe, E.O.; Sekoni, V.O. The role of pheromones and biostimulation in animal reproduction. *Anim. Reprod. Sci.* **2001**, *65*, 157–170. [CrossRef]

15. Parillo, F.; Diverio, S. Glycocomposition of the apocrine interdigital gland secretions in the fallow deer (Dama dama). *Res. Vet. Sci.* **2009**, *86*, 194–199. [CrossRef] [PubMed]

16. Nowak, R. Neonatal survival: Contributions from behavioural studies in sheep. *Appl. Anim. Behav. Sci.* **1996**, *49*, 61–72. [CrossRef]

17. Nowak, R.; Porter, R.H.; Lévy, F.; Orgeur, P.; Schaal, B. Role of mother–young interactions in the survival of offspring in domestic mammals. *Rev. Reprod.* **2000**, *5*, 153–163. [CrossRef] [PubMed]

18. Vince, M.A.; Ward, T.M. The responsiveness of newly born Clun forest lambs to odour sources in the ewe. *Behaviour* **1984**, *89*, 117–127. [CrossRef]

19. Schaal, B.; Orgeur, P.; Arnould, C. Olfactory preferences in newborn lambs: Possible influence of prenatal experience. *Behaviour* **1995**, *132*, 5351–5365. [CrossRef]

20. Vince, M.A.; Lynch, J.J.; Mottershead, B.E.; Green, G.C. Interactions between normal ewes and newly born lambs deprived of visual, olfactory and tactile sensory information. *Appl. Anim. Behav. Sci.* **1987**, *19*, 119–136. [CrossRef]

21. Vince, M.A.; Billing, A.E. Infancy in the sheep: The part played by sensory stimulation in bounding between ewe and lamb. In *Advances in Infancy Research*; Lipsitt, L.P., Rovee-Collier, C., Eds.; Ablex Noorwood: New York, NY, USA, 1986; Volume 4, pp. 1–37.

22. Logan, D.W.; Brunet, L.J.; Webb, W.R.; Cutforth, T.; Ngai, J.; Stowers, L. Learned Recognition of Maternal Signature Odors Mediates the First Suckling Episode in Mice. *Curr. Biol.* **2012**, *22*, 1998–2007. [CrossRef] [PubMed]

23. Schaal, B.; Coureaud, G.; Doucet, S.; Delaunay, E.l.; Allam, M.; Moncomble, A.-S.; Montigny, D.; Patris, B.; Holley, A. Mammary olfactory signalisation in females and odour pocessing in neonates: Ways evolved by rabbits and humans. *Behav. Brain Res.* **2009**, *200*, 346–358. [CrossRef] [PubMed]

24. Morrow-Tesch, J.; McGlone, J.J. Sources of maternal odors and the development of odor preferences in baby pigs. *J. Anim. Sci.* **1990**, *68*, 3563–3571. [CrossRef] [PubMed]

25. Schaal, B. Mammary Odor Cues and Pheromones: Mammalian Infant-Directed Communication about Maternal State, Mammae, and Milk. In *Vitamins and Hormones*; Litwack, G., Ed.; Elsevier Inc.: San Diego, CA, USA, 2009; Chapter 4; Volume 83, pp. 83–136.

26. Vaglio, S.; Minicozzi, P.; Bonometti, E.; Mello, G.; Chiarelli, B. Volatile signals during pregnancy: A possible chemical basis for mother-infant recognition. *J. Chem. Ecol.* **2009**, *53*, 131–139. [CrossRef] [PubMed]

27. Doucet, S.; Sooussignan, R.; Sagot, P.; Schaal, B. The secretion of areolar (Montgomery's) glands from lactating women elicits selective, unconditional responses in neonates. *PLoS ONE* **2009**, *4*, e7579. [CrossRef] [PubMed]

28. Legendre, A.; Faure, P.; Tiesset, H.; Potin, C.; Jakob, I.; Sicard, G.; Schaal, B.; Artur, Y.; Coureaud, G.; Heydel, J.M. When the Nose Must Remain Responsive: Glutathione Conjugation of the Mammary Pheromone in the Newborn Rabbit. *Chem. Sens.* **2014**, *39*, 425–437. [CrossRef] [PubMed]

29. Hudson, R.; Rojas, C.; Carolina Rojas, L.; Martínez-Gómez, M.; Distel, H. Rabbit Nipple-Search Pheromone versus Rabbit Mammary Pheromone Revisited. In *Chemical Signals in Vertebrates*; Muller-Schwarze, D., Ed.; Springer Link: London, UK, 2008; pp. 315–324.

30. Coureaud, G.; Charra, R.; Datiche, F.; Sinding, C.; Thomas-Danguin, T.; Languille, S.; Hars, B.; Schaal, B. A pheromone to behave, a pheromone to learn: The rabbit mammary pheromone. *J. Comp. Physiol. A Neuroethol. Sens. Neural. Behav. Physiol.* **2010**, *196*, 779–790. [CrossRef] [PubMed]

31. Brus, M.; Meurisse, M.; Franceschini, I.; Keller, M.; Lévy, F. Evidence for cell proliferation in the sheep brain and its down-regulation by parturition and interactions with the young. *Horm. Behav.* **2010**, *58*, 737–746. [CrossRef] [PubMed]

32. Lindsay, D. *Breeding the Flock: Modern Research and Reproduction in Sheep*; Inkata Pr.: Berlin, Germany, 1988; ISBN 0909605459.

33. Vince, M.A. Newborn lambs and their dams: The interaction that leads to suckling. *Adv. Study Behav.* **1993**, *22*, 239–268.

34. Sell, C. *A Fragrant Introduction to Terpenoid Chemistry*; The Royal Society of Chemistry: Cambridge, UK, 2003; ISBN 0-85404-681-X.

35. Poindron, P.; Lévy, F. Physiological, sensory and experiential determinants of maternal behaviour in sheep. In *Mammalian Parenting: Biochemical, Neurobiological and Behavioral Determinants*; Krasnegor, N.A., Bridges, R.S., Eds.; Oxford University Press: New York, NY, USA, 1990; pp. 133–156, 485–488. ISBN1 0195056000. ISBN2 9780195056006.

36. Poindron, P.; Nowak, R.; Lévy, F.; Porter, R.H.; Schaal, B. Development of exclusive mother-young bonding in sheep and goats. In *Oxford Reviews of Reproductive Biology*; Michigan, S.R., Ed.; Oxford University Press: Oxford, UK, 1993; Volume 15, pp. 311–364.

37. Foitzik, K.; Krause, K.; Conrad, F.; Nakamura, M.; Funk, W.; Paus, R. Human scalp hair follicles are both a target and a source of prolactin, which serves as an autocrine and/or paracrine promoter of apoptosis-driven hair follicle regression. *Am. J. Pathol.* **2006**, *168*, 748–756. [CrossRef] [PubMed]

38. Egli, M.; Leeners, B.; Kruge, TH. Prolactin secretion patterns: Basic mechanisms and clinical implications for reproduction. *Reproduction* **2010**, *140*, 643–654. [CrossRef] [PubMed]

39. Foitzik, K.; Krause, K.; Nixon, A.J.; Ford, C.A.; Ohnemus, U.; Pearson, A.J.; Paus, R. Prolactin and its receptor are expressed in murine hair follicle epithelium, show hair cycle-dependent expression, and induce catagen. *Am. J. Pathol.* **2003**, *162*, 1611–1621. [CrossRef]

40. Foitzik, K.; Langan, E.A.; Paus, R. Prolactin and the skin: A dermatological perspective on an ancient pleiotropic peptide hormone. *J. Investig. Dermatol.* **2009**, *129*, 1071–1087. [CrossRef] [PubMed]

41. Freeman, M.E.; Kanyicska, B.; Lerant, A.; Nagy, G. Prolactin: Structure, function, and regulation of secretion. *Physiol. Rev.* **2000**, *80*, 1523–1631. [PubMed]

42. Kikuyama, S.; Yamamoto, K.; Iwata, T.; Toyoda, F. Peptide and protein pheromones in amphibians. *Comp. Biochem. Physiol. B Biochem. Mol. Biol.* **2002**, *132*, 69–74. [CrossRef]

43. González-Mariscal, G. Neuroendocrinology of Maternal Behaviour in the Rabbit. *Horm. Behav.* **2001**, *40*, 125–132. [CrossRef] [PubMed]

44. Shaaban, A.M.; Green, A.R.; Karthik, S.; Alizadeh, Y.; Hughes, T.A.; Harkins, L.; Ellis, I.O.; Robertson, J.F.; Paish, E.C.; Saunders, P.T.; et al. Nuclear and cytoplasmic expression of ERbeta1, ERbeta2, and ERbeta5 identifies distinct prognostic outcome for breast cancer patients. *Clin. Cancer. Res.* **2008**, *14*, 5228–5235. [CrossRef] [PubMed]

45. Revankar, C.M.; Cimino, D.F.; Sklar, L.A.; Arterburn, J.B.; Prossnitz, E.R. A Transmembrane Intracellular Estrogen Receptor Mediates Rapid Cell Signaling. *Science* **2005**, *307*, 1625–1630. [CrossRef] [PubMed]

46. Robert, X.-D.; Song, R.X.-D.; Zhang, Z.; Santen, R.J. *Estrogen Rapid Action via Protein Complex Formation Involving ERα and Src*; Cell Press: Cambridge, MA, USA, 2005; Volume 16, pp. 347–353.

47. Simoncini, T.; Rabkin, E.; Liao, J.K. Membrane estrogen receptor interaction with ph osphatidylinositol 3-kinase in endothelial cells. *Arterioscler. Thromb. Vasc. Biol.* **2003**, *23*, 198–203. [CrossRef] [PubMed]

48. Hammes, S.R. The further redefining of steroid-mediated signalling. *Proc. Nat. Acad. Sci. USA* **2003**, *100*, 2168–2170. [CrossRef] [PubMed]

49. Cato, A.; Nestl, A.; Mink, S. Rapid actions of steroid receptors in cellular signaling pathways. *Science's STKE* **2002**. [CrossRef] [PubMed]

50. Levin, E.R. Cell localization, physiology and nongenomic actions of estrogen receptors. *J. Appl. Physiol.* **2001**, *91*, 1860–1867. [PubMed]

51. Wittliff, J.L.; Hilf, R.; Brooks, W.F., Jr.; Savlov, E.D.; Hall, T.; Robert, A.O. Specific estrogen-binding capacity of the cytoplasmic receptor in normal and neoplastic breast tissues of humans. *J. Cancer Res.* **1972**, *11*, 1–73.

52. Welsh, A.W.; Lannin, D.R.; Gregory, S.; Young, G.S.; Mark, E.; Sherman, M.E.; Jonine, D.; Figueroa, J.D.; Lynn, N.; Henry, N.L.; et al. Cytoplasmic Estrogen Receptor in breast cancer. *Clin. Cancer Res.* **2012**, *18*, 118–126. [CrossRef] [PubMed]

53. Klinge, C.M. Estrogen receptor interaction with estrogen response elements. *Nucleic Acids Res.* **2001**, *29*, 2905–2919. [CrossRef] [PubMed]

54. McKenna, N.J.; O'Malley, B.W. Combinatorial control of gene expression by nuclear receptors and coregulators. *Cell* **2002**, *108*, 465–474. [CrossRef]

55. Stossi, F.; Likhite, V.S.; Katzenellenbogen, J.A.; Katzenellenbogen, B.S. Estrogen-occupied estrogen receptor represses cyclin G2 gene expression and recruits a repressor complex at the cyclin G2 promoter. *J. Biol. Chem.* **2006**, *281*, 16272–16278. [CrossRef] [PubMed]

56. Farach-Carson, M.C.; Davis, P.J. Steroid hormone interactions with target cells: Cross talk between membrane and nuclear pathways. *J. Pharmacol. Exper. Therap.* **2003**, *30*, 839–845. [CrossRef] [PubMed]

57. Alexandre-Pires, G.; Martins, C.; Miguel Galvão, A.; Correia, M.; Ramilo, D.; Quaresma, M.; Ligeiro, D.; Nunes, T.; Caldeira, R.M.; Ferreira-Dias, G. Morphological Aspects and Expression of Estrogen and Progesterone Receptors in the Interdigital Sinus in Cyclic Ewes. *Microsc. Res. Tech.* **2014**, *77*, 313–325. [CrossRef] [PubMed]

58. Zaviacic, M.; Zajickova, M.; Blazekova, J.; Donarov, L.; Stvrtina, S.; Mikuleck, M.; Zaviacic, T.; Holom, A.N.K.; Breza, J. Size, macroanatomy and histology of the normal prostate in the adult human female: A minireview. *J. Histotechnol.* **2000**, *21*, 61–69. [CrossRef]

59. Gesase, A.P.; Satoh, Y.; Ono, K. Secretagogue-induced apocrine secretion in the rat Harderian gland of the rat. *Cell Tissue Res.* **1996**, *285*, 501–507. [CrossRef] [PubMed]

60. Gesase, A.P.; Satoh, Y. Apocrine secretory mechanism: Recent findings and unresolved problems. *Histol. Histopathol.* **2003**, *18*, 597–608. [PubMed]

61. Jarret, A. *Physiology and Pathophysiology of the Skin: The Sweat Gland, Skin Permeation, Lymphatics, the Nails*; Academic Press: Salt Lake, UT, USA, 1978; ISBN: 0123806054, 9780123806055.

62. Pourlis, A.F. Functional morphological characteristics of the interdigital sinus in the sheep. *Folia Morphol.* **2010**, *69*, 107–111.

63. Cristofoletti, P.T.; Ribeiro, A.F.; Terra, W.R. Apocrine secretion of amylase and exocytosis of trypsin along the midgut of Tenebrio molitor larvae. *J. Insect. Physiol.* **2001**, *47*, 143–155. [CrossRef]

64. Craven, A.J.; Ormandy, C.J.; Robertson, F.G.; Wilkins, R.J.; Kelly, P.A.; Nixon, A.J.; Pearson, A.J.; Craven, T. Prolactin signaling influences the timing mechanism of the hair follicle: Analysis of hair growth cycles in prolactin receptor knockout mice. *Endocrinology* **2001**, *142*, 2533–2539. [CrossRef] [PubMed]

65. Craven, A.J.; Nixon, A.J.; Ashby, M.G.; Ormandy, C.J.; Blazek, K.; Wilkins, R.J.; Pearson, A.J. Prolactin delays hair regrowth in mice. *J. Endocrinol.* **2006**, *19*, 1415–1425. [CrossRef] [PubMed]

66. Goffin, V.; Bernichtein, S.; Touraine, P.; Kelly, P.A. Development and potential clinical uses of human prolactin receptor antagonists. *Endocr. Rev.* **2005**, *26*, 400–422. [CrossRef] [PubMed]

67. Horseman, N.D.; Gregerson, K.A. Prolactin actions. *Soc. Endocrinol.* **2014**, *52*, 95–106. [CrossRef] [PubMed]

68. Oftedal, O.T. The origin of lactation as a water source for parchment-shelled eggs. *J. Mammary Gland Biol. Neoplasia* **2002**, *7*, 253–266. [CrossRef] [PubMed]

69. Oftedal, O.T. The mammary gland and its origin during synapsid evolution. *J. Mammary Gland Biol. Neoplasia* **2002**, *7*, 225–252. [CrossRef] [PubMed]

70. Oftedal, O.T. The evolution of milk secretion and its ancient origins. *Animal* **2012**, *6*, 355–368. [CrossRef] [PubMed]

71. Nicoll, C.S. Prolactin: Ontogeny and evolution of prolactin's functions. *Fed. Proc.* **1980**, *39*, 2561–2566.

72. Srivastava, S.; Matsuda, M.; Hou, Z.; Bailey, J.P.; Kitazawa, R.; Herbst, M.P.; Horseman, N.D. Receptor activator of NF-κB ligand induction via Jak2 and Stat5a in mammary epithelial cells. *J. Biol. Chem.* **2003**, *278*, 46171–46178. [CrossRef] [PubMed]

73. Bole-Feysot, C.; Goffin, V.; Edery, M.; Binart, N.; Kelly, P.A. Prolactin (PRL) and its receptor: Actions, signal transduction pathways and phenotypes observed in PRL receptor knockout mice. *Endocr. Rev.* **1998**, *19*, 225–268. [CrossRef] [PubMed]

74. Horseman, N.D.; Yu-Lee, L.Y. Transcriptional regulation by the helix bundle peptide hormones: Growth hormone, prolactin, and hematopoietic cytokines. *Endocr. Rev.* **1994**, *15*, 627–649. [CrossRef] [PubMed]

75. Brooks, C.L. Molecular mechanisms of prolactin and its receptor. *Endocr. Rev.* **2012**, *33*, 504–525. [CrossRef] [PubMed]

76. Sidis, Y.; Horseman, N.D. Prolactin induces rapid p95/p70 tyrosine phosphorylation, and protein binding to GAS-like sites in the anx Icp35 and c-fos genes. *Endocrinology* **1994**, *134*, 1979–1985. [CrossRef] [PubMed]

77. Standke, G.J.R.; Meier, V.S.; Groner, B. Mammary gland factor activated by prolactin in mammary epithelial cells and acute-phase response factor activated by interleukin-6 in liver cells share DNA binding and transactivation potential. *Molecul. Endocrinol.* **1994**, *8*, 469–477. [CrossRef]

78. Schramek, D.A.; Sigl, V.; Kenner, L.; Pospisilik, J.A.; Lee, H.J.; Hanada, R.; Joshi, P.A.; Aliprantis Glimcher, L.; Pasparakis, M.; Khokha, R.; et al. Osteoclast differentiation factor RANKL controls development Leibbrandt of progestin-driven mammary cancer. *Nature* **2010**, *468*, 98–102. [CrossRef] [PubMed]

79. Cao, Y.; Bonizzi, G.; Seagroves, T.N.; Greten, F.R.; Johnson, R.; Schmidt, E.V.; Karin, M. IKKα provides an essential link between RANK signaling and cyclin D1 expression during mammary gland development. *Cell* **2001**, *107*, 763–775. [CrossRef]

80. Baxter, F.O.; Came, P.J.; Abell, K.; Kedjouar, B.; Huth, M.; Rajewsky, K.; Pasparakis, M.; Watson, C.J. IKKβ/2 induces TWEAK and apoptosis in mammary epithelial cells. *Development* **2006**, *133*, 3485–3494. [CrossRef] [PubMed]

81. Müller-Schwarze, D. The chemical ecology of ungulates. *Appl. Anim. Sci.* **1991**, *29*, 389–402. [CrossRef]

82. Luna, LG. *Histopathologic Methods and Color Atlas of Special Stains and Tissue Artifacts*, 1st ed.; Johnson Printers: Downers Grove, IL, USA, 1992.

83. Bruno, T.J.; Svoronos, P.D.N. *Handbook of Basic Tables for Chemical Analysis*, 2nd ed.; Taylor & Francis Group: New York, NY, USA, 2004; Chapter 3; pp. 177–212.

International Journal of
Molecular Sciences

MDPI

Article

Mixture Concentration-Response Modeling Reveals Antagonistic Effects of Estradiol and Genistein in Combination on Brain Aromatase Gene (*cyp19a1b*) in Zebrafish

Nathalie Hinfray [1,*], Cleo Tebby [2], Benjamin Piccini [1], Gaelle Bourgine [3], Sélim Aït-Aïssa [1], Jean-Marc Porcher [1], Farzad Pakdel [3] and François Brion [1]

[1]   Institut National de l'Environnement Industriel et des Risques (INERIS), Unité d'Ecotoxicologie in vitro et in vivo, 60550 Verneuil-en-Halatte, France; benjamin.piccini@ineris.fr (B.P.); selim.ait-aissa@ineris.fr (S.A.-A.); jean-marc.porcher@ineris.fr (J.-M.P.); francois.brion@ineris.fr (F.B.)
[2]   Institut National de l'Environnement Industriel et des Risques (INERIS), Unité de modélisation pour la Toxicologie et l'Ecotoxicologie, 60550 Verneuil-en-Halatte, France; cleo.tebby@ineris.fr
[3]   Institut de Recherche en Santé, Environnement et Travail (IRSET), Inserm U1085, Unité Transcription, Environnement et Cancer (TREC), Université de Rennes 1, 35000 Rennes, France; gaelle.bourgine@univ-rennes1.fr (G.B.); farzad.pakdel@univ-rennes1.fr (F.P.)
*   Correspondence: nathalie.hinfray@ineris.fr; Tel.: +33-344-556-969

Received: 15 March 2018; Accepted: 26 March 2018; Published: 1 April 2018

Abstract: Comprehension of compound interactions in mixtures is of increasing interest to scientists, especially from a perspective of mixture risk assessment. However, most of conducted studies have been dedicated to the effects on gonads, while only few of them were. interested in the effects on the central nervous system which is a known target for estrogenic compounds. In the present study, the effects of estradiol (E2), a natural estrogen, and genistein (GEN), a phyto-estrogen, on the brain ER-regulated *cyp19a1b* gene in radial glial cells were investigated alone and in mixtures. For that, zebrafish-specific in vitro and in vivo bioassays were used. In U251-MG transactivation assays, E2 and GEN produced antagonistic effects at low mixture concentrations. In the *cyp19a1b*-GFP transgenic zebrafish, this antagonism was observed at all ratios and all concentrations of mixtures, confirming the in vitro effects. In the present study, we confirm (i) that our in vitro and in vivo biological models are valuable complementary tools to assess the estrogenic potency of chemicals both alone and in mixtures; (ii) the usefulness of the ray design approach combined with the concentration-addition modeling to highlight interactions between mixture components.

Keywords: estradiol; genistein; mixture; aromatase B; transgenic zebrafish; U251-MG

1. Introduction

Compounds able to interact with the estrogen receptors (ER) have been extensively studied over the years due to the threat they represent to aquatic species and particularly to fish reproduction and development [1,2]. However, in the aquatic environment, fish are often exposed not only to estrogenic chemicals alone but rather to mixtures, highlighting the constant need to understand compound interactions in mixtures and to develop new assays and approaches to this end.

Several studies have evaluated the combined effects of estrogenic compounds on aquatic organisms [3–12], generally concluding on an additive effect in mixtures both in vivo and in vitro. However, most in vivo studies addressed the effects of mixtures of estrogenic compounds on gonads and other peripheral organs, while only few of them studied effects on the central nervous system despite increasing evidences that estrogenic compounds interfere in neuroendocrine regulations.

In fish, the *cyp19a1b* gene encodes the brain aromatase. In zebrafish, *cyp19a1b* is expressed in radial glial cells of the brain which are crucial neuronal progenitors [13,14]. Estrogenic compounds are known to highly up-regulate *cyp19a1b* gene expression by an ER-dependent mechanism [3,15]. By using zebrafish-specific in vitro and in vivo bioassays based on the *cyp19a1b* gene coupled to a complete modeling approach, we recently demonstrated additive effects of ethynilestradiol and levonorgestrel, a pro-estrogenic compound, on *cyp19a1b* gene in glial cells [12].

In this context, the present study aimed to investigate the effects of single and combined exposure to two estrogens with different estrogenic potencies, on the expression of the zebrafish *cyp19a1b* gene. For this purpose, two in vitro and in vivo bioassays based on the zebrafish *cyp19a1b* gene were used: (i) an ER-negative human glial cell culture (U251-MG) co-transfected with two different zebrafish ER subtypes (zfERα and zfERβ2) and a luciferase gene under the control of the zebrafish *cyp19a1b* promoter [16] and (ii) a transgenic zebrafish (*cyp19a1b*-GFP) line expressing GFP under the control of the zebrafish *cyp19a1b* promoter which is used to evaluate estrogenicity of chemicals at the embryo-larval stage [3,6]. Mixtures assessed were composed of a natural potent estrogen, i.e., 17β-estradiol (E2), and a weaker estrogen, i.e., genistein (GEN), a major phytoestrogen of the isoflavone class. GEN, a human ERβ selective activator, was shown to be estrogenic in diverse in vitro and in vivo assays (for review see [17]), including in the U251-MG and in transgenic *cyp19a1b*-GFP zebrafish bioassays [3,16]. In the end, this comparative in vitro and in vivo approach was used to (i) determine the potential interactions between E2 and GEN in mixtures and (ii) evaluate the complementarity of the in vitro and in vivo models used in these experiments. The concentration-addition (CA) model for mixtures was used as the reference no-interaction model. Deviations from the CA model were quantified in terms of antagonism or synergism by using Jonker et al. interaction models [18] and their statistical significance was tested. Thereby, the present study reports antagonistic effects of E2 and GEN in mixtures on the expression of the ER-regulated gene *cyp19a1b* in a glial cell context.

2. Results

2.1. In Vitro Effects of Single Test Compounds

The effects of E2 and GEN alone were assessed in U251-MG glial cells co-transfected with zfERs and the zf-*cyp19a1b* promoter-luciferase reporter. One concentration-response experiment was carried out for each compound and each ER with three replicate wells for each condition. Luciferase activity was induced by E2 treatment in a concentration-dependent manner (Supplementary Figure S1). The EC50s (median effective concentration) were estimated at 1.86×10^{-10} M for ERα and 1.18×10^{-10} M for ERβ2. EC50s for GEN were estimated at 5.05×10^{-8} M for ERα and 2.96×10^{-9} M for ERβ2.

2.2. In Vitro Effects of Binary Mixtures of E2 and GEN

To confirm EC50s previously obtained in single test compound experiments, EC50s for E2 and GEN were also estimated from mixture experiments. EC50s for E2 were 4.29×10^{-11} and 1.38×10^{-11} M for U251-MG glial cells transfected with ERα and ERβ2 respectively. For GEN, EC50s were 1.68×10^{-8} and 4.78×10^{-9} M for U251-MG glial cells transfected with ERα and ERβ2 respectively. The U251-MG cells were slightly more sensitive to estradiol in the mixture experiments (EC50s were 2 to ten-fold lower than in the single compound experiments), but the relative potency on ERα was relatively unchanged (1.5 factor) and the relative potency on ERβ2 changed over ten-fold, so that in the mixture experiment the relative potencies were the same on ERα and ERβ2.

In vitro, a concentration-dependent induction of luciferase activity was measured with the three different mixture ratios of E2 and GEN in U251-MG cells transfected both with ERα and ERβ2 (Figure 1). The mixture model that displayed the best fit to the in vitro data was the dose-level dependent interaction model (DL) with identical slopes for the single compounds [18]. This model showed a significant improvement of the goodness of fit as compared to the CA model (approximate F-tests: $p = 7.9 \times 10^{-4}$ for ERα and $p = 8.4 \times 10^{-3}$ for ERβ2). Interaction parameters of the DL model

($\alpha = 19.3$ and $\beta = 0.0248$ for ERα, $\alpha = 12.9$ and $\beta = 0.0158$ for ERβ2) indicated strong antagonism at low concentrations between E2 and GEN in mixtures for both estrogen receptors. This antagonism was also highlighted by the deviation of the EC50s isoboles observed in the two experiments (Figure 2).

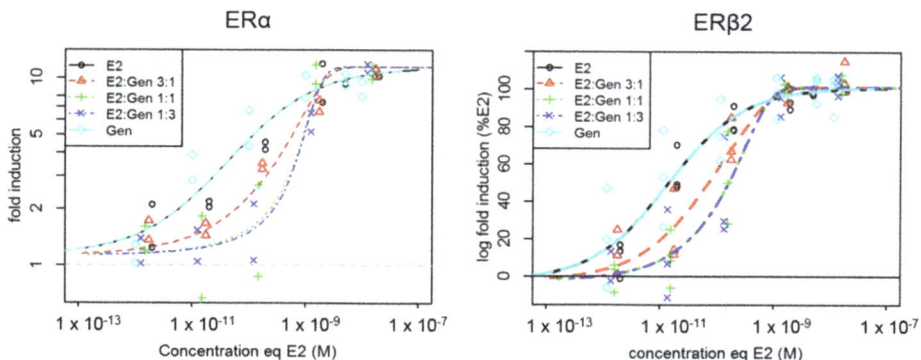

Figure 1. Concentration-response curves of luciferase activity in U251-MG cells transfected with ERα or ERβ2 after exposure to estradiol (E2) and genistein (GEN) alone or in combinations (three different ratios of substances). These data originated from 2 (ERα) or 3 (ERβ2) independent experiments. All the data were modeled by the dose-level dependent interaction model (DL). Each point represents the mean of triplicated wells. E2 and GEN concentration-response curves are superimposed because the concentration is expressed in E2-equivalents which have been calculated with the EC50s from these curves.

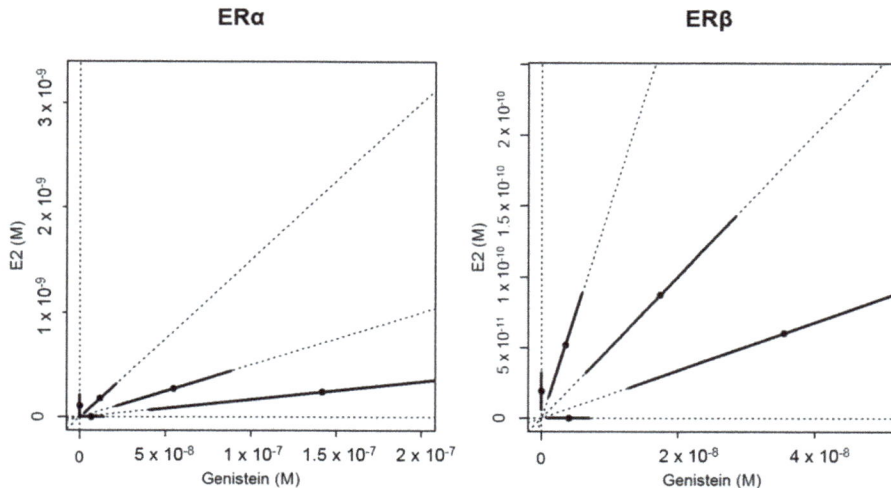

Figure 2. Illustration of the EC50 for each ray of the estradiol (E2) + genistein mixtures (EC50s isobologram). The points represent the EC50 and the bars represent the 95% confidence interval. These data originated from the in vitro assays with U251-MG cells transfected with the promoter of the zebrafish *cyp19a1b* gene coupled to the luciferase reporter gene and the zebrafish ERs (ERα or ERβ2). The isobole is the line formed when EC50s of each ray are joined. A straight isobole would indicate additivity. The deviation of the isobole to the right indicates an antagonism.

2.3. In Vivo Effects of Single Test Compounds

No effect due to chemical exposure was observed on lethality or time to hatch during any of the in vivo studies at the concentrations used in the experiments reported here.

A 96-h exposure to E2 led to a concentration-dependent induction of GFP expression in transgenic *cyp19a1b*-GFP zebrafish (Supplementary Figure S2), with an EC50 of 2.25×10^{-10} M (4 independent experiments with 8–19 transgenic zebrafish per condition). GFP expression is detected in radial glial cells of the brain (Figure 3). A concentration-dependent induction of GFP expression in radial glial cells of the brain after exposure to GEN was measured with an EC50 of 1.45×10^{-6} M (Supplementary Figure S2) (2 experiments with 8–19 transgenic zebrafish per condition).

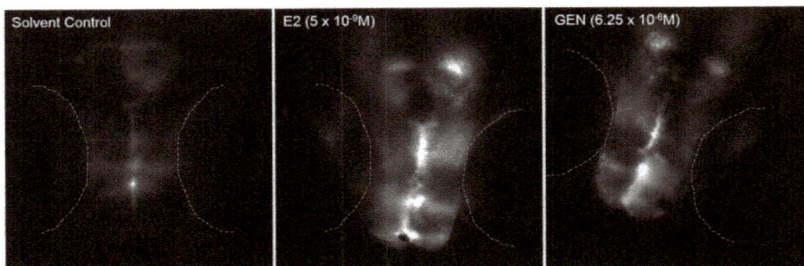

Figure 3. In vivo imaging of transgenic *cyp19a1b*-GFP zebrafish embryos (4-dpf old) exposed to solvent (DMSO), estradiol (E2) or genistein (GEN) for 96 h. Dorsal view of the brain showing GFP induction in the radial glial cells. For each chemical, the concentration used is indicated. Dotted lines delimit the eyes.

2.4. In Vivo Effects of Binary Mixtures of E2 and GEN

In the two independent mixture experiments with transgenic *cyp19a1b*-GFP zebrafish, estimated EC50s for single compounds were 2.69×10^{-10} M for E2 and 7.48×10^{-7} M for GEN. The relative potency based on these EC50s is of the same order of magnitude as in the single compound experiments.

In transgenic *cyp19a1b*-GFP zebrafish, the mixtures of E2 and GEN induced GFP expression in radial glial cells in a concentration-dependent manner (Figure 4). Observed responses were compared to the concentration-response surfaces modeled with the CA model with different slopes for each compound to improve the model goodness of fit [19] and were in good agreement in the two independent experiments (lack-of-fit *F*-test compared to the analysis of variance model: $p = 0.755$ (experiment 1) and $p = 0.195$ (experiment 2)) indicating that the CA model was not rejected. Interactions were then added to the CA model [18]. The simple interaction model (SA) showed a significant improvement of the adjustment quality of the model (approximate *F*-tests: $p = 0.00325$ for experiment 1 and $p = 0.0276$ for experiment 2) with interaction parameters of 2.19 (experiment 1) and 1.54 (experiment 2), indicating antagonism between E2 and GEN. This antagonism was also highlighted by the deviation of the isoboles observed in the two experiments (Figure 5). The simple antagonism model was the one finally accepted as none of the two-parameters models (dose-ratio (DR) and DL) improved goodness of fit compared to the SA model.

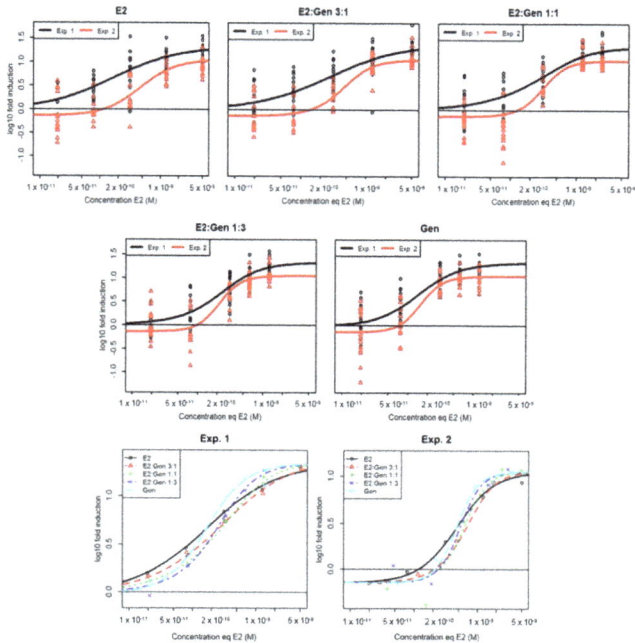

Figure 4. Concentration-response curves of GFP in *cyp19a1b*-GFP transgenic zebrafish after exposure to estradiol (E2) and genistein (GEN) alone or in combinations (3 different mixture ratios). These data originated from two independent experiments (Exp. 1 and Exp. 2). All the data were modeled by the simple interaction (SA) model. In the first five graphics, each point represents one measure of GFP in one transgenic fish brain (*n* = 8–19 fish per condition). In the last two graphics (bottom) which gather all the concentration-response curves, the points represent the means of the GFP experimentally measured for each experiment.

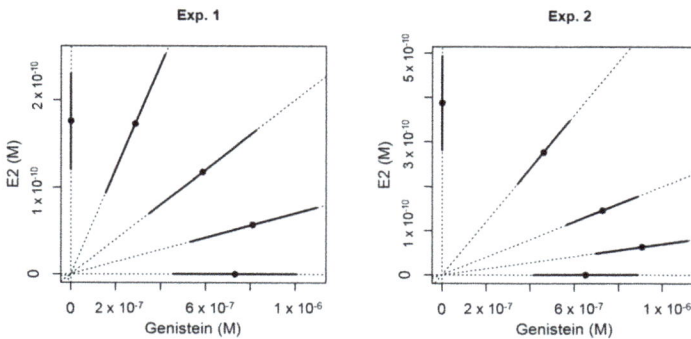

Figure 5. Illustration of the EC50 for each ray of the estradiol (E2) + genistein mixtures. The points represent the EC50 and the bars represent the standard error. These data originated from two independent exposure experiments with the *cyp19a1b*-GFP transgenic zebrafish line (described in Figure 4). The isobole is the line formed when EC50s of each ray are joined. A straight isobole would indicate additivity. The deviation of the isobole to the right indicates an antagonism.

3. Discussion

In the present study, natural estrogen (E2) and phytoestrogen (GEN) were shown to be potent inducers of the ER-regulated *cyp19a1b* gene both in vitro and in vivo. In the *cyp19a1b*-GFP transgenic zebrafish line, E2 up-regulated GFP expression in radial glial cells of the brain. In a previous study, a similar EC50 was reported for the same in vivo biological model [3]. As previously reported for EE2, EC50 reported for E2 in an ERE-GFP transgenic zebrafish line [20] is over ten times higher than that calculated in the *cyp19a1b*-GFP transgenic zebrafish line, confirming the high sensitivity of this biological in vivo model to (xeno-)estrogens. In U251-MG cells transfected with ERα or ERβ2, calculated EC50s are equivalent to those reported for the *cyp19a1b*-GFP transgenic zebrafish line and for transactivation assays using human embryonic kidney (293HEK) cells [20] and zebrafish liver cells (ZELH) [21] transfected with zfERs. However, 293HEK cells displayed a greater sensitivity to E2 when transfected with human ERs than with zfERs, supporting the relevance of using fish models to evaluate potential effects of compounds on fish. Nevertheless, our results confirm that the two biological models used in the present study are sensitive and reliable tools for the study of estrogenic potency of chemicals.

GEN is a phytoestrogen with several effects on fish reproduction and development (for a review see [22]), and is known to have a lower estrogenic potency than E2. In U251-MG cells, EC50s calculated for GEN were 270 (ERα) and 25 (ERβ2) times higher than those calculated for E2 (based on single compound experiments), confirming the higher estrogenic potency of E2. In other human cell lines (MELN with endogenous human ERs, HELN transfected with human ERs), EC50 reported for GEN were in the same range as in our U251-MG assays [23,24], while in fish cell lines transfected with fish ERs (PELN, PRTH, ZELH), EC50s were 10–100 times higher suggesting a lower sensitivity of these biological models to estrogens [21,25]. Interestingly, in our U251-MG cell model, GEN is a selective ERβ modulator as previously reported in humans [23]. Furthermore, while for E2, the sensitivity of *cyp19a1b* gene expression was the same in both *cyp19a1b*-GFP transgenic zebrafish and U251-MG cells; for GEN, the transgenic zebrafish line was less sensitive than the U251-MG cells. These differences might be related to the presence in the entire organism (transgenic zebrafish) of metabolic capacities, including phase I and II biotransformation and efflux transporter proteins. Such metabolic processes might help reducing GEN availability for the ERs in radial glial cells, leading to higher EC50s in vivo. Although high GEN concentrations are necessary to induce *cyp19a1b* expression in the *cyp19a1b*-GFP transgenic zebrafish assay, response to GEN seems to be more sensitive compared to another in vivo short-term (48 h) zebrafish embryo assay using morphological defects as endpoints (edema, head and tail malformation, reduced spontaneous movement and blood circulation) (EC50 for GEN of 427 µg/L in *cyp19a1b*-GFP transgenic zebrafish vs. 2.8 mg/L) [26]. Moreover, it is of interest to note that, in the *cyp19a1b*-GFP transgenic zebrafish assay, the effects were observed in the central nervous system, i.e., radial glial cells that are progenitor cells of the brain, in early-life stage fish exposed for very short periods.

In the present study, the effects of mixtures of E2 and GEN on the expression of zebrafish *cyp19a1b* gene were also addressed both in vitro and in vivo. By this approach, antagonistic effects of E2 and GEN in mixtures on *cyp19a1b* gene expression in a glial cell context were highlighted. In vivo in transgenic *cyp19a1b*-GFP zebrafish, these antagonistic effects were observed at all concentrations and ratios of mixtures while in vitro in U251-MG cells, the antagonism was underlined at low concentrations for both estrogen receptors. Even if mixtures of estrogenic compounds usually lead to additive effects, some deviations from the CA model (antagonisms/synergisms) have already been reported both in vitro [8,27,28] and in vivo [5,29]. As regards E2 and GEN mixtures, only additive effects were observed in MCF-7 human breast cancer cells on the ER-dependent proliferation process [30,31]. However, in MCF-7 cells transfected with an ER-reporter gene transactivation system, E2 and GEN exerted an antagonistic interaction at low concentrations of mixtures [32] just like the effects measured in our U251-MG cells model. To our knowledge, in the literature, no antagonism of E2 and GEN in mixtures was reported for in vivo experiments, however, the results obtained in the present study in the *cyp19a1b*-GFP transgenic zebrafish usefully confirmed the antagonistic effects of these compounds in a glial cell context. The origin of the antagonistic interaction observed both in vitro and in vivo is not known but it may rely on the differing abilities of E2 and GEN to recruit ERs and/or coregulators as previously showed in human cell lines [33,34]. Overall, these results clearly demonstrate the importance of our two biological models for the study of mixtures since they both highlighted the antagonistic effects of E2 and GEN in mixtures on the *cyp19a1b* gene expression.

4. Materials and Methods

4.1. Compounds

E2 (purity ≥ 98%, CAS number: 50-28-2; ref E8875) and GEN (purity ≥ 98%; CAS number: 446-72-0; ref G6649) were purchased from Sigma-Aldrich (Saint Quentin Fallavier, France).

4.2. Zebrafish Maintenance and Breeding

The *cyp19a1b*-GFP transgenic zebrafish [35] were raised in our laboratory facility at INERIS (Institut National de l'Environnement Industriel et des Risques, Verneuil-en-Halatte, France). They were maintained in a recirculation system (Zebtec, Tecniplast, Buguggiate, Italy) filled with 3.5 L aquaria. They were kept on a 14 h light:10 h dark cycle at a temperature of 27.0 ± 2.0 °C. For reproductions, 2 males and 1 female adult fish were gathered in each aquarium. Fertilized eggs were harvested and disinfected for 5 min in water supplemented with 0.1% of commercial bleach (2.6% of sodium hypochlorite).

4.3. Zebrafish Exposure to Estrogenic Compounds

Fertilized *cyp19a1b*-GFP transgenic zebrafish eggs were exposed to chemicals (alone or in mixtures) or to solvent control (DMSO, 0.02% *v/v*) according to [12]. Each experimental condition contained 20 embryos in 100 mL of water. Embryos were exposed for 96 h between 0 and 4 days post fertilization (dpf) without water renewal. At the end of the exposure period, 4-dpf old zebrafish were processed for fluorescence measurement by image analysis. Only *cyp19a1b*-GFP transgenic zebrafish larvae were photographed at the end of the experiment. Each experiment was conducted twice independently. All experimentations were performed in accordance with the European directive 2010/63/EU for animal experimentation.

4.4. In Vivo Imaging

In vivo fluorescence imaging was performed according to [3]. Each live *cyp19a1b*-GFP transgenic embryo was photographed once in dorsal view using a Zeiss AxioImager Z1 fluorescence microscope equipped with an AxioCam Mrm camera (Zeiss GmbH, Göttingen, Germany). The same exposure conditions were used to acquire each photograph (X10 objective, 134 ms of fluorescent light exposure, maximal light intensity). Fluorescence quantification was performed using Image J software (available online: http://rsb.info.nih.gov/ij/). For each picture, the integrated density (IntDen) was measured, i.e., the sum of the gray-values of all the pixels within the region of interest. All gray-values of 300 or less were considered as background values.

4.5. U251-MG Cell Bioassay

The U251-MG (ECACC, human astrocyte) are ER-negative glial cell line and were maintained at 37 °C in 5% CO_2 atmosphere in phenol red–free Dulbecco's Modified Eagle's Medium (DMEM-F12, Sigma-Aldrich, St. Louis, MO, USA) supplemented with 8% fetal calf serum (FCS), 2 mM L-glutamine, 20 U/mL penicillin, 20 µg/mL streptomycin and 50 ng/mL amphotericin B.

One day before the transfection, cells were scraped, washed and seeded at 2×10^4 cells/mL in 24-well plates in fresh medium containing 8% FCS. Transfection and luciferase assays were performed as previously described [16]. Briefly, after 24 h, the medium was replaced with fresh phenol red-free

DMEM containing 2% charcoal/dextran FCS. Cells were transfected with plasmid-DNA using JetPEI^{TM} reagent, as indicated by the manufacturer (Polyplus-transfection, Illkirch, France). The DNA templates for each well correspond to 25 ng of zfER expression plasmid [36], 150 ng of the zebrafish *cyp19a1b* promoter linked to the luciferase reporter plasmid [37] and 25 ng of the internal β-galactosidase control vector (CMV-βgal). After one night, medium was replaced with fresh DMEM-F12 containing 2% charcoal/dextran FCS and cells were exposed to vehicle (DMSO, 0.1% *v*/*v*) and various concentrations of the test compounds. After 36 h, the luciferase activities were determined using the luciferase assay system (Promega, Charbonnières-les-Bains, France) and the β-galactosidase activity was used to normalize transfection efficiency in all experiments. Each experiment was performed at least in triplicate and the results were expressed as fold induction relative to the vehicle.

4.6. Data Normalization

In the in vivo assay with *cyp19a1b*-GFP transgenic zebrafish, induction of GFP fluorescence was measured as IntDen and, since the data were obtained from several independent experiments, they were normalized by dividing by the geometric mean of the IntDen in the DMSO control group, thus expressing results as Log Fold inductions.

In the in vitro assays with U251-MG cell cultures, data normalization was performed by dividing by the geometric mean of the corresponding solvent control group. The mixture dose-response experiments were reproduced twice in vivo and in the ERα assay, and three times in the ERβ2 assay. However, due to high variability in replicate measures within each experiment, the three replicate datapoints at each concentration within each experiment were averaged, and the mixture dose-response model was based on two (ERα) or three (ERβ2) average values at each concentration. Moreover, in the ERβ2 assay, the maximum responses varied between experiments, and the data were therefore expressed as a percent of log-fold induction produced by E2 in each experiment.

4.7. Concentration-Response Modeling

The relationship between concentration and log-fold induction was modeled with a 4-parameter Hill model:

$$\Phi(c) = Min + \frac{Max - Min}{1 + \left(\frac{c}{EC50}\right)^{\beta}} \tag{1}$$

where *Min* is the minimum level of induction, *Max* is the maximum level of induction, *c* is the concentration, *EC50* is the concentration producing 50% of the maximum induction, and β is the Hill slope. The first step of the dose-response analysis in both in vivo and in vitro experiments was to estimate common values of *Min* and *Max* for both single compounds within each biological model. The models with a common *Min* and *Max* on the one hand and with freely varying *Min* and *Max* on the other hand were compared with lack-of-fit F-tests. Appropriateness of the dose-response models were also tested with a test for lack-of-fit compared to ANOVA models. As a second step, common values of *Min*, *Max* and *Slope* were estimated for E2 and GEN. Lack-of-fit F-tests were performed to check that the model did not fit less well than when the slope varied freely. The parameters were estimated by least squares, using R 3.1.1 [38] and package drc [39]. All the data and R codes used in this study are available in Supplementary materials (Supplementary Figure S3).

4.8. Mixture Experimental Designs

The mixture experimental design was developed based on relative potency of single compounds according to [12].

In the in vivo assay, relative potency of E2 and GEN were estimated with the EC50 obtained by modelling dose-response data from respectively four and two experiments with the 4-parameter Hill model. Mixture experiments were performed according to a ray design with one ray for each single chemical and three mixture ratios (3:1, 1:1, and 1:3 expressed as relative potencies). Each ray was tested with five concentrations with 4-fold serial dilutions, centered around the EC50, except for

GEN for which the highest concentrations were reduced because of their toxicity to fish (range of E2 concentrations from 4.8×10^{-12} to 5×10^{-9} M and range of GEN concentrations from 2.4×10^{-8} to 6.25×10^{-6} M) (Table 1). In theory, interactions are likely to be most visible at the equimolar mixture ratios.

The same approach was used to design the experiments for the in vitro assay with U251-MG cells. To calibrate the design, we used one experiment with E2 and GEN performed on the same plate. The ray designs built for the in vitro assays are presented in Table 2 for both ERα and ERβ2.

Table 1. Experimental ray design for the assessment of the effects of estradiol (E2) and genistein (GEN) alone and in mixtures on the expression of GFP in the brain of *cyp19a1b*-GFP transgenic zebrafish line.

Condition	[E2] (M)	[GEN] (M)	Ray
1	0	0	-
2	5.00×10^{-9}	0	1:0
3	1.25×10^{-9}	0	1:0
4	3.12×10^{-10}	0	1:0
5	7.81×10^{-11}	0	1:0
6	1.95×10^{-11}	0	1:0
7	3.75×10^{-9}	6.25×10^{-6}	3:1
8	9.37×10^{-10}	1.56×10^{-6}	3:1
9	2.34×10^{-10}	3.91×10^{-7}	3:1
10	5.86×10^{-11}	9.77×10^{-8}	3:1
11	1.46×10^{-11}	2.44×10^{-8}	3:1
12	1.25×10^{-9}	6.25×10^{-6}	1:1
13	6.25×10^{-10}	3.12×10^{-6}	1:1
14	1.56×10^{-10}	7.81×10^{-7}	1:1
15	3.91×10^{-11}	1.95×10^{-7}	1:1
16	9.77×10^{-12}	4.88×10^{-8}	1:1
17	3.12×10^{-10}	4.69×10^{-6}	1:3
18	1.56×10^{-10}	2.34×10^{-6}	1:3
19	7.81×10^{-11}	1.17×10^{-6}	1:3
20	1.95×10^{-11}	2.93×10^{-7}	1:3
21	4.88×10^{-12}	7.32×10^{-8}	1:3
22	0	6.25×10^{-6}	0:1
23	0	3.12×10^{-6}	0:1
24	0	1.56×10^{-6}	0:1
25	0	3.91×10^{-7}	0:1
26	0	9.77×10^{-8}	0:1

Table 2. Experimental ray design for the assessment of the effects of estradiol (E2) and genistein (GEN) alone and in mixtures on the luciferase activity in U251-MG cells transfected with zebrafish ERs.

Condition	ERα		ERβ2		Ray
	[E2] (M)	[GEN] (M)	[E2] (M)	[GEN] (M)	
1	0	0	0	0	-
2	2.00×10^{-8}	0	2.00×10^{-8}	0	1:0
3	2.00×10^{-9}	0	2.00×10^{-9}	0	1:0
4	2.00×10^{-10}	0	2.00×10^{-10}	0	1:0
5	2.00×10^{-11}	0	2.00×10^{-11}	0	1:0
6	2.00×10^{-12}	0	2.00×10^{-12}	0	1:0
7	1.50×10^{-8}	1.00×10^{-6}	1.50×10^{-8}	2.00×10^{-7}	3:1
8	1.50×10^{-9}	1.00×10^{-7}	1.50×10^{-9}	2.00×10^{-8}	3:1
9	1.50×10^{-10}	1.00×10^{-8}	1.50×10^{-10}	2.00×10^{-9}	3:1
10	1.50×10^{-11}	1.00×10^{-9}	1.50×10^{-11}	2.00×10^{-10}	3:1
11	1.50×10^{-12}	1.00×10^{-10}	1.50×10^{-12}	2.00×10^{-11}	3:1
12	1.00×10^{-8}	2.00×10^{-6}	1.00×10^{-8}	4.00×10^{-7}	1:1
13	1.00×10^{-9}	2.00×10^{-7}	1.00×10^{-9}	4.00×10^{-8}	1:1
14	1.00×10^{-10}	2.00×10^{-8}	1.00×10^{-10}	4.00×10^{-9}	1:1

Table 2. *Cont.*

Condition	ERα		ERβ2		Ray
	[E2] (M)	[GEN] (M)	[E2] (M)	[GEN] (M)	
15	1.00×10^{-11}	2.00×10^{-9}	1.00×10^{-11}	4.00×10^{-10}	1:1
16	1.00×10^{-12}	2.00×10^{-10}	1.00×10^{-12}	4.00×10^{-11}	1:1
17	5.00×10^{-9}	3.00×10^{-6}	5.00×10^{-9}	6.00×10^{-7}	1:3
18	5.00×10^{-10}	3.00×10^{-7}	5.00×10^{-10}	6.00×10^{-8}	1:3
19	5.00×10^{-11}	3.00×10^{-8}	5.00×10^{-11}	6.00×10^{-9}	1:3
20	5.00×10^{-12}	3.00×10^{-9}	5.00×10^{-12}	6.00×10^{-10}	1:3
21	5.00×10^{-13}	3.00×10^{-10}	5.00×10^{-13}	6.00×10^{-11}	1:3
22	0	4.00×10^{-6}	0	8.00×10^{-7}	0:1
23	0	4.00×10^{-7}	0	8.00×10^{-8}	0:1
24	0	4.00×10^{-8}	0	8.00×10^{-9}	0:1
25	0	4.00×10^{-9}	0	8.00×10^{-10}	0:1
26	0	4.00×10^{-10}	0	8.00×10^{-11}	0:1

4.9. Mixture Concentration-Response Modeling

The mixture concentration-response modeling was performed as described in [12] with some adjustments when necessary. Concentration-response surfaces were modeled with the CA model [19] under the assumption of absence of interactions, using Berenbaum's general solution [40]. This application of the CA model can be used in cases where the dose-responses of the mixture components produce same minimal and maximal effect: it does not require equal slopes in the dose-response models for single compounds. The use of the CA model when slopes differ remains a subject of controversy, because this would suggest that the single compound's modes of action are different [41,42]. In agreement with Berenbaum's general solution, the concept of Toxic Equivalent Factors, where slopes are required to be equal, is viewed as a more restrictive version of CA [43]. Prior to modelling the dose-response surface, individual rays were modelled with a Hill model with either freely varying slopes and EC50s or simply freely varying EC50s when this was not detrimental to the goodness-of-fit. On the other hand, Faust et al. [44] underline that their results do not support the idea that CA can only be applied with similar DR curves. Other authors believe that differences in Hill parameters or even differences in dose-response functions do not necessarily imply different sites of action and consider that the heterogeneity of binding sites can imply more complex dose-response functions [45]. CA has however provided adequate predictions even for mixtures where the mode of action was not identical [46].

A variety of methods have been developed for quantifying interactions based on analysis of specifically designed mixture dose-response data [47]. These include graphical methods that quantify deviations from isoboles [48,49], the widely-used Combination Index designed by Chou and Talalay [50,51], statistical methods for testing local departure from additivity [52], and modelling of the entire dose-response surface [47]. The method developed by [52] and dose-response surface modeling both allow appropriate error structure modelling [53,54] statistical tests of the significance of departures from the no-interaction model. Nonlinear response-surface analysis has the additional advantage of allowing for more complex interactions that depend on the response level or the mixture ratio [18] which could be relevant especially for endocrine disrupting compounds.

Interaction terms for simple antagonism/synergy (SA), dose-ratio dependent interactions (DR), and dose-level dependent interactions (DL) were subsequently added to the CA model [18]. These interaction models developed by Jonker et al. (2005) [18] allow for different slopes for the single compounds, but the interactions can be either calculated on toxic units based on the EC50 or on the EC at the response level under study:

$$z_i = \frac{TUx_i}{\sum\limits_{j=1}^{n} TUx_j}$$

where either $TUx_i = \frac{c_i}{EC50_i}$ or $TUx_i = \frac{c_i}{ECx_i}$. For example, for simple synergism or antagonism, these toxic units are used in the following deviation function used by Jonker et al. (2005) [18].

$$G(z_1, \ldots, z_n) = a\prod_{i=1}^{n} z_i$$

We therefore tested both implementations of the interaction definitions. Significance of the interactions was assessed using approximate F-tests on the residual sums of squares by considering that the models were nested. Acceptability of the concentration-response surface models was assessed with a lack-of-fit F-test compared to the analysis of variance model. Optimisation of parameter values for the dose-response surfaces was performed with R 3.1.1 [38], package dfoptim [55].

5. Conclusions

In the present study, we confirm (i) that our in vitro (U251-MG cells) and in vivo (*cyp19a1b*-GFP transgenic zebrafish) biological models are valuable tools to assess the estrogenic potency of chemicals both alone and in mixtures as previously stated [12]; (ii) the usefulness of the ray design approach to highlight interactions between mixture components in providing surface dose-response data and simple graphical representations. Our results show that mixture of two ER agonists, a phytoestrogen (GEN) and a natural estrogen (E2), could produce effects that deviate from the assumption of simple additivity, i.e., antagonistic effects, demonstrating the importance of considering chemical mixtures for a better understanding of the effects of ER agonists on organisms. From that point of view, both the U251-MG transactivation assay and the *cyp19a1b*-GFP transgenic zebrafish assay are reliable, flexible and simple assays, useful for the complex experimental design needed for mixture testing. Although extrapolation from the present assays to environmental situations appears difficult, as reported in our previous study [12], this assay could possibly help in determining the interactions of compounds in multi-component mixtures and/or identify compounds/mixtures that need further investigations in in vivo studies with more integrative endpoints.

Supplementary Materials: Supplementary materials can be found at http://www.mdpi.com/1422-0067/19/4/1047/s1.

Acknowledgments: This work was supported by the French National Research Program on Endocrine Disruptors (PNRPE) from the French Ministry of Environment [MIXEZ program no. 11-MRES-PNRPE-7-CVS-033] and by a fund to INERIS from the French Ministry of Environment [program 190, AP 2012-2015].

Author Contributions: Nathalie Hinfray, Cleo Tebby, Farzad Pakdel, François Brion, Sélim Aït-Aïssa conceived and designed the experiments; Benjamin Piccini raised and reproduced zebrafish in INERIS facility; Benjamin Piccini and Nathalie Hinfray performed in vivo experiments; Gaelle Bourgine performed in vitro experiments; Cleo Tebby performed the modelling of the data; Nathalie Hinfray, Cleo Tebby, Farzad Pakdel, François Brion, Sélim Aït-Aïssa, Jean-Marc Porcher revised the manuscript.

Conflicts of Interest: The authors declare no conflict of interest.

Abbreviations

CA	Concentration Addition
DL	Dose-Level dependent interaction model
E2	17β-estradiol
EC50	Effective Concentration 50%
ER	Estrogen Receptor
FCS	Fetal Calf Serum
GEN	Genistein
GFP	Green Fluorescent Protein
SA	Simple Interaction model

References

1. Jobling, S.; Beresford, N.; Nolan, M.; Rodgers-Gray, T.; Brighty, G.C.; Sumpter, J.P.; Tyler, C.R. Altered sexual maturation and gamete production in wild roach (*Rutilus rutilus*) living in rivers that receive treated sewage effluents. *Biol. Reprod.* **2002**, *66*, 272–281. [CrossRef] [PubMed]
2. Nash, J.P.; Kime, D.E.; Van der Ven, L.T.; Wester, P.W.; Brion, F.; Maack, G.; Stahlschmidt-Allner, P.; Tyler, C.R. Long-term exposure to environmental concentrations of the pharmaceutical ethynylestradiol causes reproductive failure in fish. *Environ. Health Perspect.* **2004**, *112*, 1725–1733. [CrossRef] [PubMed]
3. Brion, F.; Le Page, Y.; Piccini, B.; Cardoso, O.; Tong, S.K.; Chung, B.C.; Kah, O. Screening estrogenic activities of chemicals or mixtures in vivo using transgenic (*cyp19a1b*-GFP) zebrafish embryos. *PLoS ONE* **2012**, *7*, e36069. [CrossRef] [PubMed]
4. Kortenkamp, A. Ten years of mixing cocktails: A review of combination effects of endocrine-disrupting chemicals. *Environ. Health Perspect.* **2007**, *115* (Suppl. 1), 98–105. [CrossRef] [PubMed]
5. Lin, L.L.; Janz, D.M. Effects of binary mixtures of xenoestrogens on gonadal development and reproduction in zebrafish. *Aquat. Toxicol.* **2006**, *80*, 382–395. [CrossRef] [PubMed]
6. Petersen, K.; Fetter, E.; Kah, O.; Brion, F.; Scholz, S.; Tollefsen, K.E. Transgenic (*cyp19a1b*-GFP) zebrafish embryos as a tool for assessing combined effects of oestrogenic chemicals. *Aquat. Toxicol.* **2013**, *138–139*, 88–97. [CrossRef] [PubMed]
7. Petersen, K.; Tollefsen, K.E. Assessing combined toxicity of estrogen receptor agonists in a primary culture of rainbow trout (*Oncorhynchus mykiss*) hepatocytes. *Aquat. Toxicol.* **2011**, *101*, 186–195. [CrossRef] [PubMed]
8. Rajapakse, N.; Silva, E.; Scholze, M.; Kortenkamp, A. Deviation from additivity with estrogenic mixtures containing 4-nonylphenol and 4-*tert*-octylphenol detected in the e-screen assay. *Environ. Sci. Technol.* **2004**, *38*, 6343–6352. [CrossRef] [PubMed]
9. Thorpe, K.L.; Hutchinson, T.H.; Hetheridge, M.J.; Scholze, M.; Sumpter, J.P.; Tyler, C.R. Assessing the biological potency of binary mixtures of environmental estrogens using vitellogenin induction in juvenile rainbow trout (*Oncorhynchus mykiss*). *Environ. Sci. Technol.* **2001**, *35*, 2476–2481. [CrossRef] [PubMed]
10. Garriz, A.; Menendez-Helman, R.J.; Miranda, L.A. Effects of estradiol and ethinylestradiol on sperm quality, fertilization, and embryo-larval survival of pejerrey fish (*Odontesthes bonariensis*). *Aquat. Toxicol.* **2015**, *167*, 191–199. [CrossRef] [PubMed]
11. Song, W.T.; Wang, Z.J.; Liu, H.C. Effects of individual and binary mixtures of estrogens on male goldfish (*Carassius auratus*). *Fish Physiol. Biochem.* **2014**, *40*, 1927–1935. [CrossRef] [PubMed]
12. Hinfray, N.; Tebby, C.; Garoche, C.; Piccini, B.; Bourgine, G.; Ait-Aissa, S.; Kah, O.; Pakdel, F.; Brion, F. Additive effects of levonorgestrel and ethinylestradiol on brain aromatase (*cyp19a1b*) in zebrafish specific in vitro and in vivo bioassays. *Toxicol. Appl. Pharmacol.* **2016**, *307*, 108–114. [CrossRef] [PubMed]
13. Diotel, N.; Vaillant, C.; Gabbero, C.; Mironov, S.; Fostier, A.; Gueguen, M.-M.; Anglade, I.; Kah, O.; Pellegrini, E. Effects of estradiol in adult neurogenesis and brain repair in zebrafish. *Horm. Behav.* **2013**, *63*, 193–207. [CrossRef] [PubMed]
14. Pellegrini, E.; Mouriec, K.; Anglade, I.; Menuet, A.; Le Page, Y.; Gueguen, M.M.; Marmignon, M.H.; Brion, F.; Pakdel, F.; Kah, O. Identification of aromatase-positive radial glial cells as progenitor cells in the ventricular layer of the forebrain in zebrafish. *J. Comp. Neurol.* **2007**, *501*, 150–167. [CrossRef] [PubMed]
15. Le Page, Y.; Menuet, A.; Kah, O.; Pakdel, F. Characterization of a cis-acting element involved in cell-specific expression of the zebrafish brain aromatase gene. *Mol. Reprod. Dev.* **2008**, *75*, 1549–1557. [CrossRef] [PubMed]
16. Le Page, Y.; Scholze, M.; Kah, O.; Pakdel, F. Assessment of xenoestrogens using three distinct estrogen receptors and the zebrafish brain aromatase gene in a highly responsive glial cell system. *Environ. Health Perspect.* **2006**, *114*, 752–758. [CrossRef] [PubMed]
17. Lecomte, S.; Demay, F.; Ferriere, F.; Pakdel, F. Phytochemicals targeting estrogen receptors: Beneficial rather than adverse effects? *Int. J. Mol. Sci.* **2017**, *18*, 1381. [CrossRef]
18. Jonker, M.J.; Svendsen, C.; Bedaux, J.J.; Bongers, M.; Kammenga, J.E. Significance testing of synergistic/antagonistic, dose level-dependent, or dose ratio-dependent effects in mixture dose-response analysis. *Environ. Toxicol. Chem. SETAC* **2005**, *24*, 2701–2713. [CrossRef]
19. Loewe, S. The problem of synergism and antagonism of combined drugs. *Arzneimittelforschung* **1953**, *3*, 285–290. [PubMed]

20. Legler, J.; Zeinstra, L.M.; Schuitemaker, F.; Lanser, P.H.; Bogerd, J.; Brouwer, A.; Vethaak, A.D.; De Voogt, P.; Murk, A.J.; Van der Burg, B. Comparison of in vivo and in vitro reporter gene assays for short-term screening of estrogenic activity. *Environ. Sci. Technol.* **2002**, *36*, 4410–4415. [CrossRef] [PubMed]

21. Cosnefroy, A.; Brion, F.; Maillot-Marechal, E.; Porcher, J.M.; Pakdel, F.; Balaguer, P.; Ait-Aissa, S. Selective activation of zebrafish estrogen receptor subtypes by chemicals by using stable reporter gene assay developed in a zebrafish liver cell line. *Toxicol. Sci.* **2012**, *125*, 439–449. [CrossRef] [PubMed]

22. Jarošová, B.; Javůrek, J.; Adamovský, O.; Hilscherová, K. Phytoestrogens and mycoestrogens in surface waters—Their sources, occurrence, and potential contribution to estrogenic activity. *Environ. Int.* **2015**, *81*, 26–44. [CrossRef] [PubMed]

23. Escande, A.; Pillon, A.; Servant, N.; Cravedi, J.-P.; Larrea, F.; Muhn, P.; Nicolas, J.-C.; Cavaillès, V.; Balaguer, P. Evaluation of ligand selectivity using reporter cell lines stably expressing estrogen receptor alpha or beta. *Biochem. Pharmacol.* **2006**, *71*, 1459–1469. [CrossRef] [PubMed]

24. Pillon, A.; Servant, N.; Vignon, F.; Balaguer, P.; Nicolas, J.C. In vivo bioluminescence imaging to evaluate estrogenic activities of endocrine disrupters. *Anal. Biochem.* **2005**, *340*, 295–302. [CrossRef]

25. Cosnefroy, A.; Brion, F.; Guillet, B.; Laville, N.; Porcher, J.M.; Balaguer, P.; Ait-Aissa, S. A stable fish reporter cell line to study estrogen receptor transactivation by environmental (xeno)estrogens. *Toxicol. In Vitro* **2009**, *23*, 1450–1454. [CrossRef] [PubMed]

26. Schiller, V.; Wichmann, A.; Kriehuber, R.; Muth-Kohne, E.; Giesy, J.P.; Hecker, M.; Fenske, M. Studying the effects of genistein on gene expression of fish embryos as an alternative testing approach for endocrine disruption. *Comp. Biochem. Physiol. C Toxicol. Pharmacol.* **2013**, *157*, 41–53. [CrossRef] [PubMed]

27. Soto, A.M.; Fernandez, M.F.; Luizzi, M.F.; Oles Karasko, A.S.; Sonnenschein, C. Developing a marker of exposure to xenoestrogen mixtures in human serum. *Environ. Health Perspect.* **1997**, *105* (Suppl. 3), 647–654. [CrossRef] [PubMed]

28. Suzuki, T.; Ide, K.; Ishida, M. Response of mcf-7 human breast cancer cells to some binary mixtures of oestrogenic compounds in vitro. *J. Pharm. Pharmacol.* **2001**, *53*, 1549–1554. [CrossRef] [PubMed]

29. Kunz, P.Y.; Fent, K. Estrogenic activity of ternary UV filter mixtures in fish (*Pimephales promelas*)—An analysis with nonlinear isobolograms. *Toxicol. Appl. Pharmacol.* **2009**, *234*, 77–88. [CrossRef] [PubMed]

30. Van Meeuwen, J.A.; ter Burg, W.; Piersma, A.H.; van den Berg, M.; Sanderson, J.T. Mixture effects of estrogenic compounds on proliferation and pS2 expression of MCF-7 human breast cancer cells. *Food Chem. Toxicol.* **2007**, *45*, 2319–2330. [CrossRef] [PubMed]

31. Zhu, Z.; Edwards, R.J.; Boobis, A.R. Proteomic analysis of human breast cell lines using SELDI-TOF MS shows that mixtures of estrogenic compounds exhibit simple similar action (concentration additivity). *Toxicol. Lett.* **2008**, *181*, 93–103. [CrossRef] [PubMed]

32. Charles, G.D.; Gennings, C.; Zacharewski, T.R.; Gollapudi, B.B.; Carney, E.W. Assessment of interactions of diverse ternary mixtures in an estrogen receptor-α reporter assay. *Toxicol. Appl. Pharmacol.* **2002**, *180*, 11–21. [CrossRef] [PubMed]

33. Routledge, E.J.; White, R.; Parker, M.G.; Sumpter, J.P. Differential effects of xenoestrogens on coactivator recruitment by estrogen receptor (ER) α and ERβ. *J. Biol. Chem.* **2000**, *275*, 35986–35993. [CrossRef] [PubMed]

34. Chang, E.C.; Charn, T.H.; Park, S.-H.; Helferich, W.G.; Komm, B.; Katzenellenbogen, J.A.; Katzenellenbogen, B.S. Estrogen receptors α and β as determinants of gene expression: Influence of ligand, dose, and chromatin binding. *Mol. Endocrinol.* **2008**, *22*, 1032–1043. [CrossRef] [PubMed]

35. Tong, S.K.; Mouriec, K.; Kuo, M.W.; Pellegrini, E.; Gueguen, M.M.; Brion, F.; Kah, O.; Chung, B.C. A *cyp19a1b*-GFP (aromatase B) transgenic zebrafish line that expresses GFP in radial glial cells. *Genesis* **2009**, *47*, 67–73. [CrossRef] [PubMed]

36. Menuet, A.; Pellegrini, E.; Anglade, I.; Blaise, O.; Laudet, V.; Kah, O.; Pakdel, F. Molecular characterization of three estrogen receptor forms in zebrafish: Binding characteristics, transactivation properties, and tissue distributions. *Biol. Reprod.* **2002**, *66*, 1881–1892. [CrossRef] [PubMed]

37. Menuet, A.; Pellegrini, E.; Brion, F.; Gueguen, M.M.; Anglade, I.; Pakdel, F.; Kah, O. Expression and estrogen-dependent regulation of the zebrafish brain aromatase gene. *J. Comp. Neurol.* **2005**, *485*, 304–320. [CrossRef] [PubMed]

38. R Core Team. *R: A Language and Environment for Statistical Computing*; R Foundation for Statistical Computing: Vienna, Austria, 2014.

39. Ritz, C.; Streibig, J.C. Bioassays analysis using R. *J. Stat. Softw.* **2005**, *12*, 1–22. [CrossRef]

40. Berenbaum, M.C. The expected effect of a combination of agents: The general solution. *J. Theor. Biol.* **1985**, *114*, 413–431. [CrossRef]

41. Gennings, C.; Carter, W.H., Jr.; Carchman, R.A.; Teuschler, L.K.; Simmons, J.E.; Carney, E.W. A unifying concept for assessing toxicological interactions: Changes in slope. *Toxicol. Sci.* **2005**, *88*, 287–297. [CrossRef] [PubMed]

42. Backhaus, T.; Arrhenius, A.; Blanck, H. Toxicity of a mixture of dissimilarly acting substances to natural algal communities: Predictive power and limitations of independent action and concentration addition. *Environ. Sci. Technol.* **2004**, *38*, 6363–6370. [CrossRef] [PubMed]

43. Webster, T.F. Mixtures of endocrine disruptors: How similar must mechanisms be for concentration addition to apply? *Toxicology* **2013**, *313*, 129–133. [CrossRef] [PubMed]

44. Faust, M.; Altenburger, R.; Backhaus, T.; Blanck, H.; Boedeker, W.; Gramatica, P.; Hamer, V.; Scholze, M.; Vighi, M.; Grimme, L.H. Predicting the joint algal toxicity of multi-component s-triazine mixtures at low-effect concentrations of individual toxicants. *Aquat. Toxicol.* **2001**, *56*, 13–32. [CrossRef]

45. Goldoni, M.; Johansson, C. A mathematical approach to study combined effects of toxicants in vitro: Evaluation of the bliss independence criterion and the loewe additivity model. *Toxicol. In Vitro* **2007**, *21*, 759–769. [CrossRef] [PubMed]

46. Thienpont, B.; Barata, C.; Raldùa, D. Modeling mixtures of thyroid gland function disruptors in a vertebrate alternative model, the zebrafish eleutheroembryo. *Toxicol. Appl. Pharmacol.* **2013**, *269*, 169–175. [CrossRef] [PubMed]

47. Greco, W.R.; Bravo, G.; Parsons, J.C. The search for synergy—A critical-review from a response-surface perspective. *Pharmacol. Rev.* **1995**, *47*, 331–385. [PubMed]

48. Gessner, P.K. Isobolographic analysis of interactions: An update on applications and utility. *Toxicology* **1995**, *105*, 161–179. [CrossRef]

49. Kortenkamp, A.; Altenburger, R. Synergisms with mixtures of xenoestrogens: A reevaluation using the method of isoboles. *Sci. Total Environ.* **1998**, *221*, 59–73. [CrossRef]

50. Chou, T.C. Drug combination studies and their synergy quantification using the Chou-Talalay method. *Cancer Res.* **2010**, *70*, 440–446. [CrossRef] [PubMed]

51. Chou, T.C.; Talalay, P. Quantitative-analysis of dose-effect relationships—The combined effects of multiple-drugs or enzyme-inhibitors. *Adv. Enzym. Regul.* **1984**, *22*, 27–55. [CrossRef]

52. Gennings, C.; Carter, W.H. Utilizing concentration-response data from individual components to detect statistically significant departures from additivity in chemical mixtures. *Biometrics* **1995**, *51*, 1264–1277. [CrossRef]

53. Boik, J.C.; Newman, R.A.; Boik, R.J. Quantifying synergism/antagonism using nonlinear mixed-effects modeling: A simulation study. *Stat. Med.* **2008**, *27*, 1040–1061. [CrossRef] [PubMed]

54. Sørensen, H.; Cedergreen, N.; Skovgaard, I.; Streibig, J.C. An isobole-based statistical model and test for synergism/antagonism in binary mixture toxicity experiments. *Environ. Ecol. Stat.* **2007**, *14*, 383–397. [CrossRef]

55. Varadhan, R.; Borchers, H.W.; ABB Corporate Research. *Dfoptim: Derivative-Free Optimization*; Ravi Hopkins University: Baltimore, MD, USA, 2011.

International Journal of
Molecular Sciences

MDPI

Article

Triclosan Lacks (Anti-)Estrogenic Effects in Zebrafish Cells but Modulates Estrogen Response in Zebrafish Embryos

Hélène Serra [1,2], François Brion [1], Jean-Marc Porcher [1], Hélène Budzinski [2] and Selim Aït-Aïssa [1,*]

[1] Institut National de l'environnement Industriel et des Risques (INERIS),
Unité d'Ecotoxicologie in vitro et in vivo, UMR-I SEBIO 02, BP 2, 60550 Verneuil-en-Halatte, France;
serra.helene@gmail.com (H.S.); francois.brion@ineris.fr (F.B.); jean-marc.porcher@ineris.fr (J.-M.P.)
[2] Université de Bordeaux, EPOC-UMR 5805, Laboratoire de Physico- et Toxico-Chimie de
l'environnement (LPTC), 33405 Talence, France; helene.budzinski@u-bordeaux.fr
* Correspondence: selim.ait-aissa@ineris.fr; Tel.: +33-344-556-511

Received: 9 February 2018; Accepted: 10 April 2018; Published: 12 April 2018

Abstract: Triclosan (TCS), an antimicrobial agent widely found in the aquatic environment, is suspected to act as an endocrine disrupting compound, however mechanistic information is lacking in regards to aquatic species. This study assessed the ability of TCS to interfere with estrogen receptor (ER) transcriptional activity, in zebrafish-specific in vitro and in vivo reporter gene assays. We report that TCS exhibits a lack of either agonistic or antagonistic effects on a panel of ER-expressing zebrafish (ZELH-zfERα and -zfERβ) and human (MELN) cell lines. At the organism level, TCS at concentrations of up to 0.3 µM had no effect on ER-regulated brain aromatase gene expression in transgenic cyp19a1b-GFP zebrafish embryos. At a concentration of 1 µM, TCS interfered with the E2 response in an ambivalent manner by potentializing a low E2 response (0.625 nM), but decreasing a high E2 response (10 nM). Altogether, our study suggests that while modulation of ER-regulated genes by TCS may occur in zebrafish, it does so irrespective of a direct binding and activation of zfERs.

Keywords: in vitro; in vivo; estrogen receptor; zebrafish; triclosan; brain aromatase

1. Introduction

Triclosan (TCS) is a chlorinated phenolic chemical used as a wide spectrum antimicrobial agent in many personal care products, such as cosmetics. Around 400 tons of TCS was used in Europe in 2007, reflecting its extensive application over the last few decades [1]. As a household chemical, TCS enters the aquatic environment mainly through wastewater treatment plant (WWTP) effluent releases, after incomplete removal by adsorption from sewage sludge [2]. TCS is commonly detected in WWTP effluents [3] and in river water, where it was reported to occur at concentrations ranging from the ng/L range up to 2.3 µg/L [4–6].

TCS toxicity for mammalian and several aquatic species has been well documented, as thoroughly reviewed by Dann and Hontela [2]. Among its different potential effects, concern has been raised regarding its endocrine-disrupting activity in aquatic species. TCS was reported to alter endocrine-regulated processes, such as thyroid hormone homeostasis and reproduction. In fish, different outcomes were reported. Foran et al. (2000) reported a sex ratio weakly biased toward males after exposure of medaka fry to 100 µg/L, however not at lower or higher concentrations, suggesting that TCS has a slight androgenic effect [7]. The induction of vitellogenin (VTG) expression in male mosquitofish [8] and medaka [9] was observed after exposure to 100 µg/L TCS, while it did not alter VTG in male fathead minnow [6]. A recent study using medaka reported dual effects, as

TCS was shown to decrease VTG expression in males and increase it in females after exposure to concentrations of 174 µg/L and above of TCS [10]. Overall, studies on aquatic vertebrates suggest that TCS has an endocrine-disrupting capacity, however the mechanisms by which it interferes with sex steroid-regulated pathways remain unclear [2].

Several in vitro studies using mammalian cell systems have addressed the ability of TCS to interfere with the human estrogen receptor (hER), with contrasting outcomes. For instance, in one study, TCS was found to weakly transactivate hERα in transiently transfected CV-1 cells [11], and to promote MCF-7 cell growth up to 80% of positive control at 75 µM [12]. In other studies, TCS was reported to lack agonistic effects in similar human ER- and androgen receptor (AR)-dependent reporter gene assays [13–15], but to have ER and AR antagonistic activity [13]. To date, most in vitro studies on the anti-estrogenic effects of TCS originate from mammalian test systems, and only a few mechanistic studies have focused on ER signaling pathways in fish. A recent study examined the ability of TCS to activate the ERα ligand-binding domain of different fish species using a transient reporter gene assay in human cells, and showed only weak activation at high µM concentrations of TCS on some fish ERα, but not on others [16]. Despite information obtained so far suggesting the involvement of ER signaling in TCS response, the underlying mechanism of its estrogen-related effects remains unclear, notably in aquatic species.

In this context, the aim of the present study was to assess TCS interaction with ER signaling pathways in a model fish species, the zebrafish (zf), by using recently established in vitro and in vivo zf-specific reporter gene assays. Integrative strategy based on complementary screening bioassays has been shown to be relevant for characterizing the estrogenic potency of chemicals [17], and quantifying estrogenic activity in the environmental matrices [18] in fish. In the present study, we used a panel of in vitro assays based on (1) zf liver cell lines stably transfected to express zfERα (ZELHα), zfERβ1 (ZELHβ1) or zfERβ2 (ZELHβ2) [19]; (2) human breast cancer MCF-7 cells expressing endogenous hERα (MELN) [20]; and an in vivo assay based on (3) transgenic cyp19a1b-GFP zebrafish embryos, which express *cyp19a1b* gene under the strict regulation of ER in radial glial cells (RGCs) [21]. The in vitro and in vivo effect of serial concentrations of TCS was assessed in the presence or absence of the endogenous steroidal estrogen 17β-estradiol (E2), in order to investigate agonistic and antagonistic activity.

2. Results

2.1. Triclosan Does Not Alter ZfER Transactivation In Vitro

The different reporter cell lines responded to E2, the reference compound, in the expected range of sensitivity (Figure 1), with EC50s ranging from 0.01–0.02 nM in MELN, ZELHβ1, and ZELHβ2 cells, to 2 nM in ZELHα cells. This sensitivity and different affinity of ERs for E2 is in perfect line with our previous studies using the same cell lines [18,19], which confirms the reproducibility and robustness of established cell models.

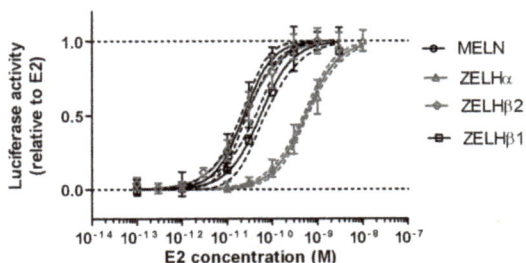

Figure 1. Luciferase induction by E2 in human (MELN) and zebrafish (ZELHα, ZELHβ1, and ZELHβ2) cell lines. Data were normalized to solvent control and to E2 maximal effect. Data are mean values ±SD (technical triplicates) and Hill fitting curve with 95% confidence interval belt.

Across all ER-expressing reporter cell lines, exposure to between 0.01 and 10 μM TCS did not induce any ER transactivation (Figure 2A–D, light bars). A slight decrease in luciferase activity was noted at 30 μM, where a cytotoxic event was also evidenced, as measured by the MTT test (Figure 2A–D, light circles). When cells were co-exposed to E2, E2-induced luciferase remained unaffected by the addition of up to 10 μM of TCS in the MELN and ZELHα cells (Figure 2A,B, dark bars), while a slight but significant decrease in E2-induced luciferase activity was observed in ZELHβ1 and ZELHβ2 cells at 10 μM (Figure 2C,D, dark bars). Above 10 μM TCS, luciferase activity significantly decreased in all the reporter cell lines, in both the presence and absence of E2. The ER-independent decrease of luciferase activity and cell viability at 30 μM TCS was confirmed in the parental ZFL and ZELH cell lines that do not express any functional zfER (Figure 2E,F). Altogether, our results suggest that TCS did not interact with ER in zebrafish liver cells, either as an ER agonist or an ER antagonist.

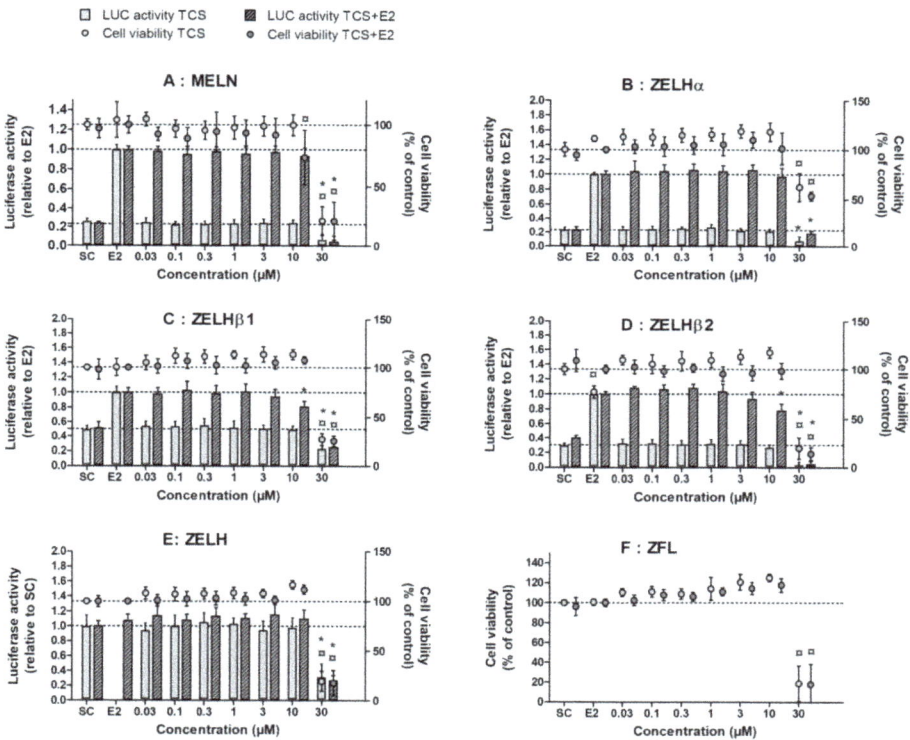

Figure 2. Luciferase response (LUC, bars) and cell viability (circles) of MELN (**A**), ZELHα (**B**), ZELHβ1 (**C**), ZELHβ2 (**D**), ZELH (**E**) and ZFL (**F**) cells after exposure to TCS (0.03 to 30 μM) for 24 h (MELN) or 72 h (ZELH-zfERs), in presence (dark symbols) or absence (light symbols) of E2. MELN, ZELHβ1 and ZELHβ2 cells were co-exposed to TCS + E2 0.1 nM, and ZELHα, ZELH and ZFL cells to TCS + E2 1 nM. Data represent the mean (+/− SD) of a minimum of 3 independent experiments using technical triplicates. Data were normalized to E2 positive control for luciferase activity. SC: solvent control ¤: response significantly different from control cell viability *: response significantly different from control luciferase activity (Mann-Whitney, $p < 0.05$).

2.2. Effect of Triclosan on Brain Aromatase Expression Using the Cyp19a1b-GFP Transgenic Zebrafish Embryo Assay (EASZY Assay)

Representative patterns of GFP expression in the developing brain of transgenic cyp19a1b-GFP zebrafish embryos are shown in Figure 3. Typically, a low GFP signal is observed in the control fish (Figure 3A) while in the E2-treated fish, a strong GFP signal is observed in the radial glial cells (Figure 3C,E). To investigate the potential effects of TCS on the expression of ER-regulated genes, transgenic cyp19a1b-GFP zebrafish embryos were exposed to graded concentrations of TCS, either alone or in combination with E2, from 3 h post fertilization (hpf) to 96 hpf. The quantification of the GFP signal after chemical treatment is presented in Figure 4.

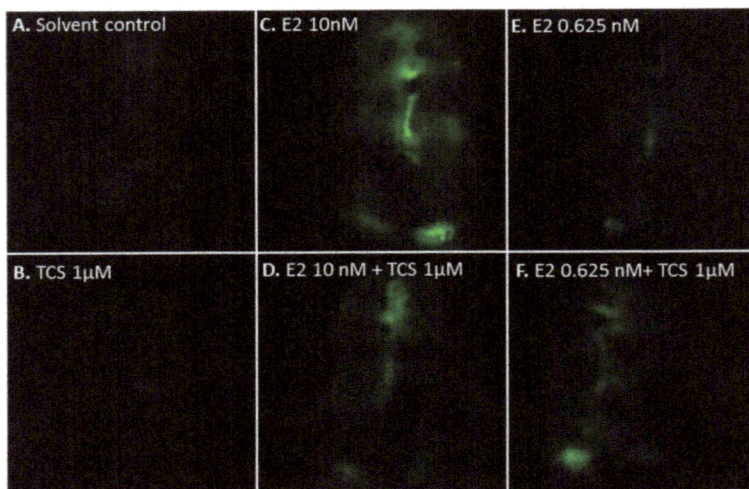

Figure 3. Dorsal view of the developing brain showing in vivo GFP expression (green signal) in the radial glial cells (RGCs) of transgenic cyp19a1b-GFP zebrafish (zf) embryos (4-dpf old) exposed to (**A**) solvent (DMSO); (**B**) TCS (1 μM); (**C**) and (**E**) E2 (10 nM and 0.625 nM); and (**D**,**F**) E2 + TCS for 96 h.

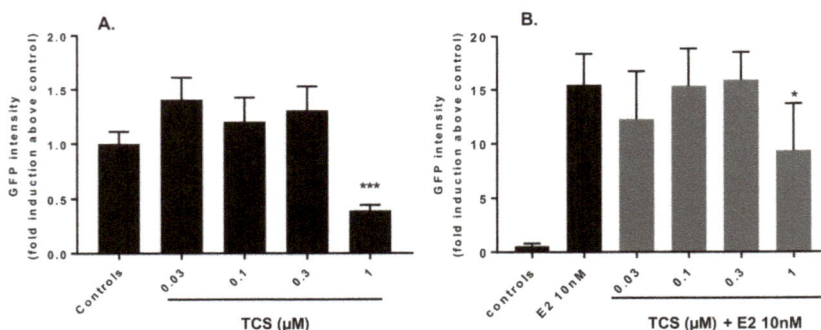

Figure 4. Effect of TCS on cyp19a1b expression in transgenic zebrafish larvae. Zebrafish embryos were exposed to TCS from 0.03 to 1 μM alone (**A**) or in co-exposure to E2 10 nM (**B**); from 3 h post fertilization (hpf) for 96 h with daily renewal of medium. GFP intensity was measured on day 4 on living organisms by fluorescence microscopy. Data are mean values ± SEM of a minimum of 2 experiments. $N = 2$ independent experiments with $n = 20$ embryos per condition per experiment. * and ***: significantly different from (**A**) control or (**B**) E2 groups ($p < 0.05$ and $p < 0.001$, respectively, Mann–Whitney test).

TCS alone did not induce any GFP expression in the EASZY assay (Figure 3A). In contrast, TCS 1 µM consistently and significantly reduced GFP intensity by 60% of the control group ($p = 0.0002$) (Figures 3B and 4A). To further assess the interference of TCS with the ER-regulated expression of the cyp19a1b gene, zf embryos were exposed to different concentrations of TCS in the presence of E2 10 nM. At this concentration, E2 alone strongly induced GFP intensity by a factor of 15 as compared to controls (Figures 3C and 4B). Co-exposure to graded concentrations of TCS had no significant effects on E2-induced cyp19a1b expression from 0.03 to 0.3 µM, but significantly decreased the E2-induced GFP intensity by 30% at 1 µM TCS (Figures 3D and 4B, $p = 0.034$).

We then further investigated the ability of TCS 1 µM to modulate the E2 concentration-dependent induction of GFP (Figure 5). E2 alone induced GFP in a concentration-dependent manner with an $EC_{50} = 0.98$ nM. In the presence of TCS 1 µM, modulation of the E2-induced response was observed, resulting in a reduction of the EC_{50} of E2 to 0.52 nM. This was the consequence of a significant 2.8-fold increase in GFP intensity at 0.625 nM E2 in the presence of TCS 1 µM, as compared to E2 alone ($p = 0.0095$) (Figure 3E,F). Furthermore, TCS reduced the response induced by E2 at 10 nM ($p = 0.061$) by 20%, a trend which confirmed the inhibitory effect of 1 µM TCS, as noted in Figure 4.

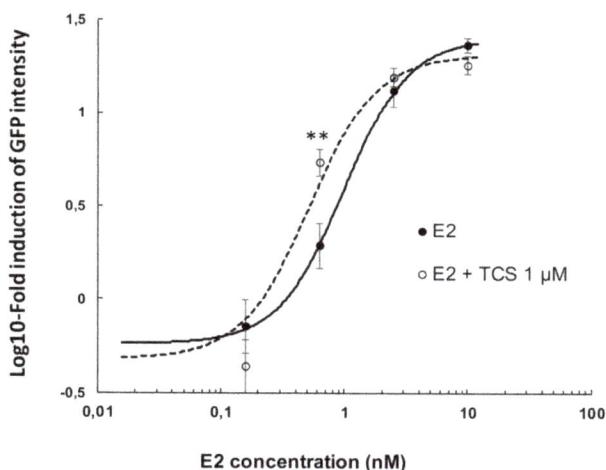

Figure 5. Effect of E2 from 0.16 to 10 nM alone (solid line) or in the presence of TCS 1 µM (dashed line) on cyp19a1b expression in transgenic zebrafish embryos. Data represent mean of log-10 transformed fold induction values ± SEM (**: significant difference at $p < 0.01$, Mann–Whitney test). $N = 2$ independent experiments with $n = 20$ embryos per condition per experiment.

3. Discussion

The in vitro assay data in ZELHs and MELN reporter cell lines demonstrate the lack of agonistic and antagonistic activity of TCS toward zfERs and hERα, in the range of tested concentrations (0.03–30 µM as nominal concentrations). A small but significant decrease in E2-induced luciferase activity was observed at TCS 10 µM in ZELHβ1 and ZELHβ2 cells. However, the decrease was minimal, and occurred close to cytotoxic concentrations, suggesting an unspecific effect on luciferase activity rather than a direct effect mediated by zfERβs. There is a lot of information regarding the estrogenic and anti-estrogenic effects of TCS in vitro. The absence of ER transactivation by TCS alone in the zebrafish reporter cell lines is in accordance with previous studies based on human cells transfected with recombinant fish [16] or human ERα [13,14]. To our knowledge, no information on the implication of zfERβ1 and zfERβ2 subtypes in TCS responses is available so far, impeding any comparison. In contrast to the lack of estrogenic effects, the absence of anti-estrogenic activity observed in the zf and human reporter cell lines differs from the findings of previous reports. With the exception of one study, which showed no effect of TCS on 17α-ethynylestradiol (EE2)-induced hERα activation in T47D-kbLuc breast cancer cell line [22], several in vitro studies have described TCS as anti-estrogenic in transactivation and proliferative assays using human cells [12–14]. Differences between test systems, e.g., organism (fish vs. human), tissue (breast cancer cells vs. hepatocytes), and ER subtype (ERα vs. ERβ1/β2), may have contributed to the differences in observations.

In the present study, we noticed a decrease in reporter gene activity at 30 μM TCS irrespective of E2 in all the cell lines, which paralleled a marked decrease in cell viability. TCS was shown to induce a caspase-dependent apoptosis in neuronal cells [23,24], and to reduce cell viability by targeting cell proliferation in mouse embryonic stem cells [25] and MCF-7 cells [26]. Thus, the decreased cell viability observed in ZFL-derived cells and in MELN cells at 30 μM is in line with the broad cellular toxicity of TCS reported in the high μM range.

The (anti-)estrogenic effects of TCS were further investigated in vivo using transgenic zf embryos, which express GFP under the control of cyp19a1b promoter in RGCs [21]. Cyp19a1b codes for the brain aromatase (aromatase B)—it is responsible for the conversion of androgens into estrogens, and its expression is highly inducible by ER agonists at the embryonic stage, by way of an autoregulatory loop [27,28]. In the present study, we showed that TCS alone did not induce the expression of cyp19a1b. These results are in line with the in vitro data obtained in this study in the same model species.

At 1 μM, TCS alone significantly decreased GFP intensity in comparison to controls. The significantly decreased GFP observed in the presence of TCS alone is unlikely due to its acute toxic effect on embryos, as no significant mortality was observed under this exposure condition. Furthermore, among the more than 100 substances tested in the EASZY assay so far [17,21,29,30], none has been reported to decrease basal cyp19a1b expression in zf embryos, even those that are known to act as anti-estrogens, such as ICI 182,780. It is therefore unlikely that the observed decrease reflects TCS anti-estrogenic activity through direct interaction with ER, as also supported by our in vitro data.

TCS is known to target the thyroid axis in many organisms, including, in fish, at the larval and adult stages [31,32]. Thyroid hormones are essential to central nervous system development and function [33]. For instance, triiodothyronine (T3) is necessary for the differentiation of glial cells onto oligodendrocytes and astrocytes [34]. TCS was shown to alter the thyroid system of *Cyprinodon variegatus* larvae, including T3 levels [32]. Thus, we may hypothesize that the observed decrease in GFP intensity observed at 1 μM TCS in zf embryos may result from neuro-developmental toxicity, including effects on RGCs expressing cyp19a1b. Further investigation at the embryonic stage in fish would be warranted, in order to study the potential direct or indirect effects of TCS on RGCs, and its consequences for brain development.

Our study suggests that TCS acts in a complex manner in the presence of E2. On the one hand, TCS at 1 μM significantly potentiates the GFP expression induced by E2, but only at a low concentration (i.e., 0.625 nM), resulting in a shift of E2 EC50. On the other hand, TCS tends to decrease the effect of E2, but only at the highest E2 concentrations (i.e., 10 nM). The potentiation of EE2 effects by TCS has been observed in the uterotrophic assay following a 21-day exposure period, while TCS alone lacked estrogenic effect [35]. Overall, our results show that TCS can act on the ER-regulated brain aromatase as an ambivalent substance, since its effects appear to be dependent on exposure conditions and concentration ratios between E2 and TCS. Interestingly, our observations support recent data regarding adult medaka, where TCS exposure led to either the inhibition of VTG synthesis in males, or an increased VTG synthesis in reproductive active females [10]. Other studies of adult carp (*Cyprinus carpio*) reported the ability of a 42-day exposure to TCS to increase VTG levels through non-ER pathways [36,37]. This induction resulted from increased E2 synthesis in gonads, i.e., gonadal aromatase (Cyp19a1a) induction. The authors also noticed aromatase (Cyp19a1b) induction in the hypothalamus, likely due to an increase in plasmatic E2-levels following induction of gonadal aromatase. The hypothalamus-pituitary gonadal (HPG) axis is not functional at the investigated embryonic stage, impeding the assessment of such effects of TCS in the EASZY assay. Altogether, our data provide complementary information supporting the hypothesis that steroid levels may influence the endocrine-disrupting effects of TCS in mammalian and non-mammalian models, through non-ER mediated but still undefined mechanisms.

In summary, among the possible TCS toxicity pathways, our study addressed the very specific effect of TCS on zfER transactivation and the ER-regulated expression of brain aromatase in zebrafish reporter gene assays. Our results demonstrated that TCS lacks agonistic and antagonistic activities towards zfER α, β1 and β2 subtypes, and hERα in vitro. At the organism level, TCS interfered with the expression of the ER-regulated brain aromatase in an ambivalent manner in zebrafish, irrespective of direct binding to and transactivation of zfERs.

4. Materials and Methods

4.1. Chemicals and Reagents

17β-estradiol (E2, CAS No. 50-28-2) and triclosan (TCS, CAS No. 3380-34-5) were purchased from Sigma-Aldrich (Saint-Quentin Fallavier, France). Dimethylsulfoxide (DMSO), Leibovitz 15 culture medium (L-15), fetal calf serum (FCS), 4-(2-hydroxy-ethyl)-1-piperazineethanesulfonic acid (HEPES), epidermal growth factor (EGF), G418, 3-[4,5-dimethylthiazol-2-yl]-2,5-diphenyltetrazolium-bromide (MTT), and D-luciferin were purchased from Sigma Aldrich. Dulbecco's Modified Eagle Medium High Glucose (DMEM HG) powder, F-12 nutrient mixture (Ham's F12) powder, penicillin, and streptomycin were purchased from Gibco. Insulin, hygromycin B, and sodium bicarbonate were purchased from Dominique Dutscher (Issy-les-Moulineaux, France).

4.2. In Vitro Assays: Cell Culture, Luciferase and Cell Viability Assays

The zf in vitro assays were derived from the zf liver (ZFL) cell line, which was stably transfected, firstly by an ERE-driven luciferase gene, yielding the ZELH cell line, secondly by zfERα subtype, yielding the ZELH-zfERα, or by zfERβ1 subtype yielding the ZELH-zfERβ1 cell lines, or by zfERβ2 subtype yielding the ZELHβ2 cell lines. The establishment of these cell models, and their response to different classes of well-known xeno-estrogens has been previously described [18,19]. In addition, the ZFL and ZELH cell lines were used as ER-negative controls. In addition to the zebrafish cell models, we used the human-derived MELN cell line [20], kindly provided by Dr. Patrick Balaguer (INSERM, Montpellier, France). The MELN cells are derived from the MCF-7 cells, which endogenously express the hERα, but no functional hERβ (P. Balaguer, personal communication).

Conditions for routine cell culture and exposure to chemicals have been detailed previously [19,20]. Briefly, ZELH-zfERs cells were seeded in 96-well white opaque culture plates (Greiner CellStar™, Dominique Dutscher, Brumath, France) at 25,000 cells per well in phenol red free LDF-DCC medium (containing 50% of L-15, 35% of DMEM HG, 15% of Ham's F12, 15 mM of HEPES, 0.15 g/L of sodium bicarbonate, 0.01 mg/mL of insulin, 50 ng/mL of EGF, 50 U/mL of penicillin and streptomycin antibiotics, and 5% *v/v* stripped FCS). MELN were seeded at 80,000 cells per well cell line in steroid-free DMEM medium. Cells were left to adhere for 24 h. Following this, they were exposed in triplicates to serial dilutions of test compound for either 72 h at 28 °C for zebrafish cells, or 16 h at 37 °C for MELN cells. After exposure, luciferase activity was measured. The medium was removed and replaced by 50 μL per well of medium containing 0.3 mM luciferin. The luminescence signal was measured in living cells using a microtiter plate luminometer (Synergy H4, BioTek Instruments, Luzern, Switzerland). The effect of test chemicals on cell viability was assessed by using the 3-(4,5-dimethyl-thiazol-2-yl)-2,5-diphenyltetrazolium bromide (MTT) assay [38]. After cell exposure, culture medium was removed and replaced by 100 μL of medium containing 0.5 mg/mL MTT. Cells were incubated for 3 h. In metabolically active cells, MTT was reduced onto a blue formazan precipitate, which was dissolved by adding 100 μL of DMSO after removal of MTT containing medium. Plates were then read at 570 nm against a 640 nm reference wavelength on a microplate reader (KC-4, BioTek Instruments, Colmar, France), and results were expressed as absorbance relative to control cells.

4.3. *In Vivo Zebrafish Bioassays*

In vivo anti-estrogenicity of TCS was assessed using the transgenic cyp19a1b-GFP zebrafish line, previously developed [39] and well characterized with different classes of estrogens and xeno-estrogen compounds [21]. The assay procedure for individual chemical testing has been described in detail by Brion et al. [21]. To briefly summarize, 20 fertilized transgenic eggs were selected for each experimental group and exposed for 96 h in 25 mL of acclimated water in glass crystallizers. Serial dilutions were tested with a final volume of solvent (DMSO) of 0.01% v/v, a concentration without any effects on embryo development or GFP expression. In each experimental series, positive (EE2 0.05 nM) and DMSO controls were included as separate experimental groups. Exposed embryos were incubated at 28 °C, under semi-static conditions with complete renewal of the medium daily. After the exposure period, each zebrafish larva was photographed using a Zeiss Axio Imager.Z1 microscope equipped with an AxioCam Mrm camera (Zeiss GmbH, Gottingen, Germany) to measure GFP expression in the brain. Image analysis was performed using the ImageJ software, and fluorescence data was treated exactly as previously described [21].

Animal maintenance was performed under strict respect of animal welfare. The assays based on zebrafish embryos used in this study are not subjected to animal experiments according to the European Directive 2010/63/EU, and are to be considered as alternative methods for animal experiments.

4.4. *Data Analysis*

Dose-response curves were fitted to the experimental data using the Hill equation as provided in the RegTox 7.5 Microsoft Excel™ macro [40]. Significant effects on luciferase induction, GFP intensity, and cell viability in exposed conditions were determined by comparing each condition to solvent control (estrogenic effects), or to E2 control (antiestrogenic effects), applying a non-parametric bilateral Mann–Whitney statistical test with α set at 5%. Furthermore, DMSO and water controls of zebrafish cyp19ab-GFP larvae were pooled together when no significant differences were observed, applying a non-parametric bilateral Mann–Whitney statistical test with α set at 5%.

Acknowledgments: This study was supported by the French Ministry of Environment (grant P181-DRC50), the SOLUTIONS project from the European Union Seventh Framework Program (FP7-ENV-2013-two-stage Collaborative project) under grant agreement 603437, and a doctoral fellowship from INERIS to Hélène Serra. We would like to thank Emmanuelle Maillot-Maréchal and Benjamin Piccini for their precious technical support as well as anonymous reviewers for their valuable comments that helped in improving the manuscript.

Author Contributions: Hélène Serra, François Brion, Jean-Marc Porcher, Hélène Budzinski and Selim Aït-Aïssa conceived and designed the experiments; Hélène Serra performed the experiments; Hélène Serra, François Brion, and Selim Aït-Aïssa analyzed the data; Hélène Serra, François Brion, and Selim Aït-Aïssa wrote the paper. All authors have read and approved the final manuscript.

Conflicts of Interest: The authors declare no conflict of interest.

References

1. SCCS (Scientific Committee on Consumer Safety). *Opinion on Triclosan (Antimicrobial Resistance)*; SCCS: Brussels, Belgium, 2010.
2. Dann, A.B.; Hontela, A. Triclosan: Environmental exposure, toxicity and mechanisms of action. *J. Appl. Toxicol.* **2011**, *31*, 285–311. [CrossRef] [PubMed]
3. Loos, R.; Carvalho, R.; Antonio, D.C.; Cornero, S.; Locoro, G.; Tavazzi, S.; Paracchini, B.; Ghiani, M.; Lettieri, T.; Blaha, L.; et al. EU-wide monitoring survey on emerging polar organic contaminants in wastewater treatment plant effluents. *Water Res.* **2013**, *47*, 6475–6487. [CrossRef] [PubMed]
4. Brausch, J.M.; Rand, G.M. A review of personal care products in the aquatic environment: Environmental concentrations and toxicity. *Chemosphere* **2011**, *82*, 1518–1532. [CrossRef] [PubMed]
5. Kolpin, D.W.; Furlong, E.T.; Meyer, M.T.; Thurman, E.M.; Zaugg, S.D.; Barber, L.B.; Buxton, H.T. Pharmaceuticals, hormones, and other organic wastewater contaminants in US streams, 1999–2000: A national reconnaissance. *Environ. Sci. Technol.* **2002**, *36*, 1202–1211. [CrossRef] [PubMed]

6. Schultz, M.M.; Bartell, S.E.; Schoenfuss, H.L. Effects of Triclosan and Triclocarban, Two Ubiquitous Environmental Contaminants, on Anatomy, Physiology, and Behavior of the Fathead Minnow (*Pimephales promelas*). *Arch. Environ. Contam. Toxicol.* **2012**, *63*, 114–124. [CrossRef] [PubMed]

7. Foran, C.M.; Bennett, E.R.; Benson, W.H. Developmental evaluation of a potential non-steroidal estrogen: Triclosan. *Mar. Environ. Res.* **2000**, *50*, 153–156. [CrossRef]

8. Raut, S.A.; Angus, R.A. Triclosan has endocrine-disrupting effects in male western mosquitofish, *Gambusia affinis*. *Environ. Toxicol. Chem.* **2010**, *29*, 1287–1291. [CrossRef] [PubMed]

9. Ishibashi, H.; Matsumura, N.; Hirano, M.; Matsuoka, M.; Shiratsuchi, H.; Ishibashi, Y.; Takao, Y.; Arizono, K. Effects of triclosan on the early life stages and reproduction of medaka Oryzias latipes and induction of hepatic vitellogenin. *Aquat. Toxicol.* **2004**, *67*, 167–179. [CrossRef] [PubMed]

10. Horie, Y.; Yamagishi, T.; Takahashi, H.; Iguchi, T.; Tatarazako, N. Effects of triclosan on Japanese medaka (*Oryzias latipes*) during embryo development, early life stage and reproduction. *J. Appl. Toxicol.* **2018**, *38*, 544–551. [CrossRef] [PubMed]

11. Huang, H.Y.; Du, G.Z.; Zhang, W.; Hu, J.L.; Wu, D.; Song, L.; Xia, Y.K.; Wang, X.R. The in vitro estrogenic activities of triclosan and triclocarban. *J. Appl. Toxicol.* **2014**, *34*, 1060–1067. [CrossRef] [PubMed]

12. Henry, N.D.; Fair, P.A. Comparison of in vitro cytotoxicity, estrogenicity and anti-estrogenicity of triclosan, perfluorooctane sulfonate and perfluorooctanoic acid. *J. Appl. Toxicol.* **2013**, *33*, 265–272. [CrossRef] [PubMed]

13. Ahn, K.C.; Zhao, B.; Chen, J.; Cherednichenko, G.; Sanmarti, E.; Denison, M.S.; Lasley, B.; Pessah, I.N.; Kultz, D.; Chang, D.P.Y.; et al. In vitro biologic activities of the antimicrobials triclocarban, its analogs, and triclosan in bioassay screens: Receptor-based bioassay screens. *Environ. Health Perspect.* **2008**, *116*, 1203–1210. [CrossRef] [PubMed]

14. Gee, R.H.; Charles, A.; Taylor, N.; Darbre, P.D. Oestrogenic and androgenic activity of triclosan in breast cancer cells. *J. Appl. Toxicol.* **2008**, *28*, 78–91. [CrossRef] [PubMed]

15. Tarnow, P.; Tralau, T.; Hunecke, D.; Luch, A. Effects of triclocarban on the transcription of estrogen, androgen and aryl hydrocarbon receptor responsive genes in human breast cancer cells. *Toxicol. Vitro* **2013**, *27*, 1467–1475. [CrossRef] [PubMed]

16. Miyagawa, S.; Lange, A.; Hirakawa, I.; Tohyama, S.; Ogino, Y.; Mizutani, T.; Kagami, Y.; Kusano, T.; Ihara, M.; Tanaka, H.; et al. Differing Species Responsiveness of Estrogenic Contaminants in Fish Is Conferred by the Ligand Binding Domain of the Estrogen Receptor. *Environ. Sci. Technol.* **2014**, *48*, 5254–5263. [CrossRef] [PubMed]

17. Le Fol, V.; Ait-Aissa, S.; Sonavane, M.; Porcher, J.M.; Balaguer, P.; Cravedi, J.P.; Zalko, D.; Brion, F. In vitro and in vivo estrogenic activity of BPA, BPF and BPS in zebrafish-specific assays. *Ecotoxicol. Environ. Saf.* **2017**, *142*, 150–156. [CrossRef] [PubMed]

18. Sonavane, M.; Creusot, N.; Maillot-Marechal, E.; Pery, A.; Brion, F.; Ait-Aissa, S. Zebrafish-based reporter gene assays reveal different estrogenic activities in river waters compared to a conventional human-derived assay. *Sci. Total Environ.* **2016**, *550*, 934–939. [CrossRef] [PubMed]

19. Cosnefroy, A.; Brion, F.; Maillot-Marechal, E.; Porcher, J.M.; Pakdel, F.; Balaguer, P.; Ait-Aissa, S. Selective Activation of Zebrafish Estrogen Receptor Subtypes by Chemicals by Using Stable Reporter Gene Assay Developed in a Zebrafish Liver Cell Line. *Toxicol. Sci.* **2012**, *125*, 439–449. [CrossRef] [PubMed]

20. Balaguer, P.; Francois, F.; Comunale, F.; Fenet, H.; Boussioux, A.M.; Pons, M.; Nicolas, J.C.; Casellas, C. Reporter cell lines to study the estrogenic effects of xenoestrogens. *Sci. Total Environ.* **1999**, *233*, 47–56. [CrossRef]

21. Brion, F.; Le Page, Y.; Piccini, B.; Cardoso, O.; Tong, S.K.; Chung, B.C.; Kah, O. Screening Estrogenic Activities of Chemicals or Mixtures In vivo Using Transgenic (*cyp19a1b*-GFP) Zebrafish Embryos. *PLoS ONE* **2012**, *7*, e36069. [CrossRef] [PubMed]

22. Louis, G.W.; Hallinger, D.R.; Stoker, T.E. The effect of triclosan on the uterotrophic response to extended doses of ethinyl estradiol in the weanling rat. *Reprod. Toxicol.* **2013**, *36*, 71–77. [CrossRef] [PubMed]

23. Park, B.K.; Gonzales, E.L.T.; Yang, S.M.; Bang, M.J.; Choi, C.S.; Shin, C.Y. Effects of Triclosan on Neural Stem Cell Viability and Survival. *Biomol. Ther.* **2016**, *24*, 99–107. [CrossRef] [PubMed]

24. Szychowski, K.A.; Sitarz, A.M.; Wojtowicz, A.K. Triclosan induces fas receptor-dependent apoptosis in mouse neocrotical neurons in vitro. *Neuroscience* **2015**, *284*, 192–201. [CrossRef] [PubMed]

25. Chen, X.J.; Xu, B.; Han, X.M.; Mao, Z.L.; Chen, M.J.; Du, G.Z.; Talbot, P.; Wang, X.R.; Xia, Y.K. The effects of triclosan on pluripotency factors and development of mouse embryonic stem cells and zebrafish. *Arch. Toxicol.* **2015**, *89*, 635–646. [CrossRef] [PubMed]

26. Liu, B.Q.; Wang, Y.Q.; Fillgrove, K.L.; Anderson, V.E. Triclosan inhibits enoyl-reductase of type I fatty acid synthase in vitro and is cytotoxic to MCF-7 and SKBr-3 breast cancer cells. *Cancer Chemother. Pharmacol.* **2002**, *49*, 187–193. [CrossRef] [PubMed]

27. Le Page, Y.; Menuet, A.; Kah, O.; Pakdel, F. Characterization of a cis-acting element involved in cell-specific expression of the zebrafish brain aromatase gene. *Mol. Reprod. Dev.* **2008**, *75*, 1549–1557. [CrossRef] [PubMed]

28. Menuet, A.; Pellegrini, E.; Brion, F.; Gueguen, M.M.; Anglade, I.; Pakdel, F.; Kah, O. Expression and estrogen-dependent regulation of the zebrafish brain aromatase gene. *J. Comp. Neurol.* **2005**, *485*, 304–320. [CrossRef] [PubMed]

29. Cano-Nicolau, J.; Garoche, C.; Hinfray, N.; Pellegrini, E.; Boujrad, N.; Pakdel, F.; Kah, O.; Brion, F. Several synthetic progestins disrupt the glial cell specific-brain aromatase expression in developing zebra fish. *Toxicol. Appl. Pharmacol.* **2016**, *305*, 12–21. [CrossRef] [PubMed]

30. Neale, P.A.; Altenburger, R.; Ait-Aissa, S.; Brion, F.; Busch, W.; Umbuzeiro, G.d.A.; Denison, M.S.; Du Pasquier, D.; Hilscherova, K.; Hollert, H.; et al. Development of a bioanalytical test battery for water quality monitoring: Fingerprinting identified micropollutants and their Contribution to effects in surface water. *Water Res.* **2017**, *123*, 734–750. [CrossRef] [PubMed]

31. Pinto, P.I.S.; Guerreiro, E.M.; Power, D.M. Triclosan interferes with the thyroid axis in the zebrafish (*Danio rerio*). *Toxicol. Res.* **2013**, *2*, 60–69. [CrossRef]

32. Schnitzler, J.G.; Frederich, B.; Dussenne, M.; Klaren, P.H.M.; Silvestre, F.; Das, K. Triclosan exposure results in alterations of thyroid hormone status and retarded early development and metamorphosis in *Cyprinodon variegatus*. *Aquat. Toxicol.* **2016**, *181*, 1–10. [CrossRef] [PubMed]

33. Di Liegro, I. Thyroid hormones and the central nervous system of mammals (Review). *Mol. Med. Rep.* **2008**, *1*, 279–295. [CrossRef] [PubMed]

34. Noda, M. Possible role of glial cells in the relationship between thyroid dysfunction and mental disorders. *Front. Cell. Neurosci.* **2015**, *9*. [CrossRef] [PubMed]

35. Stoker, T.E.; Gibson, E.K.; Zorrilla, L.M. Triclosan Exposure Modulates Estrogen-Dependent Responses in the Female Wistar Rat. *Toxicol. Sci.* **2010**, *117*, 45–53. [CrossRef] [PubMed]

36. Wang, F.; Guo, X.; Chen, W.; Sun, Y.; Fan, C. Effects of triclosan on hormones and reproductive axis in female Yellow River carp (*Cyprinus carpio*): Potential mechanisms underlying estrogen effect. *Toxicol. Appl. Pharmacol.* **2017**, *336*, 49–54. [CrossRef] [PubMed]

37. Wang, F.; Liu, F.; Chen, W.; Xu, R.; Wang, W. Effects of triclosan (TCS) on hormonal balance and genes of hypothalamus-pituitary-gonad axis of juvenile male Yellow River carp (*Cyprinus carpio*). *Chemosphere* **2018**, *193*, 695–701. [CrossRef] [PubMed]

38. Mosmann, T. Rapid colorimetric assay for cellular growth and survival —Application to proliferation and cyto-toxicity assays. *J. Immunol. Methods* **1983**, *65*, 55–63. [CrossRef]

39. Tong, S.K.; Mouriec, K.; Kuo, M.W.; Pellegrini, E.; Gueguen, M.M.; Brion, F.; Kah, O.; Chung, B.C. A *cyp19a1b*-GFP (Aromatase B) Transgenic Zebrafish Line That Expresses GFP in Radial Glial Cells. *Genesis* **2009**, *47*, 67–73. [CrossRef] [PubMed]

40. Vindimian, E. REGTOX: Macro Excel™ Pour Dose-Réponse. Available online: http://www.normalesup.org/~vindimian/fr_index.html (accessed on 29 January 2018).

MDPI

St. Alban-Anlage 66

4052 Basel

Switzerland

Tel. +41 61 683 77 34

Fax +41 61 302 89 18

www.mdpi.com

International Journal of Molecular Sciences Editorial Office

E-mail: ijms@mdpi.com

www.mdpi.com/journal/ijms

www.ingramcontent.com/pod-product-compliance
Lightning Source LLC
Chambersburg PA
CBHW051717210326
41597CB00032B/5514